VLSI CUSTOM MICROELECTRONICS

VLSI CUSTOM MICROELECTRONICS
DIGITAL, ANALOG, AND MIXED-SIGNAL

STANLEY L. HURST
Faculty of Technology (Retired)
The Open University
Milton Keynes, England

MARCEL DEKKER, INC. NEW YORK • BASEL

Library of Congress Cataloging-in-Publication Data

Hurst, S. L. (Stanley Leonard)
VLSI custom microelectronics : digital, analog, and mixed-signal / Stanley L. Hurst.
p. cm.
Includes bibliographical references and index.
ISBN 0-8247-0220-4 (alk. paper)
1. Integrated circuits—Very large scale integration—Design and construction—Data processing. 2. Computer-aided design.
I. Title
TK7874.75.H87 1998
621.39′5—dc21 98-31682
CIP

This book is printed on acid-free paper.

Headquarters
Marcel Dekker, Inc.
270 Madison Avenue, New York, NY 10016
tel: 212-696-9000; fax: 212-685-4540

Eastern Hemisphere Distribution
Marcel Dekker AG
Hutgasse 4, Postfach 812, CH-4001 Basel, Switzerland
tel: 44-61-261-8482; fax: 44-61-261-8896

World Wide Web
http://www.dekker.com

The publisher offers discounts on this book when ordered in bulk quantities. For more information, write to Special Sales/Professional Marketing at the headquarters address above.

Current printing (last digit):
10 9 8 7 6 5 4 3 2 1

PRINTED IN THE UNITED STATES OF AMERICA

Preface

In its short life span, microelectronics has become the most complex of our everyday technologies, embracing as it does physics, chemistry, materials, thermodynamics, and micromechanical engineering, as well as electrical and electronic engineering and computer science. No one person can hope to be expert in all these diverse aspects.

Yet in spite of all this complexity and sophistication, we often take the end products for granted. Nowadays, our homes contain tens of thousands of transistors in domestic appliances, communications, and entertainment equipment, to add to the vast range of applications in industrial, military, space, and commercial products. To make all this possible has required the simultaneous evolution of not only the ability to fabricate the microelectronic circuits themselves, but also the ability to design them without errors in the first place and to test them appropriately during and after production.

This is not a text detailing silicon design and fabrication methods. Instead, it is principally concerned with the important branch of microelectronics dealing with custom circuits, whereby specific circuit designs required by original equipment manufacturers can be realized rapidly and economically in possibly small production quantities. The term *application-specific IC* (ASIC) has been widely used up to now, but the more accurate term *user-specific IC* (USIC) is now increasingly used. Custom circuits have only become viable with the maturity of both the circuit fabrication methods and the computer-aided design resources necessary for their design and test, thus releasing the circuit designer from personal involvement in the range and depth of detail involved.

However, having said this, it is the hallmark of a good engineer to be aware of all the aspects involved even though he or she may not need—or indeed, not be able—to do anything about them. This text adopts this approach, and attempts to give a comprehensive overview of all aspects of custom electronics, including the very important but difficult managerial decisions of when or when not to use it.

The concept of custom microelectronic circuits is not new; it has been almost three decades since its simple conception, but it probably did not achieve very great prominence until the early 1980s. The introduction of the microprocessor as a readily available standard off-the-shelf part delayed the wider adoption of custom microelectronics and still is a major force in all the initial ''how-shall-we-make-it'' product design decisions. Nevertheless, custom ICs offer their particular technical or economic advantages, and every product designer should be aware of the strengths as well as the weaknesses of both custom and other design styles. These are the matters we shall consider in the pages of this text.

It is assumed that the reader will be familiar with the principles of microelectronic devices, semiconductor physics, basic electronic circuits, and system design as covered in many standard teaching texts. For readers requiring a more detailed treatment than is given here, specific texts dealing solely with such areas as device physics, computer-aided design, or other aspects, should be consulted. Regrettably, no one book (and no one author!) can cover all these aspects in depth, but it is hoped that the references contained in these pages will point to such additional in-depth information.

The principles of custom microelectronics are now well established, but in common with the whole of the microelectronics industry, there is a continual evolution in complexity of devices, in obsolescence of particular components and CAD resources, and in the rise and fall of industrial companies that make up the custom circuits field. It is therefore inevitable that some of the commercial products described in this text will no longer continue to represent the status quo, but the basic educational points being made using these illustrations should remain valid. It is therefore hoped that this book will enable readers to appreciate the broad spectrum of custom microelectronics and will serve as an appropriate reference to whatever newer products and design strategies evolve in this field.

Stanley L. Hurst

Contents

VLSI CUSTOM MICROELECTRONICS

1

Introduction: The Microelectronics Evolution

The development of microelectronics spans less than one half century—less than one person's life expectancy—but during this short period it has become the most pervasive technology that has yet been developed. It touches all aspects of our lives, embracing communications, transportation, entertainment, medical matters, comfort and safety, and yet many professional design engineers have still to become involved in the design of original products using the full advantages of microelectronics.

In this introductory chapter we will survey the developments which have led to the present levels of microelectronic expertise. As noted in the Preface, it will be assumed that the reader is already familiar with basic electronic components, particularly bipolar and MOS transistors, and also with basic electronic circuit configurations and semiconductor physics. We will not attempt to deal with these important fundamentals in any depth, but instead we will attempt to cover the subject in a way that is relevant to a design engineer who incorporates microelectronics as a means of producing new and innovative company products. The commercial hardware and software which may be mentioned in this text must not be taken as representative of what may be available now and in the future, since both are in a continuous dynamic state of rapid change and development, but rather are used in the following pages as illustrations of the principles and practice of the areas being discussed.

1.1 INITIAL HISTORY

The development of the first transistor in 1948 was the start of the microelectronics evolution [1], although it was not until about the mid-1950s that suitable discrete devices became available for use in industrial equipments. Early commercial devices were germanium junction transistors, where the collector and emitter regions were diffused from opposite sides into the base region to form the pnp or npn construction. Although germanium has a higher hole and electron mobility than silicon—0.39 and 0.19 $m^2 V^{-1} s^{-1}$, respectively, for germanium compared with 0.14 and 0.05 $m^2 V^{-1} s^{-1}$ for silicon, thus making it easier to achieve high-frequency performance—the poorer temperature characteristics and the inability to protect the critical transistor perimeters (the 'metallurgical junctions') of germanium meant that silicon would become the dominant microelectronic technology from the 1960s onwards.

Details of germanium versus silicon and other III/IV compounds may be found in many standard textbooks [2–5]. The gradual dominance of silicon technology was principally due to the development of planar fabrication, whereby all the fabrication steps are made on one surface (plane) of the silicon wafer (see Chapter 2) silicon dioxide providing the key to the protection of the metallurgical junctions as well as providing good isolation between areas where required. Silicon was initially more difficult to process than germanium [6], which accounted in part for germanium being the first commercially used technology.

The first person to perceive the possibility of a fully interconnected monolithic circuit, rather than the fabrication of discrete semiconductor devices which were then interconnected by wires or other separate means, was probably G. W. A. Dummer of the (then) Royal Radar Establishment at Great Malvern, England in 1952 [7–9]; however, the bulk of the development work was pioneered in the United States, particularly by Jack Kilby of Texas Instruments, followed by other companies such as Fairchild and Sprague. By 1962 small scale integrated (SSI) packages were becoming widely available for industrial use.

1.2 THE CONTINUING EVOLUTION

Since the 1960s the history of silicon technology has been one of continuous and rapid evolution based upon planar techniques. During the following two decades chip complexity increased from the initial small scale integration (SSI) capability through medium scale integration (MSI) and large scale integration (LSI) to the present status of very large scale and ultra large scale integration (VLSI and ULSI). This is indicated in the classic illustration shown in Figure 1.1 [10,11]. Since the silicon area required for transistors is less than that required for resistors and capacitors, the impact of this evolution on electronic circuit design has been to encourage the use of active rather than passive devices. This change of emphasis has in its turn given rise to greatly improved circuit performance, since much more overall gain is readily available for circuit designs than was the case when separate and relatively expensive active devices were necessary. The operational

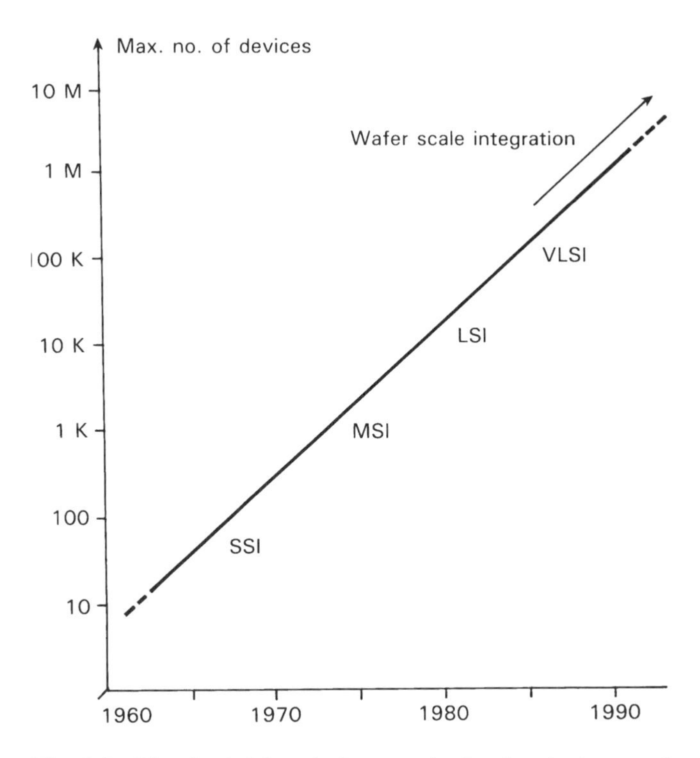

Fig. 1.1 The classic Moore's Law graph, showing the increase in maximum possible capability per single IC chip over the three decades since 1960. SSI is usually taken as tens of transistors, MSI as hundreds of transistors, LSI as thousands, and VLSI as tens of thousands of transistors upwards.

amplifier and the microprocessor (see Chapter 3) are both good examples of this evolution.

The increase in capability shown in Figure 1.1 combines two effects, namely the growth in possible wafer size and chip area and the decrease in on-chip feature size. The improvements in these two parameters together combine to give the overall increase in capability per IC, with over 10^7 transistors now available on a single chip. At the same time performance and costs have improved, as indicated in Figure 1.2. These developments are still continuing, but possibly at a slower rate as device geometries become deep submicron and the limits of well-established photolithographic techniques approach their theoretical limit. The escalating cost of fabrication lines to manufacture state-of-the-art microelectronic circuits may also have a future slow-down effect on the rate of increase of larger and more complex ICs.

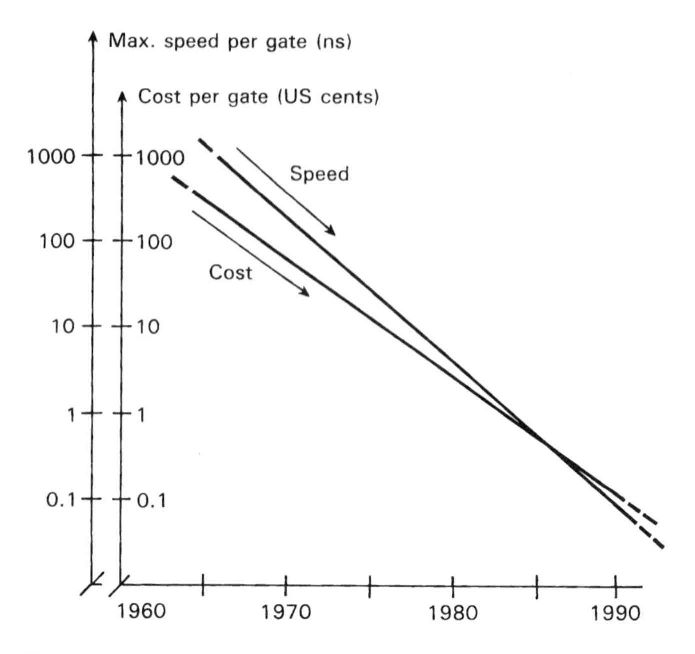

Fig. 1.2 The reduction in propagation speed per logic gate and the reduction in cost over the same period as shown in Figure 1.1.

The starting point of fabrication remains the growing of pure silicon, which has increased from the original pulled-crystal capability which gave wafers of only one or two inches diameter to the present wafer sizes of six inch (150 cm) or eight inch (200 cm) diameter. Crystal defects and discontinuities in the silicon which affect the subsequent manufacture of fault-free circuits have also been reduced (see Appendix B), and consistency of processing across the whole wafer area, which is vital for good yeild, has also been improved.

The overall characteristics such as those shown in Figure 1.2 usually represent the best technology for each parameter; the fastest available speed per gate at any time, for example, is likely to be bipolar, whereas the maximim chip area will be MOS. Nevertheless these overall characteristics represent the trend in capability of all the possible silicon technologies.

The two technologies which have been continuously developed over this period are of course bipolar and MOS (metal-oxide-silicon) [2–5,12]. As will be noted in Chapter 2 the terminology 'MOS' is something of a misnomer since the original metal-gate transistor fabrication has now been completely superseded by polysilicon-gate fabrication, but unfortunately the terminology 'MOS' continues as the device abbreviation. Very briefly, the significant features of each technology may be summarized as follows:

1. Bipolar
   - faster than MOS, but usually taking more power
   - a higher transconductance *gm* than is generally available from MOS (see Appendix D), thus giving better signal amplification performance
   - fewer fabrication steps required compared with high-performance MOS/CMOS fabrication techniques
2. MOS
   - Generally much higher impedance levels than in bipolar, thus reducing on-chip power dissipation except at very high operating speeds, but limiting the drive capability and increasing the time taken to charge and discharge large circuit capacitances
   - device performance very strongly dependent upon the device geometry, particularly the length L and the width W of the transistors (see Appendix D), thus making it readily possible to tailor device performance during the chip layout stage, a facility completely absent in bipolar design.

Bipolar remains the fastest technology [13,14], but MOS is the technology with the superior on-chip packing density. MOS, and specifically nMOS, has been and will remain the dominant technology for memory circuits, both read-only memory (ROM) and random-access memory (RAM), with feature sizes now approaching about 0.1 μm which is currently considered to be about the limit of optical lithography [12,15–17].

Outside these mainstream technologies are the others which have been actively pursued. These include:

- BiCMOS, which attempts to combine the advantages of bipolar and MOS on one chip without incurring an intolerable increase in fabrication complexity and hence cost [18,19]
- silicon-on-insulator technologies, which attempt to replace the bulk silicon semiconductor substrate with an insulating substrate; sapphire has been the most widely researched, giving silicon-on-sapphire (SOS) devices which have the particular attribute of possessing high radiation immunity ('hardness') [11,20]
- the use of III/IV compounds instead of silicon, particularly gallium arsenide (GaAs), which offers a higher speed and higher temperature performance than silicon [21,22]

- monolithic microwave integrated circuits, frequently using GaAs, but where great emphasis is placed on transmission line interconnect in order to handle signals in the GHz region without reflection problems [23,24]

More recent research has also considered the possibilities of GaAs-on-silicon, which attempts to overcome the cost and difficulty of growing perfect GaAs substrates from which the planer circuits are made by using instead conventional silicon substrates, and forming by epitaxial growth a GaAs working layer on this silicon substrate. Unfortunately the lattice constant of GaAs is 0.563 nm compared with 0.543 for silicon, and this 4% mismatch in crystal dimensions means that an unacceptably high number of crystal dislocations occur in the planar working surface. To reduce this difficulty some intermediate processing using other compounds to 'interleave' the silicon and GaAs have been investigated [25].

We shall have no occasion to refer to the last two areas again in this text, but further mention of BiCMOS, SOS and GaAs will be found in subsequent chapters. However none of these compete on financial terms with conventional bipolar and MOS silicon technologies, and are therefore likely to remain technologies for very specialized applications rather than mainstream microelectronic applications.

1.3 CAD DEVELOPMENTS

The ability to fabricate chips containing hundreds of thousands of transistors would be meaningless without the ability to design the circuits correctly. In the very early days of MSI and LSI, where possibly only five or six different layers of fabrication were involved, the chip layout could be done by hand on large drawing boards, the layouts usually being 100–200 times full size. The mask layouts were manually cut in dimensionally stable plastic sheets (rubyliths), and were visually checked both individually and also collectively by laying one on top of the other, being viewed on a light box which provided a uniform underneath illumination.

This design effort was highly inefficient and costly, taking many man-years of design effort as chip size increased, and with many errors in the fine detail of the layouts. Correct first-time designs were almost impossible to achieve, and iteration of the design was invariably required. Automation was clearly necessary. However, up to the mid-1960s computational power, graphics capability and data storage were all inadequate to handle all the information and geometric rules necessary to translate a large circuit into a chip design. Peripherals such as large visual display units (VDUs) and large plotters had yet to be developed.

As well as difficulties in the physical layout design, in the early days there were no means of simulating the circuit layout design to check for correct functionality and performance before fabrication. Breadboard techniques could no longer be relied upon to simulate circuit behavior, since the performance of the devices on the chip and the interconnect capacitances were not the same as on a breadboard model. As chip complexity increased, all these problems escalated. It was therefore necessary for the computer industry to mature alongside the increasing maturity of the microelectronics fabrication. 'Correct-by-construction' would be the objective of the supporting computer-aided design (CAD) developments, although even now this has not been 100% achievable in all circumstances.

The first generation of CAD tools for IC design appeared in the latter part of the 1960s and early 1970s. These were aimed directly at the chip layout design; they were graphics machines with large memory capacity which could manipulate geometric shapes and sizes on VDU screens, and provide design rule checking (DRC) of the resulting layouts to ensure that all the geometric design rules of minimum widths, spacings, overlaps, etc. were correct. These early machines did not, however, provide any electrical rule checking (ERC) to ensure that the circuit functionality was correct.

The two main commercial contenders in this field were Applicon and Calma, although many large industrial organizations also developed their own in-house systems based on mainframe computers. These machines grew in capability, and were sometimes known as 'polygon pushers' because of their specific duties. Their drawback was that they remained graphics engines only, not being capable of other design activities such as simulation. Thus, although they grew in size, their demise eventually came because of their limitations.

This 'dinosaur era,' as it has been unfairly called, came to a close in the late 1970s with the introduction of the workstations as we now know them. The first principal contenders in the marketplace were Daisy, Mentor and Valid. These products were smaller, cheaper and more relevant for CAD purposes than items previously available, with software to provide not only graphics capabilities but also schematic capture to enable a circuit design to be entered into the CAD system, and some simulation capability. They were the first major steps along the path of providing a truly hierarchical suite of CAD software which could link all the design phases of

- initial circuit design and simulation
- the required IC layout design
- postlayout simulation
- test vector generation

to final manufacture and test.

The principal reason for the emergence of the workstation was the availability of the microprocessor. Early microprocessors, such as the Intel 4040, were

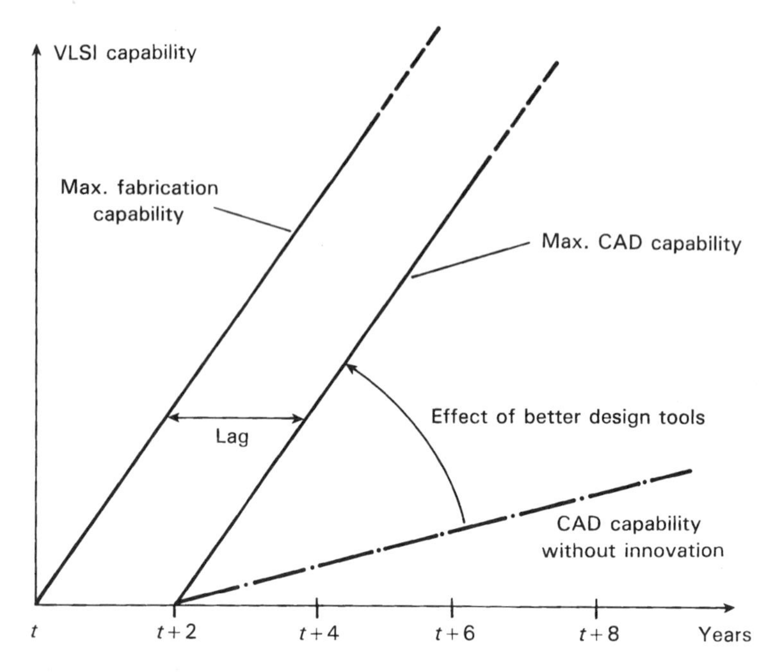

Fig. 1.3 A representation of the time lag that is present between maximum fabrication capability and the maximum design capability of the CAD resources. (*Source*: based on Ref. 26.)

largely designed by hand, and it is therefore significant that their commercial availability, at an economical price, was the means of producing the CAD tools for the more rapid and successful design of the next generation of microprocessors and VLSI circuits. This hand-in-hand evolution is still continuing: new CAD tools lead to better or more comprehensive chips, which in turn power more comprehensive CAD tools. However, the fabrication capability has up to now always managed to keep ahead of the demands of the design engineer and the capability of the CAD, as illustrated in Figure 1.3.

The later entry into this CAD market was, of course, the ubiquitous personal computer (PC). Early PCs did not have the graphics or simulation capability to be very useful, being largely confined to schematic capture purposes to enter a circuit design into a database. Subsequent geometric and simulation activities could be transferred to a more powerful workstation or mainframe computer. However the capabilities of the PC have continually increased so that there is now a blurring of capability and cost between the PC and the workstation. The latter remains the principal tool for complex design activities, but the PC now

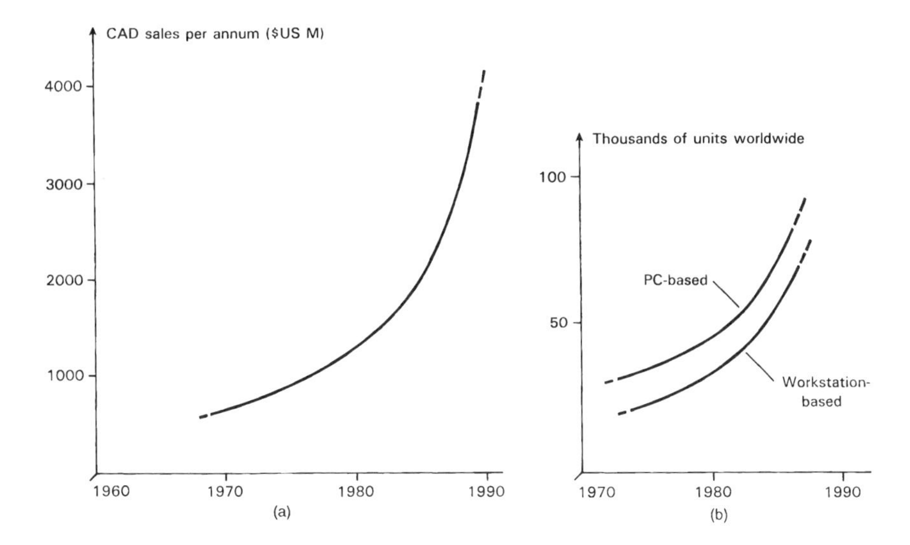

Fig. 1.4 The early increase in CAD resources with the introduction of the workstation and the personal computer: (a) worldwide sales of CAD systems; (b) the take-off in the total number of workstations and PCs in use for CAD purposes.

has the capability for the basic design requirements of the majority of original equipment design engineers.

The early exponential growth in CAD resources is shown in Figure 1.4. This growth, particularly in PC production, still continues, but it is now considered that the supporting CAD software is the major area requiring the most important continuing development for IC design. In particular, the ability of being able (a) to handle automatically all the hierarchical levels of IC design, (b) to handle mixed analog/digital design and simulation, and (c) to be technology and device independent, all in a user-friendly manner is still sought by many design engineers. Further information may be found elsewhere [26–31]; we will return to these considerations in greater depth in Chapter 5.

1.4 WHY CUSTOM MICROELECTRONICS?

Every new product designed by an original equipment manufacturer (OEM) is novel in some way. If it is an electronic product, then the circuit requirements will be unique to that product, and provided that the circuit requirements are within the capabilities of the existing off-the-shelf components, then these requirements can always be made from an appropriate assembly of standard parts. So why custom microelectronics?

The fundamental and only reason for the adoption of custom microelectronics is to satisfy the needs of the product designer who would like—or has to have—one IC package to do exactly what he or she wants, rather than having to assemble together a miscellany of standard packages to do the job. This gives rise to the following potential advantages:

- fewer packages to assemble in the final product
- a smaller final product
- improved product reliability due to fewer IC package interconnections
- lower power dissipation due to fewer ICs

and also possibly some sales advantages over competitors. In simple terms this is the *raison d'etre* for custom microelectronics, although for specialized applications such as portable or aerospace equipment the size or weight advantages may be manditory. As we will see in Chapter 4, other technical advantages such as improved overall performance may also accrue.

It has always been possible to commission an IC design which contains exactly what the product designer requires, no more or no less. This is known as 'full custom' design, with the chip design being undertaken in every detail by expert designers so as to produce the fastest or the smallest circuit possible within the constraints of the existing technology. The problems with this approach are the cost and time it takes to design. Before the advent of good CAD tools, a full custom design could take two or three years to design and finally manufacture, with a low guarantee of correct circuit operation without one or two design iterations. Hence, this design route was only relevant for military or other noncommercial applications where cost was not a prime consideration, and where possibly nondelivery of the final design was not unexpected! For the commercial market, much more viable means of production were needed, with a high guarantee of working chips within a given time scale.

It was, however, clear at a very early stage that it was not always necessary to design every individual IC from scratch. This was evident in the development of the range of off-the-shelf standard 7400-series TTL circuits, where proven basic logic configurations were used in the expanding product range, with a confidential in-house library of circuit designs being continuously built up and improved. The concept of a *library of standard circuit parts ('cells')* being available for custom design activities was therefore obvious.

Another early concept of the late 1960s was that if a number of components or basic circuits were made available on a standard chip which could be mass-produced as a standard production line item up to but not including the final interconnections between all these elements, then all that would be needed to produce a unique chip would be the design and fabrication of the chip interconnect. In essence, all the discrete components and/or logic circuits with which a product designer was familiar could be on the chip, leaving the interconnection design as the principal custom design activity.

Both of the above custom approaches are directly targeted at saving design time and therefore cost by using as far as possible already-designed data or already-designed chip layouts. It was only by using such predesigned resources that custom ICs became economically viable for the OEM designer in a competitive market, but to gain these advantages supporting CAD was also vital.

Figure 1.5 illustrates these interlocking relationships which were involved in this evolution [32–34]. This currently leaves us with the following three categories of custom microelectronics, namely:

- full custom ICs, where the chip layout is fully optimized ('handcrafted') for its particular duty, and does not rely exclusively upon predesigned circuit elements—this very expensive design methodology is fast disappearing with the increase in capabilities of the following two categories
- standard cell ICs, where the chip layout is compiled from predesigned standard circuit elements ('cells' or 'building blocks') held in the CAD memory—recent developments have allowed the CAD system to trim ('fine tune') these cells in order to optimize their specification for the specific application
- the prefabricated but uncommitted array, usually but possibly incorrectly referred to as a 'gate array,' where only the final interconnection design on a fixed chip layout has to be custom designed and made

Figure 1.6 illustrates the last two categories, with the regular structure which typifies the latter being evident.

1.5 SUMMARY

The maturity of both the semiconductor industry and the digital computer has given us not only standard off-the-shelf microelectronic parts of increasing complexity, but also the means whereby custom ICs may be designed and fabricated in relatively small quantities at a viable overall cost. The following chapters will deal in greater detail with the technical aspects of this subject, and finally an attempt will be made to quantify the many parameters involved in the possible cost equations.

Custom microelectronics has received wide publicity over many years. Many surveys of products and potential uses have been made in this period, and will doubtless continue to be made [27,35–40]. However, with some noticeable exceptions the majority of the custom circuits developed up to now have been

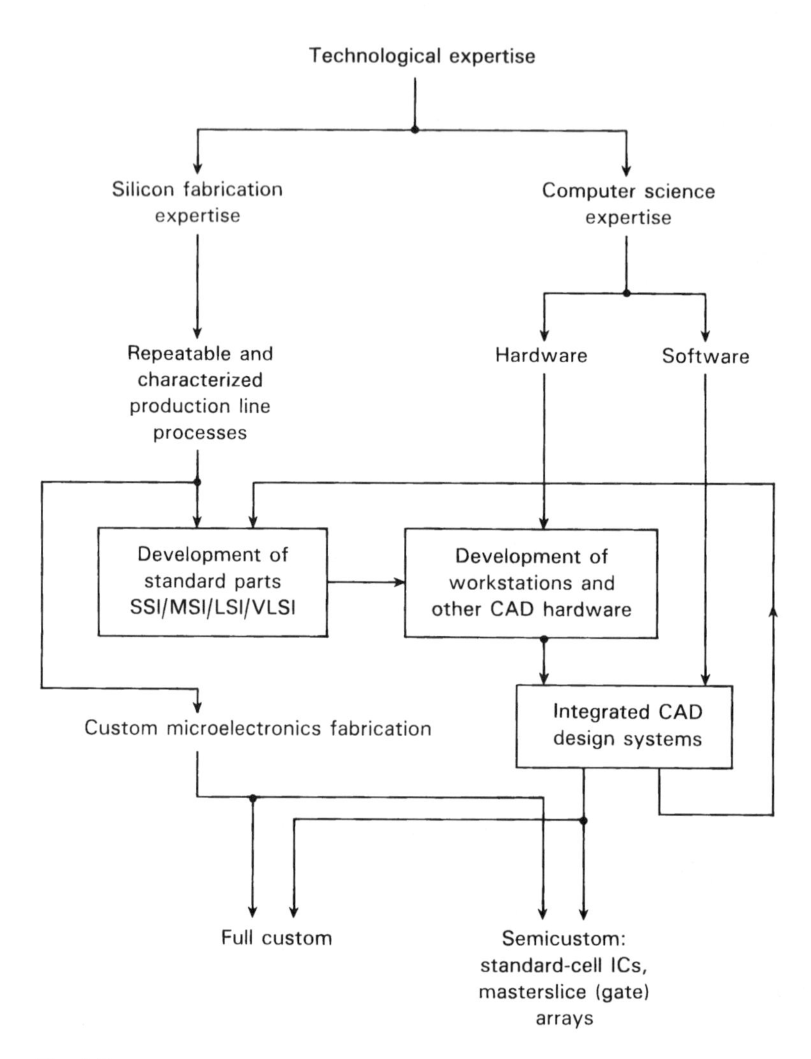

Fig. 1.5 The relationships between silicon technology and computer science which have influenced the developments and the viability of custom microelectronics. Full-custom, however, remains a minority area except where very high production volumes or great sophistication is involved.

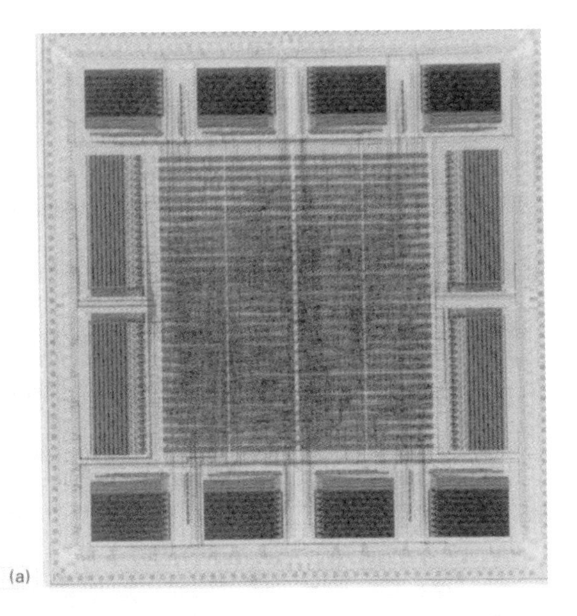

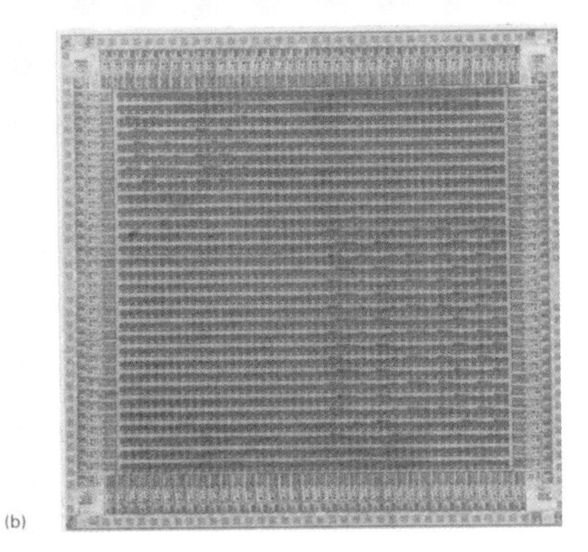

Fig. 1.6 The chip architecture of two commercial custom ICs, both from the same vendor for comparison purposes: (a) a 1.5 μm CMOS standard cell circuit with large standard blocks of ROM and RAM around the perimeter of the layout and 6000 logic gates in the center region; (b) a 1.5 μm CMOS gate array circuit with 3140 logic gates arranged in equispaced rows. (Courtesy of NEC Electronics, Mountain View, CA.)

entirely digital, with no analog capability. Analog has been a slow starter in this field, but now an increasing number of vendors are beginning to address this area [41]. Nevertheless, analog capability and its adoption remains the greatest challenge in the custom microelectronics field, and perhaps its greatest need, one that we shall have particular occasion to address in later chapters.

As a final introductory observation, there are a number of encyclopedic reference books available which give exceedingly good overviews of technological subjects [42]. Readers are recomended to these publications, which have a very high standard of both acumen and accuracy.

REFERENCES

(Note: many surveys of commercial resources are subject to periodic revision and reprinting, for example, biannually. It is therefore suggested that readers check the available literature for possibly later published survey data than that listed in this section.)

1. Kilby, J.S., Invention of the integrated circuit, *IEEE Trans. Electronic Devices*, Vol. ED.23, July 1976, pp. 648–654
2. Sparkes, J.J., *Semiconductor Devices: How They Work*, Van Nostrand Reinhold, UK, 1987
3. Till, W.C. and Luxton, J.T., *Integrated Circuits: Materials, Devices and Fabrication*, Prentice Hall, NJ, 1982
4. Neudeck, G.W. and Pierret, R.F., *Modular Series on Solid State Devices*, Vols. 1–4, Addison-Wesley, MA, 1988, 1989
5. Sze, S.M., *Physics of Semiconductor Devices*, Wiley, NY, 1981
6. Thomas, R.C., Czochralski silicon, *IEEE Trans. Electronic Devices*, Vol. ED.31, 1984, pp. 1547–1549
7. Noyce, R.N., Microelectronics, *Scientific American*, Vol. 237, September 1977, pp. 63–69
8. Dettmer R., Prophet of the integrated circuit, *IEE Electronics and Power*, Vol. 30, 1984, pp. 279–281
9. Roberts, D.H., Silicon integrated circuits: a personal view of the first 25 years, *IEE Electronics and Power*, pp. 282–284
10. Moore, G., VLSI: Some fundamental challenges, *IEEE Spectrum*, Vol. 16, 1979, pp. 30–37
11. Dilinger, T.E., *VLSI Engineering*, Prentice Hall, NJ, 1988
12. Saucier, G. and Trile, J. (Eds.), *Wafer Scale Integration*, North Holland, Amsterdam, 1986
13. Andrews, W., New generation ECL challenges GaAs in speed and power, *Computer Design*, Vol. 28, January 1989, pp. 38–39
14. Vora, M., Fairchild's radical process for building bipolar VLSI, *Electronics*, Vol. 53, September 4, 1986, pp. 55–59

15. Pugh, E.W., Critchlow, D.L., Henle, R.A. and Russell, L.A., Solid state memory developments at IBM, *IBM Journal of Research and Development*, Vol. 25, 1981, pp. 585–602
16. Dennard, D.H., Evolution of the MOSFET dynamic RAM—a personal view, *IEEE Trans. Electronic Devices*, Vol. ED.31, 1984, pp. 1549–1555
17. Meindl, J.D., Ultra-large scale integration, *IEEE Trans. Electronic Devices*, 1555–1561
18. Dettmer, R., BiCMOS—Getting the best of both worlds, *IEE Electronics and Power*, Vol. 33, 1987, pp. 499–501
19. Cole, B.C., Is BiCMOS the next technology driver?, *Electronics*, Vol. 61, 4 February 1988, pp. 55–57
20. Maly W., *Atlas of IC Technologies: an Introduction to VLSI Processes*, Benjamin-Cummings, CA, 1987
21. Morgan, D.V., Gallium arsenide: a new generation of integrated circuits, *IEE Review*, Vol. 35, September 1988, pp. 315–319
22. Greiling, P., The historical development of GaAs FET digital technology, *IEEE Trans. Microwave Theory and Techniques*, Vol. MTT.32, 1984, pp. 1144–1156
23. Howe, H., Microwave integrated circuits; a historical perspective, *IEEE Trans. Microwave Theory and Techniques*, 1984, pp. 991–996
24. McQuiddy, D.N., Wassel, J.W., Lagrange, J.B. and Weisseman, W.R., Monolithic microwave integrated circuits: a historical perspective, *IEEE Trans. Microwave Theory and Techniques*, pp. 997–1008
25. Dettmer, R., GaAs on silicon: the ultimate wafer?, *IEE Review*, Vol. 35, April 1989, pp. 136–137
26. Russell, G. (Ed.), *Computer Aided Tools for VLSI System Design*, IEE Peter Peregrinus, UK, 1987
27. Editorial Staff, Directory of silicon compilers, *VLSI System Design*, Vol. 9, March 1988, pp. 52–56
28. Ayres, R.F., *VLSI Silicon Compilation and the Art of Automatic Microchip Design*, PrenticeHall, NJ, 1983
29. Horbst, E., (Ed.), *VLSI Design of Digital Systems*, North Holland, Amsterdam, 1985
30. Rubin, S.V., *Computer Aids for VLSI Design*, Addison-Wesley, MA, 1987
31. Russell, G., Kinniment, D.J., Chester, E.G., and McLauchlan, M.R., *CAD for VLSI*, Van Nostrand Reinholt, UK, 1985
32. Freeman, W.J. and Freund, V.J., A history of semiconductor design at IBM, *Semicustom Design Guide*, Summer 1986, pp. 14–22
33. Rappaport, A.S. Technical evolution of the semi-custom design process, *Semicustom Design Guide*, Summer 1986, pp. 40–42
34. Open University, *The Management Perspective; Microelectronic Matters*, PT504 Microelectronics for Industry, The Open University, Milton Keynes, UK, 1987
35. Hinder, V.L. and Rappaport, A.S., Employing semicustom: a study of users and potential users, *VLSI System Design*, Vol. 8, May 1987, pp. 6–25
36. Editorial Staff, Gate array directory, *EDN*, Vol. 32, 25 June 1987, pp. 134–201
37. Editorial Staff, Survey of gate array and cell libraries, *VLSI Systems Design*, Vol. 8, November 1987, pp. 76–101

38. Editorial Staff, Buyers' guide to ASICs and ASIC design tools, *Computer Design*, Vol. 26, August 1987, pp. 67–136
39. Editorial Staff, Survey of CAE systems, *VLSI Systems Design*, Vol. 8, June 1987, pp. 51–77
40. Editorial Staff, Survey of IC layout systems, *VLSI Systems Design*, Vol. 8, December 1987, pp. 63–77
41. Editorial Staff, Survey of analog semicustom ICs, *VLSI Systems Design*, Vol. 7, May 1987, pp. 89–106
42. Muroga, S., Very large scale integrated design, in Vol. 14 of the *Encyclopedia of Physical Science and Technology*, Academic Press, NY, 1987

2

Technologies and Fabrication

The semiconductor technologies used in custom microelectronic circuits are essentially the same as those used in standard off-the-shelf ICs. Indeed, it would be strange if this was otherwise since the custom IC market is not as large as the global demand for standard parts, and hence the costs of semiconductor production lines are largely recouped from sales of the latter. There may have been a few exceptions to this, for example, very large high-tech companies engaged in the computer or instrumentation industry who require components at the leading edge of performance, but in general the normal customer for user specific ICs (USICs) will be using well-proven technology in the ICs which he or she commissions or purchases from a vendor. (Note, in this text we will use the terminology ''*user specific*'' for a custom IC rather than the term ''*application specific*'' [ASIC], which has been widely used up to now. The reason is that the term ''*ASIC*'' has also been applied to off-the-shelf VLSI ICs made for specific applications [see Chapter 3, Section 3.7], which are truly application specific but available to all. A further confusion in terminology is the increasing use of the term *application specific standard part* (ASSP) for the latter, which is much clearer, thus leaving the term ASIC as a rather inexplicit early term in this field.)

Chapter 1 looked back on the broad developments in germanium, silicon and other semiconductor technologies, from which it is evident that silicon is the dominant technology both now and in the foreseeable future. Germanium technology, with its poor temperature characteristics compared with silicon due to its lower energy gap (0.72 eV for Ge compared with 1.12 eV for Si), will not be mentioned further.

The depth of treatment in the following sections is intended for readers to whom custom microelectronics may be a part of their system design activity, and who should therefore be aware of the methods of fabrication and the characteristics of the various alternative technologies. Full details of the physics and chemistry of fabrication and basic device design is not our purpose; for this in-depth information, which is necessary for those engaged in the manufacture of the ICs or in design at the silicon level, reference should be made to more specialized texts in the specific subject areas [1–8].

2.1 BIPOLAR SILICON TECHNOLOGIES

By definition, bipolar technology involves both hole and electron flow in the action of the active devices, unlike unipolar (MOS) technology which is based upon the controlled flow of majority carriers only [1,3,4,7,8]. The three principal devices involved in bipolar technology are resistors, diodes and npn or pnp junction transistors. The basic fabrication process involved is a planar process, involving the diffusion or the implantation of areas of doped silicon into a silicon substrate so as to form the appropriate p-type or n-type areas. ''Planar'' means that all the stages of fabrication are performed on one surface (plane) of the silicon wafer.

Figure 2.1 shows the basic fabrication method of creating doping areas in silicon. The substrate may be either n-type or p-type material. Between the two types of silicon a pn junction is created, which, if the applied voltages are appropriate, will conduct across the junction in one direction or act as a nonconductor (barrier) between the two regions.

Resistors and diodes may be formed by making connections as shown in

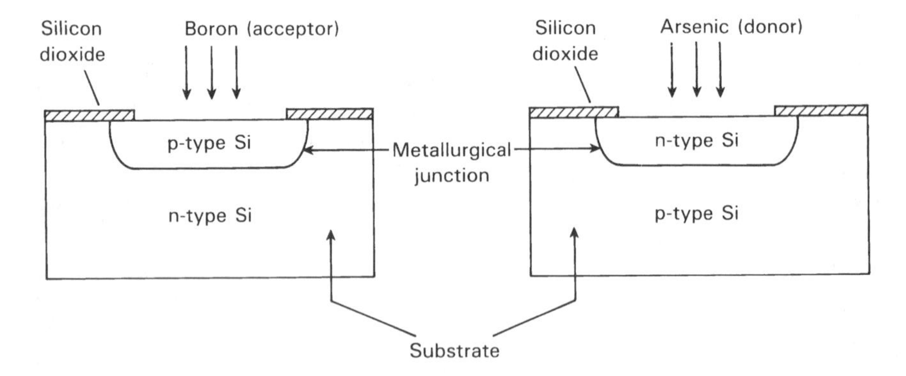

Fig. 2.1 The planar fabrication process with the diffusion of areas of doped silicon in the substrate made through a 'window' in the protective silicon dioxide layer so as to form appropriate p-type or n-type areas.

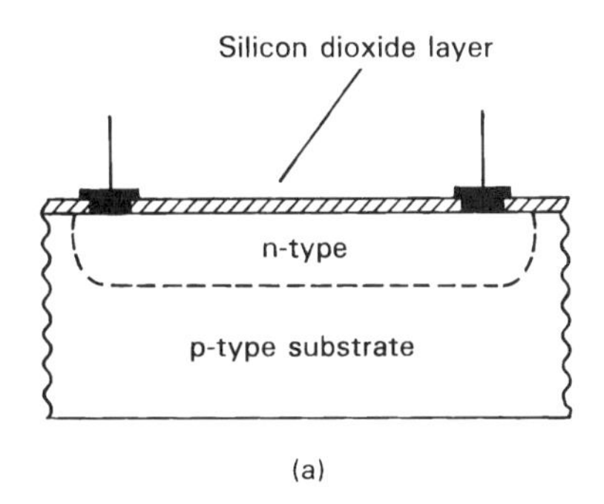

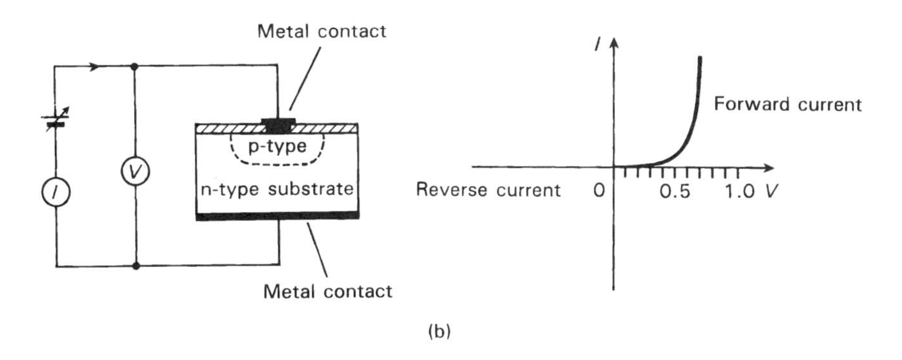

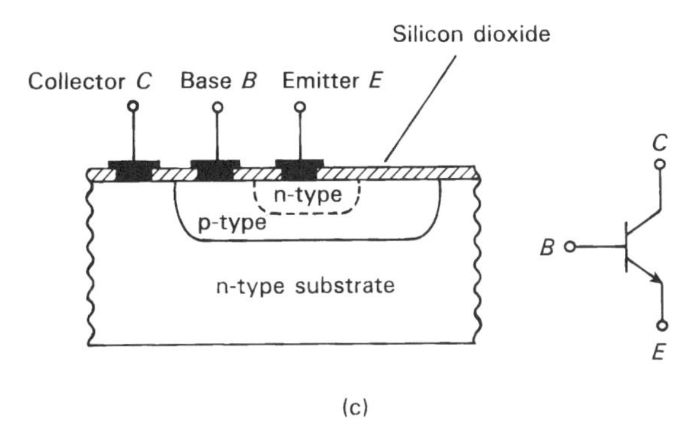

Fig. 2.2 Basic concepts of integrated resistor, diode and bipolar junction transistor planar construction: (a) resistor—the length, area and doping density determine the effective resistance; (b) simple pn junction, with connections made to each region; (c) npn transistor, with connections made to all three regions.

Figures 2.2(a) and (b). Junction transistors require a further window and diffusion stage to form the third of the pnp or npn regions, as indicated in Figure 2.2(c). In the case of the npn transistor, the flow of electrons from emitter to collector is determined by the positive voltage applied between the base and emitter, which injects holes into the base region.

These are the basic fabrication methods used in all bipolar technologies. One further step may be involved in modern bipolar technologies and in most MOS technologies, an epitaxial process which is first performed on the basic silicon substrate [1,2]. This involves the growing of a very precisely doped epitaxial layer of silicon, the ''epi. layer,'' on top of the starting substrate; it is in this thin precise epi. layer—usually n-type silicon on a p-type substrate—that all the principal steps of device fabrication and isolation are subsequently done, rather than in the bulk substrate. This feature is shown in most of the following illustrations, and is generally necessary where maximum performance is required. This epi. layer should also contain fewer crystal defects than the bulk substrate, hence improving production yield as well as providing more accurately controlled doping levels.

The present range of silicon bipolar technologies is given in Figure 2.3. The use of discrete transistors and other devices has not been included, although for many very high power and/or high voltage applications discrete semiconductors rather than monolithic circuits still remain appropriate.

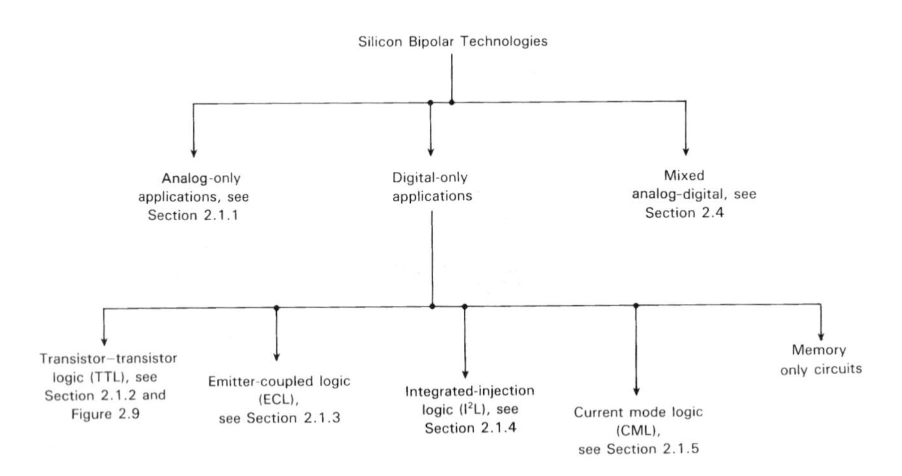

Fig. 2.3 The bipolar silicon technologies: the mixed analog/digital area, however, is subject to continuing evolution. Memory technologies will be separately reviewed in Section 2.3.

2.1.1 Analog ICs

Analog bipolar circuits employ both pnp and npn transistors, together with junction diodes, resistors and sometimes small-value capacitors. Precise fabrication and layout details depend very largely upon the individual IC manufacturers, and although not generally disclosed in detail, are based upon the fundamentals introduced in Figure 2.2.

There are, however, two distinct methods of achieving isolation between the devices fabricated on a wafer. The first, and oldest, method is the *junction-isolated* process, which relies upon the presence of a reverse-biased pn junction barrier between adjacent devices to achieve isolation. The second and more recent method is *silicon dioxide isolation*, which provides an insulating barrier between devices.

Figure 2.4 illustrates the cross-sectional arrangement of a bipolar transistor with junction isolation. This process is sometimes termed the Standard Buried Collector (SBC) process, since it requires a buried area of n^+ silicon to be first made below each transistor on the p-type substrate, which will form a low resistance subcollector contact area for each transistor. The seven process steps are as follows; appropriate masks are used at each step to define each window, the silicon dioxide protecting layer at the surface being reformed between steps when necessary:

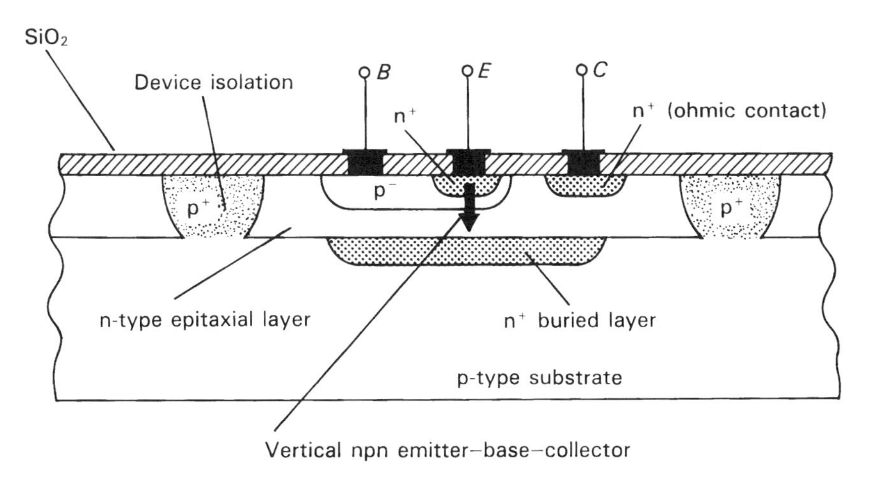

Fig. 2.4 The cross-sectional fabrication details of an SBC bipolar npn junction transistor with diode isolation (not to scale).

- Using mask No. 1, open window in the silicon dioxide (SiO_2) protecting layer on the lightly doped p-type substrate, and diffuse in a n^+ buried subcollector.
- Remove the remaining SiO_2 and grow a lightly doped n-type silicon epitaxial layer over the whole wafer.
- Regrow the SiO_2, and open a new window using mask No. 2 to diffuse in a deep p^+ trench (''moat'') around the perimeter of the device so as to form an isolated island.
- Regrow the SiO_2 and open a new window using mask No. 3 to diffuse in the p-type base area, leaving a thin n-type collector width in the epi. layer between base and buried subcollector.
- Regrow the SiO_2 and open a new window using mask No. 4 to diffuse in the n^+ transistor emitter area and also the collector contact.
- Regrow the SiO_2 and open a new window using mask No. 5 for contacts.
- Cover with interconnection metal, and then etch away using mask No. 5 so as to form the final required interconnection pattern.

With the substrate of this structure connected to the most negative voltage of the circuit, the pn junction around each device formed by the p-type moat will be reverse-biased, and therefore will provide device isolation. The n^+ buried layer also prevents the formation of parasitic vertical pnp transistors which may otherwise occur between base, collector and substrate regions.

While still ocasionally used, this SBC process has drawbacks in that the active area of the transistor is only the area below the emitter, which is within the base area, and the base area is itself within the collector area. Also due to lateral diffusion, the minimum width of the p^+ isolation moat will be about twice the depth of the epitaxial layer, so that in total the useful active area of the transistor is often less than 5% of the total device area. Figure 2.5 illustrates this shortcoming, the active area under the emitter being only 2.67% of the total device area.

The use of silicon dioxide isolation instead of pn junction isolation is an example of MOS fabrication techniques being applied to bipolar technology. This allows a considerable decrease in transistor area compared with junction isolation, and hence an increase in device density and operating speed.

The cross-section of a typical oxide-isolated npn transistor is shown in Figure 2.6. The first fabrication steps of making the buried n^+ region and the growth of the epitaxial layer remain as previously described. The next three steps, however, to create the SiO_2 isolation boundaries are as follows:

- A thin SiO_2 layer followed by a thick protecting Si_3N_4 (silicon nitride) layer is grown over the whole surface, the latter then being lightly oxidized. (If Si_3N_4 is formed directly on the epitaxial surface it causes

Minimum feature size	5 μm
Worst-case alignment tolerance between levels	2 μm
Epitaxial-layer thickness	10 μm
Collector-base junction depth	5 μm
Emitter-base junction depth	3 μm
Minimum emitter-to-collector spacing at surface	5 μm
Minimum base-to-isolation spacing at surface	5 μm
Minimum collector contact n^+ diffusion to isolation spacing	5 μm
Minimum collector contact n^+ diffusion to base spacing	5 μm
Buried-layer diffusion (both up and down)	2 μm
Buried layer to isolation spacing	5 μm
Lateral diffusion = vertical diffusion	

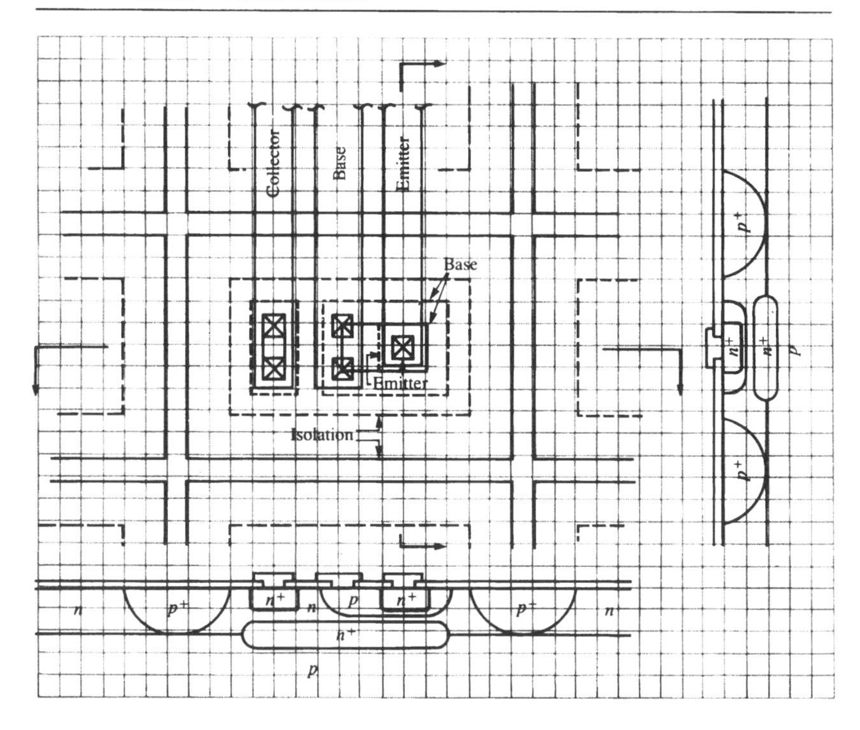

Fig. 2.5 An example of minimum area bipolar transistor layout on a 5 μm grid with a given feature size of 5 μm. The solid lines in the plan view represent the edges of the step-and-repeat mask patterns, with the dotted lines representing the final lateral spread of emitter, base and collector regions assuming equal vertical and lateral diffusion distances. Note that the isolation area occupies over 60% of the total transistor area, illustrating the inefficiency of isolation in this early fabrication process. (Reprinted with permission from Ref. 2.)

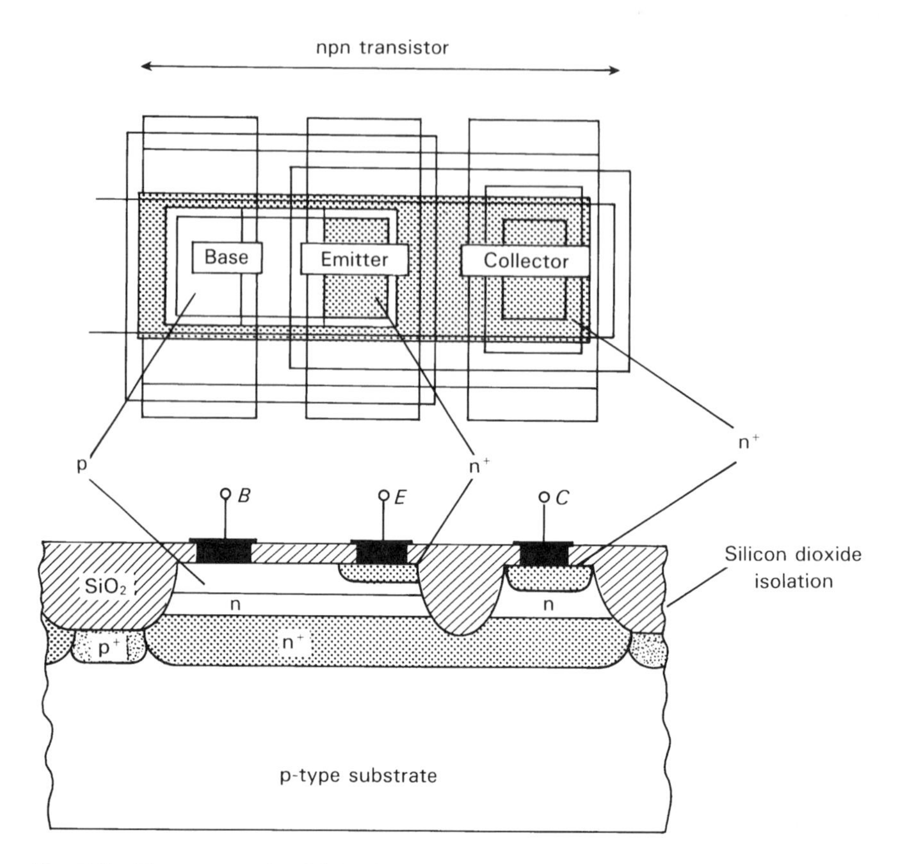

Fig. 2.6 The cross-section fabrication details of a typical oxide-isolated bipolar junction transistor [9]. The base and emitter regions may be interchanged as shown here by some manufacturers (not to scale).

surface damage due to differing thermal expansions, and hence the need for the thin interleaving SiO_2 layer.)

- Windows where oxide-isolation boundaries are required are cut through the Si_3N_4 and the SiO_2, followed by an etch which is allowed to dissolve away about half the thickness of the exposed epitaxial layer.
- A boron implant to form a p^+ region is then made on the surface of all these etched boundaries, followed by a long high-temperature cycle to grow a thick SiO_2 layer in these boundaries.

Since SiO_2 occupies about twice the volume of the silicon from which it is produced, the effect of the last step is to cause the SiO_2 to grow deeper into and fill

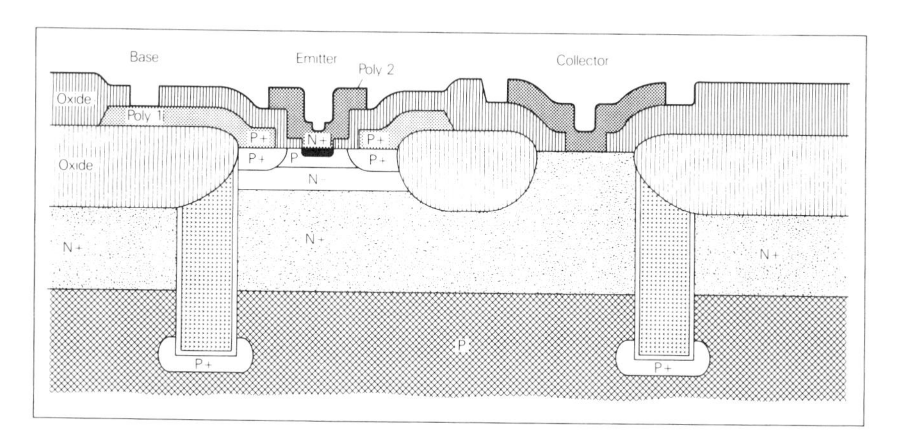

Fig. 2.7 The HE bipolar fabrication process, with 1 μm double-polysilicon and deep trench isolation. (Courtesy of GEC Plessey Research, UK.)

the cuts in the epitaxial layer. The SiO_2 growth finally reaches the silicon substrate, with the p^+ implant being driven ahead of it.

The subsequent processing stages window and implant the n^+ and p^+ regions into the epitaxial layer after removal of the thick Si_3N_4 layer. However, unlike the SBC process where the emitter region is within the base and collector regions, the base, emitter and collector regions of the oxide-isolated junction transistor are side by side. Also in the detailed fabrication steps, the emitter, base and contact regions are self-aligned, that is, they are positioned by the windows made in the SiO_2 rather than relying upon the accuracy and positioning of separate masks of the areas required—this is a carry-over from CMOS processing and will be mentioned again later. The total effect of this method of bipolar fabrication is to make the transistor size considerably smaller than the SBC process, giving better performance and much higher packing density.

Different manufacturers have different variations of this oxide-isolated process. Polysilicide, a low resistance form of polysilicon, may be used instead of metal for the interconnections into the base, emitter and collector regions, another technique which has been copied from MOS technology.

Yet another way that has been pursued to provide device isolation in bipolar technology is 'deep-trench' isolation. This is illustrated in Figure 2.7. The deep trenches, which reach down to the substrate, are cut by a reactive-ion etching process, which produces a sharp etch with minimum side-spreading. The surfaces of these trenches are then oxidized to produce the isolating SiO_2, and the remaining volume is then filled with polysilicon. This isolation process is done after a Si_3N_4 protective layer has been applied and windowed, as in the oxide-isolated process described above.

The above illustrations each show npn transistors. The alternative polarity pnp devices are available, but inherently do not have the same maximum performance as npn since the majority carriers in a pnp device are holes rather than electrons [7]. In the SBC process the pnp transistors are frequently lateral devices, with the pnp action lying along the plane of the device, rather than vertically as in Figure 2.4; in oxide-isolated and deep-trench fabrication the n^+ buried layer forms part of the base rather than the collector.

The remaining devices required for analog circuit designs, namely diodes, resistors and possibly capacitors, are all available in silicon technology. Diodes are frequently and conveniently formed by using one of the pn junctions of a npn transistor, either the collector-to-base junction with the base possibly shorted to the emitter, or the base-to-emitter junction with possibly the collector shorted to the base [1]. In general, therefore, the silicon area occupied by a diode is the same as for a transistor.

Resistors and capacitors are formed by appropriate areas in the monolithic structure. Resistors may be made in the polysilicon layer, with values from about 20 Ω to 20 kΩ with good matching but poor absolute value tolerances. Alternatively, a p-type diffusion can be used. (In MOS technology a permanently biased-on field-effect transistor may be used as a resistor [see Section 2.2], which occupies far less silicon area than these bipolar technology methods.) Capacitors may be fabricated using polysilicon and metal as the two plates, or alternatively the depletion layer capacitance of a reverse biased pn junction may be employed. The latter is very voltage sensitive, and can only provide very small capacitance values, generally less than 1 pF. However, the ready availability of voltage gain in monolithic circuits using operational amplifiers means that the professional analog circuit designer can magnify the value of any capacitance value which is physically present to greater values when required, and can also generate inductance which is not directly possible in semiconductor circuits by appropriate negative impedance circuits and gyrators. Hence very efficient analog circuits, including complex filters, can be designed in silicon.

At present, bipolar technology is dominant for analog applications because of its superior performance compared with MOS technologies. For this reason we have introduced the basic bipolar fabrication details here under analog considerations. Further fabrication details may be found in many more specialized texts [1–3,9,10]; specific analog custom microelectronic products using bipolar technology will be covered in Chapter 4.

2.1.2 Transistor-Transistor Technology

Transistor-transistor logic (TTL) represents the original and the most widely used technology for digital logic applications, being the standard off-the-shelf product for original-equipment-manufacturer (OEM) use from the late 1950s onwards. It has, however, evolved from its original 'standard' form into many variants which provide enhanced performance capabilities. Off-the-shelf CMOS logic circuits have also had to become compatible with standard bipolar logic. Figure 2.8 illustrates these TTL developments.

Figure 2.9(a) shows the original standard TTL logic configuration, employing a multiple-emitter transistor as the input stage, and a totem-pole output stage giving active pull-up to V_{CC} and active pull-down to 0 V for the logic 1 and 0 output states, respectively. The principal disadvantage with this standard circuit is its high power consumption, about 10 mW per gate, with a sharp current spike when switching from one output logic level to the other because of the momentary conducting path from V_{CC} to 0 V while the two output transistors are changing state [1,7].

Variations of this circuit were developed to provide higher speed or lower power consumption by varying the resistor values. However, the principal modi-

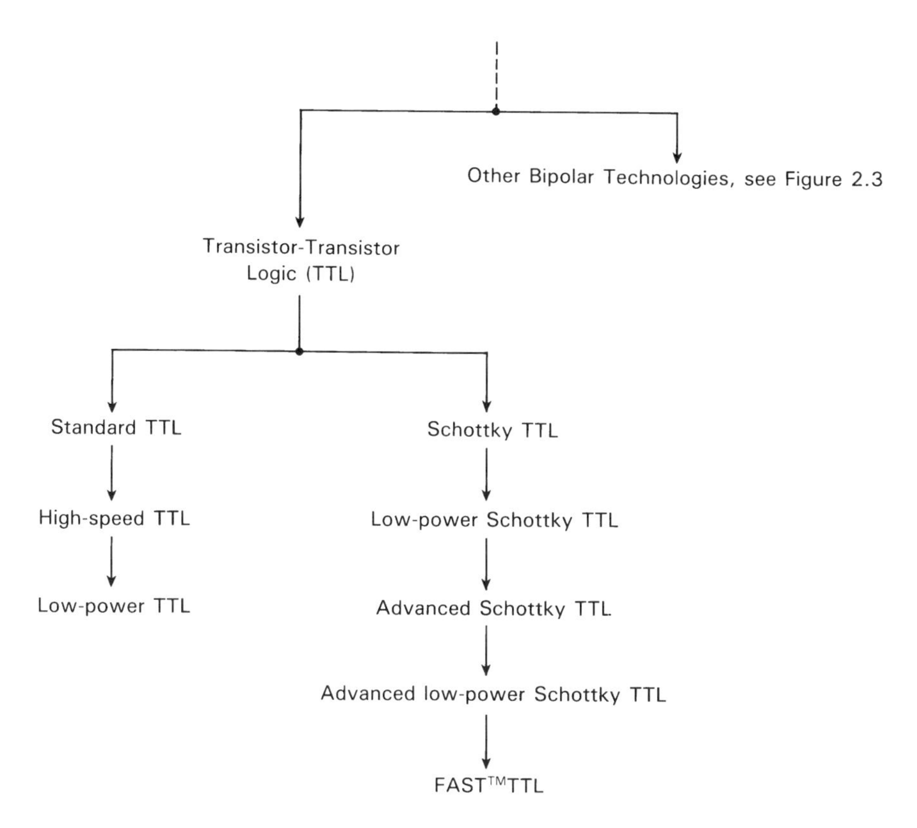

Fig. 2.8 The main families of transistor-transistor logic (TTL) digital circuits.

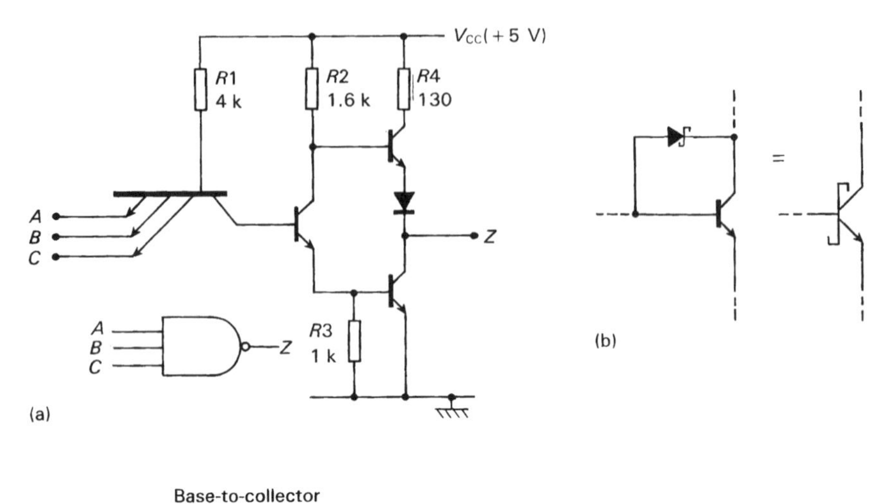

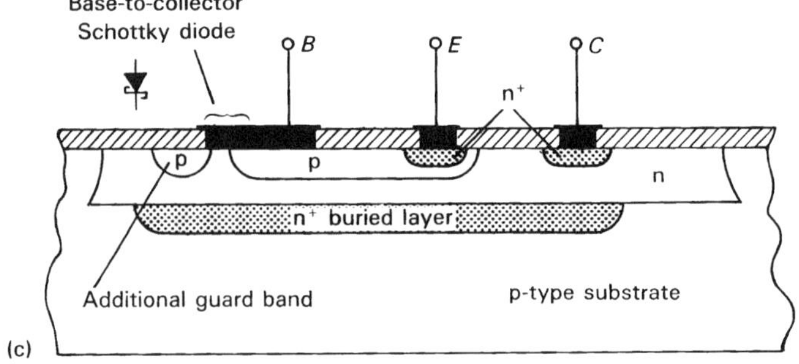

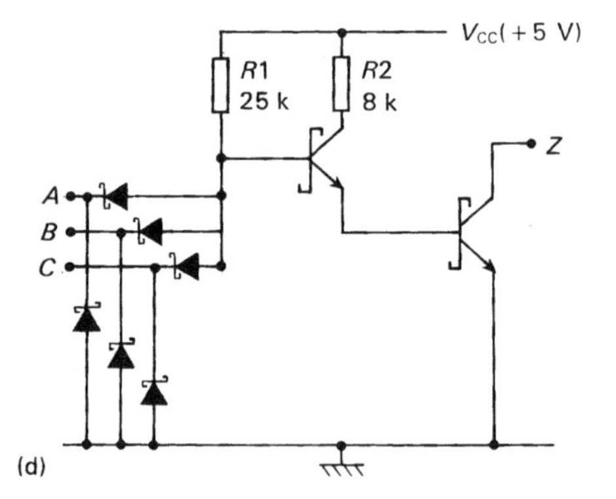

Fig. 2.9 TTL circuit details: (a) the original TTL, with a propagation delay of about 10 ns are a maximum frequency of operation of about 10–15 MHz; (b) the Schottky diode used to prevent collector saturation; (c) the fabrication details of (b); (d) typical low-power Schottky TTL NAND gate, cf. the resistor values with those of (a).

fication that improved efficiency came with the introduction of Schottky diodes to eliminate transistor saturation and hence increase switching speeds. These variants are collectively known as Schottky-clamped TTL.

A Schottky diode is a rectifying junction formed between aluminium and lightly doped n-type silicon [1,3,7]. The aluminium may be considered to act as weak p-type dopant at the surface of the silicon, resulting in a junction with pn properties. However, conduction is almost entirely due to electron emission from the silicon into the aluminium, which gives rise to junction characteristics that differ from normal pn junctions, particularly in the negligble storage time and low forward volt-drop, typically 0.35 V compared with 0.6–0.7 V for normal silicon pn junctions. However, its very low reverse breakdown voltage precludes this form of junction diode from being used as a normal rectifying device. Note that a junction between aluminium and more heavily doped n-type silicon becomes a normal ohmic connection and not a pn junction, since the p-type dopant effect of the aluminium is now swamped by the higher n-type dopant.

The incorporation of Schottky diodes in bipolar logic circuits is now universal, both as separate diodes for isolation and unidirectional purposes and also between the base and collector of npn transistors to prevent saturation where they can be incorporated with a very low silicon area overhead. Figure 2.9(b) shows the latter, with Figure 2.9(c) showing typical fabrication details. As the collector voltage falls and approaches the saturation value, the Schottky diode between base and collector will begin to conduct, thus preventing the collector potential from dropping further to its $V_{CE(sat)}$ value, excess input current being diverted from the transistor base into the collector as soon as the Schottky diode conducts.

A typical low-power Schottky TTL circuit is illustrated in Figure 2.9(d), which shows Schottky diodes used as the gate inputs rather than the multiple emitters shown in Figure 2.9(a).

The other TTL families noted in Figure 2.8 generally follow the principals shown in Figure 2.9, with individual manufacturer's variations. FAST™ TTL, for example, was a second generation, low power Schottky family developed by Fairchild, which introduced oxide isolation to increase the maximum operating speed and the packing density of the logic circuits.

2.1.3 Emitter-Coupled Technology

Emitter-coupled logic (ECL) represents the fastest available silicon-based technology for digital circuits, surpassed only by modern gallium-arsenide (GaAs) developments (see Section 2.5). However, hand-in-hand with this very high speed capability goes high power dissipation, which limits the number of gates per IC, and also small voltage swings between logic 0 and 1, which causes interfacing complexity to other types of circuit. The high power dissipation often results in very sophisticated packaging and cooling in products such as state-of-the-art computers. In general, ECL is not regarded as a mainstream VLSI technology, and does not feature very significantly in custom microelectronics.

Figure 2.10 shows the circuit topology of a typical commercial ECL logic gate. Transistors *T*1 and *T*2 form the main component, an emitter-coupled ('long-tailed-pair') circuit, where constant current is switched from one side to the other depending upon whether an input voltage is less than or greater than the reference voltage on the base of transistor *T*2. The resistor values *R*1, *R*2 and R_E are chosen such that all transistors are working in the nonsaturated (active) mode, so that maximum switching speeds result. The small output voltage swings, typically about 0.8 V, also mean that charging and discharging times of the circuit capacitances are kept to a minimum, which, with the low circuit impedance values,

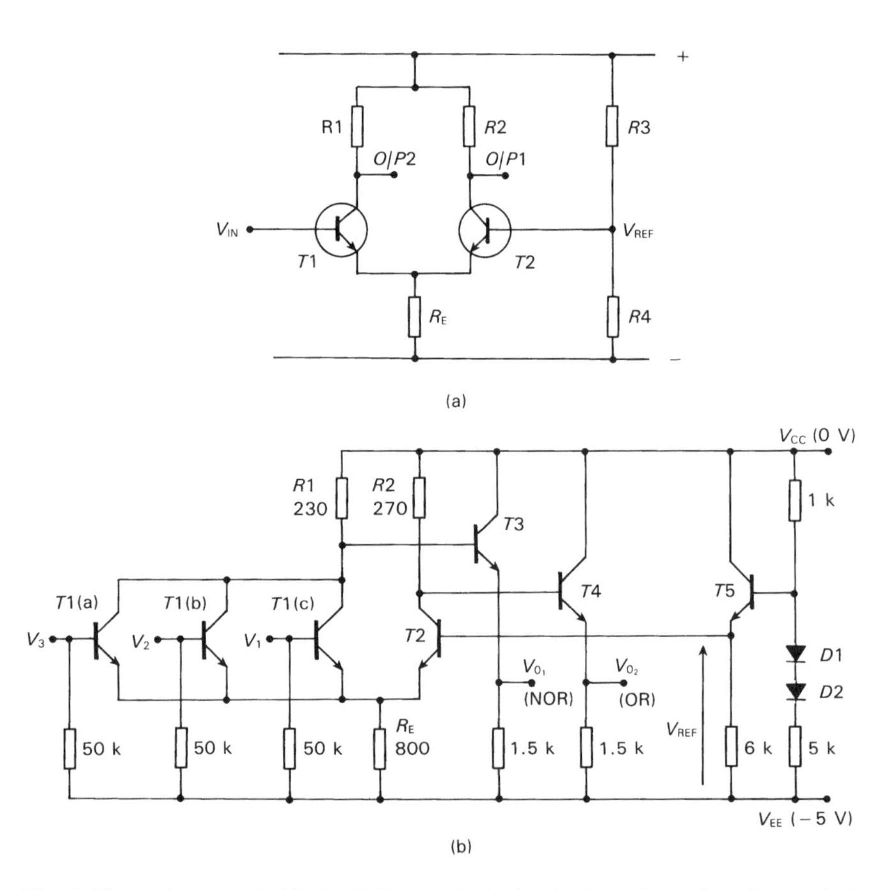

Fig. 2.10 Emitter-coupled logic (ECL): (a) the basic circuit configuration; (b) a typical commercial 3-input gate with NOR and OR outputs.

help to give the very high speed switching times. The emitter follower transistors *T*3 and *T*4 provide the low output impedance drive capability, and adjust the output voltage levels so as to be appropriate to drive further similar circuits.

Because of its small voltage swings and high power dissipation, ECL is not widely used except in the very fastest mainframe computers and in extremely specialized high-speed applications.

2.1.4 Integrated-Injection Logic Technology

The original integrated-injection logic (I^2L), sometimes referred to as 'merged transistor logic' (MTL), was first developed in the 1970s in an attempt to develop a bipolar logic family with reduced power consumption and better packing density than the available TTL circuits. The full range of developments included oxide-isolated integrated-injection logic, termed I^3L, integrated Schottky logic, termed ISL, and Schottky transistor logic, termed STL. These bipolar technologies have never been adopted for standard off-the-shelf products, but they have been used in a small number of custom IC products.

The fabrication techniques used in I^2L are similar to those already covered, but the circuit configurations are completely different [1,3,11,12]. Figure 2.11(a) shows the basic I^2L gate circuit. Transistor *T*1 is a lateral pnp device and *T*2 is a vertical multiple-collector device, the fabrication of these two transistors being merged into one, as shown in Figure 2.11(b). Junction isolation is used between separate gates.

The action of I^2L is as follows. Transistor *T*1, termed the 'injector,' provides a constant current which acts as the base input for *T*2 unless this current is sunk by a low voltage (logic 0) signal applied to the gate input terminal. When acting as the *T*2 base input current, this current switches *T*2 on, and allows each of the individual and separate collector outputs to act as a sink (low resistance) to 0 V. In the absence of this base input, each collector effectively becomes open-circuit. Thus, the circuit configuration inherently has an inverting action, acting on current sinks rather than our more familiar voltage signals. It is also a single-input, multiple-output logic gate, requiring specific logic functions to be made by commoning of outputs, for example, as shown in Figures 2.11(c) and (d).

The I^2L fabrication follows standard planar epitaxial methods. However, the vertical transistor is operating in reverse mode compared with the usual mode shown in earlier diagrams, with the n-type epitaxial layer acting as the emitter of *T*2 rather than the collector. This allows the multiple collectors to be formed by separate n-type diffusions into the p-type diffusion region.

The elimination of all resistors apart from R_E means that the packing density ol I^2L gates can be high. Speed performance can be varied by increasing or decreasing the value of the injector current I_B (varying the value of R_E), trading

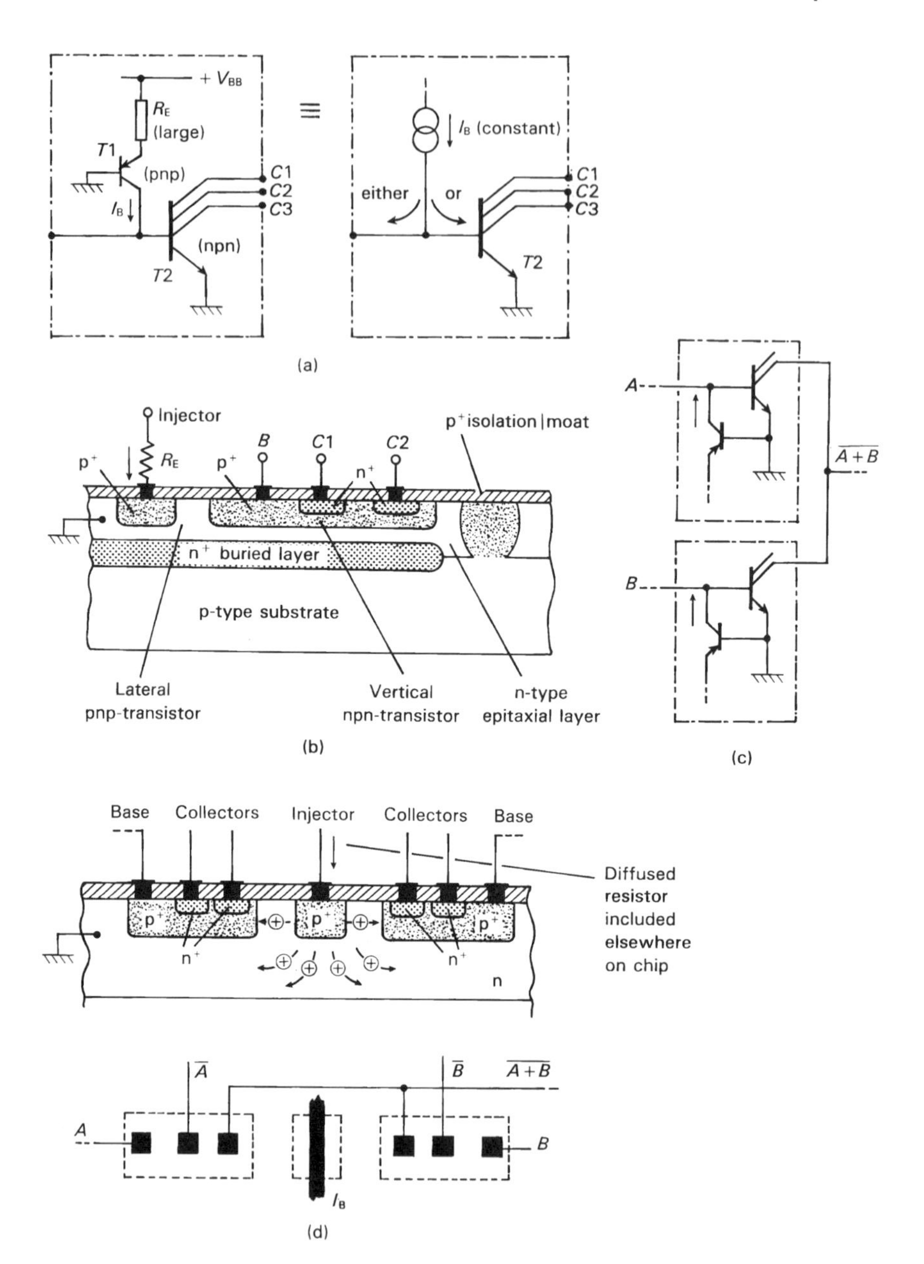

Fig. 2.11 Integrated-injection logic (I^2L): (a) the basic circuit configuration; (b) fabrication details; (c) wiring of collector outputs to provide NOR logic; (d) merged fabrication for (c) using one common pnp transistor.

off speed against power dissipation, for example, 0.01 μs gate delay at 100 μW dissipation or 10 μs delay at 0.1 μW dissipation.

However, although the packing density and speed-power product of I^2L is good, its maximum speed performance is not as good as other bipolar technologies. This is due to saturation of the npn transistor plus its low gain due to using the lightly doped epi. layer as the emitter, and also to limitations in the lateral pnp transistor. One improvement was to use oxide-isolation (see Figure 2.6) around each device rather than the pn junction isolation shown in Figure 2.11; the terminology 'isoplanar I^2L' or 'I^3L' was used for this development. Additional advantages were the self-aligned base and collector regions, which meant that smaller devices and hence better performance could be achieved [3]. Nevertheless, performance remained generally inferior to Schottky-clamped TTL.

Integrated Schottky logic (ISL) was an attempt to combine the best features of both I^2L and Schottky TTL. Figure 2.12(a) shows the basic circuit configuration, with the fabrication details being shown in Figure 2.12(b). Unlike the original integrated-injection logic, the current to the base of the npn transistor *T*2 is now supplied through some external polysilicon resistor rather than through the merged lateral transistor *T*1. The latter now acts as a clamp across the collector of *T*2 to prevent it from saturating. Additionally, the vertical transistor *T*2 has been reinverted compared with the I^2L fabrication, so that the epi. layer is once again the collector of the npn transistor. Finally, Schottky diodes have been added to the multiple collectors to improve the isolation between each collector output.

The effect of these changes in fabrication was to maintain the low power consumption of I^2L and I^3L, but to improve maximum speed performance by a factor of about two or three. A trade-off between speed and power consumption

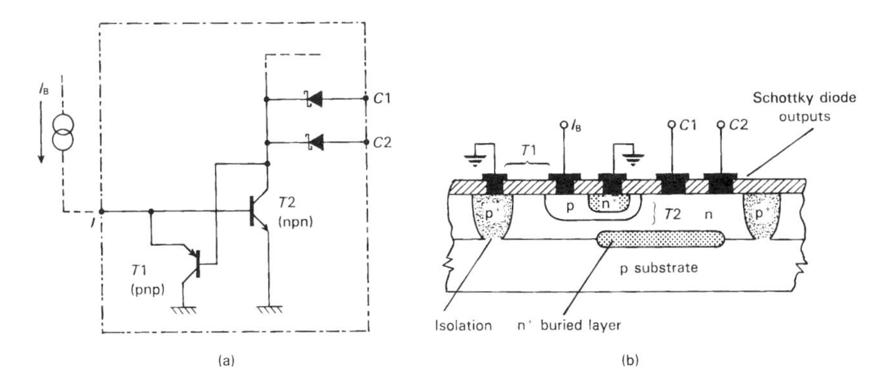

Fig. 2.12 Integrated-Schottky logic (ISL): (a) the basic circuit configuration; (b) fabrication details.

was still present but packing density was not as good, although still better than normal Schottky TTL. The Schottky-isolated output logic swings were reduced by about 250 mV by the presence of these Schottky diodes in comparison with corresponding I^2L and I^3L circuits.

The last of these closely related families of bipolar circuits which attempted to improve upon the characteristics of TTL was the Schottky transistor logic (STL), which in some respects can be viewed as a very close relative of Schottky TTL, thus completing a circle of bipolar fabrication techniques. Figure 2.13 shows the basic fabrication details of STL. The current input comes from a resis-

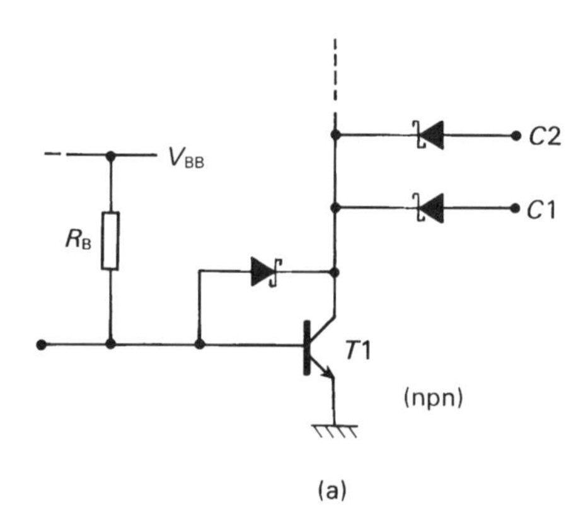

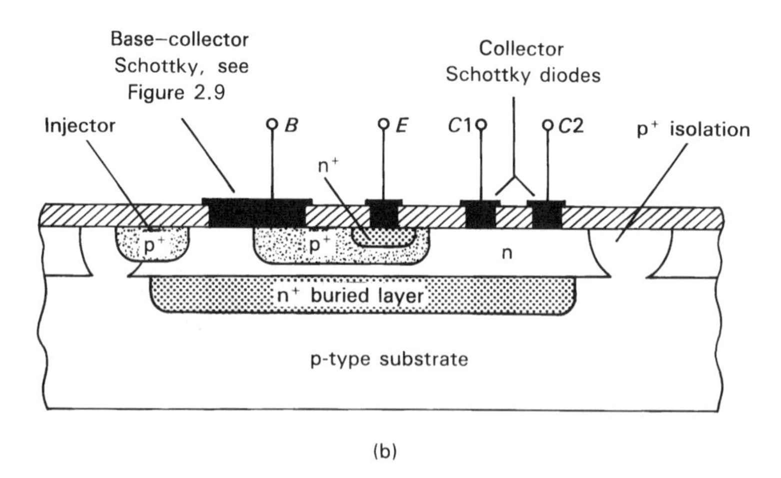

Fig. 2.13 Schottky-transistor logic (STL): (a) the basic circuit configuration; (b) fabrication details.

tor source, and the multiple collectors have individual Schottky-diode isolation as in ISL, but now the pnp lateral transistor clamp to prevent saturation of the main npn transistor has been replaced by a Schottky diode as used in normal Schottky-clamped TTL (see Figure 2.9(b)).

The final result is a logic configuration which maintains the small size of ISL, but offers about twice the maximum gate propagation speed. Two extra mask stages are, however, required. Hence we have in total the development of I^2L, I^3L, ISL and STL, all of which can have better packing densities than TTL and have therefore been considered as candidates for large scale integration. All use the same basic fabrication steps as TTL, and it is therefore theoretically possible to fabricate more than one type on the same chip if required. However, none of these alternative bipolar technologies has proved to have any lasting place in the custom field, although they may still possibly be encountered in some long-lasting custom equipments; CMOS technology is now *the* dominant technology for digital custom circuits, challenged by bipolar only when analog or mixed analog/digital circuits are required.

2.1.5 Collector Diffusion Technology

A bipolar technology which has, however, had almost three decades of use in custom microelectronic ICs is collector-diffusion isolation (CDI) technology. It also has now been largely superseded by CMOS, but it remains possibly the most successful bipolar technology to have been used in custom circuits, with a large variety of products still in use incorporating such circuits.

The CDI process was originally invented for digital applications by Bell Laboratories, but was further developed for both digital and mixed analog/digital applications by Ferranti Electronics, now part of GEC-Plessey Semiconductors, UK. The basic process is shown in Figure 2.14(a). The substrate is p-type silicon upon which is first formed a low-resistance n^+ buried layer. A thin p-type epitaxial layer is then grown on this buried layer to form the eventual base regions. Device isolation is then made by means of a deep n^+ diffusion through the p-type epi. layer to form isolated islands in this epi. layer. This device-isolation moat also forms the collector of the final transistor, thus surrounding the base region on all sides by the collector. A shallow n^+ diffusion or implantation into the epi. layer to form the emitter is then performed, followed by the final emitter, base and collector contacts to complete the fabrication steps [2].

This process is therefore an extremely simple fabrication process, and can produce high-performance, small-silicon-area devices. The relatively large collector-base and collector-substrate capacitances to some degree limit performance, but this is mitigated by low collector resistance values and very small vertical n-p-n distances.

The CDI process has principally been used in current-mode logic (CML)

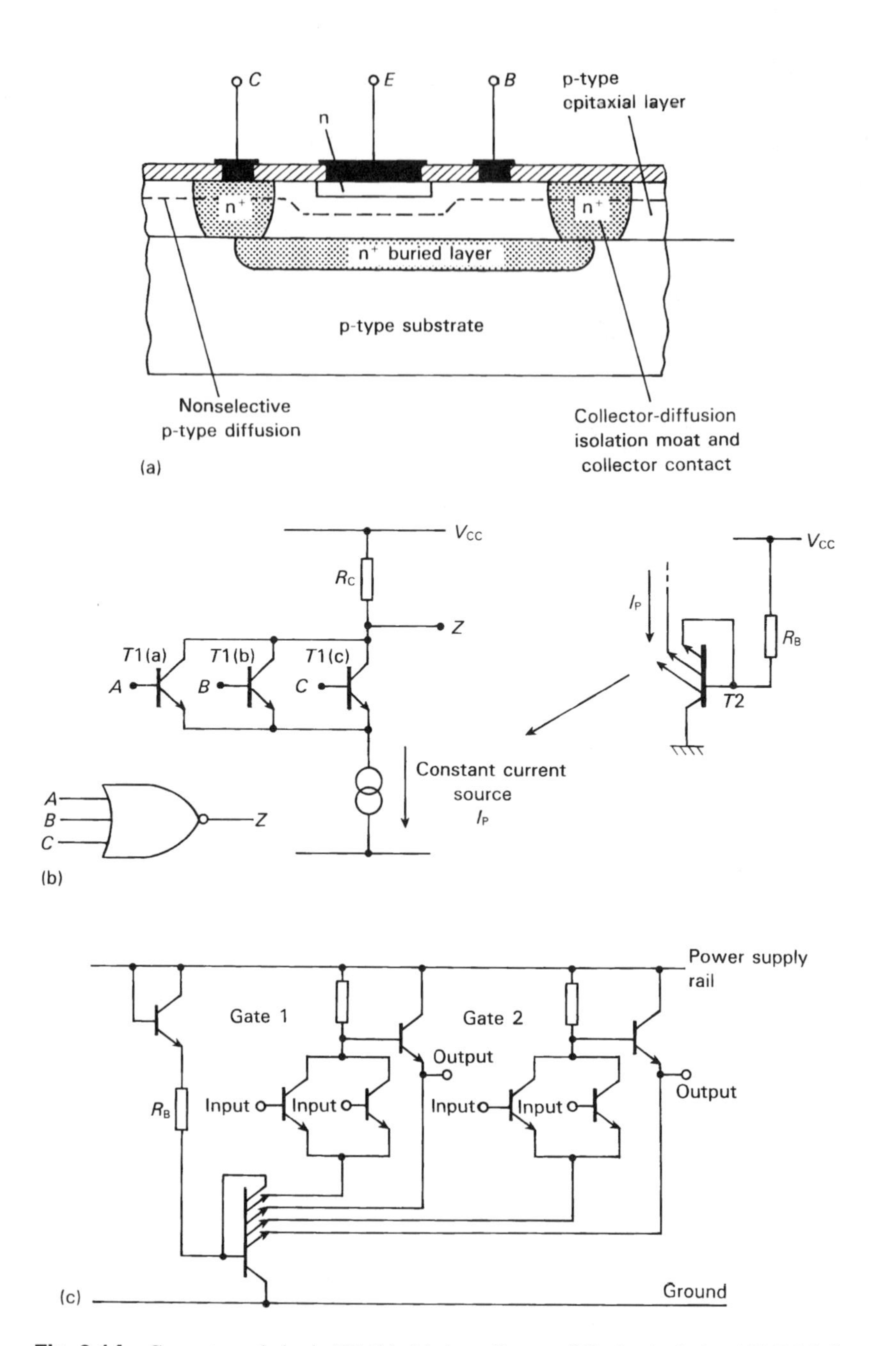

Fig. 2.14 Current-mode logic (CML): (a) the collector-diffusion isolation (CDI) fabrication; (b) the basic CML circuit configuration; (c) the multiple-emitter current source used to supply separate gates and also output emitter followers.

circuit configurations. Figure 2.14(b) illustrates the basic configuration. Unlike ECL, where a constant current is switched from one collector path to another (see Figure 2.10), in CML a preset current is allowed to flow or not to flow through the main switching transistors $T1$, the value of this current being fixed so that saturation of the switching transistors does not occur. Thus, no Schottky diodes need to be fabricated. A low output voltage swing, typically 0.8 V, with very fast switching times is achieved. Level shifting between these on-chip voltages and the outside world is required.

The preset current source in the emitter circuit is also very efficiently provided by multiple-emitter transistors, each emitter serving a separate logic gate. As shown in Figure 2.14(c), each current source is a current mirror, where the current allowed in each emitter is controlled by the value of resistor R_B. The multiple emitters are fabricated in the (normal) collector region of Figure 2.14(a), with the emitter used as the collector in this inverted mode of operation; a particular feature of the CDI process is that good inverted transistor performance is available due to the heavy p-type doping of the diffusion isolation region. The gate power can be adjusted over a very wide range by choice of the constant current value, giving a trade-off between power and speed ranging from about 10 μW per gate at a speed of a few MHz to 1 mW per gate at 100 MHz. This is comparable to the power dissipation of many standard CMOS logic circuits when toggling at very high speeds. Finally, the supply voltage tolerance of CML is extremely good since all on-chip voltages track together; operation down to as low as 1 V can be provided.

The CDI fabrication process with the resulting CML circuits has therefore been used very profitably for niche markets in the custom field, particularly where low supply voltages or other ''non-TTL''–like applications were required. We will mention them again in Chapter 4, where the analog as well as the digital capability will be noted.

2.1.6 Bipolar Summary

The basic processes used in bipolar fabrication are well established, the key steps in most modern processes being the use of the *buried layer* followed by an *epitaxial layer*. Isolation between devices, originally using reverse-biased pn junctions, has been superseded by *silicon dioxide* or *deep-trench* isolation in order to increase the packing density and speed. Collector-diffusion isolation (see Section 2.1.5) remains the odd man out, not receiving any widespread industrial support in spite of considerable advantages before CMOS was developed to its present level of sophistication.

Schottky-diode clamping, introduced in order to prevent transistor saturation, now features strongly in all bipolar logic circuits except ECL and CML. In ECL and CML saturation is inherently avoided by control of the circuit current values, thus saving the fabrication steps necessary to produce the lightly doped n-type silicon/aluminum Schottky interface. The use of *polysilicide interconnect* is increasingly evident.

The several developments of *integrated-injection logic* noted in Section 2.1.4 have not found widespread use and must now be considered as obsolete since no overriding advantages over other mainstream fabrication technologies could be attained. The principal bipolar fabrication problem which limits yield, however, continues to be collector-to-emitter leakage and short circuits. This is compounded by the desire to produce very narrow base widths in the fabrication process (see Figure 2.4), but process tolerances or imperfections can cause this width to be locally reduced or bridged. This critical fabrication feature is not present in MOS technologies, which therefore have a fundamentally higher potential yield than bipolar.

Further details of bipolar fabrication may be found in more specialized texts [1–4,6–10]. A comparison of the performance of bipolar technologies and MOS technologies will be given in Section 2.6.

2.2 UNIPOLAR SILICON TECHNOLOGIES

Unipolar technology is based upon the use of field-effect transistors (FETs) as the active devices, in which the controlled current flow is by majority carriers only. In n-channel devices the flow of carriers between source and drain is by electrons, and in p-channel devices, it is by holes. The potential applied to the third electrode, the gate, controls the number of free carriers flowing in the channel, and hence the source-to-drain current. Since electrons have a higher mobility than holes, n-channel devices are inherently faster by a factor of about 2.5 times than comparable p-channel devices.

The gate control of a FET may be achieved by the following two methods:

(a) where the gate potential controls the effective cross-sectional area and hence resistance of the source-to-drain channel by creating a reverse-biased pn junction depletion layer in the channel
(b) where the gate potential capacitively enhances or depletes the number of majority carriers in the channel

The former is usually termed a *junction FET (''JFET'')*, and can be either a n-channel JFET or a p-channel JFET. The latter may be termed an *insulated-gate FET (''IGFET'')* because the gate electrode is separated from the channel by a thin insulating layer of silicon dioxide. Since in the original IGFET construction the gate was metal, the metal-oxide-silicon cross-section of this gate gave rise to the alternative terminologies ''MOSFET'' or ''MOST,'' terms which have been erroneously maintained even when the metal gate electrode was superseded by polysilicon, as will be illustrated later. Again, both n-channel and p-channel IGFETs are possible. This complete spectrum of unipolar devices is summarized in Figure 2.15(a).

Although the fabrication of junction FETs (JFETs) is basically somewhat simpler than bipolar transistors, not requiring a buried sub-collector, their performance is not such as to justify their use except in some applications such as low-power off-the-shelf operational amplifiers. They do not feature at all in digital applications, since the insulated-gate FET has far superior advantages in logic

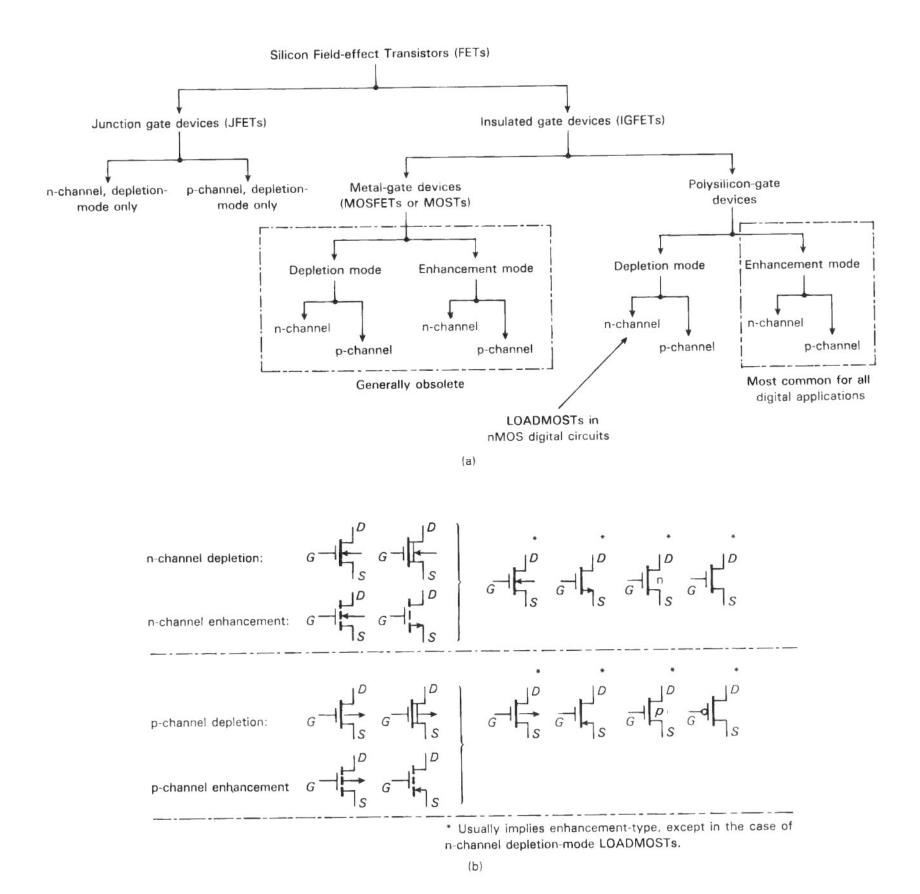

Fig. 2.15 Field-effect transistors and their symbols: (a) the FET family tree; (b) some of the insulated-gate symbols in use; several are not explicit and require the context in which they are being used to define their precise meaning. The extreme righthand symbols are commonly used in CMOS digital circuits, representing n-channel and p-channel enhancement-mode devices.

circuits. Hence, we will not refer to JFETs any further as a silicon technology, but we will encounter JFET fabrication in connection with gallium-arsenide (GaAs) technology in Section 2.5.

Therefore, confining our interests to the IGFET, there are in total four possible types:

(a) n-channel enhancement mode, in which the source-to-drain current is zero when the gate-to-source voltage V_{GS} is zero, channel conductivity being increased (enhanced) when V_{GS} is made positive
(b) p-channel enhancement mode, in which the source-to-drain current is zero when $V_{GS} = 0$, channel conductivity being increased when V_{GS} is made negative
(c) n-channel depletion mode, in which source-to-drain conduction is present when $V_{GS} = 0$, V_{GS} having to be made negative to reduce the conductivity to zero
(d) p-channel depletion mode, in which source-to-drain conduction is present when $V_{GS} = 0$, V_{GS} having to be made positive to reduce the conductivity to zero

The various symbols which may be used for these FET variants are given in Figure 2.15(b).

In most digital and other applications enhancement types are of significance, since zero source-to-drain current with $V_{GS} = 0$ is almost always required. Therefore, we shall only mention depletion mode fabrication briefly in the following sections.

2.2.1 nMOS Insulated-Gate Devices

Although complementary MOS (CMOS) technology (see Section 2.2.3) is dominant in digital custom ICs, it is appropriate to consider nMOS as a technology in its own right since it is used almost exclusively in LSI and VLSI memory circuits and certain other array-structured circuits, giving unsurpassed packing densities, as will be considered in detail in Section 2.3.

The basic structure of an n-channel (nMOS) depletion-mode FET is given in Figure 2.16. In order for source-to-drain electron flow to take place when $V_{GS} = 0$, it is necessary to implant a thin n-type region below the gate to provide majority carriers. With V_{GS} positive the number of free carriers (electrons) is increased, but with V_{GS} negative the number of carriers in the channel is decreased and source-to-drain current can be cut off. An alternative method of providing the n-type conducting channel below the gate is to grow a n-type epitaxial layer on the p-type substrate, the n^+ source and drain regions then being formed to complete the source-to-drain regions.

It should be noted that between the conducting n-type channel of an nMOS

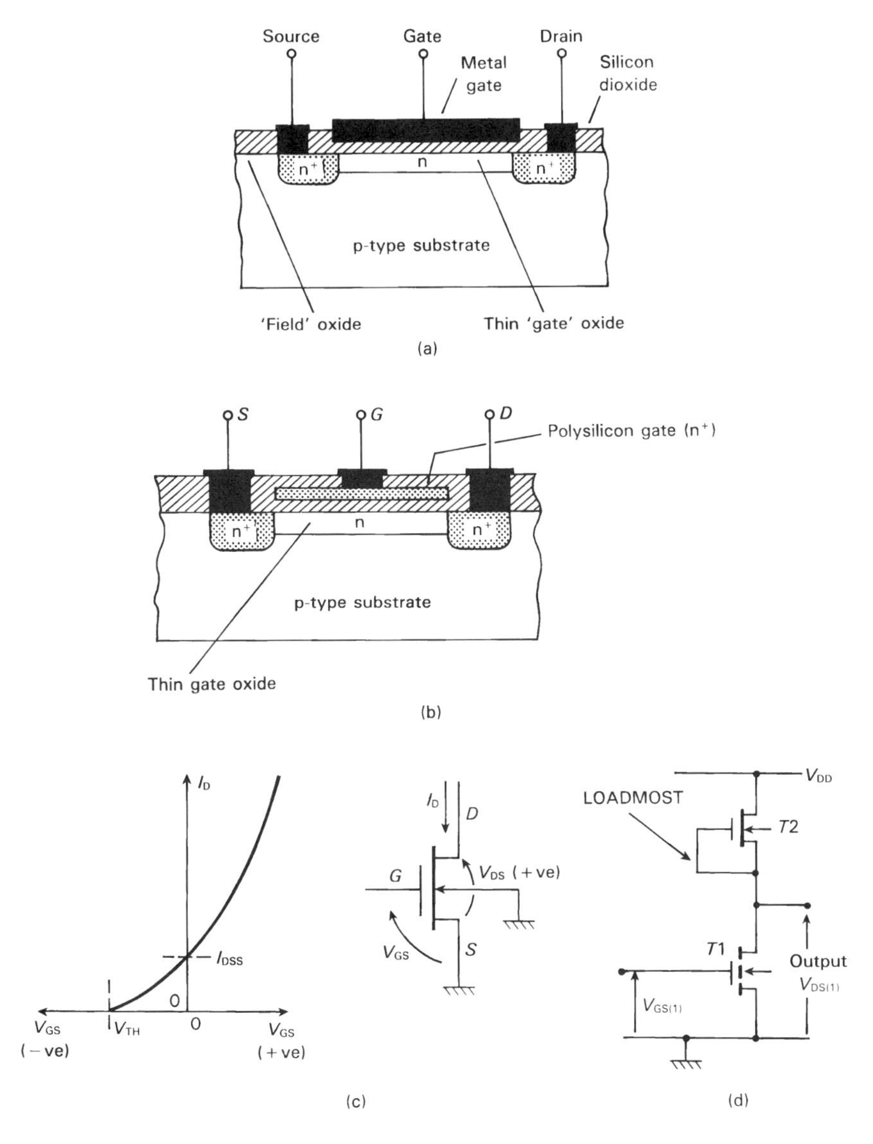

Fig. 2.16 Basic depletion-mode nMOS insulated-gate FET fabrication: (a) metal-gate fabrication, now considered obsolete; (b) silicon-gate fabrication; (c) typical I_D/V_{GS} characteristics; (d) the use of a depletion-mode FET as a load resistor.

transistor and the substrate is a pn junction, which, if the substrate is connected to the most negative potential in the circuit, means that this junction is never forward-biased. The device, including source and drain areas, is therefore isolated from the substrate, and hence is sometimes referred to as 'self-isolated.' The relatively simple controlling equations which relate the device construction and resultant characteristics may be found in many device textbooks [1,3,7,13,14], and here in Appendix D.

Depletion-mode FETs are not employed as active switching devices because of the reverse bias voltage necessary on the gate to cut off the device. However, it is extensively used as a constant-value load resistor ('loadmost' or 'loadMOST') by connecting together the gate and source electrodes (see Figure 2.16), thus establishing the fixed operating conditions of $V_{GS} = 0$ and giving roughly constant resistance between source and drain [13]. The advantage of forming a resistor in this manner is that it only occupies the area of a transistor, which is far less than resistors formed in silicon by other means.

However, enhancement-mode n-channel FETs constitute the best and most widely used FET switching device for logic gates, either in association with loadmosts or with enhancement-mode p-channel FETs in CMOS logic configurations. The basic structures of enhancement-mode devices are given in Figure 2.17. The metal-gate construction is now obsolete except in some initial families of off-the-shelf SSI and MSI CMOS logic circuits, and possibly where high-voltage operation is required, having been superseded by the higher performance, smaller silicon area silicon-gate derivatives.

The stage of fabrication of an n-channel silicon-gate FET are shown in more detail in Figure 2.18. The steps are as follows, with re-covering by photoresist being undertaken at every stage where new windows have to be formed:

1. Starting with the lightly doped p-type substrate, first form a thin layer of silicon dioxide (the 'pad' oxide), followed by a thick protecting silicon nitride layer, cf. the bipolar process given in Section 2.1.1.
2. Open window using mask No. 1 in the SiO_2 and Si_3N_4, and form a p^+ isolation border around the perimeter of each device—sometimes termed the 'channel stopper.'
3. Grow a thick SiO_2 layer, the 'field' oxide, over the p^+ channel stopper, which has the effect of driving the p^+ region deeper into the substrate as the silicon dioxide is grown.
4. Strip off the Si_3N_4 and pad oxide, and grow a very thin, very precisely controlled new SiO_2 layer (the 'gate' oxide) over the area.
5. Adjust the exact p-type doping concentration of the p-type substrate under this gate oxide by boron ion implantation, which passes freely

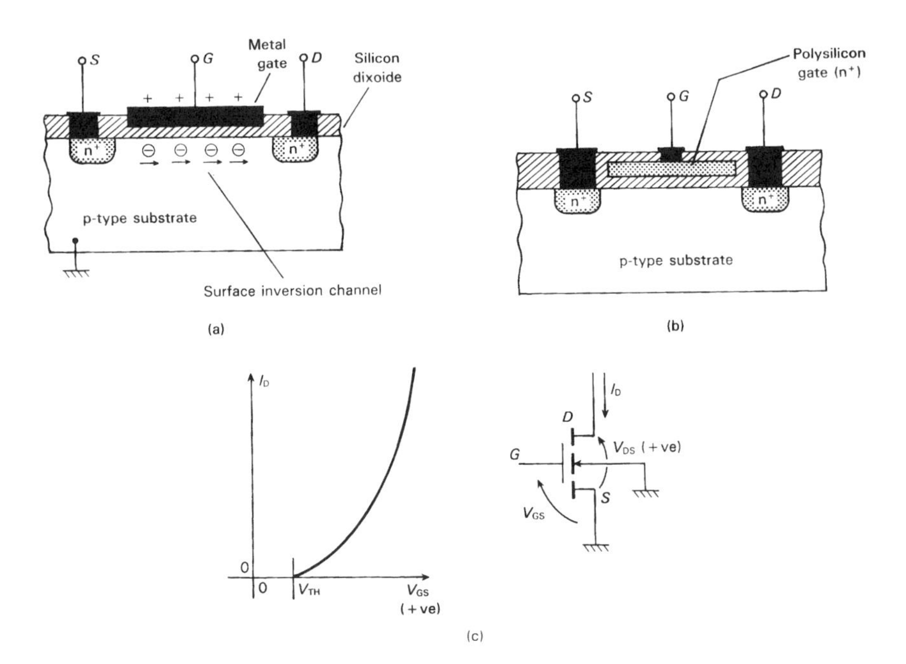

Fig. 2.17 Basic enhancement-mode nMOS insulated-gate FET fabrication: (a) metal-gate fabrication, now considered obsolete: (b) silicon-gate fabrication; (c) typical I_D/V_{GS} characteristics, V_{DS} constant.

through the thin gate oxide—this provides the threshold voltage adjustment for the final device.

6. Deposit a layer of polycrystalline silicon ('polysilicon') over the whole surface.
7. Open window (new photoresist and then mask No.2), and remove the polysilicon from everywhere except the gate area of each device—this forms the polysilicon gate for each device.
8. Form the n^+ source and drain regions of each device by arsenic ion implantation. The important feature here is that this n^+ implantation can take place through the thin gate oxide but is stopped by the thicker field oxide and polysilicon gate areas, and therefore there is no need for any separate mask to delineate the gate areas; this very significant technique is known as *self-aligning* or *self-alignment*.
9. Deposit a layer of silicon dioxide over the whole surface, thus burying each polysilicon gate.

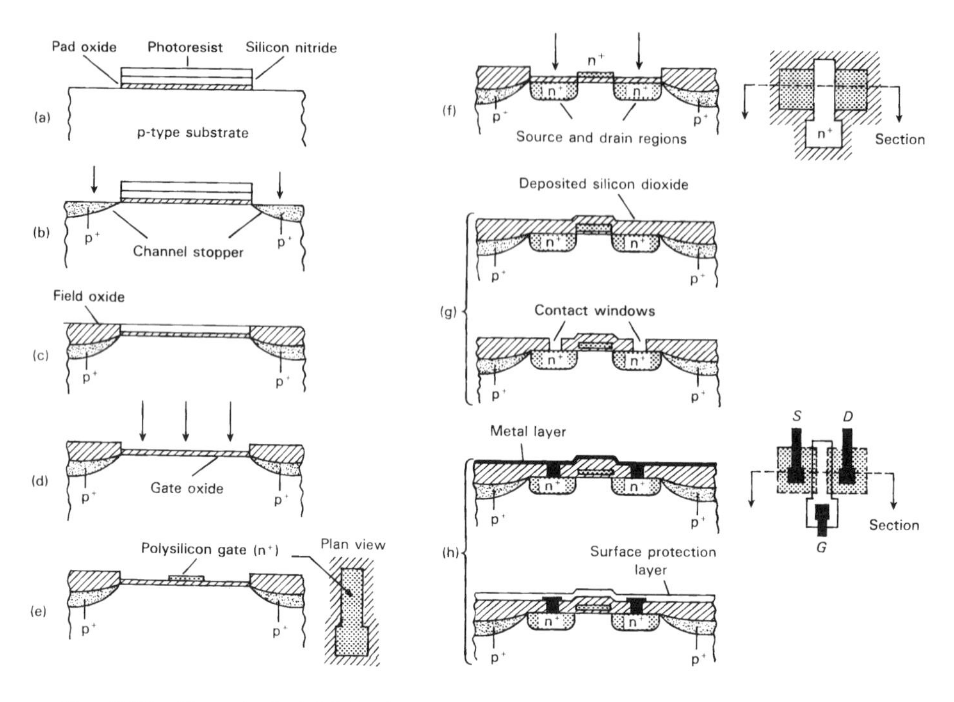

Fig. 2.18 The principal fabrication steps in enhancement-mode, self-aligned silicon-gate FET fabrication: (a) formation of pad oxide and Si_3N_4 layers; (b) p^+ implantation or diffusion; (c) growth of field oxide isolation; (d) removal of pad oxide and Si_3N_4 and growth of gate oxide and adjustment of channel doping; (e) deposition and etching of polysilicon to form gate; (f) formation of source and drain regions; (g) silicon oxide covering, with contact windows then cut through; (h) final metalization interconnect patterning and surface protection.

10. Open contact windows, known as 'vias,' through this SiO_2 layer (new photoresist and then mask No.3) to meet all source, drain and gate areas, and then cover the whole surface with metal—usually aluminium with 1% silicon.
11. Etch away all unwanted metal (new photoresist and then mask No. 4) so as to leave only the required interconnection pattern between the individual devices on the water.
12. Apply a final surface protection over the whole wafer.

This is the basic fabrication process which gives the very high performance now available from field-effect transistors. The breakthrough to high performance came with the development of the self-aligned polysilicon gate, which eliminated the problems of alignment between source, gate and drain areas inherent in metal-gate devices. With the latter there was always some overlap between the gate and the source and drain areas, which produced much higher gate-to-source and gate-to-drain capacitance values than with the nonoverlapping, self-aligned polysilicon gates. These capacitance values, accentuated by the Miller effect when the circuit is operating, severely limited metal-gate performance compared with the silicon-gate FET [1,14]. Self-alignment and the oxide isolation techniques shown in Figure 2.18 thus represent the key characteristics in modern MOS fabrication.

A further feature of the polysilicon layer of this insulated gate construction is that the polysilicon is also used as an interconnect level in the final IC to provide underpass connections below the overlaying metal layer interconnections. Although polysilicon is not as good a conductor as a metal it is adequate for underpasses and other short interconnections. This will be extensively noticed in the custom ICs covered in Chapter 4. More recent developments may use silicide rather than polysilicon, which provides a somewhat lower resistance for such connections.

The fabrication details for n-channel depletion-mode devices can be basically the same as detailed above, except that arsenide ion implantation to form the thin n-type source-to-drain conducting channel is made through the polysilicon and gate oxide layer. If both depletion mode and enhancement mode n-channel devices are required, as in nMOS memory circuits, then the channel region of each type of device has to be separately implanted to give the two distinct threshold voltages, which therefore requires an additional fabrication mask.

Certain other nMOS fabrication methods have been developed, but are now mainly applied to discrete devices for high-voltage or high-current applications rather than monolithic circuits. Figure 2.19 illustrates two such fabrication methods. In the DMOS structure shown in Figure 2.19(a) the effective channel length is the narrow p-channel dimension between the n^+ source and the intrinsic π-type region surrounding the drain. A current-carrying capacity of many amps with a switching time of a few nanoseconds is possible. The p^+ and n^+ diffusions under the source electrode are made through the same window, thus eliminating any problem of alignment.

In the V-groove (VMOS) construction of Figure 2.19(b), which was at one time considered for memory circuits, the substrate acts as the source, with the channel formed along the four sides of the groove. The channel width is determined by the thickness of the epitaxial layer, which in early days was shorter than could be achieved with conventional MOS fabrication methods using lithography—this is no longer the case. Switching times of a few nanoseconds and a current-carrying capacity of one amp or more are possible with discrete VMOS devices.

It is clear that more difficult fabrication steps are involved in these devices

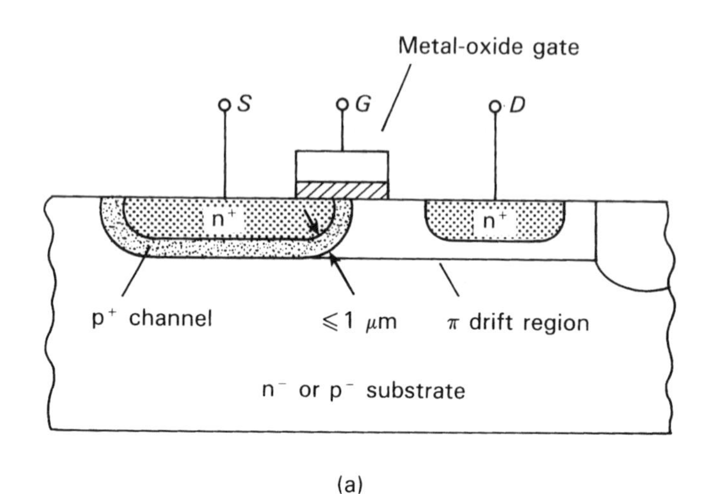

(a)

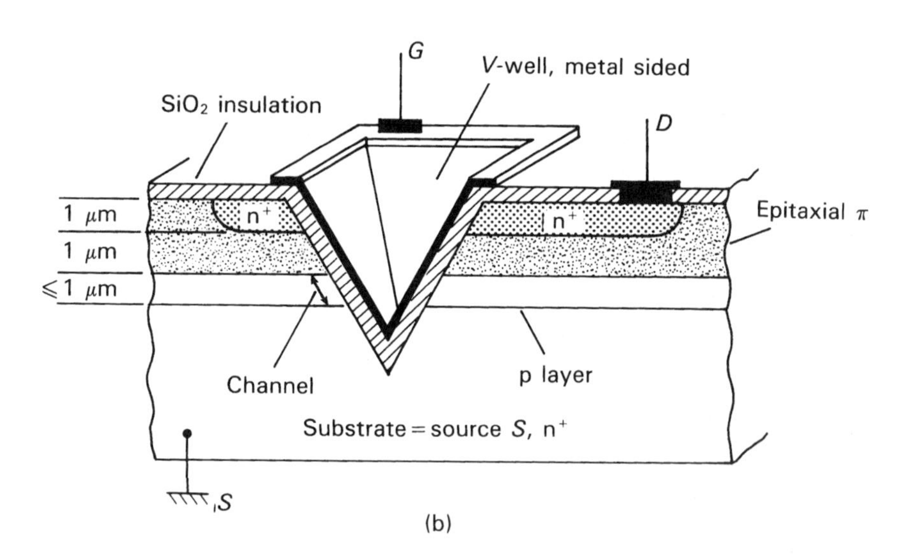

(b)

Fig. 2.19 Further depletion-mode nMOS fabrication techniques, now generally reserved for discrete power devices: (a) the double-diffused DMOS fabrication; (b) the V-groove VMOS fabrication.

compared with the self-aligned silicon-gate process, and thus they do not now present any attraction for mainstream LSI and VLSI applications. Further details may be found in more specialized publications [2,14,15–17].

2.2.2 pMOS Insulated-Gate Devices

Because of the inherently inferior performance of pMOS compared with nMOS, circuits using only pMOS devices are no longer used. This was not the case in the very early days with metal-oxide-silicon devices, since a problem known as 'surface inversion' occurred in nMOS enhancement-mode FETs between the lightly doped p-type channel region and the overlying dioxide (see Figure 2.17(a)), giving rise to an unwanted thin conducting channel and, unfortunately, depletion-mode rather than enhancement-mode operation. This problem was overcome by the introduction of the polysilicon gate with ion implantation to control device threshold voltage, thus making nMOS feasible and far preferable to pMOS for switching applications.

The general construction of pMOS devices is similar to nMOS, except that polarities are reversed. Figure 2.20 shows the basic structure of both the old metal-gate and the present silicon-gate FETs.

However, except for some discrete devices, the use of pMOS today is exclusively confined to complementary MOS (CMOS) digital logic circuits, where the enhancement-mode p-channel devices act as a switch to the positive (V_{DD}) supply rail and the complementary n-channel devices act as a switch to the negative (V_{SS} or 0 V) supply rail. We will therefore continue our discussions on pMOS in the context of CMOS fabrication techniques.

2.2.3 CMOS Technologies

The basic CMOS digital logic configurations are shown in Figure 2.21. The p-channel enhancement-mode FETs are configured in an arrangement which is always the dual of the n-channel enhancement-mode FETs, that is, a series configuration of one type is always associated with a parallel configuration of the other type; for any x-input logic gate there are always x FETs of each type, the p-channel devices providing a conducting path to V_{DD} and the n-channel devices providing a conducting path to V_{SS}. Because the resistance of a p-type FET is higher than that of a comparable n-type FET, it is preferable to series the n-type devices rather than the p-type devices, which implies that CMOS NAND configurations are preferable to NOR configurations. However, this principally affects the layout topology and interconnection details, and not the basic fabrication methods.

CMOS fabrication requires both pMOS and nMOS devices to be fabricated alongside each other on the same substrate, with appropriate insulation between

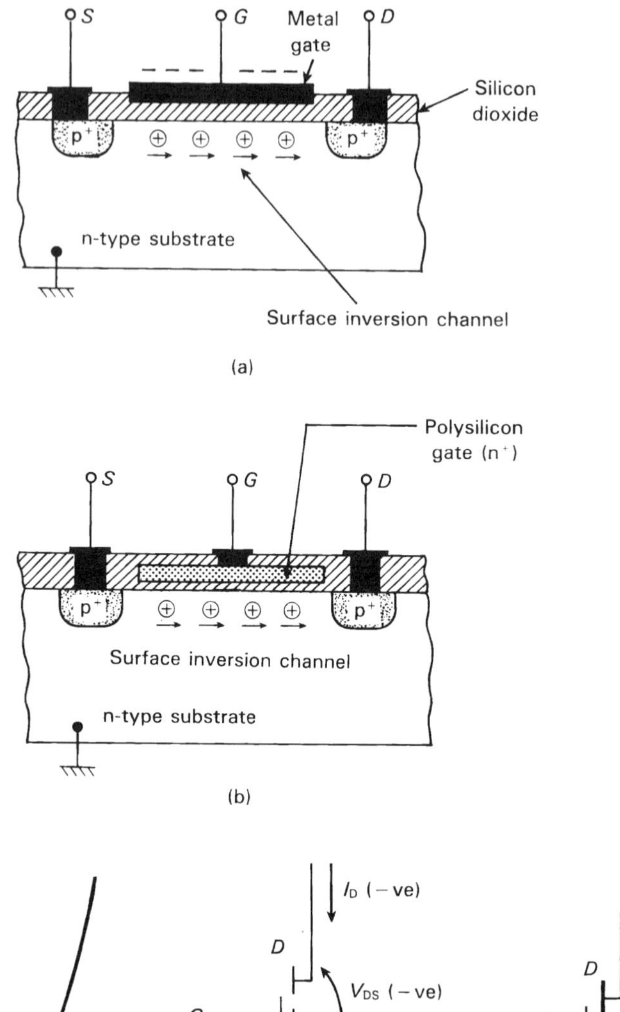

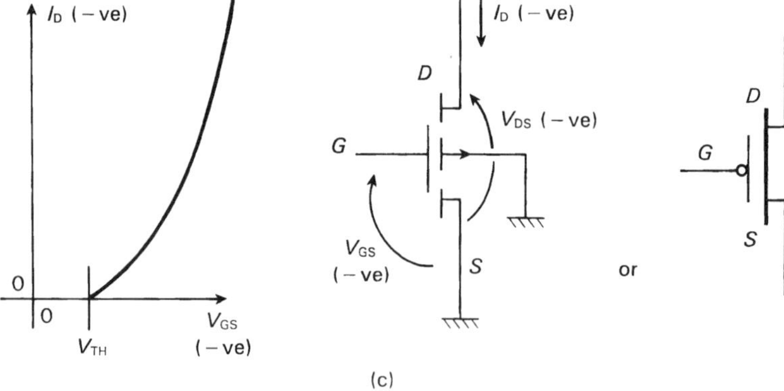

Fig. 2.20 Basic enhancement mode pMOS insulated-gate fabrication: (a) metal-gate fabrication, now considered obsolete; (b) silicon-gate fabrication; (c) typical I_D/V_{GS} characteristics, V_{DS} constant.

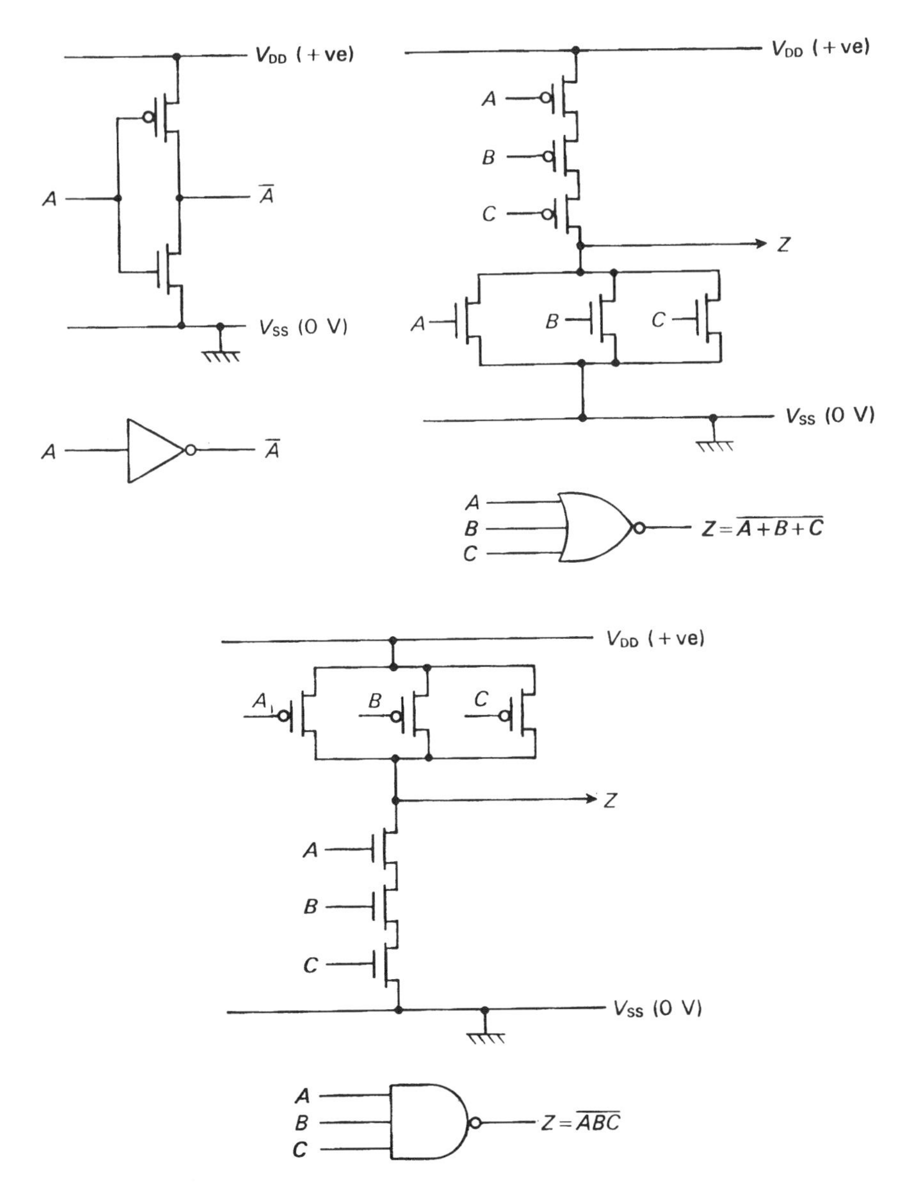

Fig. 2.21 Static CMOS logic gate configurations, with enhancement-mode p-type FETs connected to the V_{DD} rail and corresponding n-type FETs connected to the V_{SS} supply rail (cf. n-channel only logic gates where the p-channel devices are replaced by a single n-channel depletion-mode ''loadmost'').

them. The three basic methods of fabrication are shown in Figure 2.22. However, the method illustrated in Figure 2.22(a) requires that the n-wells ('n-tubs') must be rather heavily doped in order to convert the p-type substrate into n-type, and as a result the performance of the p-channel devices are degraded. Similarly, if p-wells ('p-tubs') on an n-type substrate are used, the performance of the n-channel devices is impared. The best method of fabrication is therefore the *twin-*

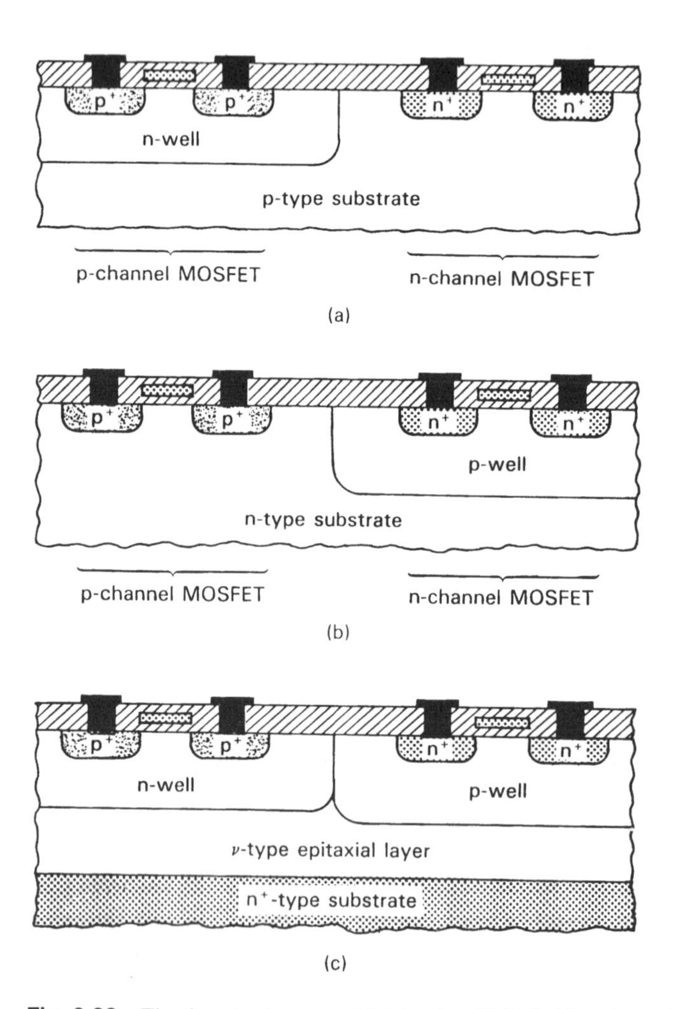

Fig. 2.22 The three basic ways of fabricating CMOS: (a) p-channel device in an n-well with p-type substrate; (b) n-channel device in a p-well with n-type substrate; (c) twin-tub epitaxial fabrication on an n-type substrate.

tub process of Figure 2.22(c), which allows both the n-wells and the p-wells that are formed in a very lightly doped epitaxial layer to be given optimum doping levels. The penalty to be paid for twin-well processing is, of course, additional processing cost.

The methods of isolation between devices are also of particular significance. One early CMOS isolation technique was to diffuse or implant a deep p^+ moat around each n-channel device and a corresponding n^+ moat around each p-channel device, thus giving pn junction isolation. This method, known as *dual guardband isolation*, was originally used with metal-gate CMOS, particularly in early off-the-shelf 4000-series logic ICs, but required a high silicon area for its implementation [14]. It did, however, allow higher voltage working than the usual 5 V if required, but it was not attractive for LSI and VLSI circuits.

Isolation techniques for CMOS are now exclusively by some form of oxide isolation. The present generation of CMOS circuits is therefore characterized by *oxide-isolated, self-aligned* silicon gate fabrication methods. Terminologies such as LOCMOS (local-oxide complementary MOS) and IsoCMOS (isoplanar CMOS) have been used by certain IC manufacturers for their particular method.

The general fabrication details of an oxide-isolated p-well CMOS process are illustrated in Figure 2.23. This is a planar, but not a planar-epitaxial, process. The various manufacturing steps are as follows, with re-covering by photoresist being done at each appropriate stage:

1. On the n-type substrate first form the thin pad oxide and thicker silicon nitride layers (see Section 2.2.1).
2. Open window in photoresist using mask No. 1, and etch away all pad oxide and silicon nitride except over the final transistor areas.
3. With new photoresist and mask No. 2, open window to define the area that will become the p-well, and then ion-implant the p-well with boron dopant—note that this boron ion implantation can be made through the pad oxide and the Si_3N_4 but is stopped by the photoresist.
4. With mask No.3 open a further window in the photoresist around the p-well perimeter and ion implant a p^+ channel-stopper border.
5. With new photoresist and mask No. 4 open window around the p-channel device and ion implant an n^+ channel-stopper border with phosphorus.
6. Grow a thick SiO_2 field oxide layer over all the areas not protected by the pad oxide and Si_3N_4 to produce the field oxide isolation around each device; the channel-stopper borders and the p-well will be driven in during this high-temperature process.
7. Strip off all remaining Si_3N_4 and pad oxide, and grow the thin SiO_2 gate oxide layer.

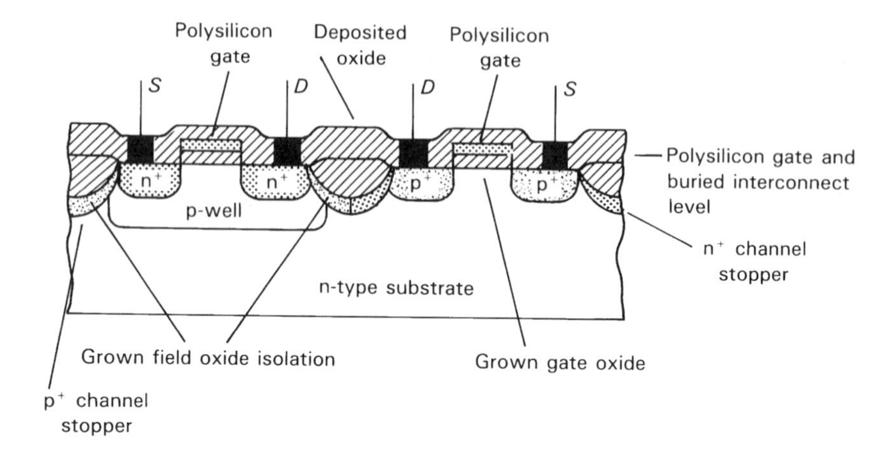

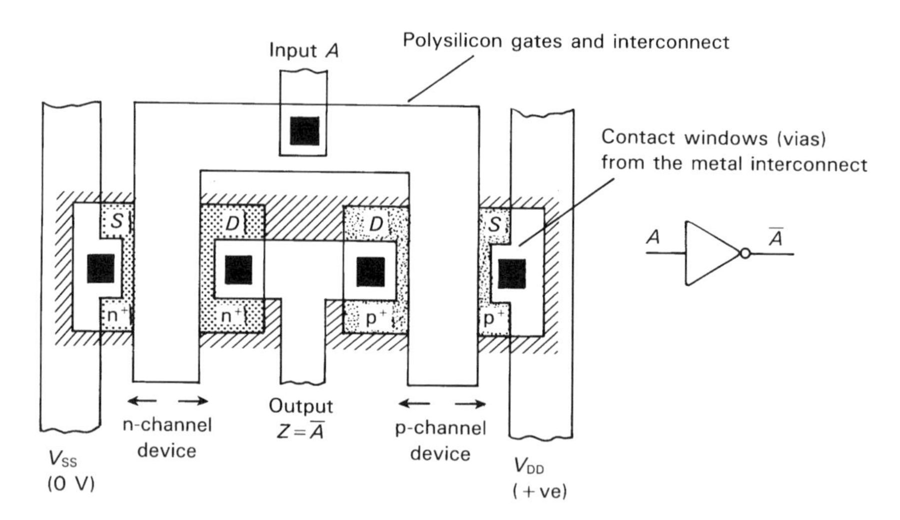

Fig. 2.23 The general fabrication details of an oxide-isolated silicon-gate CMOS inverter logic gate, with n-channel and p-channel enhancement-mode FETs. Note, for clarity the transistors are drawn here side by side; in practice they are usually located with a straight strip of polysilicon to form both gates, as shown in Figures 4.6 and 4.17.

8. Adjust the exact doping concentration of the n-channel and p-channel areas by careful ion implantation through the gate oxide so as to give the required threshold voltage values.
9. Deposit a layer of polysilicon over the whole wafer surface.
10. With new photoresist and mask No. 5, open windows and remove polysilicon from everywhere except the required gate areas.
11. With new photoresist and mask No. 6, open window and form the self-aligned n^+ source and drain regions of the n-channel device by arsenic implantation through the thin field oxide.
12. With new photoresist and mask No. 7, open window and form the self-aligned p^+ source and drain regions of the p-channel device by boron implantation through the field oxide.
13. Deposit a silicon dioxide layer over the whole wafer surface, thus burying the polysilicon gates.
14. With new photoresist and mask No. 8, open windows and cut contact vias in the SiO_2 to all source, gate and drain areas.
15. Cover the wafer with metal.
16. With new photoresist and mask No. 9, etch away all unwanted metal so as to leave the required interconnection pattern.
17. Apply the final surface protection.

CMOS fabrication will be noticeably more complex than nMOS or bipolar fabrication because of the need to fabricate the two polarity devices on the same substrate. Isolation is particularly important, as is the necessity to ensure that there is no parasitic pnpn action between the pMOS source electrode and the nMOS source electrode. If such a path with a sufficiently high current gain exists, then thyristor action can occur, with the ‘‘thyristor’’ switching on and causing a destructive short-circuit between V_{DD} and V_{SS} supply rails [3,13]. This is known as *latch-up*, and was particularly troublesome in early CMOS devices. The heavily doped p^+ and n^+ channel-stopper regions now prevent the formation of any effective pnpn latching.

In practice, CMOS fabrication may involve additional steps to those detailed above. Twin-tub fabrication, in particular, requires further processing to form the epitaxial layer and the n-well. Double- or triple-layer metal interconnections add further to the total number of masking stages required, bringing the total number up to as high as 15 or more in state-of-the-art circuits. Figure 2.24 shows the cross-sectional details of a typical double-layer-metal CMOS process.

The introduction of oxide isolation was the key feature in the first production of very high density, high performance CMOS circuits. However, the area requirements for the p^+ and n^+ channel stoppers necessary with oxide isolation are appreciable, and hence the development of the deep trench isolation (see Figure 2.7) was particularly relevant in the pursuit of better packing densities.

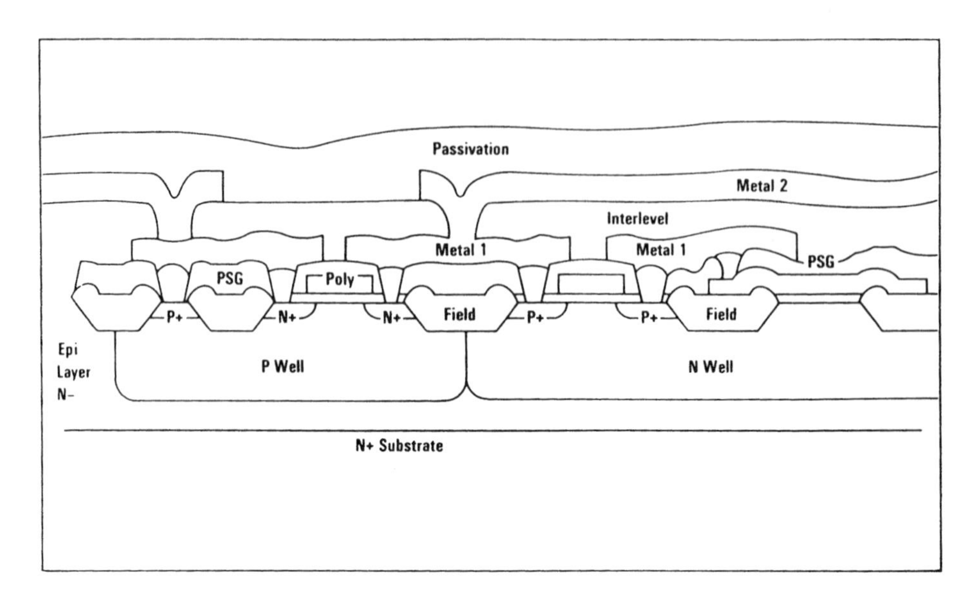

Fig. 2.24 The VO double-layer-metal interconnection CMOS process of Plessey Semiconductors, using twin-tub oxide-isolation fabrication. (Courtesy GEC-Plessey Semiconductors, UK.)

2.2.4 Silicon-On-Insulator Technologies

Devices where the bulk substrate of the device is silicon, sometimes collectively referred to as 'bulk silicon' devices, have an inherent problem in that the substrate is itself a semiconductor, which means that the action of all devices must be isolated from the substrate as well as from each other. Also, the substrate must be electrically connected in the final circuit to a supply rail, to the 0 V rail in the case of CMOS, and not left floating, and hence there will be inherent parasitic capacitance between the substrate and the active devices. Therefore the idea of planar fabrication on top of an insulating substrate has obvious advantages. This possibility is termed *silicon-on-isolator* (SOI) fabrication to distinguish it from normal bulk-silicon technologies.

The substrate insulator which has been the object of most research is artificial sapphire, giving rise to *silicon-on-sapphire* (SOS) technology. It has been considered mainly to improve the performance of CMOS, in particular to produce silicon-on-sapphire insulated-substrate CMOS (SOS-CMOS or SOSMOS) circuits [1,13,14,18]. Figure 2.25(a) shows the general fabrication details.

It will be seen that the SOSMOS fabrication is a planar epitaxial process as has been previously described, with oxide isolation between devices. The ad-

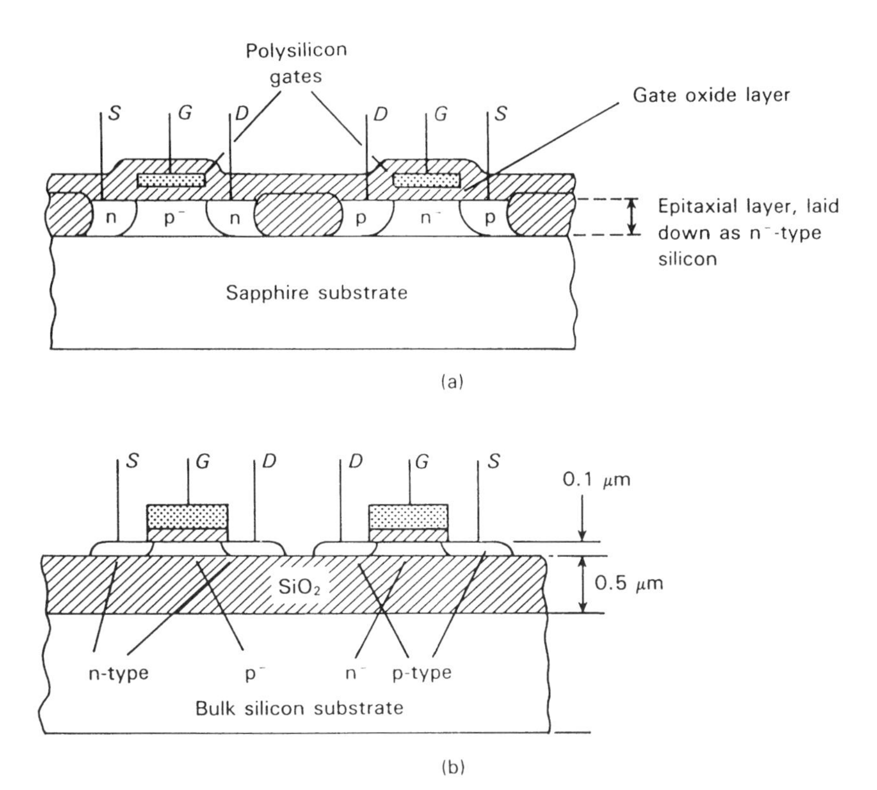

Fig. 2.25 Silicon-on-isolator (SOI) technologies: (a) silicon-on-sapphire CMOS; (b) silicon-on-oxide CMOS.

vantages claimed for this fabrication are an increase in speed by a factor of two compared with comparable bulk-silicon CMOS, and a much higher inherent resistance to radiation damage due to the sapphire rather than a silicon substrate. Against these advantages are the very much higher fabrication costs due to the manufacture of the sapphire substrate and the difficulties of forming flaw-free silicon on the surface of the sapphire substrate. Hence, this technology has largely been pursued for military and space applications, and is not a commercial competitor to normal CMOS.

An alternative fabrication method retaining bulk silicon as the substrate is shown in Figure 2.25(b). Here silicon dioxide forms a thick insulation below all devices, with the devices being built up in small isolated islands on this SiO_2 insulating layer. However, this development has not proved successful, and is no longer being researched. Other ideas are still being considered, but in spite of academic advantages there is little prospect of them challenging the present dominance of bulk-silicon CMOS.

2.2.5 MOS Summary

The basic processes involved in unipolar technologies are all well proven, with CMOS technology in particular being the most significant for custom microelectronic applications. The planar fabrication techniques used are the same as in modern bipolar technologies, with *oxide isolation* and *self alignment* being the key elements in providing high density, high performance, high yield CMOS circuits.

The buried (subcollector) layer necessary in most bipolar technologies is not required in FET fabrication; neither is the formation of any Schottky diodes for antisaturation purposes. However, an epitaxial layer and the formation of n-wells or p-wells or twin-wells is required for efficient CMOS circuits. In general, CMOS requires more mask levels than bipolar, possibly twice as many in fast circuits, this development being largely driven by the desirability of increasing circuit speeds. In this respect modern CMOS generally matches the speed of all but the fastest of bipolar technologies (see Section 2.6).

The other fundamental difference between bipolar and unipolar technologies is that the circuit performance of FETs is controlled by physical device size as well as the semiconductor parameters, in particular by the length L and the width W of the source-to-drain channel in any given fabrication process. The smaller the L/W ratio the faster the circuit. However, in bipolar technology the critical factor is the effective width of the base layer, but this is a function of processing rather than geometric layout. It is this dependence upon lengths and widths for any given processing that allows the layout of MOS and CMOS custom digital circuits to be undertaken by circuit designers outside semiconductor manufacturing companies, using *geometric design rules* supplied by the latter. It is also the reason why the majority of published textbooks and academic courses deal largely with unipolar design rather than bipolar [13,19–21], although the whole thrust of custom microelectronics should be to release the circuit and system designer from having to be involved at all with any design details at the fundamental silicon design level. Further details of FETs and unipolar design may be found in more specialized texts [1–4,6–10,13].

2.3 MEMORY CIRCUITS

The development of high capacity, low cost memory has made possible the evolution of affordable CAD systems for custom IC design work and for all other computer purposes. Memory ICs are in the vanguard of VLSI capability, and represent the highest commercially available transistor count per chip.

Memory circuits cater to two types of need, namely (a) where it is necessary to change stored data very frequently during the operation of the equipment, such as when acting as a temporary memory store, and (b) where the stored data are fixed and not altered during normal operation. The former are *random-access memories* (RAMs), with very fast read and write times normally being required, while the latter are *read-only memories* (ROMs), with fast read times being desirable.

With random-access memories the information written into the storage circuits is lost when the power is lost, and has to be rewritten after power has been restored; such circuits are therefore said to have 'volatile' memory. (The very low standby power of certain small CMOS RAMs has made it possible for an internal lithium battery to be included within the IC package so as to give continuity of memory in the event of a short power failure, but the memory itself is still volatile.) Read-only memories, on the other hand, retain their stored pattern of data under power-off conditions, and are therefore referred to as 'nonvolatile.' The permanent memory stored in ROMs may be mask-programmed into the circuit by the supplier, or supplied unprogrammed to the purchaser for his or her own electrical programming procedure. The latter are 'programmable' or 'field-programmable' ROMs, and in turn may be further divided into the two following categories:

- types which may be programmed only once with required data (PROMs)
- types where the stored data can be rewritten with required data by taking the circuit out of service and undertaking an erasing/reprogramming procedure (EPROMs)

The latter include several variants such as erasure by ultraviolet light (UV-PROMs), and electrically erasable and reprogrammable (EAPROMs or EEPROMs). There is often a limit to the recommended number of times an EPROM may be reprogrammed.

The range of memory ICs is indicated in Figure 2.26. Mask-programmable ROMs are available from very many IC vendors. Bipolar versions provide the fastest access (read) time, but are limited in capacity to 32 or 62 Kbits storage capacity; unipolar, particularly nMOS, can provide up to possibly 100 times this capacity, but with a somewhat slower access time. (Note that 'K' in this context means 1024, and should not be read as the same as 'k,' which equals 1000.) The same distinction applies also to field programmable types, where bipolar types are generally smaller but faster; for example, a representative off-the-shelf non-erasable 20-pin bipolar IC can provide 16 K of memory organized as 4096 × 4-bit words with an access time of 30 ns, whereas a comparable erasable CMOS

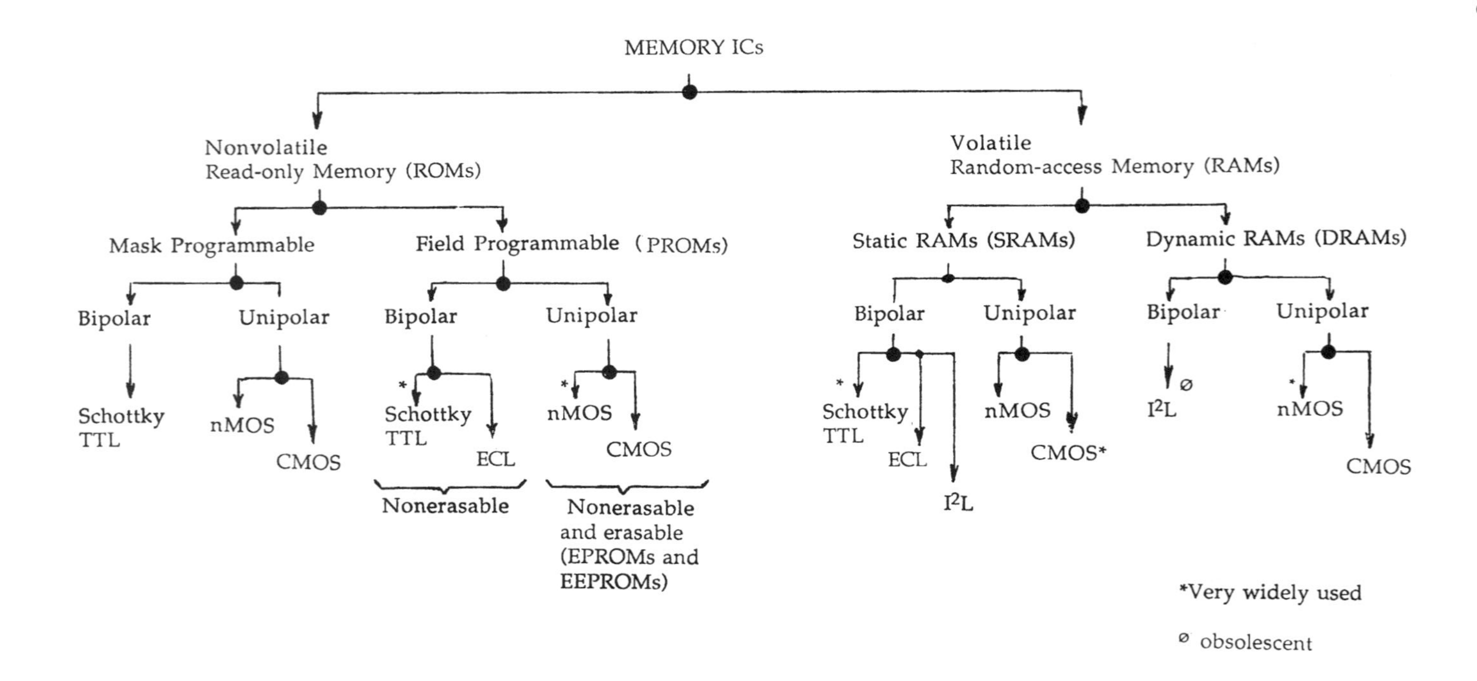

Fig. 2.26 The families of commercially available memory circuits. Note that other programmable devices such as programmable logic arrays (PLAs), which are closely related to PROMs, are not normally listed under memory devices.

EPROM can provide 1024 Kbits of memory organized as 64 K $\times$ 16 bits or 128 K $\times$ 8 bits, but with an access time of around 150 ns.

The same general distinction of maximum speed vs. capacity is also true in the range of available random-access memories. With static RAMs, usually referred to as SRAMs, Schottky TTL can for example provide 256 $\times$ 1-bit memory capacity with a 40 ns access time, whereas CMOS—by far the most widely used RAM technology—can provide 32 K $\times$ 8-bit capacity in a 28 pin package with an access time of about 100 ns. With dynamic RAMs, usually referred to as DRAMs, unipolar technologies now dominate for manufacturing reasons (see below). Up to 2048 K-bits organized as 256 $\times$ 8-bits or equivalent per IC, with access times of around 100 ns are readily available. A common feature of all these standard off-the-shelf parts is that in spite of their internal complexity, prices are usually in the range of $10 per IC.

Full details of the circuit configurations and fabrication methods of RAMs and ROMs may be found elsewhere, including the essential on-chip circuits necessary to refresh continuously the stored data in dynamic RAMs [1,3,22–24]. However, as will be seen in the following overview, the basic fabrication techniques do not involve any radically new procedures over those already covered.

ROMs

With the exception of very small memory circuits, the fundamental arrangement of all read-only memories is as shown in Figure 2.27(a). The 'memory' for each output bit consists of either forming the required connections between the horizontal 2^j row address decode lines and the 2^k column lines in the $2^j \times 2^k$ memory array matrix, or destroying those connections that are not required; for m outputs (m bits) there are m such identical matrices. The size of a ROM is defined by the examples shown in Table 2.1. The heart of all ROMs and RAMs therefore consists of the mechanism used to form the pattern of required connections in each memory array. In nonerasable types, once this connection pattern has been formed it can never be altered, but in erasable types, a pattern can be destroyed and an alternative one made.

Figures 2.27(b–d) show the basic concepts. In mask-programmable bipolar types the multiple emitters to the decode lines are individually made or omitted by the surface metallization mask pattern. In nMOS types the programming is usually performed by the formation of a thin gate oxide at the locations where an effective n-channel switch (see Figure 2.18) to 0 V is required; where no path to earth is required, a thick field oxide is left over the gate position, which means that an effective gate for this depletion-mode n-channel device is not possible, and the source-to-drain conductance therefore remains zero.

Note that both depletion-mode and enhancement-mode FETs are present in the circuit Figure 2.27(d). CMOS is not particularly relevant for programmable

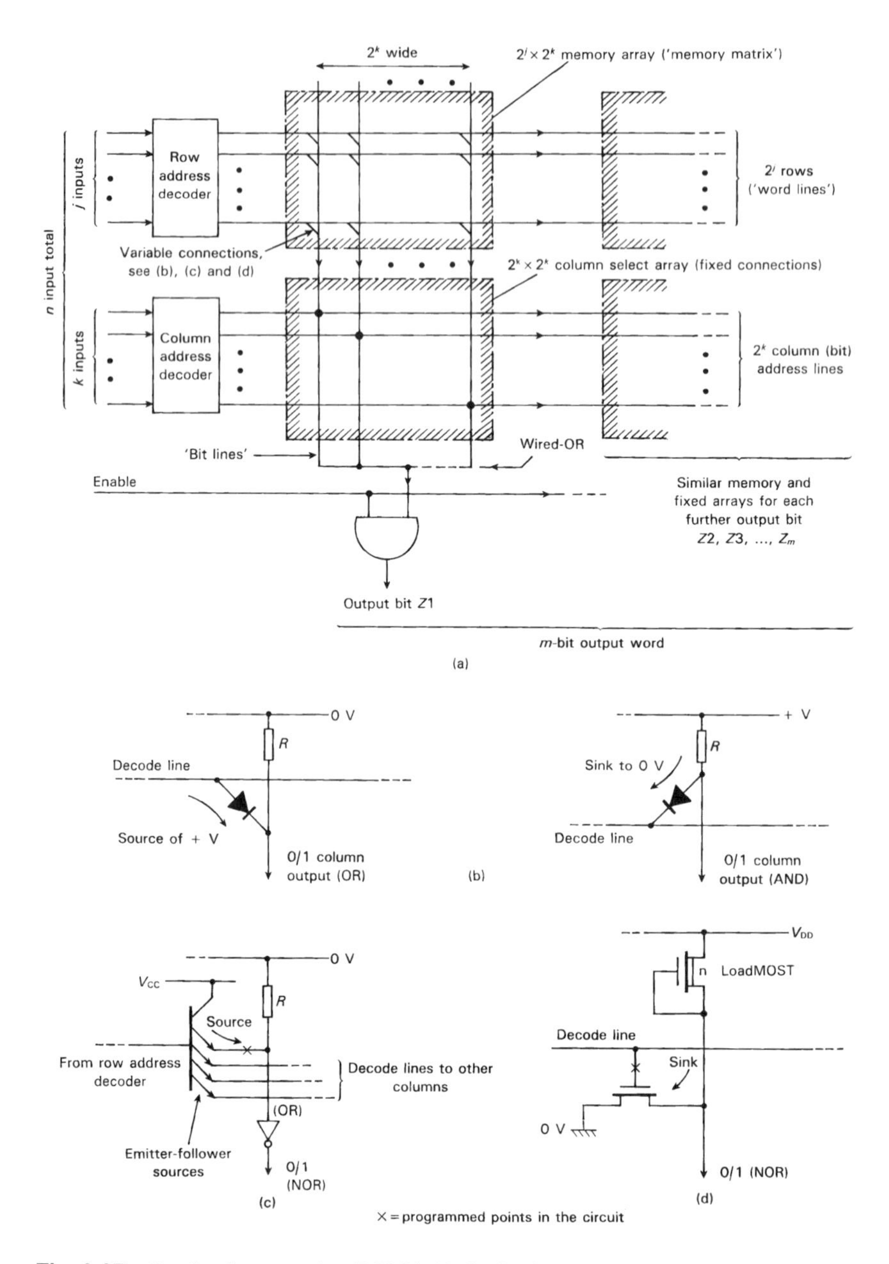

Fig. 2.27 Read-only memories (ROMs) (a) the fundamental group topology where only the read and column connections in the memory remain present in the final programmed circuit; (b) two concepts for the decoder control of the column outputs using some form of unidirectional interconnection; (c) bipolar circuits using multiple-emitter transistors, each emitter to one column line only; (c) nMOS circuit using a depletion-mode loadmost.

Table 2.1 The Relationship Between the Number of Inputs and Outputs of a ROM and the Catalogued ROM Size

No. of inputs ($n = j + k$)	No. of outputs(m)	Size of ROM
5	8	256 bits, or 32 × 8
12	4	16 K, or 4096 × 4
16	8	512 K, or 64 K × 8
18	4	1024 K, or 256 K × 4

array matrices, since it would require a much greater silicon area to use a true CMOS logic configuration where both n-channel and p-channel devices have to be switched on and off. However, CMOS may be relevant in the Enable and other nonprogrammable peripheral circuits of the ROM.

PROMs

Programmable ROMs and erasable-programmable ROMs employ the same basic fabrication techniques, but with the addition of appropriate means whereby the purchasers of the IC can select the required connections in the array. In bipolar PROMs the user programming is by means of a fusible link in each emitter connection. This link is either doped silicon or some metal such as nichrome, and the programming operation consists of applying a short-duration pulse of about 20 mA to individual circuits, which is sufficient to blow this 'fuse' link but insufficient to impair the rest of the circuit. Figure 2.28 shows the result of this method of programming; clearly, once blown, the fuse path cannot be repaired, and hence the memory is permanent (nonvolatile) and nonreprogrammable.

EPROMs

Commercially available unipolar PROMs do not employ fusible links. Other methods are used which effectively activate an FET gate by applying a short electrical voltage pulse when the device is required to be made functional, leaving it inoperative where no switching action is required. The terminology 'electrically programmable' (EPROM) is usually applied to these devices, the generic term PROM often being reserved for bipolar types only. Furthermore, since it is usually possible to erase and subsequently rewrite (alter) the programming pattern in the FET switches by opposite polarity electrical pulses or ultra-violet light, the additional designations EEPROM, EAPROM and UV-PROM are found.

Details of the methods of EPROM operation may be found in many texts [1,23–26]. Briefly, each switch in the memory array is fabricated with an isolated gate buried in the SiO_2 insulation layer, above which is the normal control gate

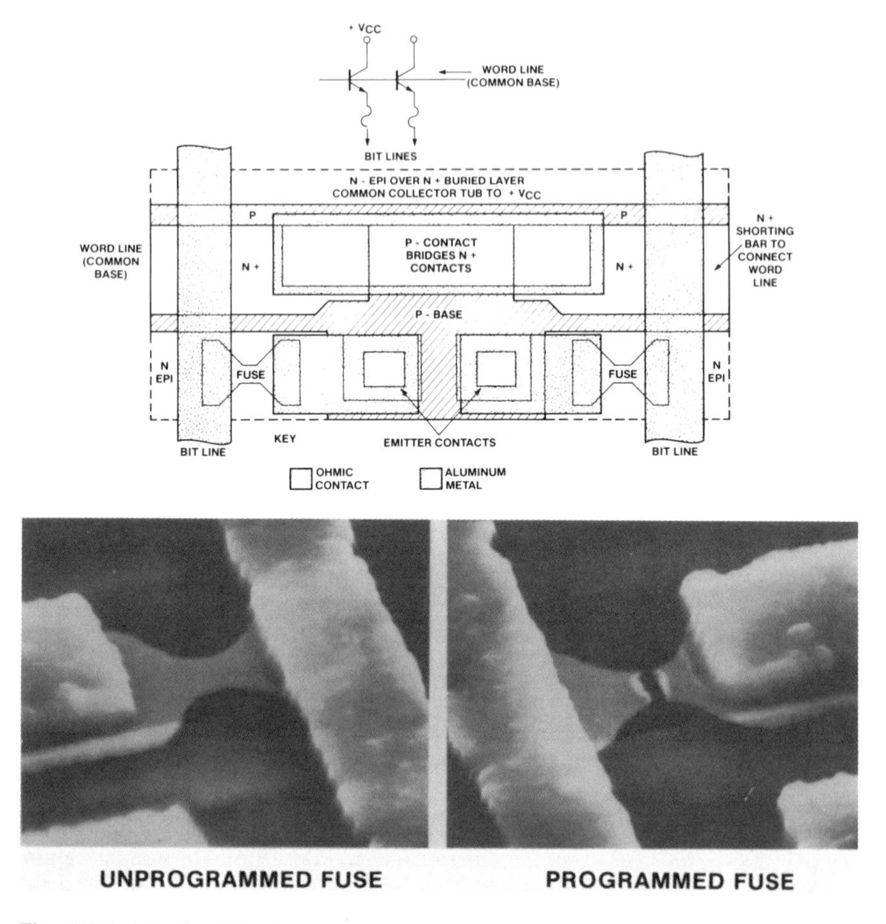

Fig. 2.28 The fuse-blowing details of a Schottky TTL PROM.

(see Figure 2.29). With no charge on this isolated (floating) gate the control gate is ineffective in controlling the source-to-drain conductance, which remains zero. However, when a short pulse of possibly 50 V is applied between the source and drain electrodes, avalanche multiplication occurs and charge becomes induced and trapped on this isolated gate, and the control gate now becomes effective in switching source-to-drain current. Because of the very high insulating properties of SiO_2, the trapped charge remains effective for a very long time, possibly more than a decade, but can always be removed, for example, by exposure to intense UV light or X-rays, which cause free carriers to be induced, which cancels the trapped charge. The device may then be reprogrammed as before with new data.

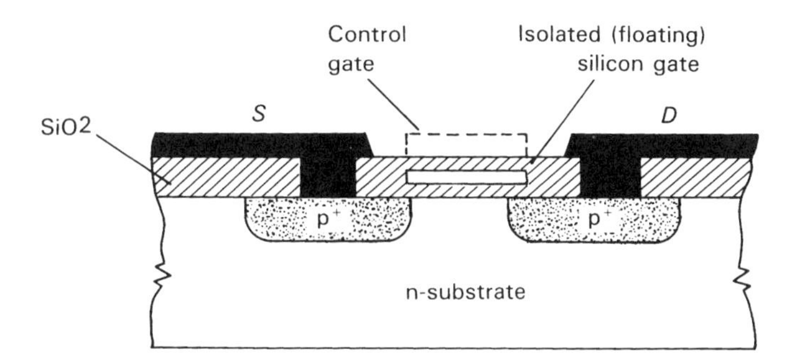

Fig. 2.29 The floating isolated gate concept of a unipolar EPROM.

The memory in EPROMs is therefore regarded as nonvolatile, but reprogrammable.

RAMs

Random-access memories by definition require circuits which can always accept new data, with no limitation on the number of times they are 'written-to' or 'read-from.' The internal circuits must therefore be read/write storage circuits, and unless very extreme high speed is required, the lower power consumption and superior packing density of nMOS and CMOS rather than bipolar is preferable.

All conventional RAM circuits are by definition volatile, since if power to the circuit is interrupted the stored data is lost. However, static RAMs retain their data as long as power is on, but dynamic RAMs require repeated internal refreshment since data is held by capacitive means rather than by any form of cross-coupled bistable circuits.

The basic topology of all RAMs whereby individual memory cells in the array can be written to and read from is given in Figure 2.30(a). Typical static memory cells are shown in Figures 2.30(b–d). In the case of nMOS, the two depletion loadMOSTs may be replaced by two high-value resistors made from lightly doped polysilicon positioned on top of the four active devices, which decreases the total silicon area of the cell but at the expense of additional processing steps and hence cost. The CMOS memory cell gives the lowest power consumption, but speed and packing density may not be as good as nMOS.

Fabrication of static RAMs does not therefore involve any basic concepts beyond those previously considered, being normal bipolar or MOS planer technologies but with great emphasis upon layout and packing density. Further details may be found elsewhere [9,23–26]. Dynamic RAMs do, however, involve some novel internal concepts which stem from the requirements to pack ever greater

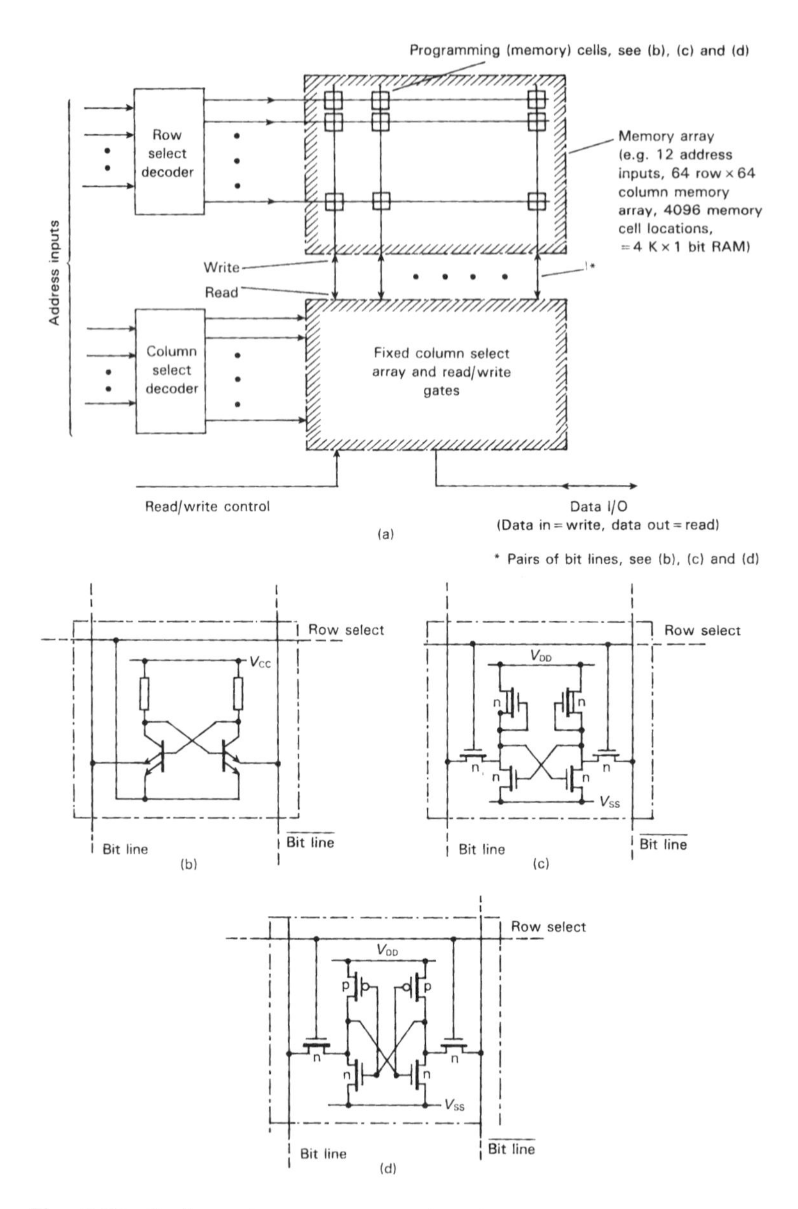

Fig. 2.30 Static random-access memories (SRAMs): (a) the basic topology of all SRAMs, one I/O bit only shown; (b) static bipolar single-bit memory cell (simplified) using multiple-emitter transistors; (c) the static 'six-transistor' single-bit nMOS cell; (d) the static CMOS single-bit memory cell. Note, in (b), (c) and (d), additional column-select logic selects the two vertical bit lines to the individual cell.

memory capacity on a single chip, which means minimizing the number of devices per memory cell. The present status is that the single-FET memory cell shown in Figure 2.31 is now the classic means of producing high-density dynamic memories.

The n-channel enhancement-mode FET shown in Figure 2.31(a) merely acts as an on/off switch to allow the storage capacity of the cell to be charged or discharged (written), and for the state of this charge to be read. However, due to circuit leakage, the charge has to be refreshed every two or three milliseconds, which requires additional on-chip circuitry to refresh all the memory cells continuously. Furthermore, the act of reading a cell discharges the capacitance, and therefore to maintain this data bit, it must be copied when read into an output buffer register and then rewritten into the same memory cell. Hence, although each memory cell is simple, considerable complexity is necessary in the peripheral circuitry of the DRAM [6,23].

The fabrication details of a double-level polysilicon memory cell are shown in Figure 2.31(b). There remains intense commercial developments in this field in efforts to provide more memory capability per chip, but as device size is reduced so also is the storage capacitance value. This results in more complex fabrication in order to provide adequate capacitance [27], for example, as illustrated in Figure 2.31(c).

FRAMs

While the preceding sections surveyed the most widely used forms of memory, a completely dissimilar technology has been developed which gives *nonvolatile* static RAM capability, thus doing away with the need for any refresh/recopy circuitry. This new technology uses a ferroelectric material which maintains a stable polarized state after the application and removal of an externally applied electric field. The market product is known as a *Ferroelectronic Random Access Memory*, or FRAM™ (FRAM is the registered trademark of Rampton Corporation, Colorado Springs, CO, and applies to their complete range of nonvolatile ferroelectronic memories).

The ferroelectric effect is the ability of the molecular structure of certain crystals to become aligned (polarized) under the influence of an electric field, and to remain in this polarized state after removal of the field. Reversal of the applied field causes polarization in the opposite direction. Hence, ferroelectric material has two stable states and can be used as a 'bistable capacitor' with two distinct threshold voltages to switch the device from one capacitive state to the other, no power being necessary to maintain the polarization of the crystal structure in either of the two states. (Note that this ferroelectric effect has nothing to do with any magnetic effect in ferrous materials, and hence is something of a historical misnomer from the early 1920s. There is, however, an analogy with

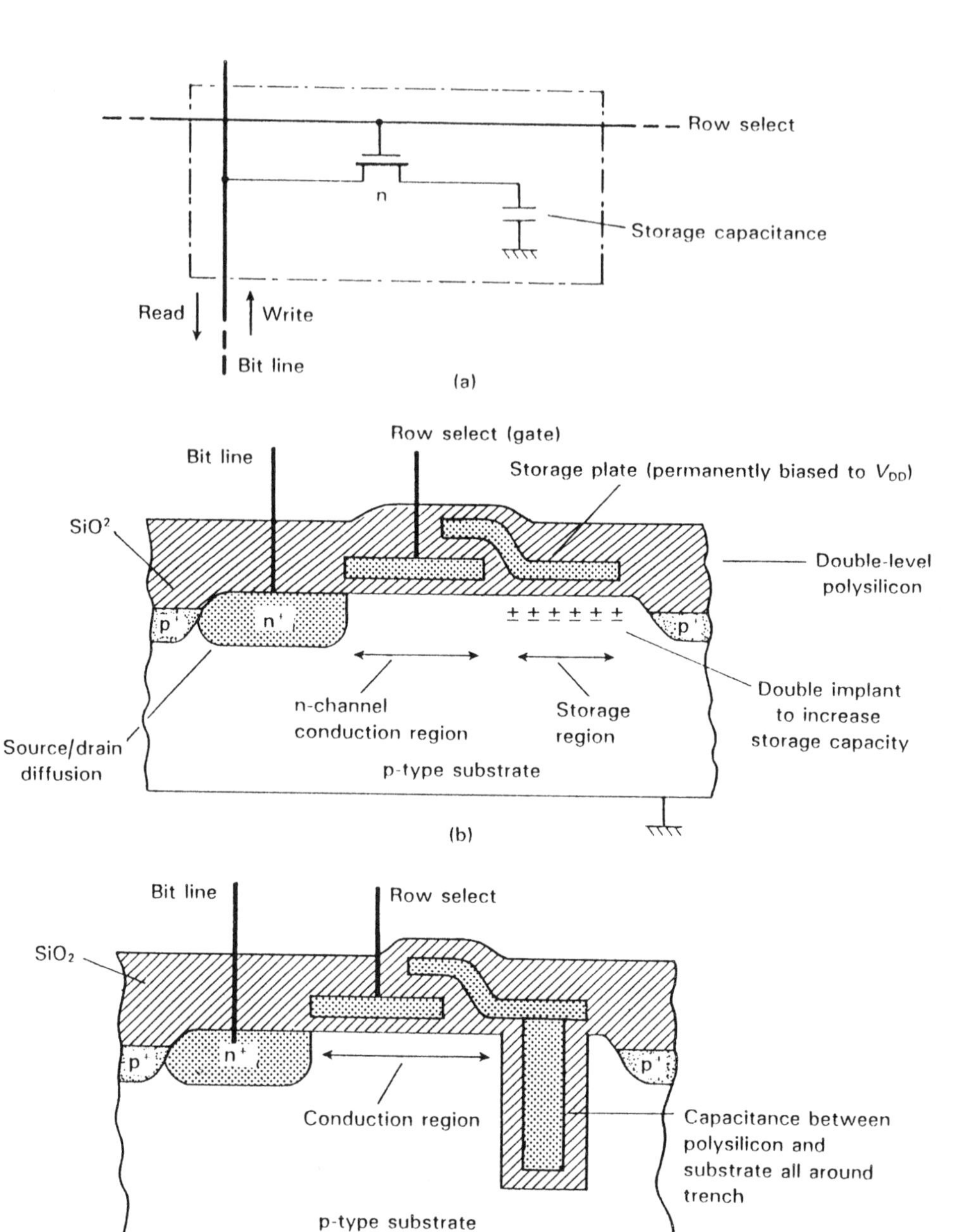

Fig. 2.31 Dynamic random-access memories (DRAMs): (a) the single-transistor memory cell; (b) fabrication details of a double-layer-polysilicon structure, giving a polysilicon gate and storage plate; (c) deep-trench construction for a 1 Mbit memory. Further developments for 4 Mbit and greater capacity involve dropping the polysilicon capacitor electrode even deeper into the substrate and stacking the gate vertically above it.

more familiar ferromagnetic materials—magnets, which maintain magnetic polarization without applied power.)

Work on ferroelectric compounds has resulted in the development of a complex ceramic material, largely composed of lead-zirconate-titanate (PZT), which is compatable with and can be deposited as a thin film over existing silicon processing. Three basic steps are required to add ferroelectric devices to convential processing, namely (i) the formation of the bottom capacitor electrode (plate), (ii) the deposition of the thin film of PZT, and (iii) the formation of the upper capacitor electrode. This is illustrated in Figure 2.32(a) [27].

Early circuits disclosed the use of PZT capacitors to convert volatile six-transistor CMOS memory cells, such as that shown in Figure 2.30(d), into nonvolatile cells by connecting one PZT capacitor to each of the two bistable circuit outputs. Under power-off conditions the two capacitors retained opposite states of polarization, depending upon which bistable output was at logic 0 or 1, and on restoration of power these dissimilar capacitive states would bias the recovery of the bistable circuit so that the original state of the memory was re-established. Additional circuit details disconnected the PZT capacitors under normal working conditions [28].

Later developments [29] produced a simpler circuit involving the two-transistor, two-capacitor memory cell configuration shown in Figure 2.32(b). In this memory cell write data to the cell on the two bit lines is used to polarize the two PZT capacitors directly. This information is permanently held by the PZT capacitors without requiring refreshment or power. Under read conditions the two capacitors are reconnected to the bit lines, and a special sense amplifier senses the difference in polarization between the two capacitors and generates 0 or 1 data output as appropriate. The action of reading, however, is destructive, and data have to be reentered into memory by repolarizing the PZT capacitors. Further developments using only one PZT capacitor per cell have also been pursued [27,28].

Figure 2.32(c) illustrates a standard commercial product [29]. Performance details in comparison with other alternatives are given in Table 2.2. The possible areas of use include both custom circuits and also computer core and disk replacements.

To summarize, all RAM and ROM circuits use well-established planar fabrication techniques, with appropriate additions or variations from normal digital logic processing to meet the particular memory requirements. The OEM digital design engineer will normally use them as standard off-the-shelf items, except where a vendor has ROM and RAM macros in his standard-cell library for custom circuits (see Section 4.2), but in any case, the precise details of the memory circuitry remains the province of the semiconductor vendor. Further details may be found in a comprehensive survey paper by Maes *et al.*, which also contains an extensive list of further references [30], and in the comprehensive text by Prince [31].

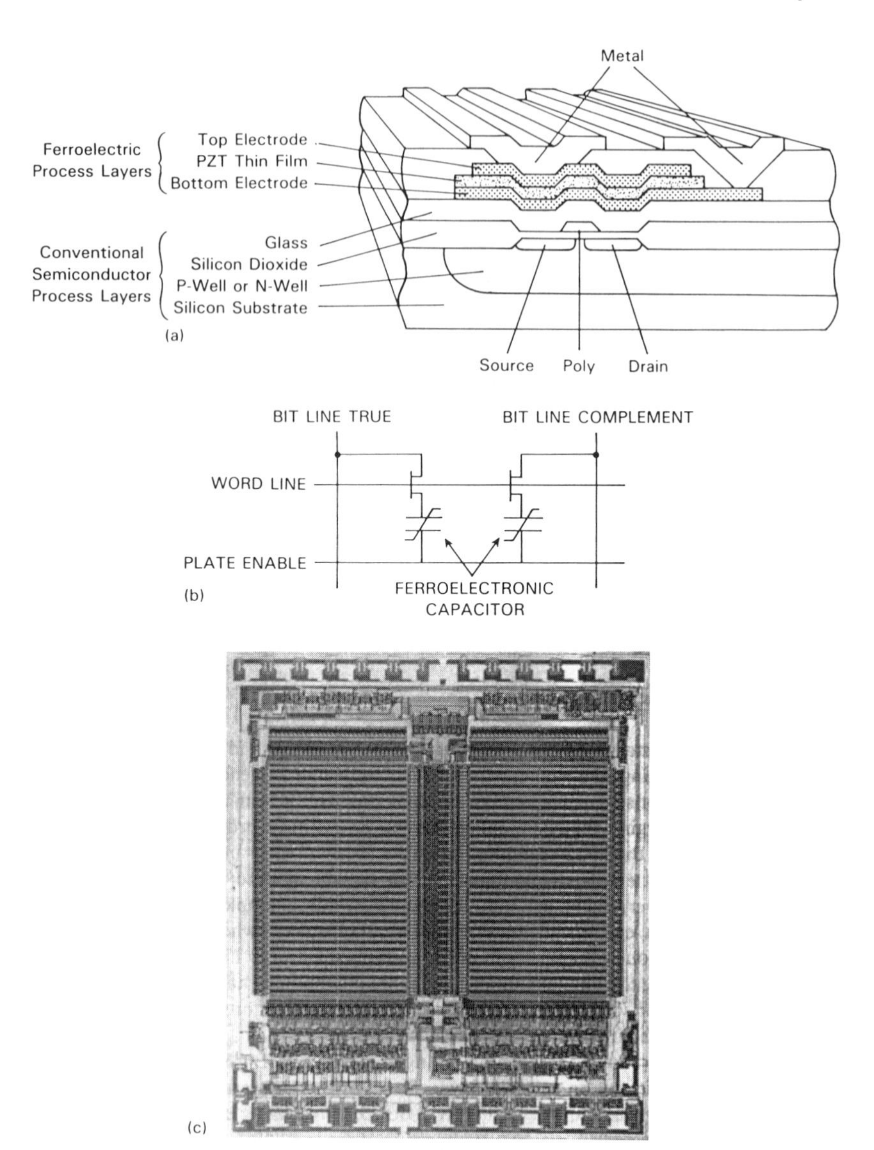

Fig. 2.32 Details of the nonvolatile ferroelectric static memory cell: (a) the addition of a PZT thin-film capacitor on conventional CMOS processing; (b) the two-capacitor, two-transistor nonvolatile memory cell; (c) a commercial product. (Courtesy of Ramtron Corporation, Colorado Springs, CO.)

Table 2.2 A Comparison of Typical FRAM Performance with Other Memory Devices

Parameter	SRAM	DRAM	EEPROM	FRAM
Read cycle (ns)	25	150	50 to 200	25 to 100
Write cycle (ns)	25	150	10^7	25 to 100
Operating voltage (V)	+5	+5	+5	+5
Programming voltage (V)	+5	+5	+12 to +21	+5
Operating temperature range (°C)	←	−55 to +125		→
Data retention at zero power (years)	0	0	10 to 100	>10
Relative cell area	3X	X	2X	X

2.4 BiCMOS TECHNOLOGY

Because of the strengths and weaknesses of both bipolar and MOS technologies—bipolar consuming more power than MOS, but MOS being, in general, not as linear or fast as bipolar or capable of driving external loads—it is an attractive proposition to consider merging the two technologies on to one chip so as to provide the best of both worlds. This is BiCMOS technology.

BiCMOS attempts to use the standard fabrication methods of both bipolar and CMOS. However, there are certain problems. For example, bipolar uses an epitaxial layer on a p-type substrate, whereas CMOS may be fabricated without an epi. layer on a n-type substrate. Hence, there is a choice between keeping the bipolar fabrication the same and adding the CMOS requirements, or vice versa.

Figure 2.33 shows the general arrangement of two BiCMOS techniques, one with and one without an epitaxial layer. Some of the process steps can be made common between the bipolar and CMOS devices, for example, in Figure 2.33(a) the emitter and collector n^+ regions can be made at the same time as the n-channel FET source and drain regions, and in Figure 2.33(b) the deep n-wells can be simultaneously fabricated. There may, however, need to be compromises between the doping levels in the bipolar and MOS regions if they are simultaneously fabricated.

The true place for BiCMOS technology in the custom microelectronic field has yet to be assured. It is not clear whether the bipolar transistors will be principally used on the output of CMOS digital logic to give better drive capability than the CMOS, or whether they will both be used to best advantage in mixed analog/digital ICs for DSP applications. With increasing interest in the latter, it is probable that DSP applications may be the most advantageous. However, up to eighteen or more mask stages can be involved in the fabrication, and hence it is likely to remain a comparatively expensive and specialized technology [9,32,33].

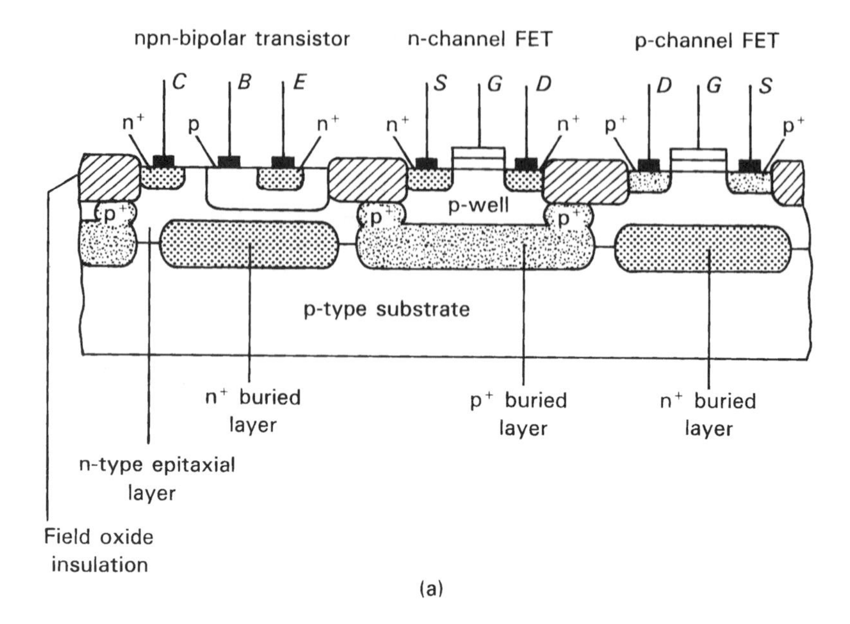

(a)

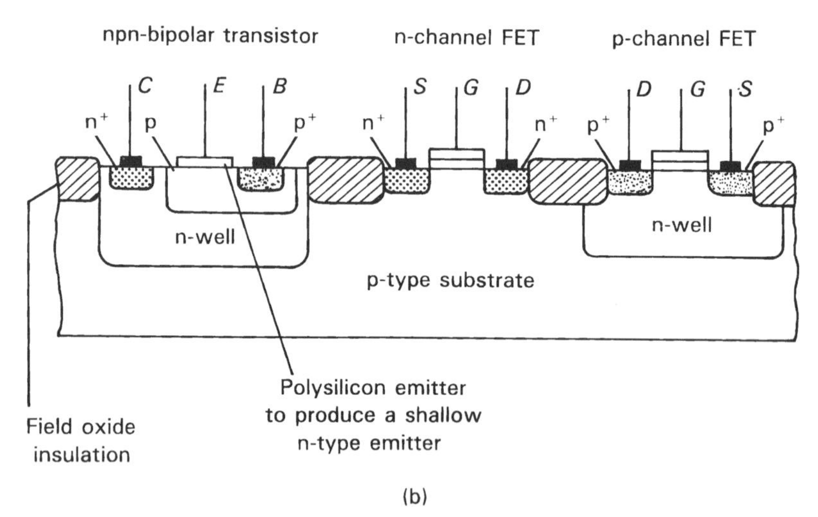

(b)

Fig. 2.33 BiCMOS fabrication methods: (a) planar epitaxia fabrication on a p-type substrate with vertical npn transistors; (b) n-well fabrication on a p-type substrate with no epitaxial layer.

2.5 GALLIUM-ARSENIDE TECHNOLOGY

Gallium-arsenide (GaAs) is a III/IV compound which can be made in a perfect crystal form as the basis for semiconductor devices. Its energy gap is higher than silicon (1.4 eV compared with 1.1 eV for silicon), and its electron mobility is higher (0.85 $m^2\ V^{-1}\ s^{-1}$ compared with 0.14 $m^2\ V^{-1}\ s^{-1}$ for silicon), thus giving it the potential for high temperature and very high speed applications. Its hole mobility, however, is no better than silicon (0.05 $m^2\ V^{-1}\ s^{-1}$), thus favoring n-channel working rather than p-channel.

The active GaAs devices used in digital applications are metal-semiconductor field-effect transistors (MESFETs), which are normally used in a conventional switching mode to 0 V rather than in any pass-transistor (transmission gate) configuration. The logic configurations are either *Schottky-diode FET logic* (SDFL), in which additional Schottky diodes perform the logic discrimination with the MESFET providing output inversion/amplification, or *buffered FET logic* (BFL), in which the MESFETs perform the logic with the Schottky diodes for output level shifting and signal isolation. Subnanosecond switching performance is available.

The fabrication technique in GaAs technology is basically extremely simple. A region of n-type material is formed by ion implantation in an almost pure GaAs substrate, and ohmic source and drain connections are made directly to it as shown in Figure 2.34. Between these two contacts a metallic gate (aluminum)

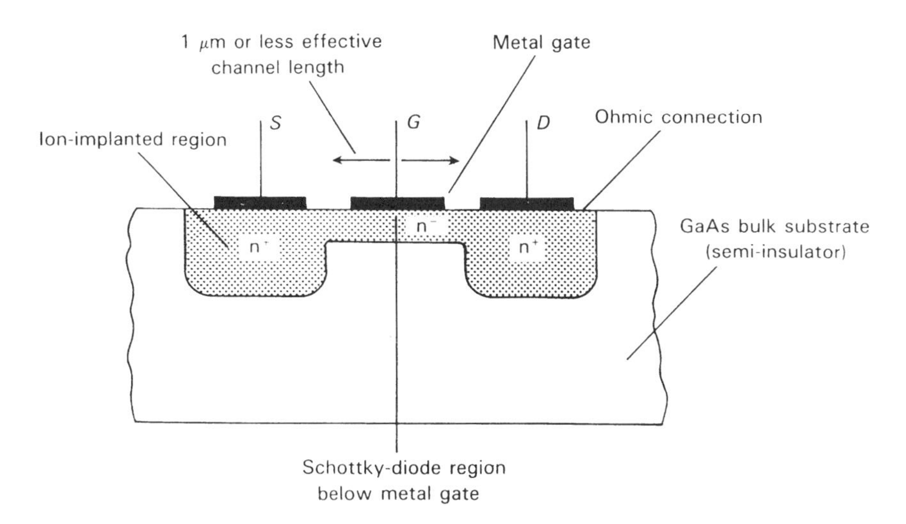

Fig. 2.34 GaAs fabrication using ion-implanted n-region and metal-gate control electrode.

is laid down, which forms a Schottky junction diode connection with the n-type material. When the gate is made negative with respect to the source, this Schottky junction is reverse-biased, and a depletion layer extends into the n-type channel, narrowing the area in which source-to-drain electrons can flow. The drain current can therefore be controlled by the value of V_{GS}. The normal JFET action has depletion-mode characteristics ($I_D \neq 0$ with $V_{GS} = 0$), but enhancement-mode characteristics can be achieved if the implanted n-type channel is sufficiently thin such that the Schottky depletion layer cuts off the channel with $V_{GS} = 0$ [1]. Terminologies D-MESFET and E-MESFET may be found for depletion mode and enhancement mode devices, respectively, with the technology sometimes termed D-mode GaAs and E-mode GaAs, or where both depletion mode and enhancement mode devices are present, E/D-mode GaAs may be encountered [33].

Because the GaAs substrate has a very high resistivity, the problems of isolating individual devices which cause complexities in silicon technology fabrication do not arise. Isolation is inherently present. Also, the negligible device-to-substrate capacitance gives excellent high-frequency performance. On the debit side, however, must be weighed the difficulty and cost of producing the defect-free GaAs substrate, and also the lack of the equivalent of CMOS since a p-channel device in GaAs would show no advantage over silicon in terms of speed.

Several variations of the fabrication methods shown in Figure 2.34 are possible, including growing an epitaxial n-type layer rather than ion implantation [1]. For discrete devices a mesa structure built as islands sitting on top of the substrate may be found. However, to date, GaAs has been principally used in microwave applications, although some GaAs digital custom products have been marketed for very specific OEM applications [34–36].

2.6 A COMPARISON OF AVAILABLE TECHNOLOGIES

Having completed this overview of the various semiconductor technologies which may be applied to custom microelectronics and their fabrication methods, an overall summary and comparison can now be made. However, it must be appreciated that all technologies rely upon similar fabrication techniques, and are subject to the same limitations in resolution and accuracy of the manufacturing steps, and hence any advance in, e.g., optical or X-ray lithography is applicable to all technologies.

The maximum digital capacity of MOS and bipolar devices over the first three decades continuously increased, as shown in Figure 2.35, with the latest possible MOS geometries now down to 0.1 or 0.2 microns and gate counts up to tens of millions. As shown in this figure, MOS has always led bipolar in packing density per chip, and this advantage seems likely to continue in spite of developments which aim to reduce bipolar transistor size and dissipation to challenge this MOS capability.

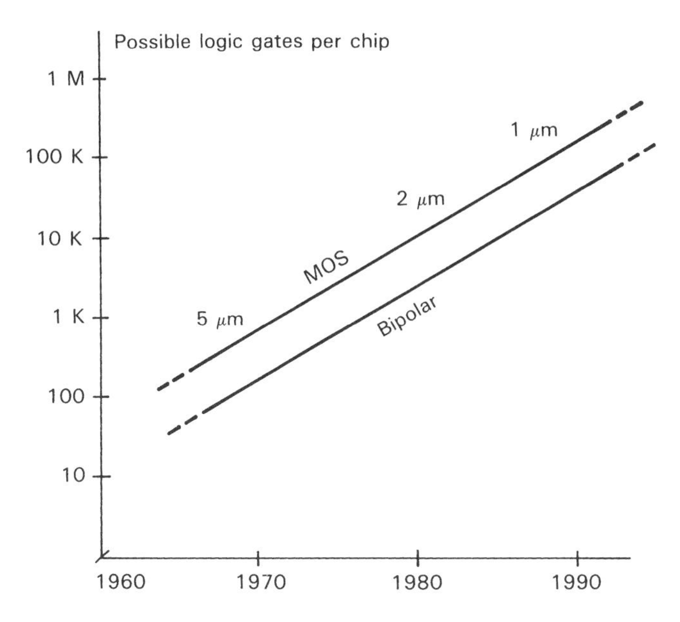

Fig. 2.35 The general evolution of maximum digital logic capability per chip.

Gallium-arsenide has the potential to exceed the packing density of normal bipolar technology, with in excess of 100 k logic gates per chip being currently possible [35,36]. However, cost will remain the dominant disadvantage of this technology for the vast majority of OEM applications, leaving MOS and particularly CMOS as the technology of choice for the vast majority of eqcustom ICs. Silicon-on-insulator does not feature in current custom technologies.

A comparison between the principal technologies when used for digital applications is given in Table 2.3. To some extent, speed and power may be traded off against each other, and therefore a broader graphical representation of these two parameters as indicated in Figure 2.36 may be more significant, with continuing developments tending to reduce both power dissipation and propagation delay and therefore the picojoule speed-power product performance.

From an applications point of view, parameters other than those considered above may be important. For example, in analog applications available voltage gain may be of greater significance than speed-power product. Table 2.4 therefore lists the relative significance of a number of parameters which may be of importance to the OEM circuit and system designer.

The following two chapters will illustrate where these various technologies are employed in both standard off-the-shelf products and custom microelectronic circuits.

Table 2.3 Typical Data for the Main Logic Technologies

Technology	Gate delay (ns)	Gate power dissipation (mW)	Power-delay product (pJ)	Max. toggle frequency (MHz)	Gate packing density (2-input gates mm^{-2})	No. of fabrication masks required
Standard TTL	10	10	100	35	20	7
Low-power Schottky TTL	10	2	20	40	20	8–9
Advanced low-power Schottky TTL	5	1	5	50	50	8–9
I^2L	15	0.5	7.5	15	100	8
ISL	4	0.25	1.0	40	75	10–12
STL	2	0.5	1.0	50	75	10–12
ECL	0.5	50	25	200+	20	8
High-speed CDI CML	3	0.5	1.5	50	100	8
Silicon-gate nMOS	10	0.5	5	35	400	7
Silicon-gate CMOS	15	0.1[a]	1.5	25	300	10–15
Silicon-on-sapphire CMOS	10	0.1[a]	1.0	35	300	10–15
BiCMOS	0.75	1	0.75	100+	200	12–18
GaAs	0.15	5	0.75	300+	50	6–7

[a] Power dissipation proportional to the number of switching operations per second; these figures are typical for 1 MHz switching at 5 V.

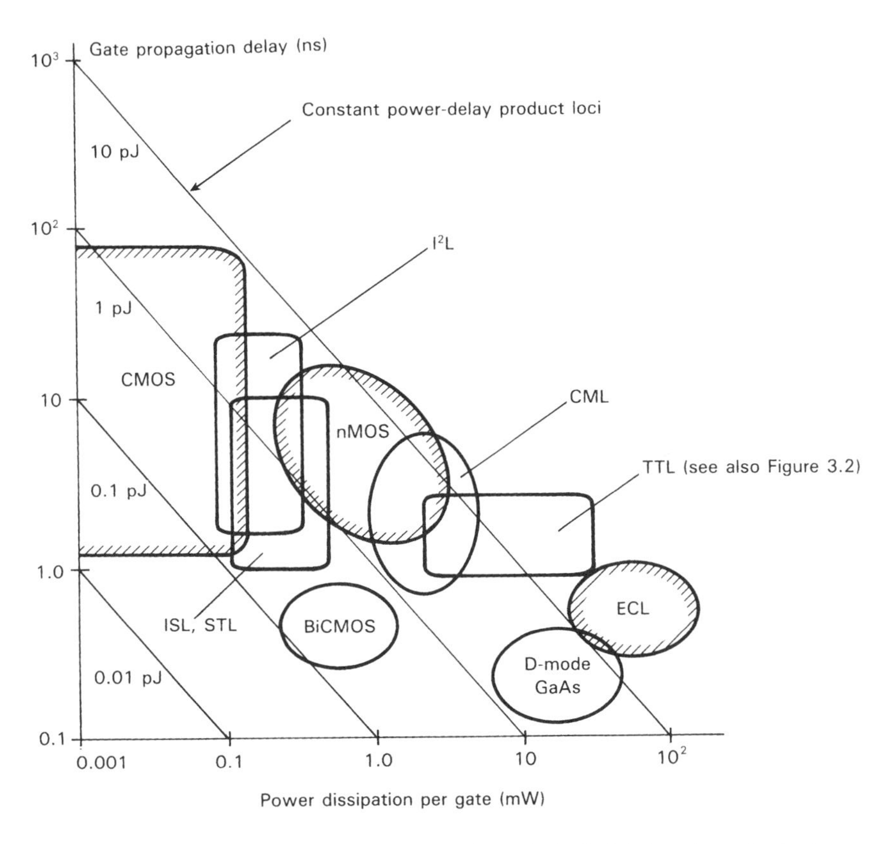

Fig. 2.36 The speed-power performance of the principal digital logic technologies.

Table 2.4 Relative Attributes of Common Bipolar and MOS Circuit Capabilities

	Bipolar npn	Bipolar I^2L/ISL	Bipolar CDI	nMOS	CMOS
General					
Supply voltage tolerance	2	3	4	3	4
Power dissipation	1/2	2/3	2/3	3	4
Speed	4	2	3	3	3
Transconductance g_m	4	4	4	1/2	1/2
Drive capability	4	2/3	2/3	1/2	2
Analog					
Gain per stage	4	[a]	3/4	1	1/2
Bandwidth	4	–	3/4	1	1/2
Input impedance	1/2	–	1/2	4	4
Output swing	4	–	2/3	2	2
Linearity	4	–	4	2	2
Precision passive devices	4	–	4	2	2
Transmission switches	1	–	1	2	4
Digital					
Max. switching speed	4	2	3	2/3	2/3
Logic swing	2	1	1	3	4
Voltage noise margin	2	1	1	3	4
Source and sink drive capability	4	3/4	3/4	1/2	2

[a] Not often employed for analog duties.

4 = superior relative performance, 3 = relatively good, 2 = very similar, and 1 = relatively poor.

REFERENCES

1. Goodge, M., *Semiconductor Device Technology*, Macmillan Press, UK, 1983
2. Jaeger, R.C., *Introduction to Microelectronic Fabrication*, Addison-Wesley Modular Series on Solid State Devices Vol. V, MA, 1988
3. Sze, S.M. (Ed.), *VLSI Technology*, McGraw-Hill, NY, 1983
4. Till, W.C. and Luxton, J.T., *Integrated Circuits: Materials, Devices and Fabrication*, Prentice Hall, NJ, 1982
5. Hodges, D.A. and Jackson, H.G., *Analysis and Design of Digital Integrated Circuits*, McGraw-Hill, NY, 1983
6. Dillinger, T.E., *VLSI Engineering*, Prentice Hall, NJ, 1988
7. Sparkes, J.J., *Semiconductor Devices: How They Work*, Van Nostrand Reinhold, UK, 1987
8. Pierret, R.F., *Semiconductor Fundamentals*, Addison-Wesley Modular Series on Solid State Devices Vol. I, MA, 1988

9. Maly, W., *Atlas of IC Technologies: An Introduction to VLSI Processes*, Benjamin/Cummings, CA, 1987
10. Ruska, W.S., *Microelectronics Processing: An Introduction to the Manufacture of Integrated Circuits*, McGraw-Hill NY, 1987
11. Hart, K. and Slob, A., Integrated-injection logic: a new approach to VLSI, *Proc. IEEE Int. Solid State Circuits Conf.*, 1972, pp. 92–97
12. Hewlett, F.W., Schottky I^2L, *IEEE J. Solid State Circuits*, Vol. SC.10, 1975, pp. 343–351
13. Uyemura, J.P., *Fundamentals of MOS Digital Integrated Circuits*, Addison-Wesley, MA, 1988
14. Hurst, S.L., *Custom-Specific Integrated Circuits: Design and Fabrication*, Marcel Dekker, NY, 1983
15. Rogers, T.J. and Meindl, D.J., VMOS, high speed compatable MOS logic, *IEEE J. Solid State Circuits*, Vol. SC.9, 1974, pp. 239–250
16. Rogers, T.J., Jenne, F.B., Frederick, B., Barnes, J.J., Hitpold, W.R. and Trotter, J.D., VMOS memory techniques, *IEEE ISSCC Digest*, February 1977, pp. 74–75
17. Baliga, B.J. and Chen, D.Y., *Power Transistors: Device Design and Application*, IEEE Press, NY, 1984
18. Davis, R.D., The case for CMOS, *IEEE Spectrum*, Vol. 20, No. 10, 1983, pp. 26–32
19. Mead, C. and Conway, L., *Introduction to VLSI Systems*, Addison-Wesley, MA, 1980
20. Glasser, L.A. and Dobberpuhl, D.W., *The Design and Analysis of VLSI Circuits*, Addison-Wesley, MA, 1985
21. Weste, N. and Eshraghian, K., *Principles of CMOS VLSI Design*, Addison-Wesley, MA, 1985
22. Texas Instruments, *MOS Memory Data Book*, Texas Instruments, Houston, 1986
23. Schroder, D.K., *Advanced MOS Devices*, Addison-Wesley Modular Series on Solid State Devices Vol. VI, MA, 1987
24. Glazier, A.B. and Subak-Sharpe, G.E., *Integrated Circuit Engineering: Design, Fabrication and Applications*, Addison-Wesley, MA, 1979
25. Howes, M.J. and Morgan, D.V. (Eds.), *Large Scale Integration*, Wiley, UK, 1981
26. Cirovic, M.M., *Handbook of Semiconductor Memories*, Reston Publishing Co., VA, 1981
27. Technical Report, *Nonvolatile Ferroelectric Technology and Products*, Ramtron Corporation, Colorado Springs, CO, 1988
28. Weber S., A new memory technology, *Electronics*, 18 February 1988, pp. 91–94
29. Ramtron Corporation, *FRAM Data Sheets FM1008/1108/1208/1408*, Ramtron Corporation, Colorado Springs, CO, 1989
30. Maes, H.E., Groeseneken, G., Lebon, H. and Witters, J., Trends in semiconductor memories, *Microelectronic Journal*, Vol. 20, Spring 1989, pp. 5–58
31. Prince, B., *Semiconductor Memories*, Wiley, UK, 1995
32. Dettmer, R., BiCMOS—getting the best of both worlds, *IEE Electronics and Power*, Vol. 8, 1987, pp. 499–501

33. Zimmer, G., Esser, W., Fichtel, J., Hostika, B., Rothermal, A. and Schardein, W., BiCMOS: technology and circuit design, *Microelectronic Journal*, Vol. 20, Spring 1989, pp. 59–75
34. Howes, J.J. and Morgan, D.V. (Eds), *Gallium-Arsenide: Materials, Devices and Circuits*, Wiley, UK, 1985
35. Special Report, Inside technology: gallium-arsenide, *Electronics*, June 1988, pp. 65–85
36. Morgan, D.V., Gallium-arsenide: a new generation of integrated circuits, *IEE Review*, Vol. 35, September 1988, pp. 315–319

3

Standard Off-the-Shelf ICs

Although the principal aim of this text is the coverage of custom electronics, it is appropriate to include information on standard off-the-shelf ICs for several reasons.

First, the use of standard ICs still heavily outweighs the use of custom circuits both in volume and financial value, and is likely to continue to do so. However, every OEM designer has to consider the pros and cons of using (a) only standard parts, (b) custom ICs, or (c) some combination of the two, in every new product design, and thus knowledge of all types of product is necessary. These engineering and also financial considerations will form the subject of Chapter 7.

Second, off-the-shelf ICs generally set the standards of capability and performance of the semiconductor industry. The design of such products is invariably state of the art at the time of design, so that each product has a market edge over preceding designs or competitors' products. Custom ICs may be tailored to perform specific duties more efficiently than with an assembly of standard ICs, but the basic performance of the individual circuits cannot be better than that of a contemporary standard product. It follows that if there is an already available off-the-shelf LSI or VLSI product which does everything a design engineer requires, for example, a video-display control IC or a keyboard controller, then there is no technical justification for designing the custom equivalent of this item.

The range of standard ICs may be divided into the following:

- technology (bipolar or MOS)
- duty performed (analog, digital, nonprogrammable or programmable)
- size and capability

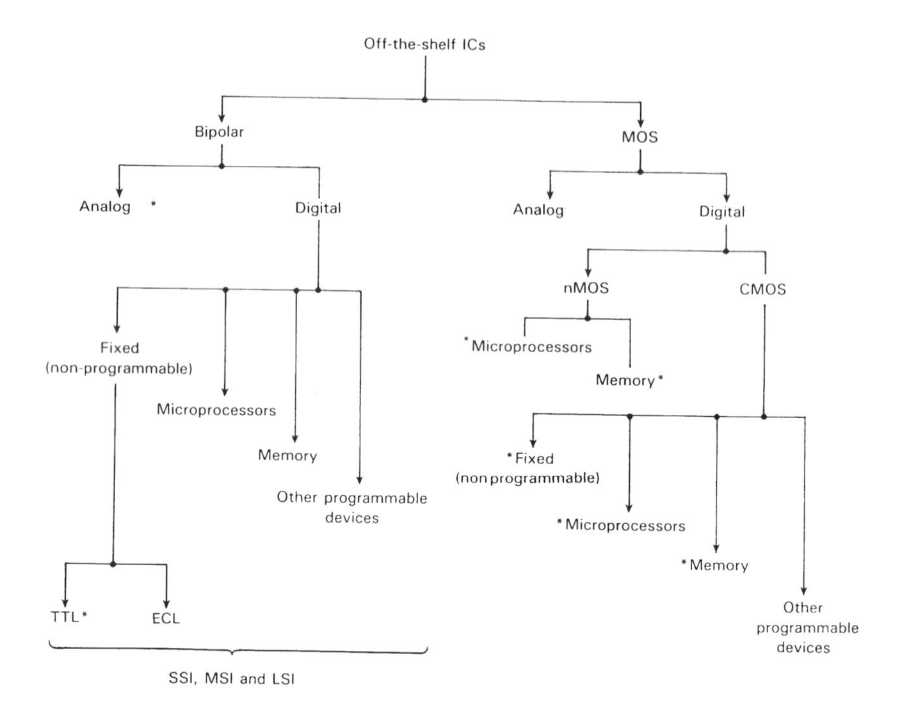

Fig. 3.1 The range of standard off-the-shelf microelectronic components, excluding discrete divices. The catagories marked with an asterisk are particularly prominent.

This is shown in Figure 3.1. To this listing must be added the vast range of discrete semiconductor devices such as individual transistors, thyristors, zener diodes, light-emitting diodes (LEDs), liquid crystal displays (LCDs), etc., which form an essential part of many systems, but which are not considered here.

3.1 NONPROGRAMMABLE SSI, MSI AND LSI DIGITAL ICs

The greatest global use of microelectronic circuits still remains standard SSI, MSI and LSI digital ICs. Both bipolar and silicon-gate CMOS digital ICs are universally available, particularly in the 74** series products. The older metal-gate CMOS 4000 series products are now obsolete, but may still be required if a higher voltage working is necessary, having a 3–15 V operating range compared with the usual 5 V of the standard. 74** series. We will consider standard VLSI processor and memory ICs later.

The eight principal categories of circuits in the standard 74** series are as follows, roughly in order of introduction on the market:

(a) standard 74** series TTL
(b) low power Schottky 74LS series TTL
(c) advanced low power Schottky 74ALS series TTL
(d) isoplanar high-speed FAST™ 74 series TTL
(e) high speed 74HC series CMOS
(f) high speed 74HCT series CMOS
(g) advanced high speed 74AC series CMOS
(h) advanced high speed 74ACT series CMOS

The four bipolar families are all designed for +5 V working with an operating temperature range of 0° to +70°C; the CMOS families allow an operating temperature range of −40° to +85°C or more, with +5 V working for the 74HCT series. The latter two CMOS families provide an improved speed performance over the the HC and HCT series. The general electrical performance is shown in Figure 3.2. (Individual manufacturers of 74** series ICs may use slightly different designations from those cited here.)

This range of ICs covers all the standard logic building blocks, ranging from the most simple SSI 7400 quad 2-input NAND package through to MSI and LSI arithmetic functions, encoders, decoders and other functions [1–8]. Fabrication details are as discussed in the preceding chapters. Because of the familiarity of these circuit functions, it is significant that in custom microelectronics, most vendors have standard circuits ('cells') for incorporation in custom ICs which mirror these standard functions; indeed, certain efforts in the past to introduce gates other than Boolean logic gates into custom circuits (see Section 4.3.1) have not been very successful due to the conservative nature of many digital system designers.

The use of off-the-shelf emitter-coupled logic (ECL) is much more specialized, and is only considered when the required performance cannot be obtained from other circuits. Hence, although off-the-shelf ECL ICs are available, the need to consider very carefully correctly-terminated transmission line interconnections in order to maintain the speed advantages means that ECL circuit design is more difficult than is otherwise the case, and hence tends to be a very specialized area of logic design [9,10].

3.2 STANDARD ANALOG ICs

The detailed design of very high performance analog circuits in microelectronic form largely remains a specialist area, since a detailed knowledge of the fabrication process and its parameters is necessary in order to achieve maximum perfor-

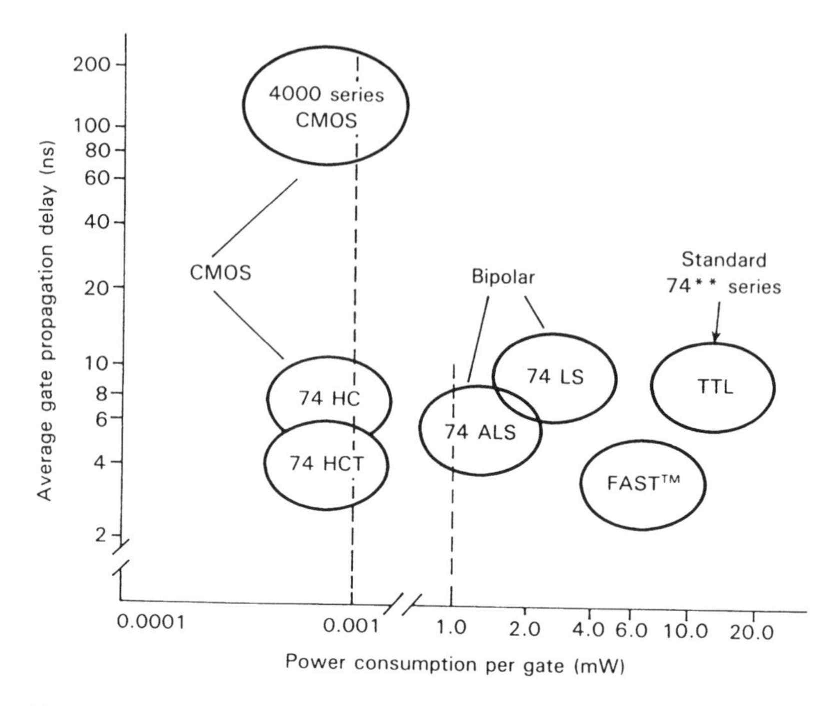

Fig. 3.2 The power consumption/propagation delay of off-the-shelf 74** series bipolar and CMOS ICs, and the now-obsolescent metal-gate 4000 series CMOS circuits.

mance. For this reason it is unlikely that any custom-designed analog circuit will have a better performance than a vendor's standard part, although in many applications this will not be important. Also, because standard analog ICs are made in very large volumes, the cost per circuit is usually far less than can be achieved in a custom IC assembly.

The range of off-the-shelf analog ICs is indicated in Figure 3.3. The operational amplifier forms the core of many analog systems, and is available in very many versions from many vendors; a main distributor's catalog may list as many as 250 or more variants in bipolar, mixed bipolar FET and CMOS technologies [11–15]. All these products contain relatively few transistors compared with a digital IC, and thus are not usually classified by the designations SSI, MSI, LSI, etc. Strictly speaking, analog circuits lie outside LSI and VLSI discussions. However, since the transistors in analog circuits are operating in their linear region rather than as on/off switches, the power dissipation per chip is inherently higher than in similar size digital circuits.

The bipolar 741 operational amplifier was one of the first commercially

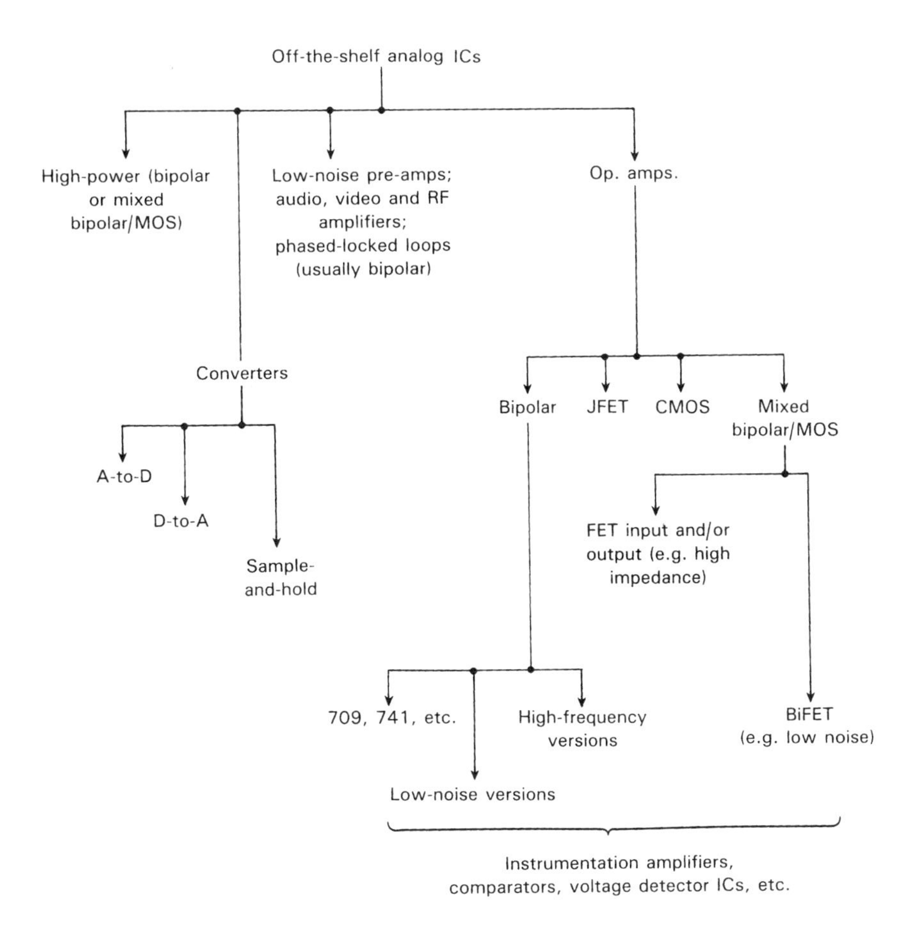

Fig. 3.3 The range of off-the-shelf analog circuits.

available unconditionally stable op. amps. on the market. Its basic circuit diagram is shown in Figure 3.4, from which the extensive use of matched transistor-pairs can be noted. There are now a number of variants on this design from many vendors, which may offer improved slew rate, bandwidth or other parametric improvements [9,15].

CMOS op. amps. are also widely available, particularly the 7600 series. In general, these do not provide such a high voltage gain as bipolar, but can have other advantages such as higher input impedance, lower cross-over distortion in output stages and lower total power consumption. Further commercial op. amp. families combining the best features of bipolar and MOS technologies are also available [12–18]. Representative performance figures are shown in Table 3.1.

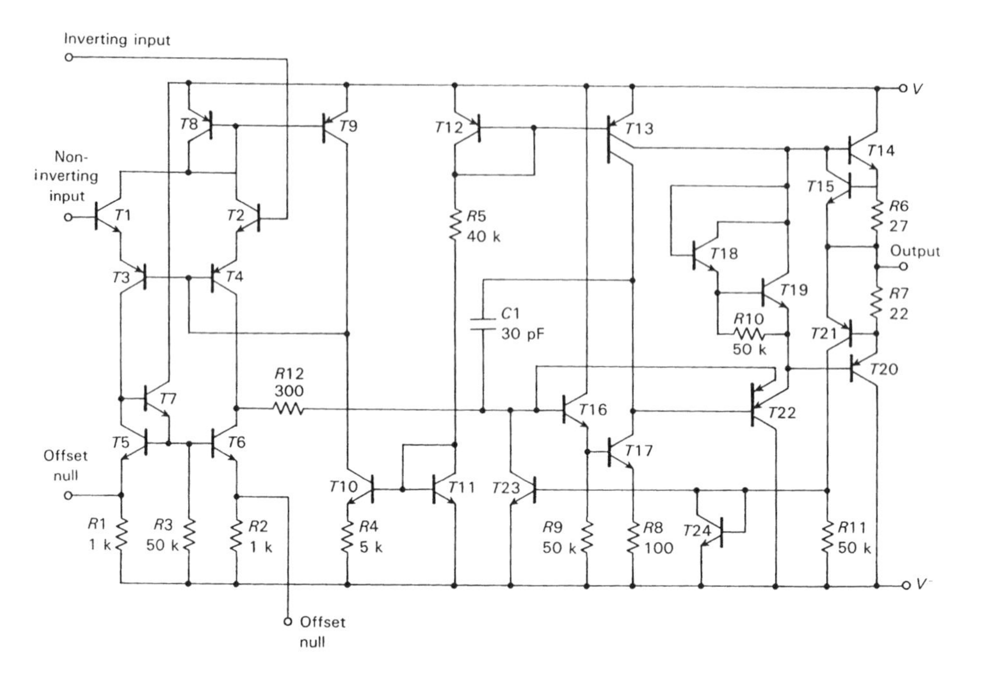

Fig. 3.4 The circuit of the classic original 741 operational amplifier.

3.3 MICROPROCESSORS

Although the specific duties which a microprocessor has to perform are dictated by the user-specific software program, the microprocessor IC itself is a standard off-the-shelf product, the purchaser being unable to change the internal circuit configuration.

The microprocessor is basically a single-chip central processing unit (CPU) (see Figure 3.5(a)). To form a complete processing or computing system requires the addition of further circuits, mainly memory to store the software instructions and the operating data, plus input/output (I/O) circuits to interface with the outside peripherals, as indicated in Figure 3.5(b). All of these circuits can be on one chip, with external bulk memory as required; we will, however, look at standard memory ICs in the following section.

A wide range of microprocessors and their supporting ICs are available from many sources. They may be characterized by:

- the technology, which may be bipolar or MOS
- the width of the data bus which carries the system instructions and

Table 3.1 Performance Figures for Widely Available Off-the-Shelf Operational Amplifiers

Op. amp. type	Supply voltage range (V)	Max. differential input voltage (V)	Max. output swing (V)[a]	Operating temperature range (°C)	Max. power dissipation (mW)	Large-signal open-loop gain (dB)	Input resistance	CMR (dB)	Slew rate ($V\,\mu s^{-1}$)	Band width (kHz)
741N (bipolar)	±5 to ±18	±15	±13	0–70	500	106	2×10^6	90	0.5	10
741S (bipolar)	±5 to ±18	±15	±13	0–70	625	100	1×10^6	90	20	200
5539 (bipolar)	±8 to ±12	–	± 2.5	0–70	550	52	1×10^5	80	600	48 000
TL081 (BiFET)	±3 to ±18	±18	±13.5	0–70	680	106	1×10^{12}	76	13	150
LF351N (J-FET)	±5 to ±18	±18	±13.5	0–70	500	110	1×10^{12}	100	13	150
7611 (CMOS)	±1 to ±8	±8	± 4.5	0–70	250	102	1×10^{12}	90[b]	0.16[b]	500[b]

[a] At max. supply voltage.
[b] Dependent upon operating point.

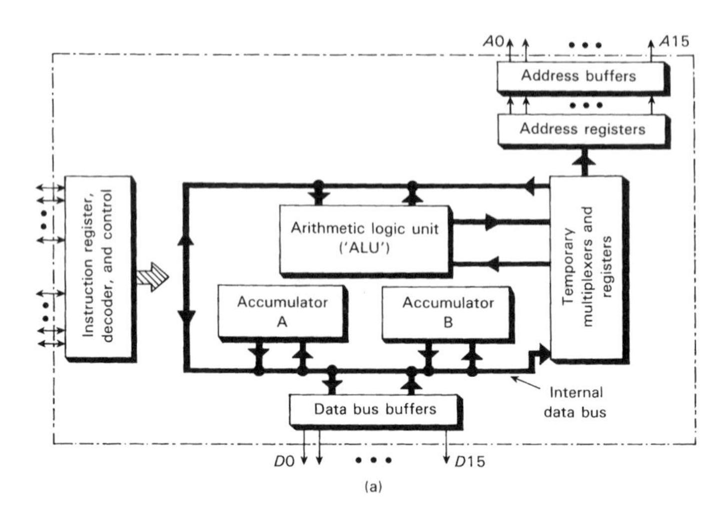

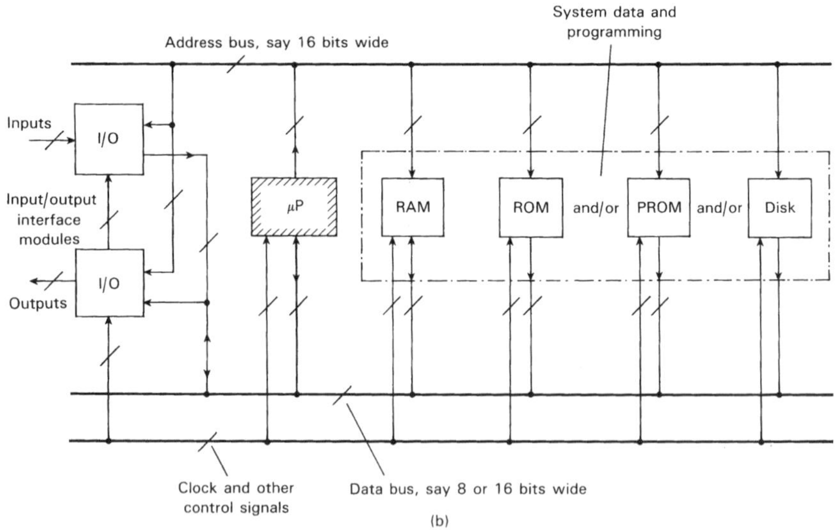

Fig. 3.5 The standard microprocessor (μP): (a) the usual μP architecture; (b) the μP with typical associated peripheral circuits. In many products some or all of the required memory can be on the μP chip itself instead of being external.

system data, and which may be 4-bits, 8-bits, 16-bits, 32-bits or 64-bits wide.

While Schottky TTL, ECL, I^2L and CDI bipolar have all been used in the past for commercially-available processors, the increasing on-chip complexity and the requirement to reduce power consumption has meant that bipolar has now been largely superseded by MOS technologies. Some bipolar products are still available for simple duties (see below) but nMOS and CMOS now dominate the market, with an increasing range of other compatible CMOS and nMOS standard peripherals [19–24]. Fabrication technologies are as covered in Chapter 2, the microprocessor and its associated ICs being in the forefront of all VLSI fabrication developments due to the extremely large mass market for such products.

For many simple domestic and industrial control systems, 4-bit processors such as the Texas Instruments' 1000, 2000 and 3000 series are appropriate; the global sales of these and similar simple processors has been reported to exceed that of the larger size products. However for more general-purpose industrial and scientific work it is the 16, 32 or 64-bit IC which is relevant. Note that the term *microcontroller* may be used for microprocessors which are specifically relevant for industrial control duties, possibly with specific on-chip resources such as A-to-D and/or D-to-A conversion, but we will not distinguish here between these and more general-purpose processors.

In between the very simple 4-bit processors and the state-of-the-art microprocessors such as the Intel Pentium, PentiumPRO and Motorola Power-PC604 and 620 designs are a range of 8-bit and 16-bit ICs which have had wide sales. These have included the following:

- 8-bit Z80 series, both nMOS and CMOS technology
- 8-bit 6500 series, both nMOS and CMOS technology
- 8-bit 8600 series, both nMOS and CMOS technology
- 8-bit 8080 series, nMOS technology
- 8-bit 8085 series, both nMOS and CMOS technology
- 16-bit 8086 series, both nMOS and CMOS technology
- 16-bit 68000 series, both nMOS and CMOS technology
- 16-bit 9900 series, nMOS technology

It will be noted that both nMOS and CMOS versions of certain types have been marketed, with the CMOS versions generally being developments of the earlier nMOS ICs, giving somewhat better performance and lower power consumption. However, as size increases towards 32 and 64 bit words, nMOS currently remains the chosen technology, largely because it is easier and more compact to make large array structures in MOS rather than in CMOS.

Microprocessors form an essential element for many systems. However, there is no question of any individual OEM designer ever designing a specific version for some custom product, as in no way could the low cost and high performance of a commercial product be matched. However, the microprocessor can find a place in custom microelectronics by using a vendor's design in a standard-cell form (a *megacell*), where the processor forms a part—possibly the major part—of the assembly of predesigned building blocks that make up the unique custom chip. These considerations will be addressed in the following chapter.

3.4 MEMORY

Section 2.3 of Chapter 2 gave an overview of the technologies, circuit configurations and fabrication used in digital memory circuits. As noted, memory ICs represent the pinnacle of transistor count and device packing density per IC, aided by the regular layout possible in memory circuits.

3.4.1 Read-Only Memory (ROM)

Mask-programmable ROMs (see Figure 2.26) are not true off-the-shelf components because the IC manufacturer has to pattern the interconnections required for a particular application. Original equipment manufacturers must therefore liaise with vendors for the supply of such parts, and known volume requirements must be such as to justify their use.

Field-programmable ROMs, however, are widely available for OEM dedication, for example, Schottky TTL bipolar ICs ranging from 256 bits (32 × 8-bit) in a 16-pin package to 32 Kbits (4 K × 8-bit) in a 24-pin package, with access times on the order of 20 ns being available. Larger sizes, particularly in surface-mount packages, are equally available.

In the range of erasable-reprogrammable read-only memories (EPROMs), MOS technology using the floating gate principle shown in Figure 2.29 is used in all standard products. In the type of EPROM where bulk erasure of all the chip data is by exposure to ultraviolet light (the UV-PROM), the available standard parts range from around 8 Kbits (1 K × 8-bit) in a 24-pin package to 4 Mbits (512 K × 8-bit) in a 32-pin package. Both nMOS and CMOS technologies with access times as fast as 120 ns are represented, but with the CMOS versions providing considerably lower power dissipation and with a static current drain of around 1 mA at 5 V compared with perhaps 50 or more times this current for comparable nMOS circuits. Figure 3.6 illustrates a typical UV-PROM. It is also possible to purchase small single-chip microprocessor ICs complete with UV-PROM, a typical example being an 8-bit CMOS processor plus 4 K × 8-bits of UV-PROM together with RAM and I/O circuitry, all on the one chip.

Electrically erasable ROMs are also widely available. Currently available

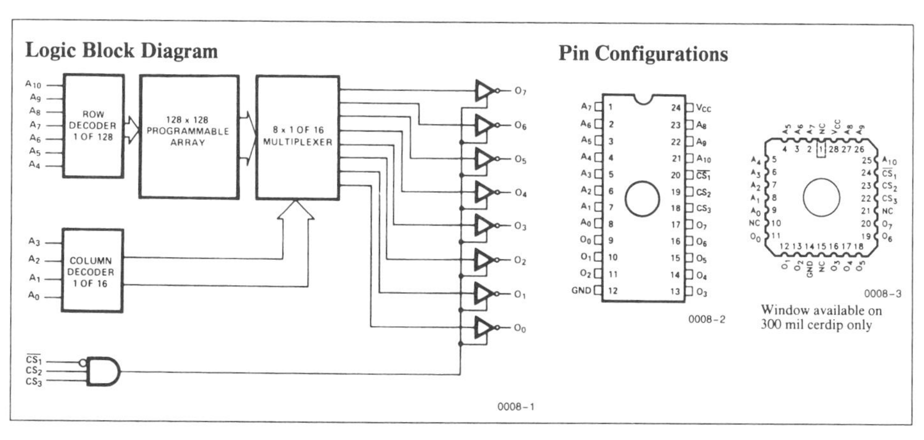

Fig. 3.6 A CMOS technology 2048 × 8-bit UV-PROM, with an access time of 25 ns. (Courtesy of Cypress Semiconductor Corporation, CA.)

standard products, termed EE-PROMs or E^2PROMs, use either nMOS or CMOS technology, and supersede earlier much slower versions which were generally known as electrically alterable-programmable ROMs (EAPROMs). EEPROMs are not available in such large sizes or with the access time performance of EPROMs; 64 Kbits (8 K × 8-bit) with an access time of 250 ns is widely available, with later developments up to greater than 1 Mbits.

The advantages and disadvantages of UV and EE reprogrammable PROMs are as follows:

(i) The UV-PROM cannot be selectively erased, and the bulk erasure process requires removal of the IC from the circuit and irradiation for a considerable time—possibly over 30 minutes—to fully erase the previously stored data.

(ii) The EEPROM, however, can have selective word erasure—this may be done in-circuit if the voltage pulses can be safely applied to the complete circuit.

(iii) The number of possible erasures/reprogramming cycles for EE-PROMs is usually higher than for UV-PROMs. However, as previously noted, the performance of EEPROMs is not as good as that available from UV-PROMs.

3.4.2 Random-Access Memories

Off-the-shelf random-access memory circuits are available in both static and dynamic form. Static RAMs ranging from 1 Kbit (256 × 4-bit) to 256 Kbit (32 K × 8-bit) are very widely used. CMOS technology dominates the larger sizes, with access times on the order of 25 to 30 ns for the high-speed versions. A dominating characteristic of these CMOS RAMs is their competitive cost, currently less than $10 for a 256 Kbit IC, and their low power, as low as 5 to 10 μW static dissipation.

Commercially available dynamic RAMs have an even greater capacity than static RAMs. 1M × 1-bit configurations in 18-pin packages are widely available, with 100 ns access time and a cost of about $15. Still larger versions are available, with 64 Mbits per IC being advertised. Note that there are periodic shortages of large memory ICs due to fluctuations in commercial demand—in times of recession, production is often reduced, and then if demand suddenly recovers there is a global shortage. It seems that production and market demand often become out of step, which places a premium on large memory circuits.

Like the microprocessor, both RAM and ROM designs are available as megacells from custom microelectronic vendors for incorporation in standard-cell custom ICs, although it may often be more satisfactory and economical to

Table 3.2 Representative Data on a Small Sample of Off-the-Shelf Memory ICs

Type no.	Device	Technology	Organization	Programmable means	Access time (ns)	Packaging (pins)
SN74371	ROM	Bipolar	256 × 8-bit	Mask	30	20
SN74387	PROM	Bipolar	256 × 4-bit	Fuse	30	16
23256	ROM	nMOS	32 K × 8-bit	Mask	200	28
53128	ROM	CMOS	16 K × 8-bit	Mask	250	28
82HS321	PROM	Bipolar	4 K × 8-bit	Fuse	35	24
27512	EPROM	nMOS	64 K × 8-bit	Electrical, UV erasable	200	28
27C1024	EPROM	CMOS	64 K × 16-bit	Electrical, UV erasable	150	40
28C64	EEPROM	CMOS	8 K × 8-bit	Electrically programmable/ reprogrammable	200	28
43256	SRAM	CMOS	32 K × 8-bit	–	100	28
50464	DRAM	nMOS	64 K × 4-bit	–	120	18
42456	DRAM	CMOS	256 K × 4-bit	–	100	20
27C4001	EPROM	CMOS	512 K × 8-bit	Electrical, UV erasable	150	32

keep very large memory requirements as separate ICs which are in large volume and proven production.

Table 3.2 summarizes some representative information on medium size commercial memory ICs. Further circuit and technical details may be found in the published literature [4,25–30].

3.5 PROGRAMMABLE LOGIC DEVICES

The term *programmable logic device* (PLD) theoretically covers any off-the-shelf IC which can be programmed after purchase by the customer to perform some specific logic duty. Hence, PROMs (see Section 3.4) should come under this heading, although due to their principal use in computer systems they are invariably considered as memory devices rather than PLDs.

Programmable logic devices are capable of realizing any required logic function within the capacity of the device. It will be recalled that any Boolean logic expression can be written in a sum-of-products form, for example, as shown in Figure 3.7. The product (AND) terms may be minterms, that is, containing all the input variables in either true or complemented form (the literals), or minimized if possible so as to contain fewer terms in each AND function [31–36].

The direct realization of a Boolean expression or truthtable therefore involves AND gates followed by an OR level. All programmable devices are based on this, giving the architectural arrangement shown in Figure 3.8. More than one output is always present. However, the distinction between the many types of programmable logic devices is in the details of the input decoding and in the

Inputs			Minterm designation	Output function $f(x)$
x_1	x_2	x_3		
0	0	0	m_0	0
0	0	1	m_1	1
0	1	0	m_2	0
0	1	1	m_3	1
1	0	0	m_4	1
1	0	1	m_5	1
1	1	0	m_6	0
1	1	1	m_7	0

(a)

x_3 \ x_1x_2	00	01	11	10
0	m_0 0	m_2 0	m_6 0	m_4 1
1	m_1 1	m_3 1	m_7 0	m_5 1

(b)

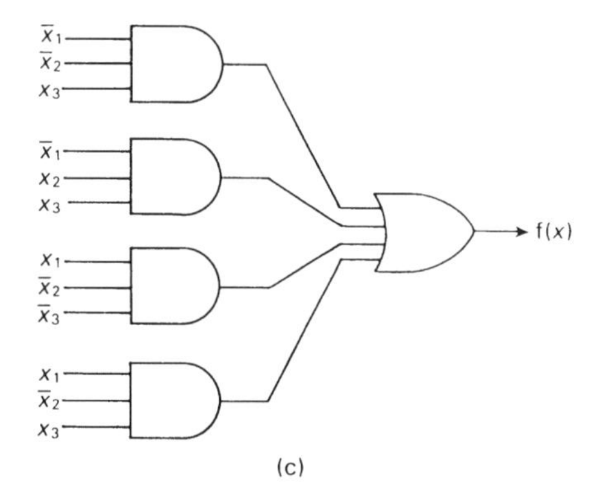

(c)

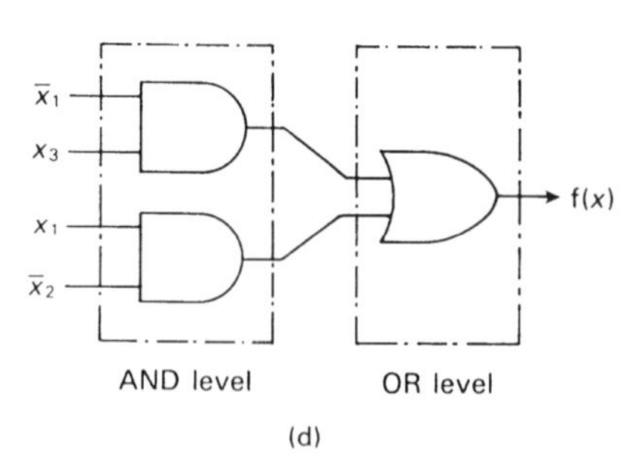

(d)

Fig. 3.7 Example realization of a simple 3-variable Boolean function $f(X)$: (a) the truth-table for $f(X)$; (b) Karnaugh map representation; (c) sum-of-minterms representation, $f(X) = \bar{x}_1\bar{x}_2x_3 + \bar{x}_1x_2x_3 + x_1\bar{x}_2\bar{x}_3 + x_1\bar{x}_2x_3$; (d) minimized sum-of-products realization, $f(X) = \bar{x}_1x_3 + x_1\bar{x}_2$.

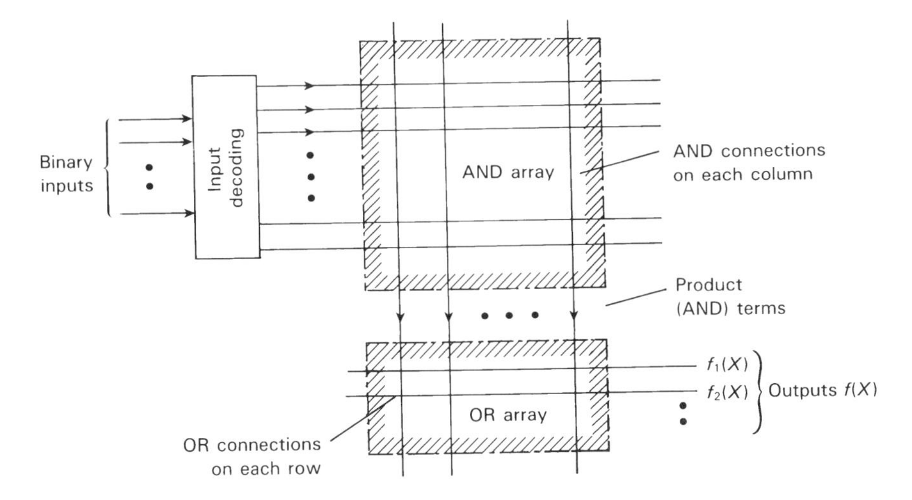

Fig. 3.8 The basic architecture of programmable logic devices.

AND and OR arrays. The OR array, for example, may be fixed rather than programmable. Also, although the two arrays are invariably referred to as 'AND' and 'OR,' respectively, they may in practice both be made as either NAND or NOR arrays. This is readily shown to be possible by applying De Morgan's law [31–36], which can algebraically transform any sum-of-products Boolean expression into the logically equivalent all-NAND or all-NOR form.

Referring back to Figure 2.27(a), it will be appreciated that programmable read-only memories (PROMs) are programmable devices where the n binary input variables are decoded by the input decoders, and then passed through the arrays to form the required true minterms. These minterms are then wired-ORed together to form the output function Z, and hence the OR array (see Figure 3.8), is fixed and nonprogrammable in the PROM case. The four or eight or more separate AND arrays of the PROM form the 4-bit, 8-bit or more output word of the PROM, every input combination ('address') being programmed at minterm level to give the corresponding output word. Each PROM output is therefore a direct copy of the full input/output truthtable.

3.5.1 Programmable Logic Arrays (PLAs)

While the PROM can be used as a powerful general-purpose device for combinational logic design, alternative programmable devices specifically designed for random logic applications are generally more appropriate. These can include sequential as well as combinational capability in the same package.

The principal distinction between PROMs and programmable logic arrays is that the PLA input variables $x_1, \ldots, x_n$ are not decoded into all the possible 2^n minterms. Instead they are first decoded into the $2n$ literals, that is xi and $*\bar{x}_i$ for each binary input x_i, $i = 1, \ldots, n$. The product (AND) array can then make product terms containing as few or as many literals as necessary. This is illustrated in Figure 3.9.

The programming of the AND array thus provides the direct realization of the product terms for any Boolean function. The OR array combines these terms as required to produce each output function $f(X)$, completing the synthesis of a sum-of-products expression.

The programming means in PLAs and the variants to be described shortly are the same as were discussed in Chapter 2 when considering memory circuits. Fusible links are employed in bipolar circuits, and floating gates in MOS circuits

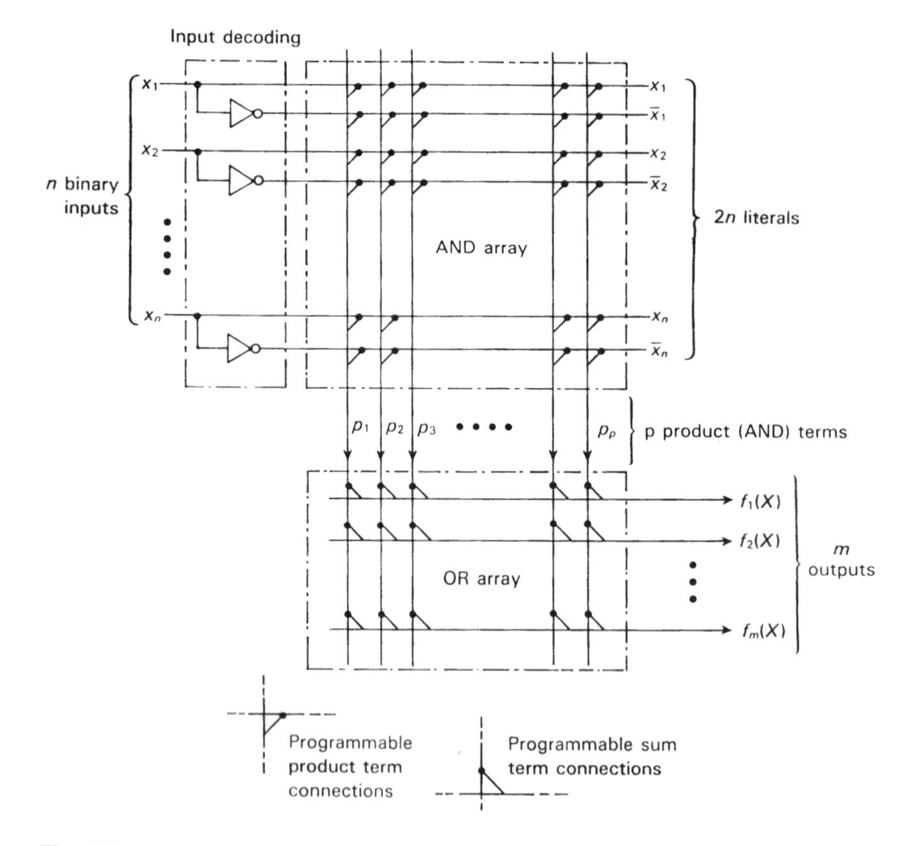

Fig. 3.9 The general-purpose PLA, with n inputs, p product terms, and m outputs, and programmable AND and OR arrays.

(see Figures 2.28 and 2.29, respectively). Bipolar devices are therefore one-time-only programmable, but some CMOS devices are available which are erasable/reprogrammable. However, there is generally not such a wide need for reprogrammability in PLAs as there is in PROMs, since the logic requirements are fixed for any given product once the design has been finalized. Reprogrammable PROMs may advantageously be used in the design development of a product, which when finalized can be transferred to a PLA for production purposes.

The conventions that are often used with programmable logic devices include the following:

- The buffered true and complement of each input is shown by a single symbol.
- The effective (programmed) connections in each array are shown by a dot where the horizontal and vertical lines cross.
- The product and sum lines are defined by single AND and OR symbols, respectively.

These conventions are shown in Figure 3.10(a). Notice that when the PLA is formed by NOR or NAND arrays, the AND and OR array terminology is slightly misleading, but the logical meaning is correct. Additional features may be found in commercial products, including:

- the ability to feed back some or all of the output lines to act as inputs or to allow latches to be constructed
- enable and/or inversion facilities on the output lines to give additional flexibility

These enhancements are shown in Figure 3.10(b). Additionally some I/Os may be programmable so as to act as either inputs or outputs as required.

The PLA and its variants provide very useful off-the-shelf packages for many small logic design duties. The principal technical limitations are in capacity, specifically:

- the number of inputs n available—clearly one 8-input PLA cannot be used if product terms containing more than eight inputs are required
- the number of outputs available
- the number of internal product lines p available—if the m outputs collectively involve more than p different product (AND) terms, then there will be insufficient internal capacity to synthesize all the required output functions

The use of PLAs, therefore, usually involves multiple-output Boolean minimization design procedures in order to realize the several outputs with a minimum of product terms. Absolute minimization may not be necessary provided the number of product terms is not greater than the device capacity. All vendors of programmable devices have appropriate commercial CAD software available

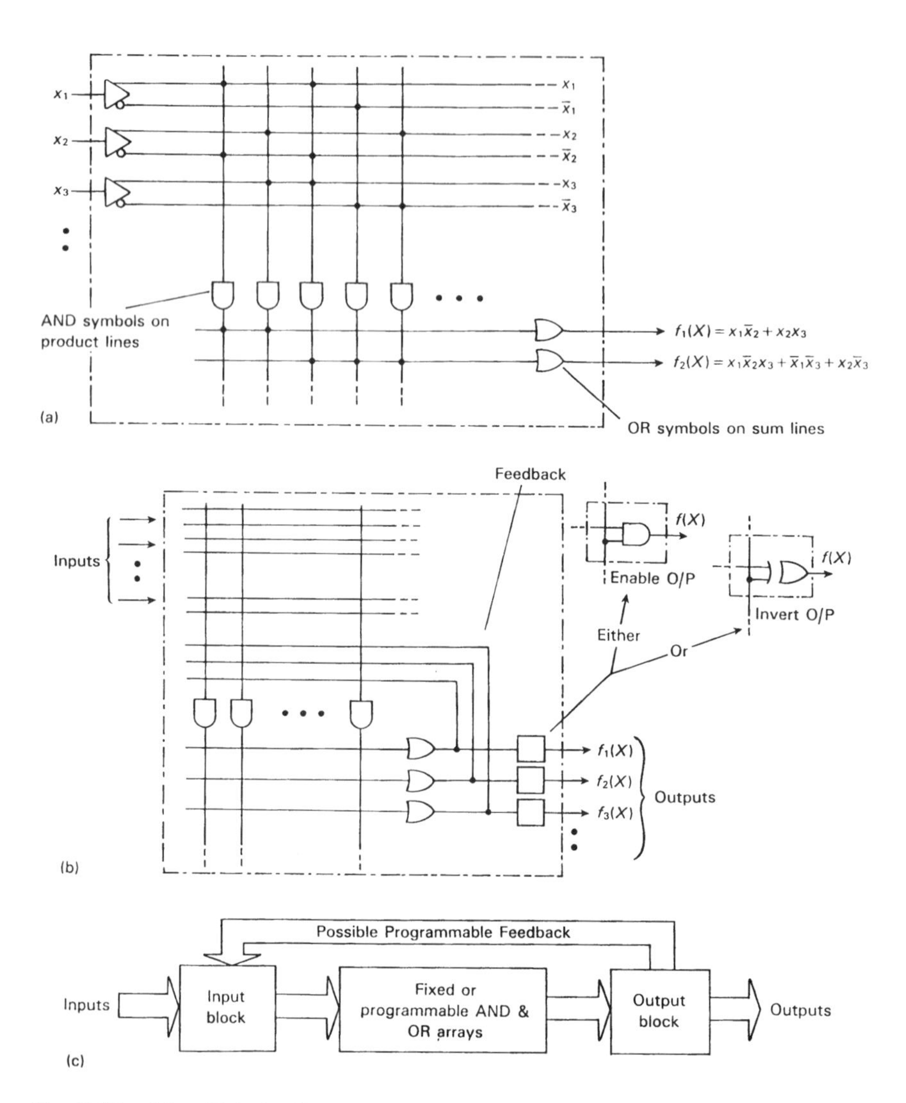

Fig. 3.10 Other PLA details: (a) input and other circuit conventions, with two simple functions illustrated; (b) additional facilities which may be incorporated; (c) the block schematic which may be used by some vendors.

to undertake this design requirement automatically from, e.g., truthtables or non-minimum Boolean expressions. What the CAD software finds difficult to do is to partition a given logic requirement which will not fit into one PLA, so that multiple PLAs in some optimum parallel or series/parallel combination may be used. This is a decomposition problem for which no simple algorithm exists, and which in practice often relies upon human experience or intuition.

Further details of the capabilities of and design techniques for PLAs may be found in many sources [4,29,37–42]. The designation FPLA—field programmable logic array—is also used by certain vendors.

3.5.2 Programmable Array Logic (PALs)

A widely available variant on the PLA is the programmable array logic (PAL™) device, whose basic architecture is similar to Figure 3.9 except that the programmable OR array is replaced by fixed OR gates. (PAL is a registered trademark of Monolithic Memories, who first introduced this type of PLD to the marketplace.) Additionally, the outputs from some of the OR gates are usually fed back to the AND array (see Figure 3.10(b)) in order to increase device capabilities. Selective output inversion through Exclusive-OR gates may also be available [37–39].

Figure 3.11 shows a typical small PAL product. The fixed number of inputs to each internal OR gate is a disadvantage of the PAL in comparison with many PLAs, although the feedback loops may be used to augment this fan-in at the expense of increased propagation times.

One additional feature with certain vendors' PAL devices may be noted. This is the availability of the equivalent PAL device with mask-programming of the custom requirements. Such devices are known as HALs—*hard array logic* devices—and may be more economical where large quantities of a particular PAL circuit commitment are found to be needed. As with ROMs, the vendor rather than the OEM has to manufacture these dedicated devices.

3.5.3 Programmable Gate Arrays (PGAs)

The programmable gate array, frequently termed a field-programmable gate array or FPGA, is a yet simpler version of the architecture shown in Figure 3.9. In PGAs, no OR gates are provided, but comprehensive feedback and inversion facilities are provided instead. A sum-of-products expression such as $\bar{x}_1x_2x_3 + x_2\bar{x}_3x_4$ is available in the alternative form of

$$\overline{\{\overline{(\bar{x}_1x_2x_3)} \cdot \overline{(x_2\bar{x}_3x_4)}\}}$$

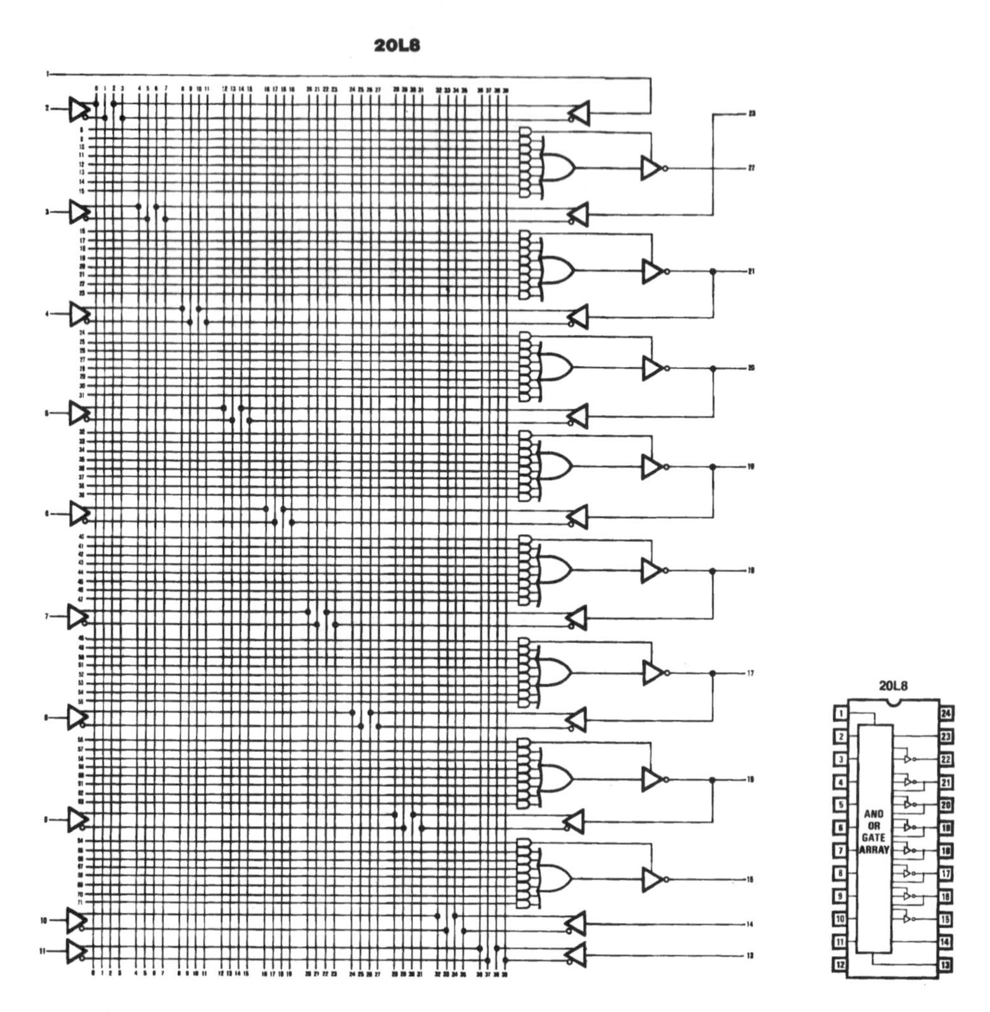

Fig. 3.11 A bipolar technology PAL, with fourteen dedicated inputs, two dedicated outputs, six programmable inputs/outputs, and 15 ns typical propagation delay. (Copyright Advanced Micro Devices, Inc., 1988. Reprinted with permission of copyright owner. All rights reserved.)

by using the feedback and inversion resources. Figure 3.12 illustrates a typical commercial architecture.

3.5.4 Programmable Logic Sequencers (PLS)

The programmable devices so far considered do not contain any dedicated storage (memory) circuits. This additional facility will be found in programmable logic sequencers, sometimes termed field-programmable logic sequencers (FPLS).

Normally, both AND and OR programmable arrays are present, with clocked D-type, RS-type or JK-type storage circuits at the outputs. The Q and $\bar{Q}$ outputs of these circuits are usually fed back into the AND array, as illustrated in Figure 3.13.

Some PAL devices with the AND-only architecture shown in Figure 3.12 are also available with dedicated storage circuits, but the designation 'PAL' is retained for these commercial products rather than the designation 'PLS.'

3.5.5 Other Programmable Standard Parts

Other variants on these programmable logic devices have been marketed, including programmable diode matrix devices (PDMs) which are simply arrays of diodes from which passive diode-AND and diode-OR networks can be made. PDMs are, however, now generally obsolete. Other variants on the PLA, PGA, and PLS architectures may be encountered. For very large devices, dynamic rather than static MOS circuits may be found, mirroring the developments in ROM technologies. Table 3.3 gives a short representative listing of some off-the-shelf products, but the device range is still expanding rapidly.

Units for programming these various types of programmable devices are readily available. These can be separate self-contained microprocessor-controlled programmers, programming a range of commercial products, or programmers which are driven by normal PCs. Great sophistication may be built into the software to verify that the device programming has been correctly completed to the required logic specification.

There are, however, two further commercial products which are distinctly different from these PLD products. One is the *programmable gate array* (PGA), and the other is the *programmable logic controller* (PLC).

PGAs are unique devices falling somewhere between a conventional PLD and a custom gate array IC. They will be considered separately in the following section. The PLC, on the other hand, is not a single IC package, but instead is a very specific commercial product to realize electronic control systems for industrial applications, such as would have been made with electromechanical means in previous years. It consists of the following parts:

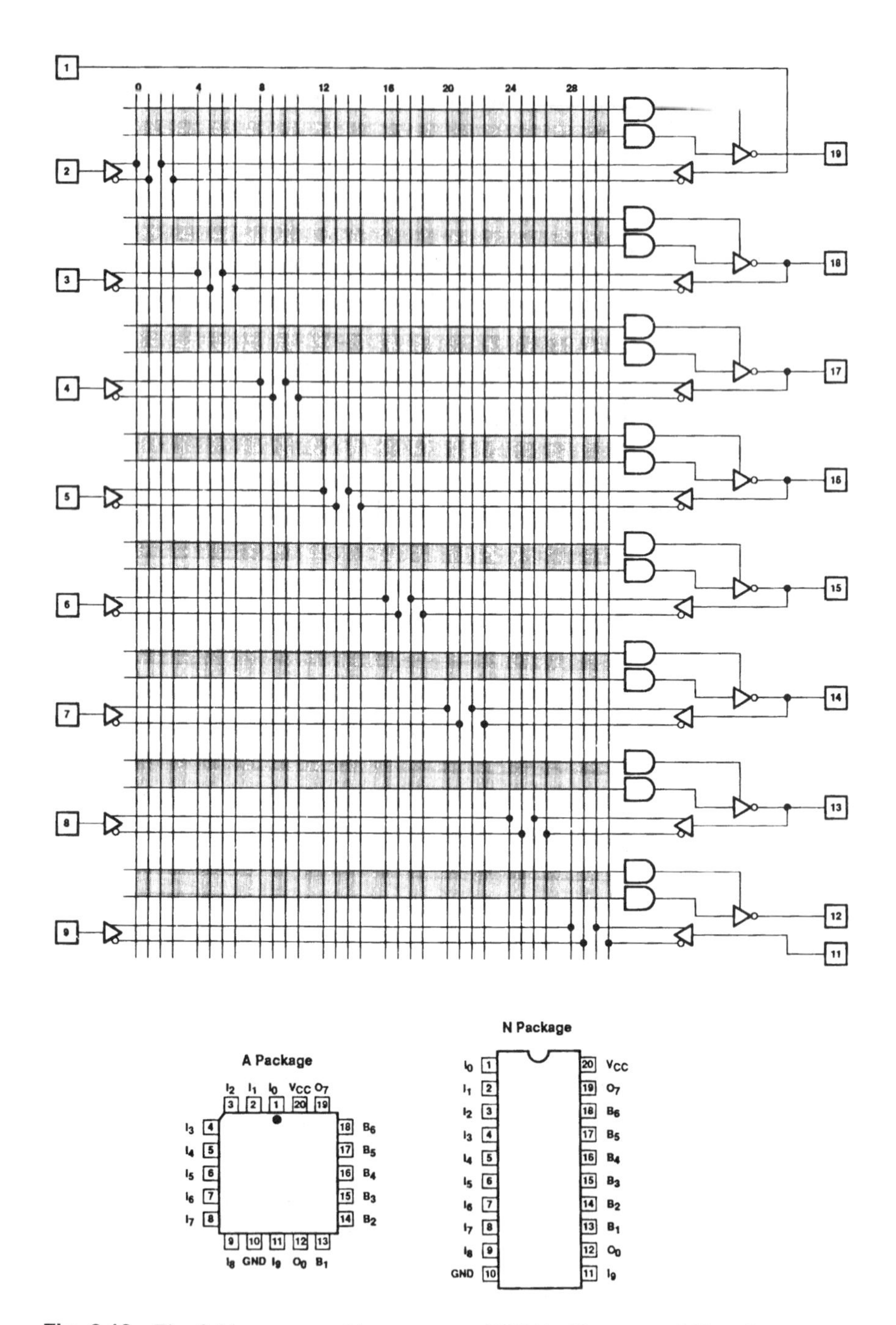

Fig. 3.12 The field-programmable gate array (FPGA). (Courtesy of Signetics/Philips Components Ltd.)

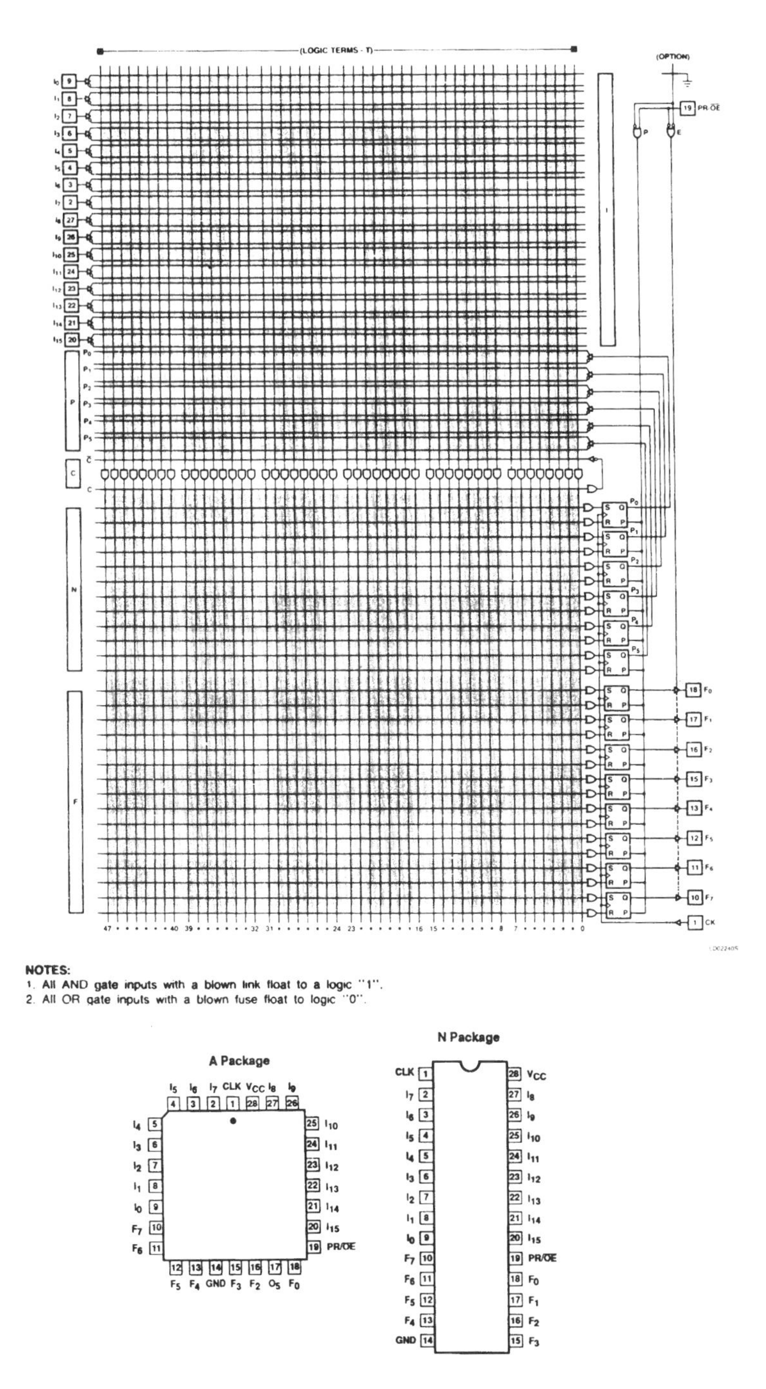

Fig. 3.13 The field-programmable logic sequencer (FPLS). (Courtesy of Signetics/Philips Components Ltd.)

Table 3.3 Some Representative Off-the Shelf MSI Programmable Logic Devices

Type no.	Device architecture	Technology	Organization $n \times p \times m$[a]	Reprogrammable	On-chip storage circuits	Access time (ns) or max. clock speed (MHz)	Packing (pins)
93459	PLA	TTL	16 × 48 × 8	No	None	25 ns	28
16L8A	PAL	TTL	10 × 64 × 8	No	None	7.5 ns[b]	20
C16L8	PAL	CMOS	10 × 64 × 8	Yes	None	25 ns	20
C22V10	PLS	CMOS	12 × 120 × 10	Yes	10	25 ns	24
PLS179N	PLS	TTL	8 × 45 × 12	No	8	45 ns	24
10H20EG8	PLS	ECL	12 × 90 × 8	No	8	3.5 ns	24
405-45	PLS	TTL	16 × 64 × 8	No	8 JK buried, 8 JK brought out	45 MHz	28
EP1810	PLS	CMOS	16 × NA × 48	Yes	48 programmable D, T, RS or JK	35 ns	68

[a] Organization $n \times p \times m = n$ dedicated inputs, p internal product terms, m dedicated outputs or reconfigurable I/Os.
[b] High-speed version.

(a) an isolating input section (an input module or modules)
(b) an electronic central processing unit (CPU)
(c) an isolating power output section (an output module or modules)
(d) a power supply unit (PSU module)

Other facilities such as timers, printer interfaces, etc., are also available as part of a comprehensive boxed commercial product.

The CPU, which contains RAM and PROM or EPROM memory, is programmed by the OEM to undertake the required control tasks, operating usually in real time using the system input data supplied via the input modules, and providing output control signals via the output modules. The PLC is thus a power electronics system with the low voltage microelectronics providing the central control, rather than an entirely low-power microelectronics system. Further details of these very useful control engineering products may be found elsewhere [43], but will not be referred to again in this text.

3.6 LOGIC CELL ARRAYS (LCAs)

Since the logic cell array (LCA), sometimes referred to as a *programmable gate array* (PGA), is fundamentally dissimilar from other PLDs, we will consider it as another type of off-the-shelf but user-programmable product. (LCA is the registered trademark of Xilinx, Inc., CA, who introduced this particular device to the commercial market.)

In comparison with many other types of PLDs, the LCA is a large chip of VLSI complexity. Its employs CMOS technology, with feature size as small as about 1 μm [44,45]. Its architecture consists of an array of identical cells separated by wiring channels, together with peripheral I/O cells, as shown in Figure 3.14(a). However, unlike a semicustom gate-array IC (see Chapter 4) where the metallic interconnections in the wiring channels have to be custom-designed and fabricated, in the LCA there are many fixed segments of metal interconnect in all the row and column wiring channels, with programmable means to form any required interconnection pattern between all cells and I/Os. The cells themselves are also programmable, each consisting of a number of logic elements as shown in Figure 3.14(b), with multiplexers which can be programmed to turn the cell into a wide range of combinational and sequential duties.

The actual programmable chip connections are performed by n-channel FET switches which are present between all adjacent interconnection segments. With a FET 'off,' adjacent segments are effectively isolated from each other, but with the FET 'on,' the segments form a connected path. Control of all the FET switches on the chip therefore configures it to a particular custom requirement.

The gates of all these controlling FET switches are in turn controlled by

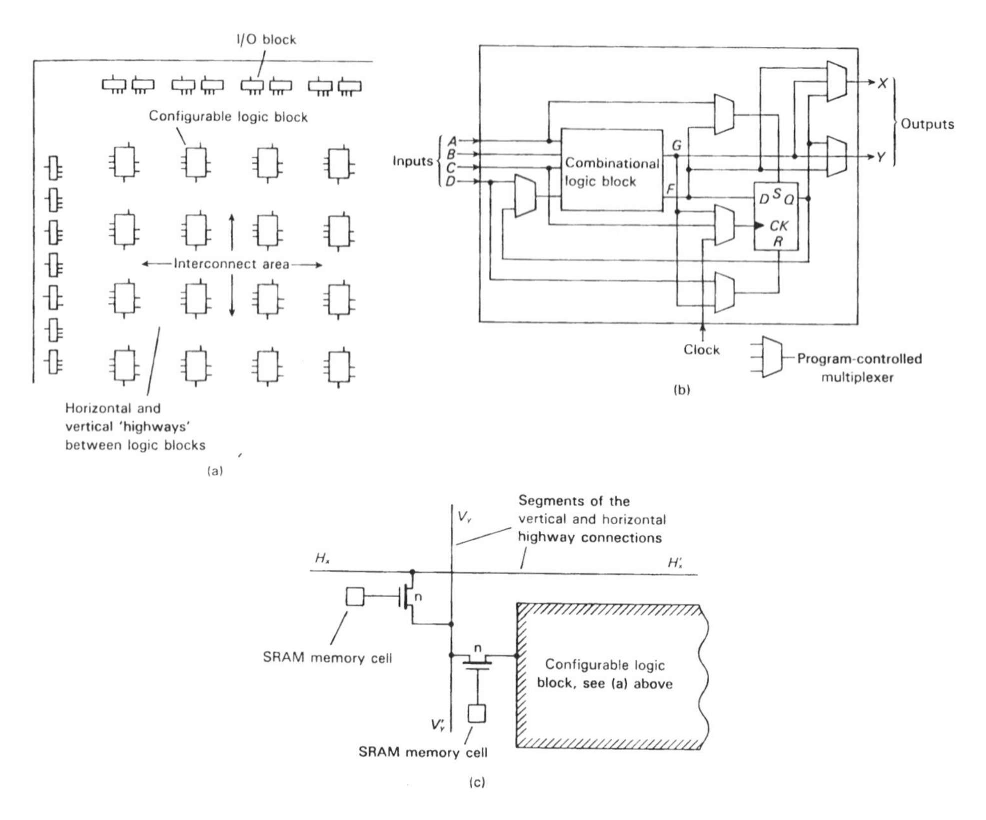

Fig. 3.14 The field-programmable logic cell array (LCA): (a) the general chip architecture; (b) the standard cell ("configurable logic block"); (c) the FET n-channel switches controlled by static RAM data storage. (Courtesy of Xilinx, Inc., CA.)

a static random-access memory (SRAM) (see Figure 3.14(c)), each SRAM cell being programmed to the on or off state by the user's program. Thus, additional to all the functional cells and their interconnect is this comprehensive SRAM array, which may be dedicated so as to control all the required interconnections of the custom circuit.

The vendors of LCAs provide unique CAD software to enable the OEM designer to design and verify the dedication program. When complete, this program may be permanently stored either on hard disk or in a separate 8-pin PROM IC. It should be appreciated that the LCA chip itself is a volatile device, being programmed by the internal SRAM which inherently loses its data on power-off. When power is restored, then the configuration program needs to be reloaded from the nonvolatile memory store into the LCA. In this respect the LCA is similar to any microprocessor system, where the volatile RAM data has to be reloaded into the system after any power-off.

It will be apparent that the LCA chip is a complex device. Nevertheless due to the absence of any area-consuming fuse elements and the use of leading-edge MOS technology, much larger capability is currently available in LCAs compared with most other PLDs. Table 3.4 gives a comparative overview, although it should be noted that the capacity of all types of device is generally still growing.

The disadvantage of the LCA being a volatile device is overcome in another company's product which employs 'antifuse' technology to configure a logic array in a nonvolatile manner. PLICE ('programmable low impedance circuit element') antifuses are two-terminal devices consisting of an extremely thin dielectric barrier between two electrodes, which breaks down to form a bidirectional ohmic path when an 18 V programming pulse is applied. (PLICE is the trademark of Actel, Inc., CA.) The resistance of such a path is about 1 kΩ compared with its previous value of about 100 MΩ. This is therefore a one-time-programmable nonvolatile technique. An off-the-shelf commercial product containing 2000 gates in its logic array contains 186,000 antifuse elements, roughly 100 per logic gate, to give the full freedom of interconnect, but only a very small percentage has to be blown to configure a chip to a given custom requirement. The terminology *application configurable technology* (ACT) will also be found used in association with these products. Some typical product specifications are

Table 3.4 Some General Comparative Characteristics of Five Ways of Making Random Logic Networks[a]

	Off-the-shelf devices			Vendor-customized devices	
	SSI and MSI ICs	Programmable logic devices (PLDs)	Logic cell arrays (LCAS)	Gate-array custom ICs	Standard-cell custom ICs
Capacity in no. of gates	Tens to few hundred	Hundreds to few thousand	Thousands to tens of thousands	Thousands to tens of thousands	Tens of thousands to hundreds of thousands
Speed	Medium to fast	Slow to medium	Slow to medium	Slow to fast	Medium to fast
Functionality defined by user	No	Yes	Yes	Yes	Yes
Time to customize	—	Seconds	Seconds	Depends upon many factors – possibly months' turn-around	
User-programmable	No	Yes	Yes	No	No

[a] Complex programmable logic devices (CPLDs) are now taking the capability of logic cell arrays to greater capacity and faster speeds. Details of vendor-customized devices follow in Chapter 4. (Source: based upon Ref. 44.)

Table 3.5 The Product Specifications for the First Generation of Actel PLICE™ Devices[a]

Device type	ACT 1010	ACT 1020	ACT 1230	ACT 1260
CMOS process	2 μm	2 μm	1.2 μm	1.2 μm
No. of logic cells	295	546	720	1404
Equiv. number of gates	1200	2000	3000	6000
Max. no. of latches	295	546	700	1400
Max. no. of D-type circuits	147	273	540	1052
No. of PLICE antifuses	112,000	186,000	333,000	666,000
No. of I/Os	57	69	108	150

[a] Second-generation devices with increased capacity and performance are now available.

given in Table 3.5. Further information may be found in published literature [46,47].

At present, it is not established whether the volatile VLSI programmable product or the non-volatile antifuse product will capture the major market share for OEM-programmable VLSI devices. The former does allow design errors to be corrected, but the nonvolatile nature of the latter may be preferred in some cases. Both currently represent a rapidly growing market, challenging to a noticable degree the custom IC approach which we will be considering in the subsequent chapters of this text. It is a pity that the terminologies such as LCA, FPGA, ACT, CPLD, etc., and 'gate array,' which are growing up around these large standard products are not always as explicit or standardized as they might otherwise be.

3.7 SPECIALIZED APPLICATION-SPECIFIC STANDARD PARTS (ASSPs)

In this final section, we are referring to off-the-shelf ICs which have been marketed to perform very specific duties, and are not general-purpose products in any respect. They are, therefore, increasingly referred to as *application-specific standard parts* (ASSPs), although this terminology has not yet received universal acclaim. (See also the comments in the first paragraph of Chapter 2 concerning the terms 'ASIC,' 'USIC,' and 'ASSP').

There is an increasing number of ICs, usually of VLSI complexity, which are becoming available for very particular purposes, having been designed and optimized for the one duty. It is an interesting evolution in that an IC designed for a special purpose may start its life as a user-specific IC (USIC), but when a supplier appreciates its global market potential, then its design may be fine-tuned to optimize its performance or capability, and marketed as a standard catalog item. The product then ceases to be the property of one customer. Some vendors who were initially suppliers of custom ICs only, for example, LSI Logic Corporation, now list standard parts in their catalog. All this makes it increasingly important to try to distinguish between USICs—specific to one OEM—and ASSPs, which are specific to an application but not to just one OEM.

Among the products which may be found in vendors' and manufacturers' catalogs are the following:

- digital clock and driver circuits
- LED and LCD drivers
- keyboard, printer, video-display unit, disk-drive and communication controllers
- speech synthesis circuits
- smoke detection circuits
- temperature, strain gauge and other instrumentation circuits
- signal processing circuits

and others. If such a product is available to match a particular OEM requirement, then in purely financial terms it would rarely be appropriate for the OEM to consider designing a custom IC for this duty. There may, however, be other considerations such as product design security which the OEM may consider, and which might influence the choice between the use of ASSPs and a more commercially secret custom design. These are considerations which will be addressed in Chapter 7.

3.8 SUMMARY

This overview of off-the-shelf parts has been considered necessary since custom microelectronics will never supersede the use of standard parts in all circumstances. The choice between the use of standard parts and custom ICs is therefore a crucial engineering decision which every design engineer must face when considering a new product design.

However, having covered this review, we may now move on to the principal theme of this text, namely *custom microelectronics*; however, in Chapter 7 we will further consider standard parts when the full pros and cons of custom ICs vs. standard parts are discussed.

REFERENCES

(Note: the many very comprehensive handbooks and data books issued by manufacturers are regularly updated, and therefore no date of publication has been cited below in such commercial references.)

1. Williams, A.B., *Designer's Handbook of Integrated Circuits*, McGraw-Hll, NY, 1984
2. Parr, A.E., *Logic Designer's Handbook*, Granada Press, 1984
3. Marston, R.M., *CMOS Circuits Manual*, Heinemann, UK, 1987
4. Texas Instruments, *The TTL Data Book, Vol. 1, Vol. 2 and Vol. 3*, Texas Instruments, Inc., Dallas, TX
5. National Semiconductor, *Logic Data Book, Vol. 1, Vol. 2 and Vol. 3*, National Semiconductor Corporation, Santa Clara, CA
6. Texas Instruments, *High-speed CMOS Logic Data Manual*, Texas Instruments, Inc., Dallas, TX
7. GE–RCA, *High-speed CMOS Logic Data Book*, General Electric–RCA, Somerville, NJ
8. Mullard-Signetics, *Technical Data Book 4, Parts 2, 4, 5, 8 and 8(a)*, Signetics Corporation, Sunnyvale, CA
9. Sedra, A.S. and Smith, K.C., *Microelectronic Circuits*, Holt, Rinehart and Winston, NY, 1987
10. Motorola, *MECL High Speed Integrated Circuits*, Motorola Semiconductor Products, Inc., Pheonix, AZ
11. Texas Instruments, *Linear Circuits Data Book*, Texas Instruments, Inc., Dallas, TX
12. National Semiconductor, *Linear Data Book*, National Semiconductor Corporation, Santa Clara, CA
13. Analog Devices, *Applications Handbook*, Analog Devices, Inc., Norwood, MA
14. Linear Technology, *Linear Data Book*, Linear Technology, Inc., Burlington, Canada
15. Gayakwad, R.A., *Op Amps and Linear Integrated Circuits*, Prentice Hall, NJ, 1988
16. Texas Instruments, *LinCMOS Design Manual*, Texas Instruments, Inc., Dallas, TX
17. Texas Instruments, *BiFET Design Manual*, Texas Instruments, Inc., Dallas, TX
18. Maxim, *New Releases Data Book, Vol. IV*, Maxim Integrated Products, Sunnyvale, CA
19. Comer, D.J., *Microprocessor-based System Design*, Holt, Rinehart and Winston, NY, 1986
20. Cassell, D.A., *Microcomputers and Modern Control Engineering*, Prentice Hall, NJ, 1973
21. McGrindle, J.A., *Microcomputer Handbook*, Collins, UK, 1985
22. Whitworth, I.R., *16-bit Microprocessors*, Collins, UK, 1986
23. Mitchell, H.J. (Ed.), *32-bit Microprocessors*, Collins, UK, 1987
24. Ciminiera, I. and Valenzano, A., *Advanced Microprocessor Architectures*, Addison-Wesley, NY, 1987
25. Goodge, M., *Semiconductor Device Technology*, Macmillian Press, UK, 1983
26. Dillinger, T.E., *VLSI Engineering*, Prentice Hall, NJ, 1988
27. Texas Instruments, *MOS Memory Book*, Texas Instruments, Inc., Dallas, TX
28. Intel, *Microprocessors Handbook*, Intel Corporation, Santa Clara, CA
29. Cypress Semiconductor, *MOS Data Book*, Cypress Semiconductors, San Jose, CA
30. *IEEE J. Solid State Circuits*, special issue on semiconductor memory annually each October
31. Muroga, S., *Logic Design and Switching Theory*, Wiley, NY, 1979

32. Lewin, D. and Protheroe, D., *Design of Logic Systems*, Chapman and Hall, UK, 1992
33. Barna, A. and Porat, D.I., *Integrated Circuits in Digital Electronics*, Wiley, NY, 1987
34. Roth, C.H., *Fundamentals of Logic Design*, West Publishing, MN, 1995
35. Hayes, J.P., *Introduction to Digital Logic Design*, Addison-Wesley, MA, 1993
36. Wakerly, J.F., *Digital Design Principles and Practices*, Prentice Hall, NJ, 1994
37. Bolton, M.J.P., *Digital System Design with Programmable Logic*, Addison-Wesley, UK, 1990
38. Bostock, G., *Programmable Logic Handbook*, Blackwell, UK, 1987
39. Hurst, S.L., *Custom-Specific Integrated Circuits: Design and Fabrication*, Marcel Dekker, NY, 1985
40. Advanced Micro Devices, *PAL Device Data Book and Design Guide*, Advanced Micro Devices, Inc., Sunnyvale, CA
41. Signetics, *PLD Data Manual*, Signetics Corporation, Sunnyvale, CA
42. Intel, *Programmable Logic Handbook*, Intel Corporation, Santa Clara, CA
43. Kissell, T.E., *Understanding and Using Programmable Logic Controllers*, Prentice Hall, NJ, 1986
44. Xilinx, *The Programmable Logic Data Book*, Xilinx, Inc., Santa Clara, CA
45. Freeman, R., User-programmable gate arrays, *IEEE Spectrum*, Vol. 25, December 1988, pp. 32–35
46. Actel, *The Beginner's Guide to Programmable ASICs*, Actel, Inc., Sunnyvale, CA
47. Actel, *FPGA Data Book and Design Guide*, Actel, Inc., Sunnyvale, CA

4

Custom Microelectronic Techniques

The custom microelectronic techniques to be discussed in this chapter, and which will continue to be the subject area of the remaining chapters of this text, now form a highly significant sector of the complete microelectronics spectrum. The technical reasons for the emergence of custom microelectronics have been covered in Chapter 1, namely the increasing maturity of both silicon fabrication and computer-aided design resources; the practical reasons for considering the use of custom circuits in preference to that of standard off-the-shelf ICs is to attempt to achieve a final product which is better in some way than can be obtained by using standard parts.

Among the advantages which can or may be acheived in adopting custom microelectronics are the following:

- the final market product is unique, so potential competitors cannot readily copy the design using standard parts
- the performance of the complete product may be improved above that possible with standard parts
- the product designer may have greater freedom to innovate if necessary, not being constrained by standard parts
- the final product's physical size may be considerably reduced
- power consumption may also be reduced, allowing smaller power supplies
- product assembly, test and documentation may be reduced in comparison with an assembly of standard parts
- product reliability may be improved due to the reduction in the number of individual parts and their interconnections

- possibly a reduced product design and development time due to appropriate computer-aided design resources
- possibly a reduced final product cost due to savings in component cost and product size

These are the principal advantages claimed for custom circuits. While there are products for which the adoption of a custom IC is obviously advantageous or indeed necessary, for example, in hand-held battery-operated products for which size is at a premium, not all the above claims are necessarily realized or indeed are significant for a particular product design. Company decisions when or when not to use custom ICs for a new product design will form the major part of the penultimate chapter of this text.

The principal subdivisions of custom microelectronic circuits are shown in Figure 4.1. As briefly introduced in the first chapter, full custom is where every detail of the custom IC is specifically designed to achieve the best end product. Standard-cell and gate-array techniques, however, rely upon predesigned circuit

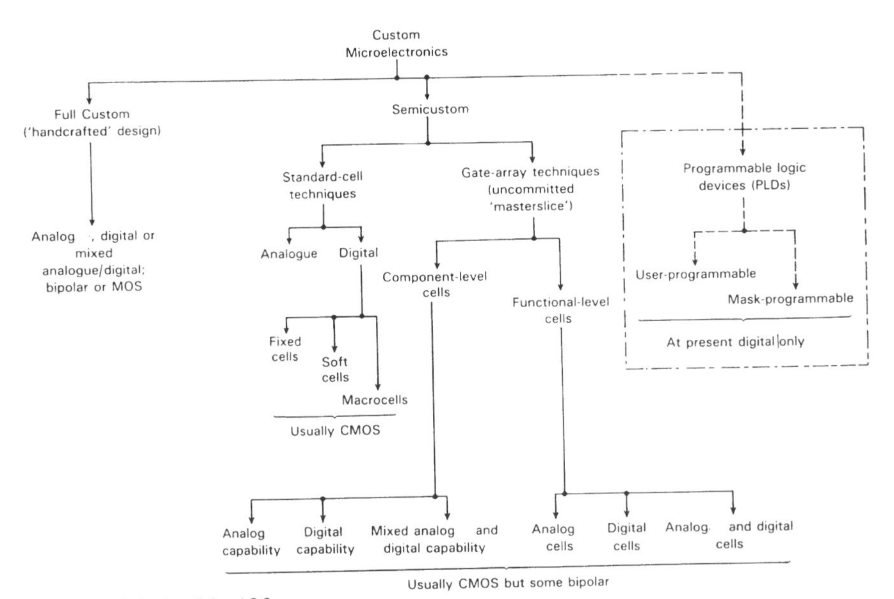

Fig. 4.1 The hierarchy of custom microelectronic ICs (USICs): at present, digital-only circuits are the majority of products on the market, but analog and mixed analog/digital products are of increasing significance. The terminology 'gate array' is ambiguous if applied to the latter products even if they have an array architecture.

details or chip layouts, respectively, in order to minimize expensive design time and cost in completing the final fabrication details. These categories and their subdivisions will be considered in the following sections.

Programmable logic devices have also been included in Figure 4.1, since they are a form of custom-dedicated device. In particular the larger sizes, the *complex programmable logic units* (CPLUs), are increasingly competitive with conventional gate arrays. However, in this chapter we shall have no need to add to the technical data on programmable logic which has already been given, but they will be referred to again in subsequent chapters dealing with CAD and with the choice of design style.

It should also be recalled that custom ICs do not involve any novel fabrication techniques. Indeed, proven stable fabrication processes are essential for the reliable and economic production of custom circuits. The fabrication details considered in Chapter 2 therefore cover all the techniques that may be involved in the custom field.

Finally, the term USIC (*user-specific IC*) rather than ASIC (*application-specific IC*) will be used in the following pages, for the reasons given earlier.

4.1 FULL HAND-CRAFTED CUSTOM DESIGN

Full-custom USICs are designed to meet the product specification in every detail, the emphasis being upon maximum possible circuit performance and minimum possible silicon area. A complete set of masks (a 'full mask set') is required for wafer production. Minimum area means the lowest eventual fabrication cost since more ICs can be made on a single wafer, but against this must be weighed the considerably higher initial design costs.

The accepted definition of full custom is that the chip is designed at the silicon level, with complete freedom to design the individual transistors and the floorplan layout without any predesigned restrictions such as those present in semicustom design techniques (see Sections 4.2 and 4.3). The designer of a full-custom chip must therefore be familiar with the following factors:

- complex circuit design techniques in order to be able to make full use of the freedom of design available
- how monolithic circuits make new circuit configurations possible, for example, by using the inherent good matching of component parameters but accepting a wide tolerance on absolute parameter values
- full details of the fabrication technology in order to be able to size all transistors and other components correctly
- detailed simulation procedures in order to verify the circuit performance as far as possible before any manufacturing costs are incurred

Thus full custom design will be appreciated as a highly skilled specialist area if maximum benifits are to be achieved. It is a task for all IC manufacturing companies when producing a new LSI or VLSI off-the-shelf circuit or family of circuits, but it is not one which a designer in a non-IC manufacturing company can easily accomplish. Indeed, the latest details of a fabrication process and its critical parameters may not be available for use outside a restricted circle, and therefore it is not possible for outside designers to become involved in any detailed device design activity at the silicon level.

One exception to this generalization has been the Mead-Conway structured approach to full-custom design. This was based upon a set of nMOS geometric design rules which were simplifications of the various mask layouts used in nMOS device fabrication, and which in their simplest form were applicable to any nMOS production line [1–4]. These design rules were therefore generalized or 'portable,' and as such did not represent the exact geometric design dimensions of any specific vendor or represent the ultimate state-of-the-art fabrication and hence performance limits.

The fundamental rationalization introduced by Mead and Conway was the normalization of the critical geometric dimensions involved in nMOS silicon layout. This involved the introduction of a basic dimensional parameter lambda (λ) by which all the critical dimensions of a nMOS layout—such as line widths, line spacings, mask alignment, overlaps between polysilicon and diffusion and between polysilicon and metal, and so on—could be expressed. This concept is illustrated in Figure 4.2. Typically, λ may be 1.5 μm, and all feature details are expressed as 2λ, 4λ, etc., for a particular design.

Lambda is therefore a conservative and possibly coarse parameter which allows independent designers to design and lay out nMOS integrated circuits, the appropriate value for lambda being incorporated immediately prior to mask making by the subsequent producer of the IC. The whole viability of this approach clearly depends upon the fundamental feature that MOS device parameters are very largely controlled by the device geometry for a given fabrication process (see Section 2.2.5 and Appendix D), and within limits a given MOS layout can be scaled up or down to meet a required specification. Such an approach is completely unavailable with bipolar technology.

Extension of these lambda-based nMOS design and layout rules to CMOS technology has been pursued [2,5–7]. There has also been some consideration of analog requirements [8]. However, because of the rationalization and generosity of these lambda-based design rules, the following disadvantages have to be accepted:

(i) The silicon area required for a given custom design is inevitably larger than can be realized with conventional full-custom design methods, possibly as much as twice as great.

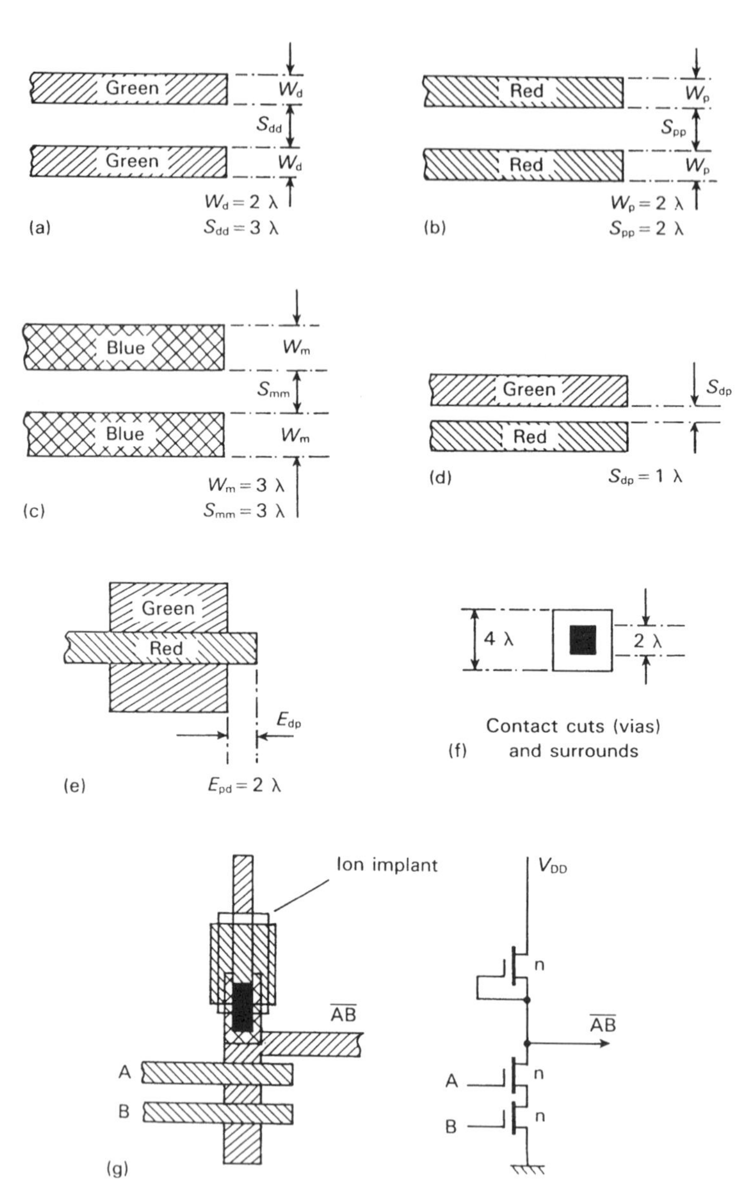

Fig. 4.2 Typical geometric layout rules in terms of lambda (λ) in the Mead-Conway nMOS full-custom design approach; green, red and blue are conventionally used in color plotters for MOS layouts: (a) diffusion widths and spacings; (b) polysilicon widths and spacings; (c) metal widths and spacings; (d) diffusion to polysilicon spacing; (e) polysilicongate overlap; (f) contact cut overlaps; (g) example of a 2-input NAND gate layout.

(ii) Performance will be inferior because of its very conservative design rules.

(iii) It is not possible to scale down all the layout dimensions linearly once actual dimensions have reached about 2 μm, since nonlinearity layout requirements then begin to dominate.

As a result of these practical disadvantages, lamba-based MOS design has received little commercial support, and must be regarded as completely inappropriate for state-of-the-art geometries. This design philosophy received its greatest support, and still does to some extent, in academic institutions, since it is an approach which gives students some practical work in the fundamentals of MOS design at the silicon level, and is both instructive and academically respectable.

To revert, therefore, to full-custom design techniques which are commercially viable, detailed design must, in practice, be undertaken by either the manufacturer of the final IC or by a specialist design house which has expert in-house knowledge both on design and also on the process parameters of chosen manufacturers. In general, the market for full custom ICs will either be where very large volume requirements are anticipated, which therefore justifies the expense of the design, or where very specialized circuits are required which cannot be acheived by using available standard parts. Complex digital signal processing circuits such as that shown in Figure 4.3 are particularly appropriate for full-custom consideration.

4.2 STANDARD CELL TECHNIQUES

Standard cell techniques rely on the existence of previously designed and fully characterized standard circuits ('building blocks'), the layout and performance details of which are held in a CAD data base. The custom design procedure is thus the assembly of the required cell designs to form the chip layout, together with simulation to ensure acceptable performance before any manufacturing costs are incurred. A full mask set is required for manufacturing purposes, and also tailor-made testing jigs, since the chip layout, number of I/Os and their positioning, and the chip size are fully flexible. This is in contrast to the gate-array design style (see Section 4.3) where only the final interconnect details are unique, thus permitting the use of a reduced mask set and standard test jigs. This distinction in mask sets is illustrated in Figure 4.4.

However, the standard-cell approach may itself be divided into subdivisions depending upon the precise nature and capabilities of the 'standard' cells. These subdivisions include the following:

- digital fixed-cell architectures, where there is no flexibility in the CAD system to alter the predesigned standard cells in any way

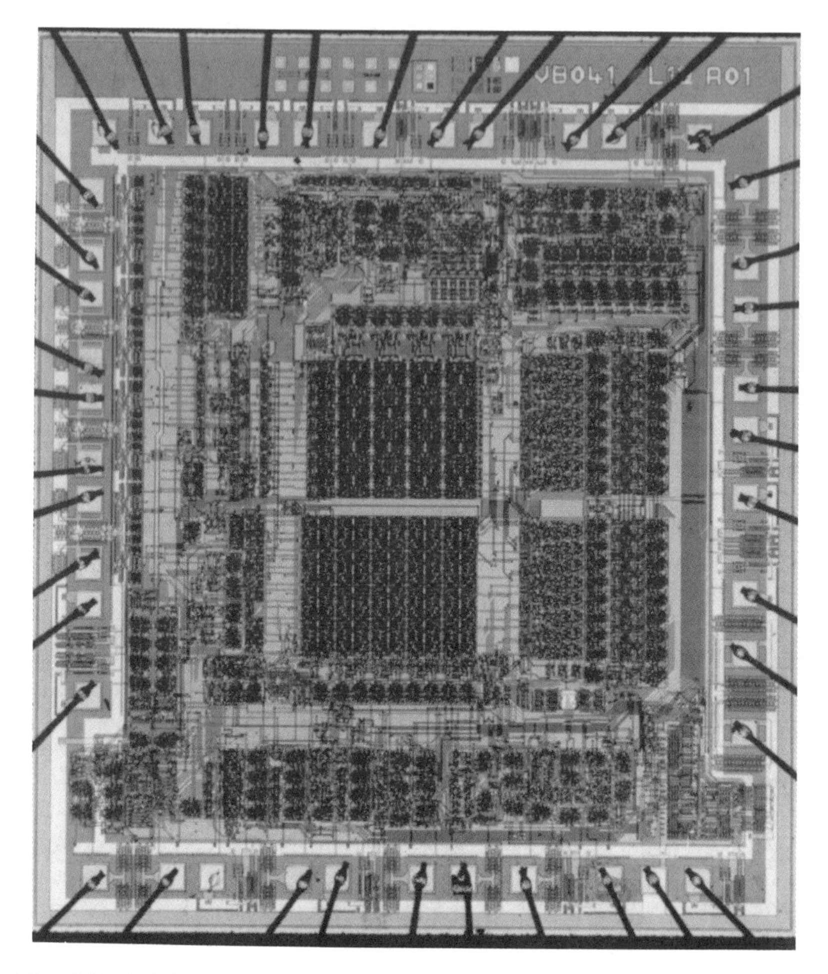

Fig. 4.3 A full-custom IC which executes all the line level functions and check sums for a 10 Mbit local area network (LAN) system. (Courtesy of Swindon Silicon Systems, UK.)

- digital soft-cell architectures, where there is some flexibility in the CAD system to alter the cell designs but not their function in some restricted way
- digital macro cells, which are large rectangular building blocks such as RAM, ROM and PLA, and which are usually flexible in size
- analog cells, which are usually flexible and not fixed so that gain and other parameters may be appropriately chosen

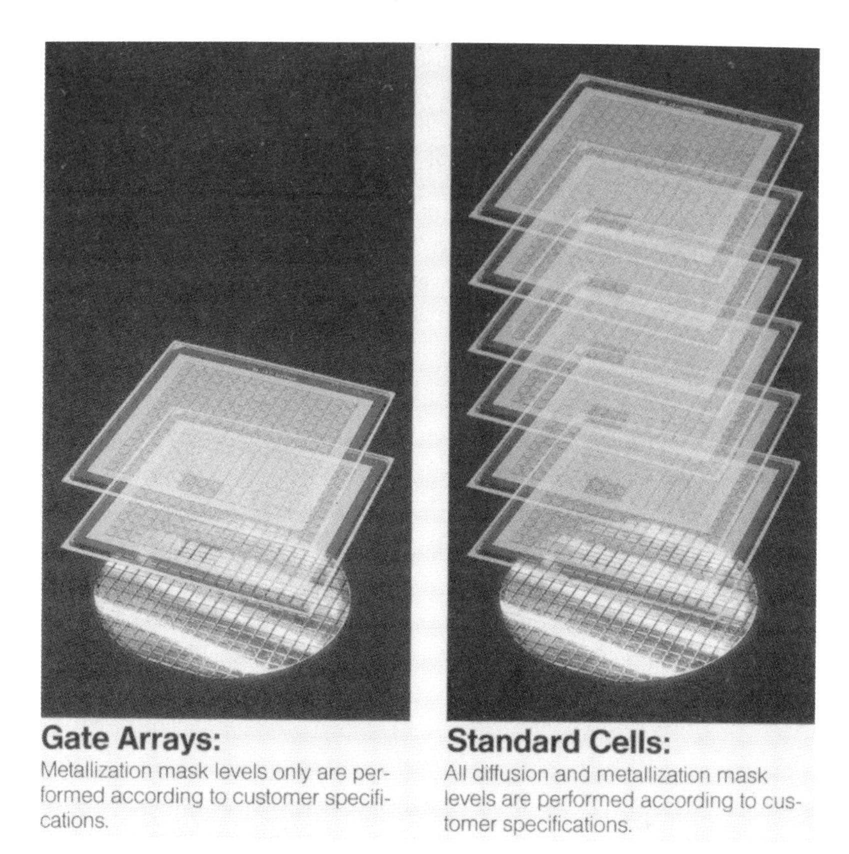

Fig. 4.4 The fundamental distinction between a standard-cell USIC and a gate-array USIC. The chip size and routing areas are fully flexible in the former but not in the latter. (Courtesy of NEC Electronics, CA.)

These distinctions will be considered further in the following sections. Section 4.5 will give a final overall summary and comparison of both standard-cell and gate-array design techniques.

4.2.1 Fixed-Cell Architectures

Fixed-cell architectures are compiled from predesigned cells in a vendor's library which cannot be altered by the CAD system in any way. These fixed-cell designs, sometimes known as 'hard' or 'hard-coded cells,' cover the range of standard logic building blocks with which all digital designers are familar, such as Boolean logic gates, latches, adders, and so on. The range of many vendors' designs often mirrors the range of off-the-shelf 74** series SSI and MSI standard parts, which promotes user confidence when transferring from standard design activities to USICs. Indeed, the analogy has been made between a product using standard 74** series parts, which involves the design of a printed-circuit board to carry and interconnect the individual ICs, and a standard-cell USIC design, which involves the placing and routing of the cell designs on the chip layout. Functionally this analogy is reasonably good, but the standard-cell design activity is more complex. As well as standard logic cells, vendors also have a number of I/O cell designs available, covering a range of input and output system requirements.

In order that the CAD system can assemble ('place') the individual cell designs in a compact manner without appreciable waste of silicon area, and then begin to route all the required interconnections between cells and I/Os, it is desirable that the original design of each cell shall conform to some dimensional standard, and not have a random size layout. This is invariably achieved by standardizing upon a fixed height of cell but varying width, thus enabling cells to be placed alongside each other in uniform rows across the chip layout. Cells which conform to this standard height layout are sometimes referred to as *polycells*. The routing between cells and to the I/Os is then done in wiring channel areas between the rows of cells, as shown in Figure 4.5.

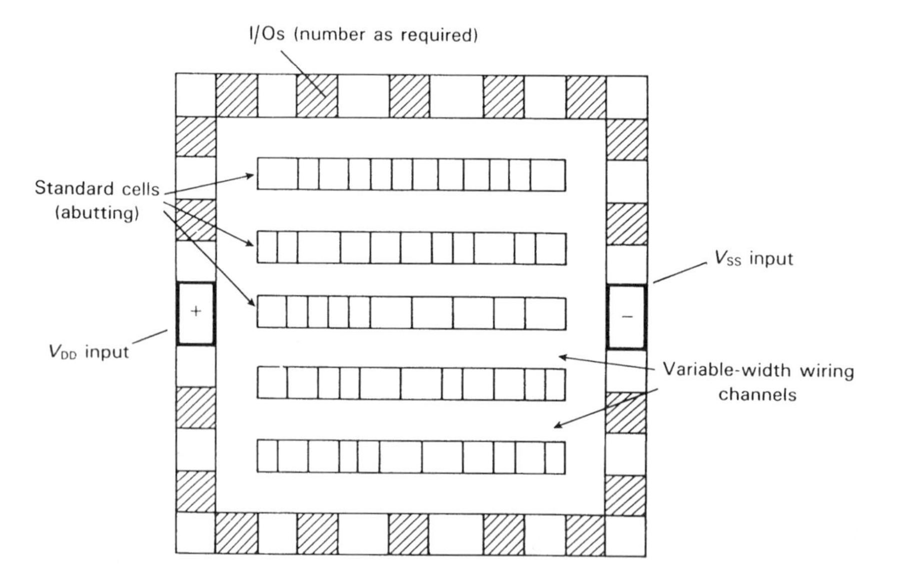

Fig. 4.5 The usual topology of a standard-cell chip design, with the required number of constant-height cells assembled in rows and appropriate width wiring channels between the rows. (cf. Figure 4.9.)

While this standardization of cell height is impractical to maintain when the capability of the cells reaches LSI and VLSI proportions (see Section 4.2.3), it is appropriate for all small SSI and MSI functions. It is particularly appropriate for CMOS technology, where all functions are composed of p-channel and n-channel transistor-pairs, allowing the height of all cells to be dictated by the dimensions required for one transistor-pair plus the V_{DD} and V_{SS} supply rails. This is illustrated in Figure 4.6. Note that a conventional circuit diagram may not mirror this physical layout arrangement. The width of the cell clearly depends upon the number of transistor-pairs involved in the logic function. Partly because of this layout convenience and partly because CMOS technology is particularly appropriate for the vast majority of custom applications, it is CMOS which dominates the digital standard-cell market. Surveys have indicated that there were fewer than ten vendors worldwide who marketed standard cells for digital applications other than CMOS; the exceptions are invariably for specialized applications, and include ECL and GaAs technology products [9,10].

Because of the fixed-height cells in a standard-cell USIC the final chip layout superficially resembles the layout of a gate array (see Figure 4.10). However, with the standard cell design it is important to appreciate the following:

- the final standard-cell chip contains only the cells required, with no unwanted or unused cells
- the width of the wiring channels between the rows of cells is only that which is necessary to contain the final required connections, not pre-chosen and fixed as with a gate-array layout
- only the exact required number of I/Os are present

The result is that a standard-cell USIC is generally smaller and more efficient than a gate array configured to perform the same duty (see Section 4.5), but the design is constrained by having to use the cells which are available in the vendor's CAD library, which may not always be exactly what the OEM designer requires.

The standard-cell design process is initially very similar to system design using standard parts. This consists of partitioning the system down to the level of the available building blocks, but because the standard cells in the vendor's library are not available as functional entities, it is not possible to make a bread-board prototype for evaluation purposes, as can be done when using off-the-shelf standard parts. Instead, reliance upon the CAD simulation is necessary to ensure that the initial design is functionally correct. Standard-cell procedures therefore are much more heavily CAD-dependent than design procedures using standard parts, which may be advantageous if the CAD capabilities are comprehensive, user-friendly and accurate.

The complete standard-cell design procedure is shown in Figure 4.7. Details of simulation, placement and routing, test vector generation and other software activities will be covered in Chapter 5, together with consideration of the possible

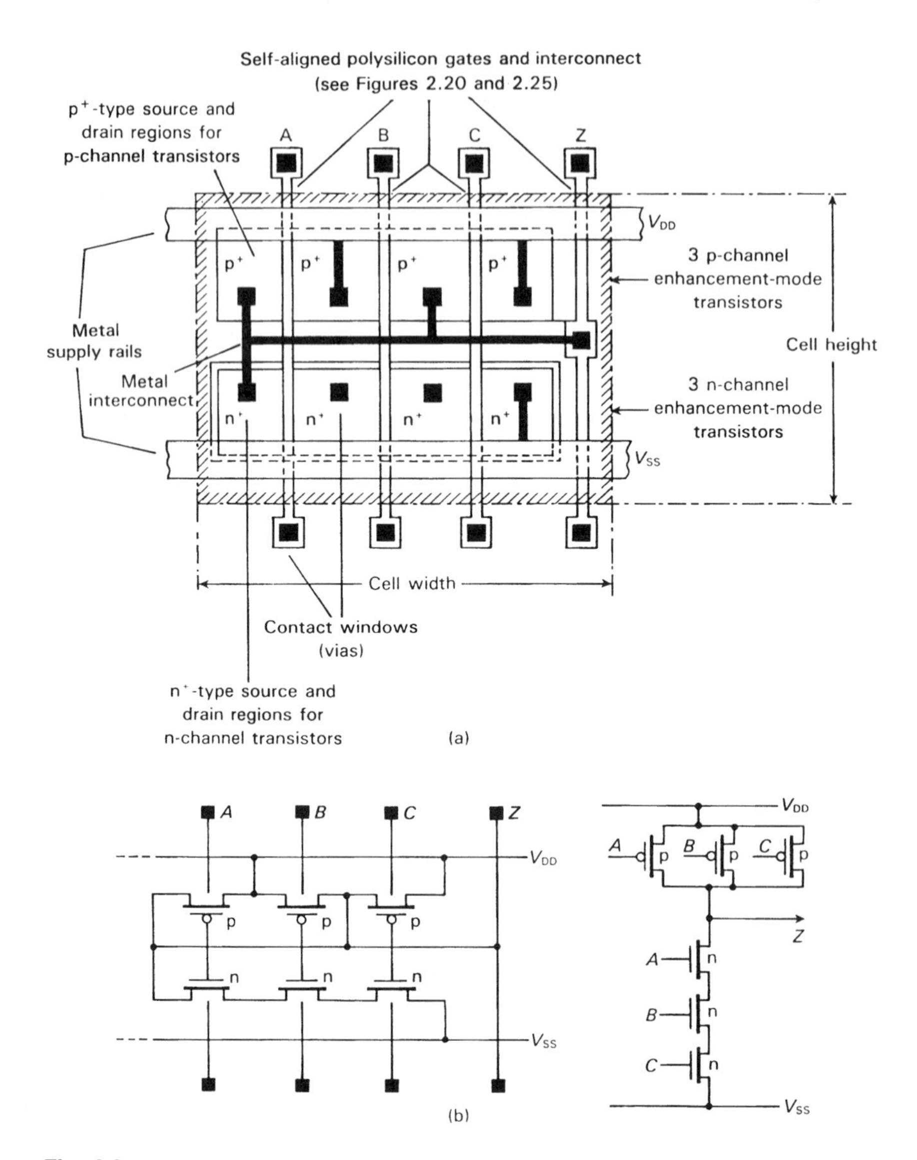

Fig. 4.6 The general method of achieving constant height but variable width standard-cell layouts in CMOS technology: a 3-input NAND gate is illustrated with inputs and outputs available at both top and bottom edges (both-sides access). The number and position of the contact windows (vias) may vary considerably between different vendors designs: (a) the basic layout; (b) the equivalent circuit diagram.

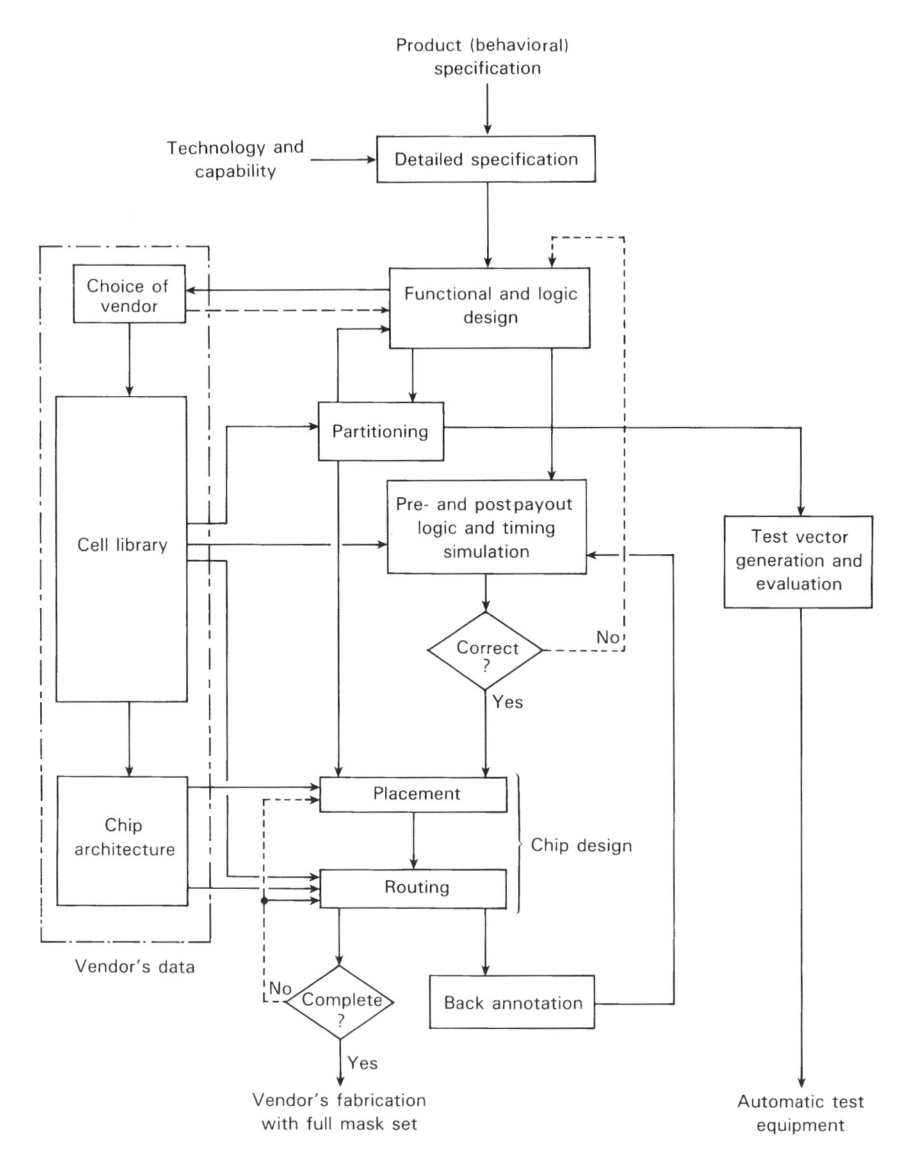

Fig. 4.7 The hierarchy of the standard-cell USIC design process, from the original product specification at the top to final fabrication at the bottom. The dotted paths are possible iterations if problems or constraints are found.

interface levels between the OEM and the IC vendor. The vendor clearly has to be involved in the final manufacturing processes of mask making, silicon fabrication, packaging and test, but the OEM may wish to do as much of the chip design work as possible before handing over the design to a vendor for manufacture. CAD costs and OEM capability enter into these interface decisions, factors which we will also consider in Chapter 7.

The precise details of the standard cells at the silicon level are not usually available to OEMs, being the intellectual property of the vendor who has hand-crafted them so as to achieve minimum silicon area and optimum performance. Such details as are available are sometimes referred to as *limited technical data* (LTD), and will include the dimensions and performance of each cell (for example, as illustrated in Figure 4.8) but not its exact internal geometric details. General data for the whole library will give supply voltages, whether single-layer or double-layer metal interconnect, and other broad information. *Full technical data* (FTD), which includes all the geometric details of the cells, may be made available to certain large OEM companies under confidentiality agreements should they wish to design their own special cells which are compatable with the vendor's standard cells, but this is not usual.

4.2.2 Soft Cell Architectures

One of the constraints with fixed standard cells is that some compromise must be present in the original cell specification with regard to performance, particularly the fan-out capability of each cell. Unlike off-the-shelf standard ICs, all of which have on-chip buffered outputs to give good fan-out capability, the provision of high fan-out on each standard cell would be extremely wasteful in silicon area, since the majority of cells used on-chip would not require this. Nevertheless, this capability may be required in some circumstances.

One way of providing different output drive capabilities is to have two or more versions of a hard-cell design in the cell library; for example, one vendor lists three variants of NAND, NOR and other gates, with cell widths in the ratio of 1:1.33:2.33 for increasing cell strength. A more sophisticated method is to allow the CAD system to modify the detailed silicon layout so as to alter the cell specification (but not the function) within limits by varying the p-channel and n-channel device areas. Cells which may be fine-tuned in this way may be known as *soft cells* or *soft-coded cells.*

It is important to note that the rules for any change in transistor details are built into the CAD software so that the designer using the system does not have to be familiar with detailed transistor design or the silicon layout rules. However, vendors of standard-cell custom ICs may still prefer to offer several variants of SSI and MSI hard cells rather than provide the facility for the customer to fine-tune them for particular applications.

CELL NUMBER 2310 (JEDEC: JXO1) REV 0.1
EXCLUSIVE OR
3 PINS
CELL WIDTH: 48 μm

SC3000
1.5 MICRON CELL LIBRARY
10 TRANSISTORS
CELL HEIGHT: 90 μm

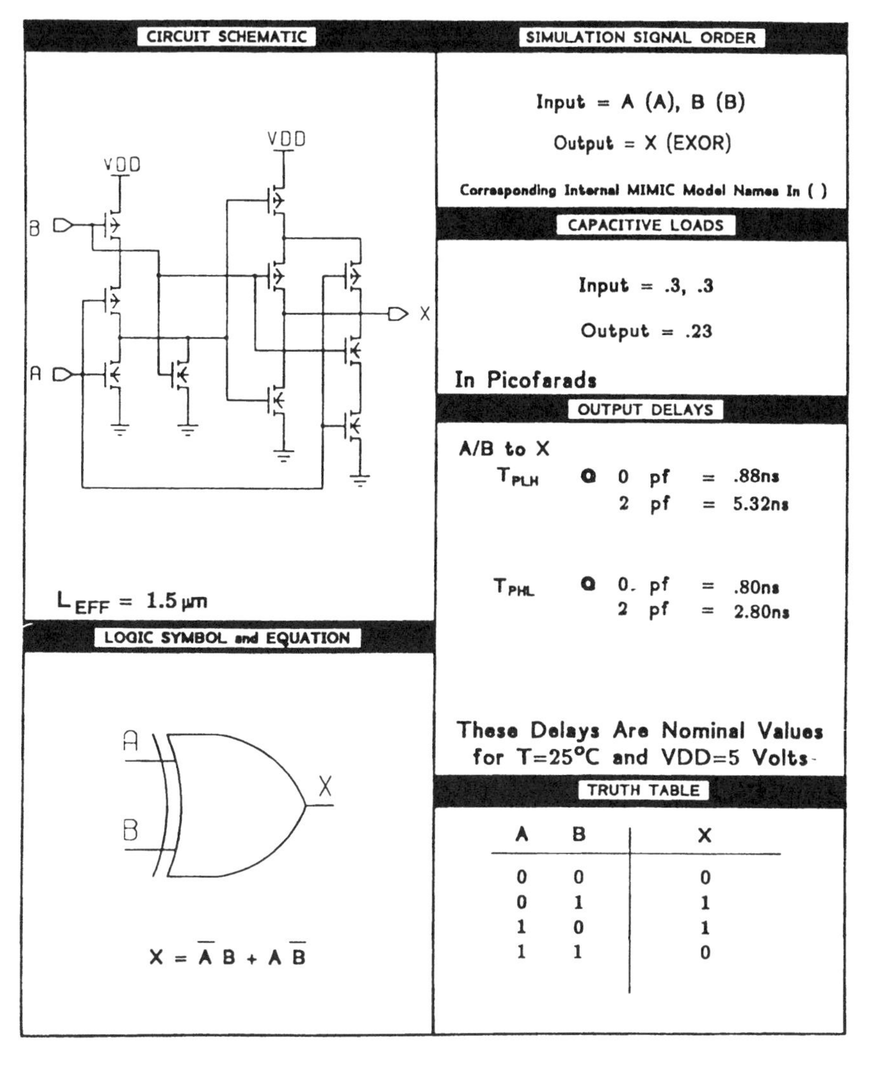

A	B	X
0	0	0
0	1	1
1	0	1
1	1	0

Fig. 4.8 Example data sheet for one cell in a vendor's standard-cell library, with cell dimensions of 48 × 90 μm. T_{PLH} and T_{PHL} are the cell propagation times when the output changes from low to high and from high to low, respectively, under given loading conditions. (Courtesy of Harris Semiconductors, FL.)

4.2.3 Macrocells

While MSI standard cells in a vendor's library may be built from a compacted assembly of fixed-height SSI cells, it is unrealistic to try to maintain this constant cell height feature for cells of LSI and VLSI complexity. Instead, large standard cells are invariably designed by the vendor with a more realistic layout, ideally square, which means that they cannot be placed in a row layout with the small size standard cells. As will be seen in Chapter 5, this involves a different concept of automatic routing in the CAD software, requiring a block router rather than a channel router.

A vendor's library may include large building blocks such as the following:

- RAM and ROM
- PLAs
- 16-bit multipliers and arithmetic units
- microprocessors and microcontrollers

and others. An increasing range of digital signal processing (DSP) blocks is also becoming available. However, with increasing size and complexity comes the problem of the increasing difficulty of meeting a customer's requirements exactly in terms of size and capability, plus the difficulties of placement and routing if only one layout of each large block is available. Hence, it is common practice for very large cells in the library to be flexible rather than fixed (hard) cells. In particular, the regularity of RAM, ROM and PLA structures affords several degrees of freedom without altering the basic transistor and other detailed dimensions, for example:

- freedom to alter the number of inputs and outputs
- freedom to scramble the inputs and the outputs in order to ease placement and routing difficulties
- freedom to mirror-image or orientate these rectangular layouts in any way

Similarly, in bus-structured architectures, arithmetic modules and the like, the freedom to increase or decrease bus widths is frequently provided.

Macrocells are therefore inherently soft-coded in order to give the necessary flexibility for a range of custom requirements. Indeed, the increasing capabilities of available CAD software have further effects such as:

- giving built-in flexibility to accommodate changes and improvements in fabrication processes, rather than having to undertake a costly redesign by hand of the cell library
- blurring the distinction between full-custom design and cell-based design, since the increasing flexibility of the latter allows optimized RAM,

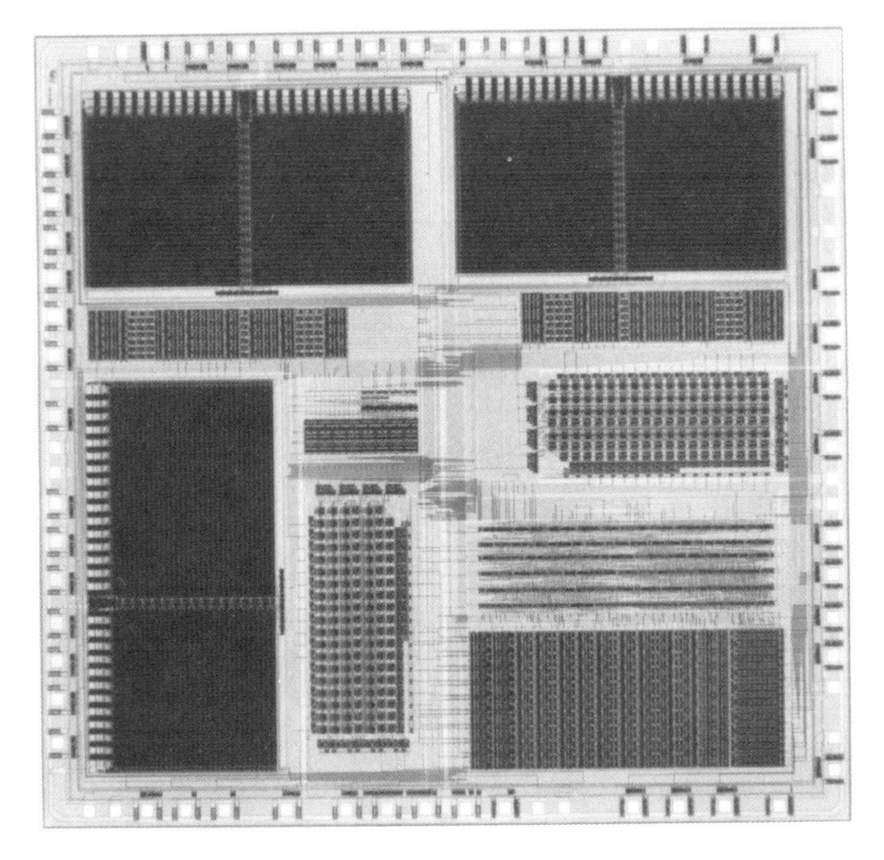

Fig. 4.9 A standard-cell chip design involving three large RAM blocks with other smaller data path and arithmetic blocks, and about 750 standard cells for random logic. (Courtesy of VLSI Technology, CA, and Evans and Sutherland, Inc., UT.)

ROM and other macrocells to be used in a full-custom design environment

- possibly allowing special macros to be readily compiled so as to augment existing designs to meet new customer requirements

Figure 4.9 illustrates the layout of a standard-cell custom IC designed for graphics image processing which contains both large macros and standard logic cells, the total chip design involving the equivalent of some 75,000 logic gates [11]. This typifies the growth in complexity which has occured since the early days of standard-cell custom design.

In summary, the libraries offered by vendors have become increasingly comprehensive and flexible, with only the smaller SSI and MSI logic cells possibly maintaining a fixed layout. The use of silicon compilers and VHDL (see Chapter 5) is a part of this still-continuing CAD evolution [12–16]. Once again, confusion in terminology may be noticed in the literature, among the terms which may be encountered when referring to the full hierarchy of increasingly complex standard-cell building blocks being:

- leaf cells
- cells, basic cells, primitives, polycells, standard cells
- macros, macrocells, macro blocks, macro functions
- composition cells, compiled cells
- megacells, megamacros, megafunctions
- supracells, gigacells

All the preceding discussions relate primarily to CMOS products, bipolar technology not being a lasting contender is this field. However, recent developments in very high speed standard cells using gallium-arsenide have been announced, although the number of vendors marketing this technology is very few [9]. The standard cells so far announced are of SSI and MSI complexity, but a comparison of performance with CMOS is as follows:

- Typical CMOS performance, 1.5 μm geometry:
  Minimum gate propagation time = 0.5 ns
  Maximum operating clock frequency = 200 MHz
- Typical GaAs performance:
  Minimum gate propagation time = 0.1 ns
  Maximum operating clock frequency = 1 GHZ

The power dissipation of a 3-input NOR gate in the GaAs technology, however, has been reported to be on the order of 2 mW, which is a penalty that has to be paid for the contemporary increase in speed over CMOS [17].

4.2.4 Analog Cells

While it is entirely straighforward to define and market a very wide range of self-contained digital building blocks suitable for most users, typified by the 74** series circuits and their equivalents, with analog the problem is much more difficult. The following are among the circuit parameters which users may specify to meet their product requirements:

- voltage, current or power gain
- input and output voltage levels
- input sensitivity
- output power

- bandwidth
- slew rate
- bandpass, bandstop and other filter characteristics

and others [18–21]. Indeed, the exact analog performance required by each new system will almost invariably be unique.

Further distinctions between analog and digital systems include the following:

- Analog circuits require the presence of resistor and capacitor elements to provide good signal amplification and frequency performance—no wide-band linear amplification can be achieved without the presence of fixed resistance.
- Analog performance needs to be accurate and stable, unlike digital circuits, where exact electrical performance is not critical.
- The preferred silicon technology for analog is bipolar, whereas MOS is more appropriate for digital applications and standard cells.
- Analog systems usually involve relatively few analog circuits in comparison with the very large number of circuits which can build up in digital systems; hundreds rather than tens of thousands of transistors may therefore be involved in an analog system, but as each transistor is usually operating in a linear mode the power dissipation may be high.

Another difficulty facing a vendor wishing to market analog standard cells is that, ideally, the cells should be capable of working alongside digital cells in order to provide a comprehensive mixed analog/digital capability, but the vendor may encounter the following problems:

- The analog cells may then need to be MOS rather than bipolar so that they can be fabricated using the same processing stages as the digital cells.
- The required operating voltages of the on-chip analog and digital cells may be different.
- The MOS processing may require modification in order to achieve acceptable analog performance, which may reflect on the performance of the digital cells.
- Great care will need to be taken in the chip floorplan layout in order that analog cells working at low signal levels do not pick up interference from the digital switching circuits on the same substrate.
- Full simulation of a large mixed analog/digital IC design is still very difficult even with the latest CAD resources.

Because of all these potential difficulties, uncommitted arrays (see Section 4.3.4) are most often used for analog and mixed analog/digital applications, most commonly up to now using bipolar technology rather than MOS. Nevertheless, an increasing number of vendors are making MOS analog cells available in their standard-cell libraries. The variety of different types listed, however, is small in comparison with the list of possibly hundreds of digital standard cells, and the analog performance may not be as good as that available from bipolar arrays. Library types include general-purpose operational amplifiers, analog-to-digital and digital-to-analog converters, voltage comparitors and timer-oscillators [22–24]. Exact circuit details are not usually published.

This area remains the least mature of standard-cell technology, and considerable work is still being undertaken to improve the technology and the capabilities of the CAD for mixed analog/digital USICs. BiCMOS technology is also a candidate for this field, but cost considerations remain a problem for products using this approach.

4.3 GATE ARRAY TECHNIQUES

Semicustom gate arrays rely upon the availability of wafers that have been processed up to but not including the final chip interconnections, and which are fabricated as stock items. Each die on an uncommited wafer contains an array of identical general-purpose cells, the subsequent interconnection design of which is the customization. The uncommitted wafers may be held in stock with a complete first-layer metal covering, the polished surface of which gives rise to the term *mirror blanks* for such stock items. However, this is only possible if all the contact windows (see Figure 2.18(g)) from the first-layer metal through the field oxide are made before the first-layer metal is applied. If these via holes need to be specified as part of the custom interconnection design, then the first-layer metal cannot be applied before the vias are subsequently manufactured, and the wafers are therefore held in stock with no metalization.

As has been noted in Chapter 1, the prime advantage of the gate-array approach is to minimize the amount of custom design and fabrication activity necessary to complete a particular custom circuit. Since only the interconnection design, manufacture of the interconnection masks (a reduced mask set) and fabrication of the final interconnection levels is customer-specific, the time and cost to make the final USIC should be less than that of a corresponding standard-cell product. However, performance penalties may have to be paid, as will be detailed later.

The very wide range of semicustom arrays which have been marketed may be classified by the following:

- the number of the general-purpose cells replicated on the uncommitted chip

- the specification of the cells
- the number of I/Os per chip
- the fabrication technology, whether MOS, bipolar or otherwise
- the number of interconnection levels which have to be customized

Simple arrays have a single layer of metal for their customization, but more complex arrays have two layers as previously illustrated in Figure 2.24. Yet more complex products have more than two interconnect levels; for example, some very high-speed ECL arrays can have one metal layer for power supply rails together with two or more for signal interconnections.

The number of customization masks in the reduced mask set therefore increases as follows:

- single-layer-metal arrays
  contact window (via) masks: none or 1
  metal patterning masks: 1
  total mask set: 1 or 2
- two-layer-metal arrays
  contact window masks: 1 or 2
  metal patterning masks: 2
  total mask set: 3 or 4
- three-layer-metal arrays
  contact window masks: 2 or 3
  metal patterning masks: 3
  total mask set: 5 or 6

In general the initial contact windows before the first layer of metal is formed have to be customized except in simple arrays, and hence the higher number above is usual.

While single-layer-metal arrays are less costly to customize than those with two or more layers of metal, routing the required interconnections using only one level is frequently impossible. The increased flexibility provided by more than one level and potentially smaller interconnect area are the principal reasons for multilayer metallization. The smaller area of interconnect may also give reduced interconnect impedance, and hence enhanced circuit performance.

Figure 4.10 shows the basic chip floorplan which is adopted for virtually all gate arrays, except for 'channel-less' architectures, which will be considered separately in Section 4.3.3. As will be seen, the layout consists of rows of identical cells separated by wiring channels, with I/Os around the perimeter of this active area. This superficially resembles the standard-cell layout shown in Figure 4.5, but now there is no flexibility to alter the number and type of cells in each row, or the width of the wiring channels, or the number of I/Os to meet the exact requirements of a customer. Hence, a major disadvantage compared with

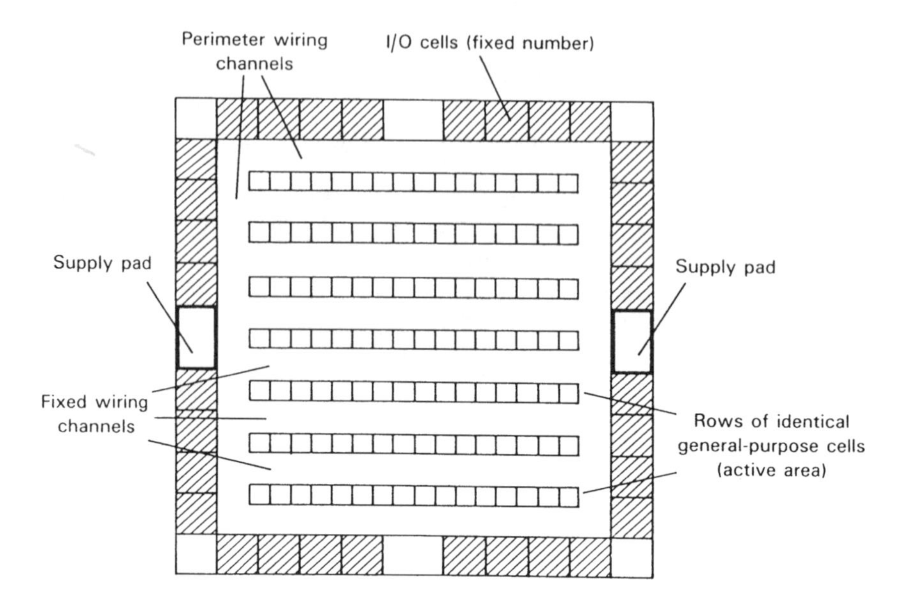

Fig. 4.10 The floorplan of a typical gate-array chip with fixed-width wiring channels, sometimes termed a channel-array architecture (cf. Figure 4.24).

a standard-cell design is that silicon area utilization will not be as good, since inevitably there will be unused parts of the fixed array structure, and performance will generally be inferior due to looser on-chip packing densities. These and other comparisons will be considered further in Section 4.5.

The vendor's choice of the general-purpose cell used in a semicustom array is a fundamental decision which has to be made when developing a gate-array chip. To a large extent this depends upon whether the array is targetted for digital-only applications, or mixed analog/digital, or solely analog purposes. The latter two categories will be considered separately in Section 4.3.4. For digital-only applications, the choice of the silicon fabrication technology will also influence the choice of the general-purpose cell; most bipolar technologies have been used, including ECL, I^2L, ISL, STL, CDI and GaAs (see Chapter 2), but CMOS is now the dominant technology. ECL and GaAS are still marketed, but the others must be considered obsolete. The following section will, however, look at all these alternatives, together with the meaning of 'cell library' within the context of gate array architectures.

The design process for a gate-array IC is very similar to that for a standard-cell design, as may be seen by comparing Figure 4.11 with Figure 4.8. However, the exact design activities are dissimilar in detail, since with a gate array the chip

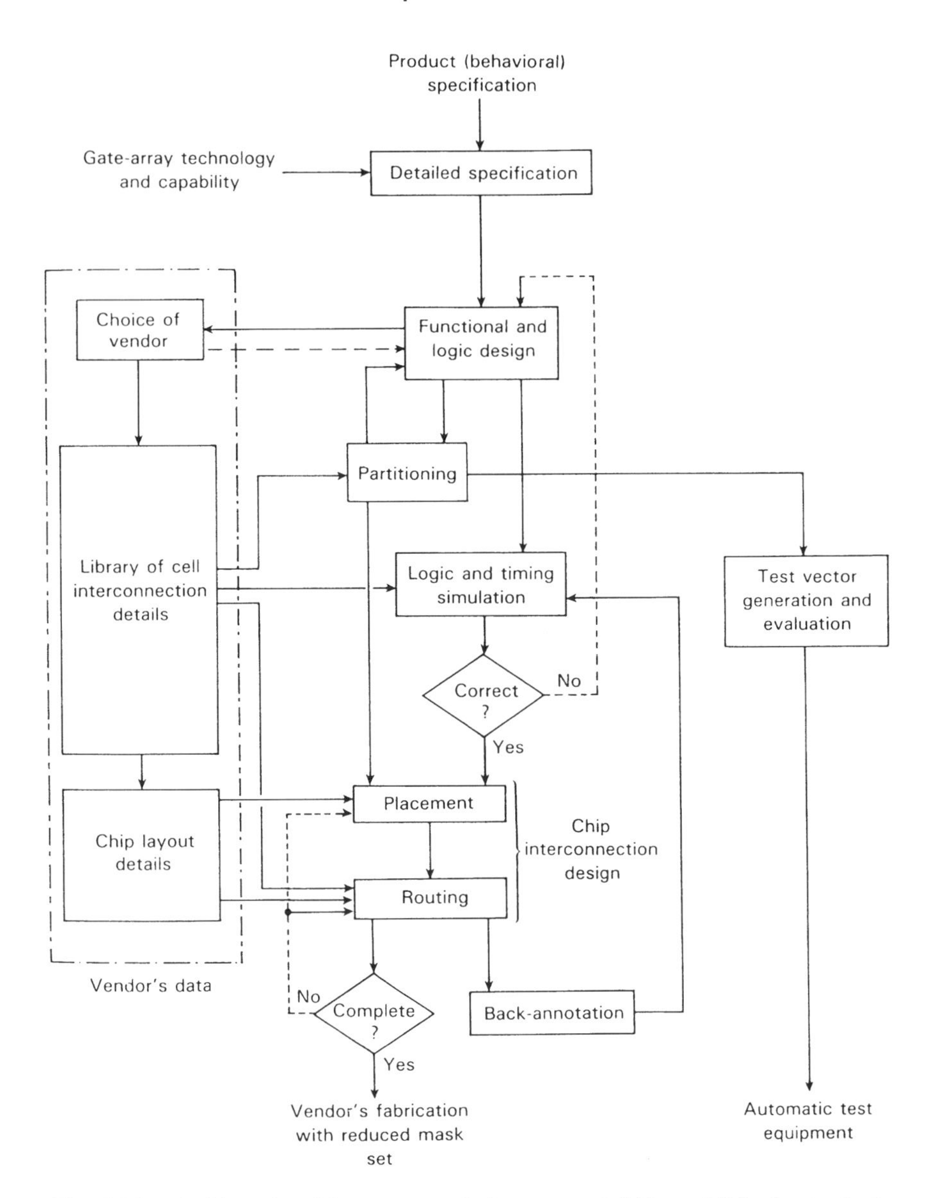

Fig. 4.11 The hierarchy of the gate-array design process (cf. Figure 4.7 for the standard-cell design equivalent).

floorplan is fixed with no flexibility to modify the rows of cells and the wiring channels in any way. Also, the 'placement' is different; with a gate array the 'placement' activity means the allocation of the fixed cells in the array to specific duties, ideally so as to ease the subsequent routing design, whereas in the standard cell design 'placement' means the physical placement of chosen cells to form the unique chip floorplan. Further details of the actual design activities will be covered later.

4.3.1 Choice of General-Purpose Cells for Digital Applications

The diversity of commercially available gate arrays arises in part from the choice of the general-purpose cell which is replicated on the chip. Whatever is chosen must be capable of being interconnected so as to form all possible types of digital logic function. Figure 4.12 shows the principal choices, with component-level cells or functional-level cells as the initial distinction. With the former, the custom

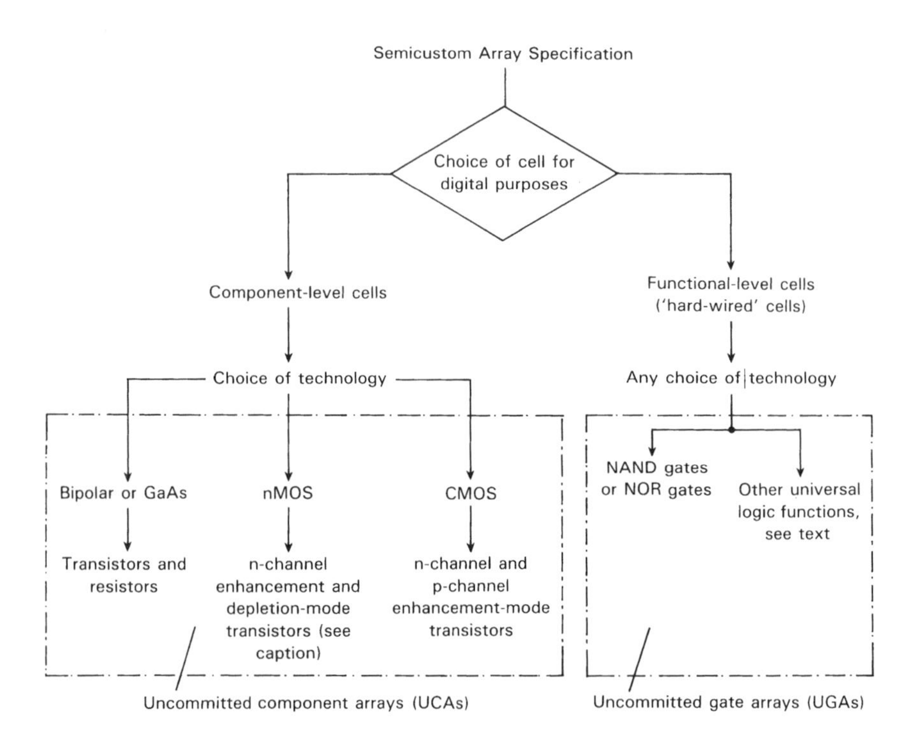

Fig. 4.12 The possible choice of a general-purpose cell for semicustom arrays. nMOS technology is now no longer employed by any vendor, having been superseded by CMOS.

interconnection design requires the internal cell connections which connect the components to be designed, as well as the interconnections between cells and I/Os; with the latter, each cell is a functional entity, sometimes termed a *hard-wired* cell, and hence only the intercell and I/O connections have to be considered.

Uncommitted Component-Level Arrays

All bipolar technologies have at some stage been considered for component-level arrays (see Section 2.1 of Chapter 2), but only ECL and possibly CDI remain commercially available [9,10]. ECL caters for the very high speed market, while CDI has particular advantages for nonstandard d.c. supply voltages and for combining with analog requirements.

As an example of a high-speed ECL component-level array, consider the product illustrated in Figure 4.13. Its specification is as follows:

- 416 cells arranged in twenty-one rows
- 76 transistors and 60 resistors per cell, arranged in four equal quadrants
- 1.5 μm minimum feature size, with two layers of metallization
- speed/power adjustable from 175 ps on-chip gate delay at 3 mW per gate to 300 ps delay at 1 mW per gate
- 256 signal I/O pins
- outputs capable of driving 25, 50 and 60 ohm transmission lines
- maximum IC dissipation 30 W
- packaging 289-pin pin-grid array or other alternatives

It is evident that this is a very sophisticated product and is not intended for general-purpose use. Like other ECL products, its applications are primarily in computers and in high-speed communications and signal processing. The vendor has a library of proven cell interconnections to cover a range of digital functions, together with CAD resources for the complete custom design activity. Great care needs to be taken with the interconnections of such circuits on a printed-circuit board in order not to degrade the very high speed capability; matched transmission line PCB interconnections, such as shown in Figure 4.14, are required.

An alternative to ECL arrays for high speed working can be gallium-arsenide products. A number of different choices of cell components for GaAs arrays have been proposed, but the simplest cell for general-purpose digital applications consists of one depletion-mode and two enhancement-mode GaAs MESFET transistors from which a 2-input NOR gate can be constructed. The specification of one of the pioneering products on the market is as follows:

- 4000 cells arranged in ten rows
- one depletion-mode and two enhancement-mode MESFETs per cell

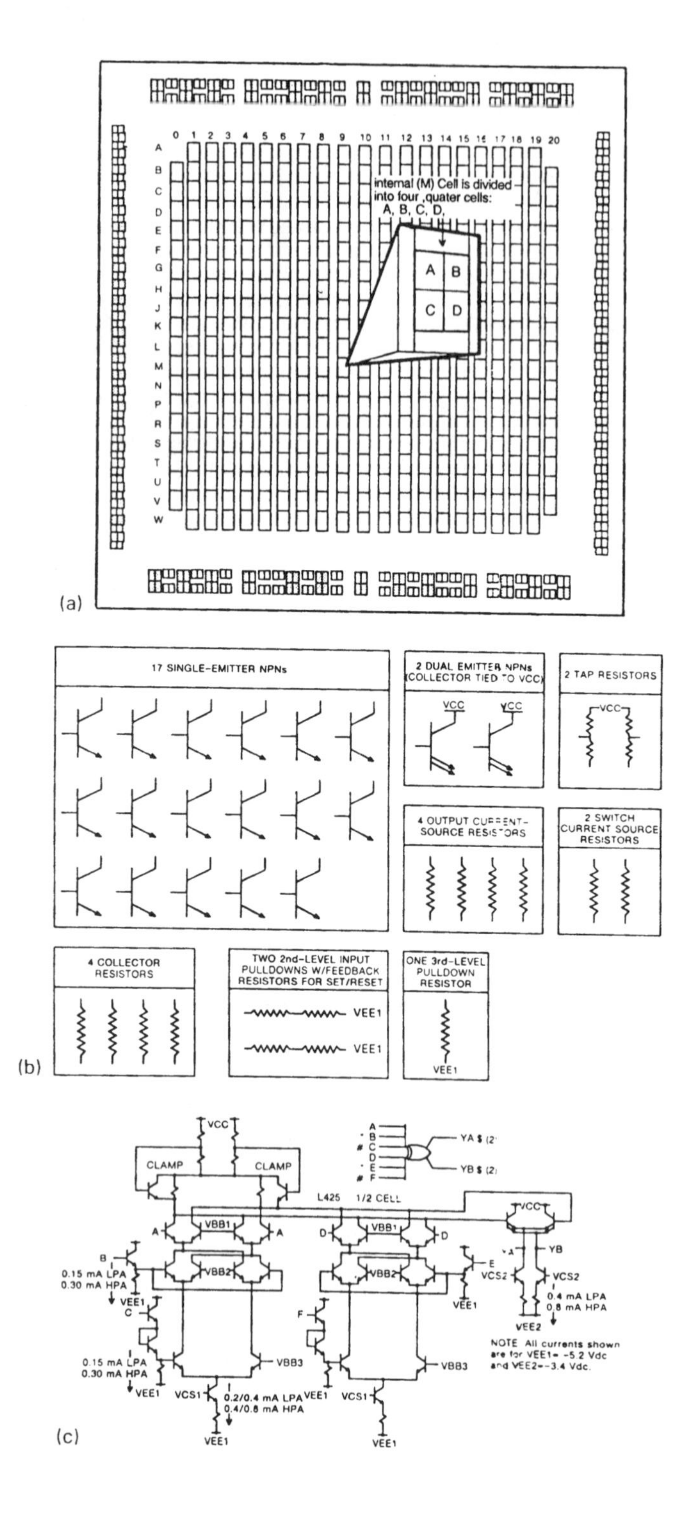

internal (M) Cell is divided into four ,quater cells: A, B, C, D,
A
B
C
D
(a)
17 SINGLE-EMITTER NPNs
2 DUAL EMITTER NPNs (COLLECTOR TIED TO VCC)
VCC
2 TAP RESISTORS
4 OUTPUT CURRENT-SOURCE RESISTORS
2 SWITCH CURRENT SOURCE RESISTORS
4 COLLECTOR RESISTORS
TWO 2nd-LEVEL INPUT PULLDOWNS W/FEEDBACK RESISTORS FOR SET/RESET
VEE1
ONE 3rd-LEVEL PULLDOWN RESISTOR
(b)
CLAMP
L425 1/2 CELL
VBB1
VBB2
VBB3
VCS1
VCS2
0.15 mA LPA
0.30 mA HPA
0.2/0.4 mA LPA
0.4/0.8 mA HPA
0.4 mA LPA
0.8 mA HPA
VEE2
NOTE All currents shown are for VEE1= -5.2 Vdc and VEE2=-3.4 Vdc.
(c)

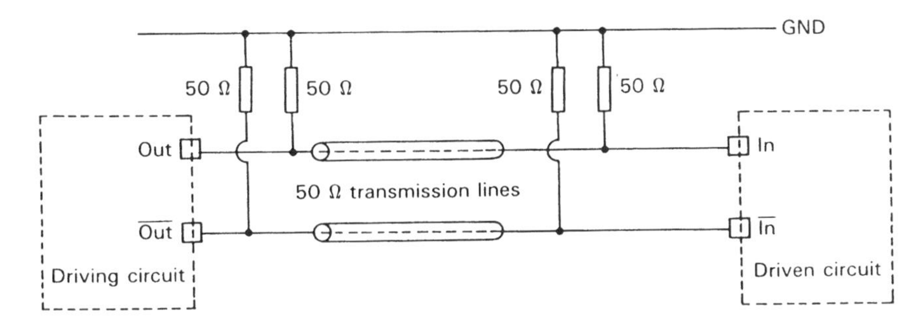

Fig. 4.14 The use of terminated balanced transmission line techniques which are mandatory for very high speed interconnection duties between ECL and other very high speed digital circuits.

- 0.8 μm minimum feature size with two layers of metallization
- typical 2-input NOR gate delay 120 ps at 340 μW
- 120 signal I/O pins
- inputs and outputs can be TTL or ECL compatible
- typical IC dissipation 2.5 watts
- 149 pin grid array package or other alternatives

Figure 4.15 illustrates this product. Larger and faster developments have subsequently been announced.

Like ECL, gallium-arsenide provides extremely high performance. In comparison with ECL the speed is slightly higher, although newer ECL technologies are giving increasing speeds, but the power consumption is less than with ECL. In general, the speed-power product of GaAs is an order of magnitude better, but ECL has a longer history of development and hence remains the major force in the very high speed logic field [25–28].

Collector-diffusion isolation (CDI) technology also has a very long history. It was pioneered by Ferranti Electronics, UK, now GEC-Plessey Semiconductors, in their range of uncommitted logic arrays (ULAs) first introduced commercially in the 1970s. (ULA™ is the trademark of Ferranti/GEC Plessey Semiconductors, and applies to all their range of semicustom array devices.) They also occupied a special niche in the custom market. It will be recalled from Chapter 2 that CDI

Fig. 4.13 The high-speed Motorola MCA III ECL uncommitted component array: (a) the channel architecture floorplan; (b) the component devices available in one quarter of a cell; (c) example of 6-input Exclusive-OR commitment, using one half of the components of a cell, i.e., two '¼' cells. (Courtesy of Motorola, Inc., AZ.)

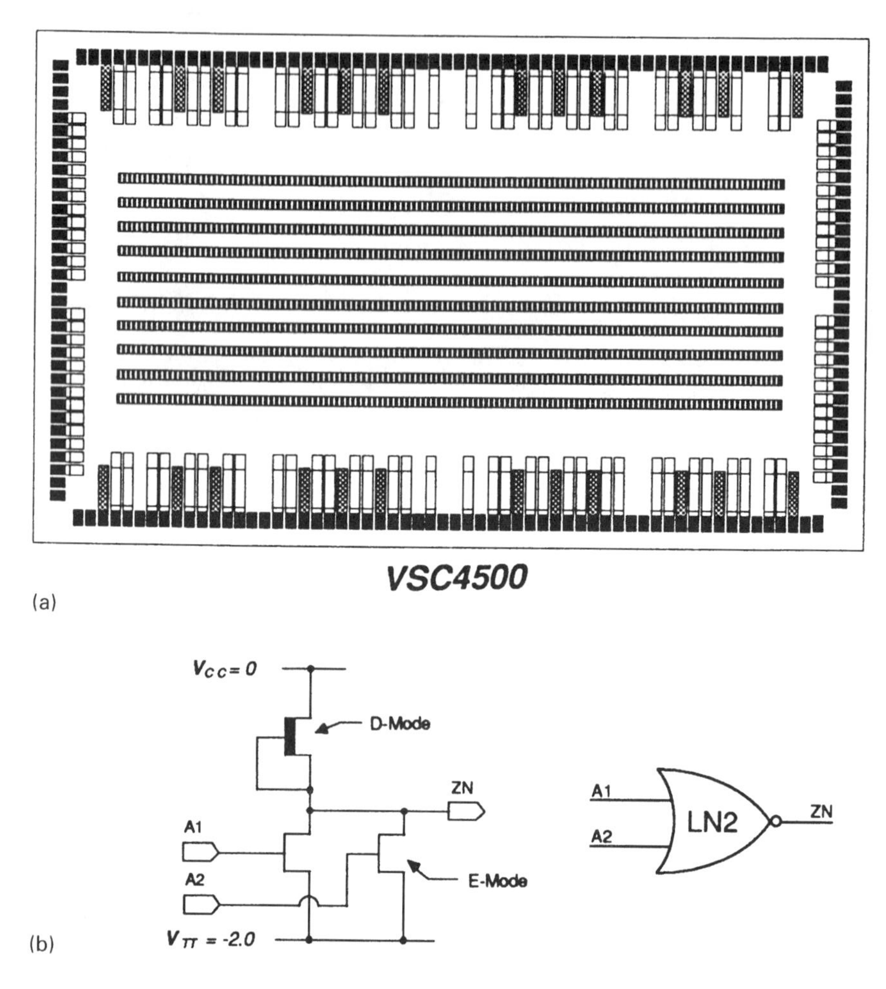

Fig. 4.15 A gallium-arsenide uncommitted component array with 4000 uncommitted cells and with programmable I/O cells for different interface requirements: (a) the chip floorplan; (b) the commitment of a cell to a 2-input NOR gate. Multiple cells provide a standard range of logic functions. (Courtesy of Vitesse Semiconductor Corporation, CA.)

is a bipolar process, which like ECL allows a trade-off between speed and power. Both speed and power are lower than ECL. The unique feature of CDI, however, is that it allows efficient analog and digital circuits to be made with the same same fabrication steps, and hence many ULA products have both digital and analog capability. We will therefore mention CDI and ULAs again in Section 4.3.4.

Early versions of the ULA series of custom ICs had component-level cells which were roughly square in layout, as shown in Figures 4.16(a) and (b). These cells were positioned on the chip with wiring access to all four sides, rather than in abuting rows as in Figure 4.10. This grid layout, sometimes referred to as a 'block-cell' layout, provided an extremely flexible means of interconnection, permitting single-layer metal customization to be used, but its flexibility proved to be its drawback since no CAD software to route the array efficiently could be developed—its sheer degree of freedom proved more than software could handle. It was, however, very efficiently routed by hand by expert designers, but this became prohibitively costly and time-consuming as array sizes grew from the early small arrays to larger products in following years.

The 1990 versions of the ULA have a cell composition as shown in Figures 4.16(c) and (d). Cells are now assembled in rows on the chip so as to give a more familiar chip layout, one which automatic CAD routing can more easily handle. A range of chip sizes up to the equivalent of 10,000 2-input NAND or NOR gates has been marketed, with a choice of speed/power options for system speeds up to 100 MHz. The average gate power is 210 μW at 100 MHz, with lower speed/power options available. For purely digital applications, this does not quite compete with the latest CMOS technology, unless the CMOS is continuously switching at very high rates.

Turning from these bipolar products to CMOS technology uncommitted component arrays, there is now no fundamental difficulty in defining a basic cell. The simplest is clearly a single p-channel and n-channel transistor-pair from which all CMOS digital logic circuits may be constructed. It is also clearly relevant to arrange such transistor-pairs in rows, giving a floorplan as was shown in Figure 4.10.

However, there is a choice of either one transistor-pair per cell or more than one, as Figure 4.17 illustrates. Commercially available CMOS arrays may have one, two or more transistor-pairs arranged within common p^+ and n^+ regions. The polysilicon cross-unders, which must be available somewhere in the layout to allow the gate inputs to be brought out from the cells under the common V_{DD} and V_{SS} supply rails, may be considered as part of a cell, or as a necessary addition between cells.

It will be apparent that these uncommitted CMOS cells closely resemble the designs available in a vendor's standard-cell library (see Figure 4.6). Indeed, exactly the same physical layout of the individual transistors may be present in a vendor's standard cell and uncommitted array products, but each standard-cell design may possibly occupy a smaller width than uncommitted transistor-pairs due to its being specifically optimized for the one functional duty.

The commitment of the cells in a CMOS array involves internal cell metallization to interconnect the p-channel and n-channel transistors in series and in parallel as required, plus the remaining interconnections between cells and I/Os.

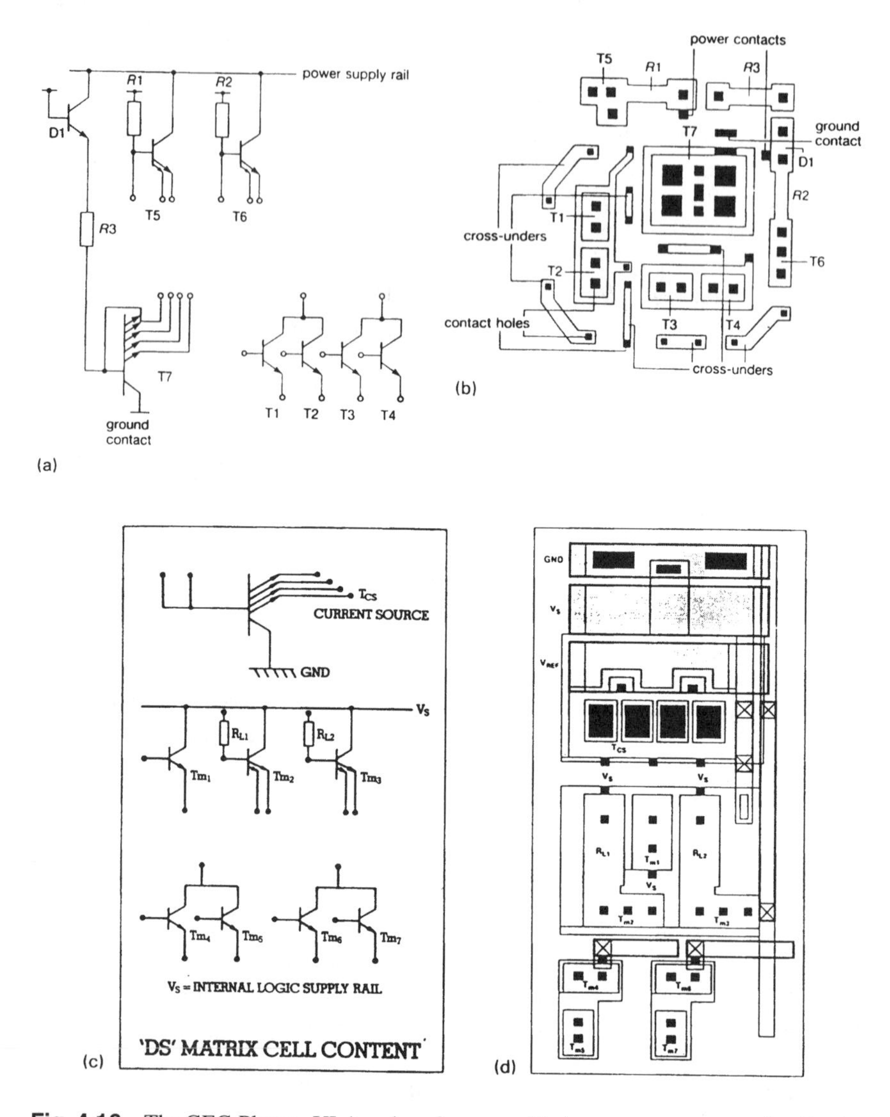

Fig. 4.16 The GEC-Plessey ULA series of uncommitted component arrays using bipolar CDI technology: (a) the component devices per cell available in early series ULAs, from which the gate configuration shown in Figure 2.14(c) or other logic functions may be constructed; (b) the physical cell layout, with wiring access to all four sides; (c) the component devices available in later series; (d) the physical cell layout of the later series, with a chip floorplan arranged in rows of abutting cells with top and bottom wiring access. (Courtesy of GEC-Plessey Semiconductors, UK.)

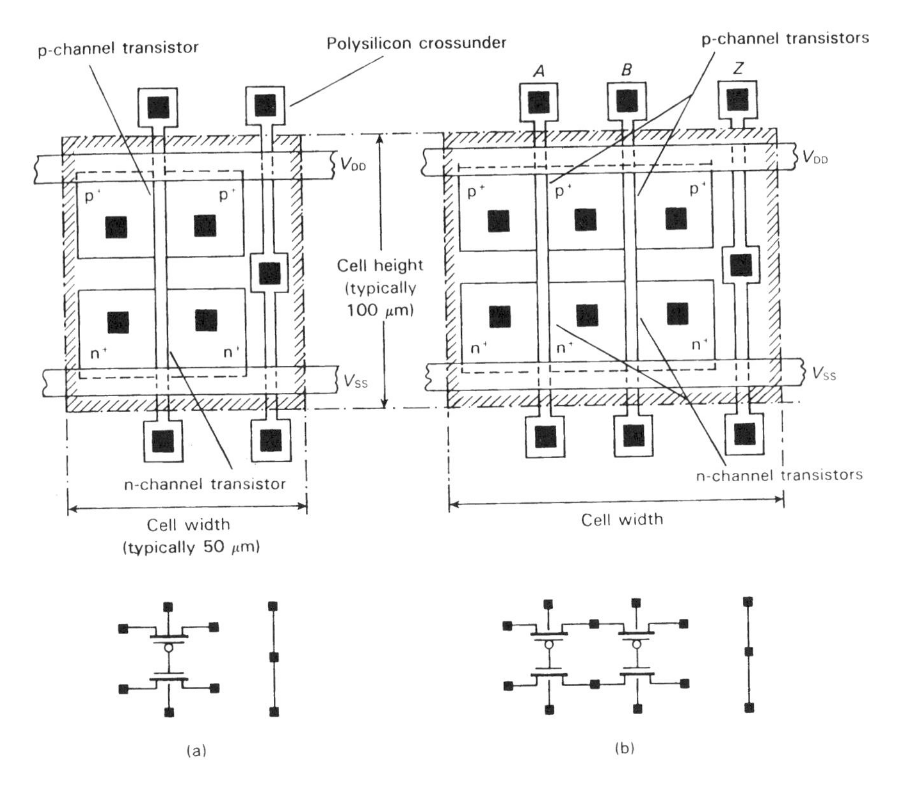

Fig. 4.17 The choice of component cells for CMOS technology uncommitted arrays, cf. the CMOS standard-cell layout of Figure 4.6: (a) a single pair of transistors per cell, with an additional polysilcon crossunder for output connections; (b) two pairs of transistors per cell, also with one polysilicon crossunder.

However, all vendors now supply the interconnection details necessary to commit their cells to a range of given logic duties, as well as details of how many cells are required to form given functions [29–38]. The terminology 'cell library' used in this context should be distinguished from the library which vendors provide for standard-cell USICs; with the latter, the complete cell fabrication details are contained in the CAD library software, whereas in the uncommitted array context, the 'cell library' is only the interconnection design details for each required function.

As an illustration of a commercial product, consider the Texas Instruments' TACH product family [29]. The chip floorplan is a normal channelled architecture, with rows of identical cells separated by wiring channels. However, the standard cell in each row illustrates the quandary facing a vendor in choosing how many transistor-pairs to place in each p^+ and n^+ region. One pair, as shown in Figure 4.17(a), may be too restrictive, but three or more pairs may be wasteful. The solution adopted in the TACH arrays is to repeat three pairs, two pairs, two pairs, three pairs in each row, with polysilicon cross-unders for output connections between each. Hence, a 'cell' in this series is defined as ten transistor-pairs, and has a total logic capacity of two 2-input and two 3-input NAND or NOR gates. This may be referred to as 'five 2-input NAND gate equivalents,' but it should be noted that it is not possible to make five separate 2-input NAND gates with the ten transistor-pairs. Figure 4.18 illustrates this product, together with the interconnection details for a 2-input Exclusive-OR gate.

Most commercially available uncommitted CMOS arrays are standardized with three transistor-pairs per cell rather than the mixed choice shown in Figure 4.18. The number of cells required to build up a range of digital logic building blocks is then as indicated in Table 4.1. The vendor's CAD system contains all the necessary interconnect information together with simulation data for the customization of the array.

Functional-Level Arrays

Because of the flexibility of the above type of CMOS array, there is little necessity to market alternative forms of uncommitted CMOS arrays containing hard-wired gates rather than component transistor-pairs. However hard-wired cells may be found in other technologies, including ISL, LSTTL, BiCMOS, ECL and GaAs. We will illustrate the first two of these in the following for comparison purposes, but they may now be regarded as obsolete.

The logic function (often termed a *primitive* in this context) chosen by a vendor as the functional cell must be capable of being used as a basic building block from which all digital functions can be assembled. It must therefore be a *universal logic primitive*. The simplest choice is either a NAND or a NOR gate. Both are universal primitives, as was noted in Chapter 3 when discussing programmable logic devices, and are widely chosen for uncommited gate arrays. A 2-input NAND or NOR gate is the smallest possible primitive, but three or more inputs are sometimes chosen in preference to 2-inputs—again, this is a compromise, since the actual required fan-in of the logic circuits in a final commitment may vary widely.

The choice of NAND or NOR gates may also be inefficient due to the limited logical discrimination of such Boolean gates; by themselves, each n-input gate can only uniquely distinguish one of the 2^n different possible binary input combinations, the remaining $2^n - 1$ input combinations producing identical gate outputs. It is this property that leads to a large number of Boolean logic gates

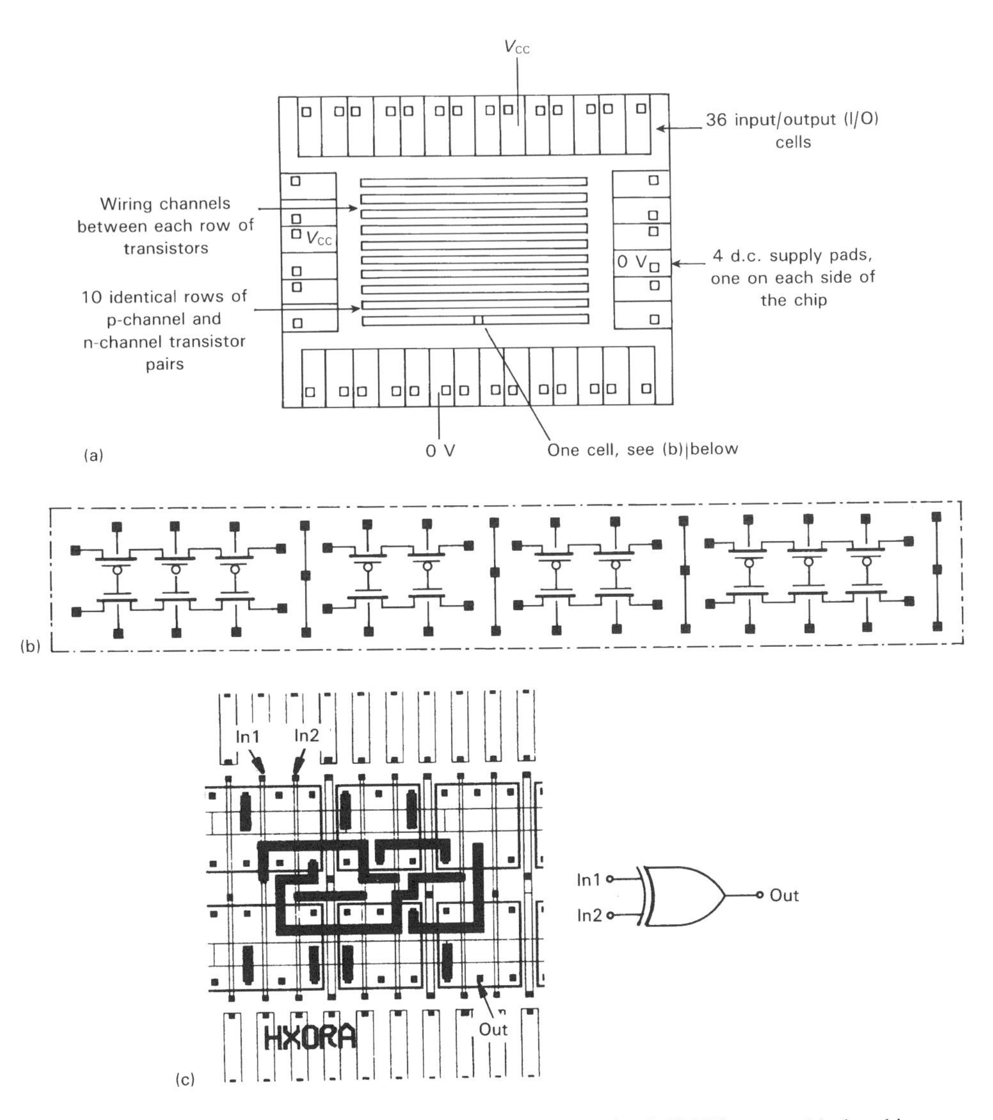

Fig. 4.18 The Texas Instruments TACH 06 uncommitted CMOS array: (a) the chip floorplan with ten rows of uncommitted cells; (b) one cell with a total of ten transistor pairs divided into four groups; (c) the cell metallization to make an Exclusive-OR function using five transistor-pairs, taken from the vendor's design manual. (Courtesy of Texas Instruments, TX.)

Table 4.1 A Short Extract from a Vendor's Library of Functions for a CMOS Gate-Array Product[a]

CMOS-4/4A gate arrays			
Function block	Block type	Function	Cells
NOR	F202	2-input NOR gate	1
	F203	3-input NOR gate	2
	F204	4-input NOR gate	2
	F208	8-input NOR gate	7
OR	F212	2-input OR gate	2
	F213	3-input OR gate	2
	F214	4-input OR gate	3
NAND	F302	2-input NAND-gate	1
	F303	3-input NAND gate	2
	F304	4-input NAND gate	2
	F305	5-input NAND gate	3
	F306	6-input NAND gate	3
	F308	8-input NAND gate	7
AND	F312	2-input AND gate	2
	F313	3-input AND gate	2
	F314	4-input AND gate	3
EX-OR	F511	2-input Exclusive-OR gate	3
EX-NOR	F512	2-input Exclusive-NOR gate	3
Full adder	F521	1-bit full adder	7
	F523	4-bit full adder	30
Multiplexer	F569	8-1 multiplexer	17
	F570	4-1 multiplexer	8
	F571	2-1 multiplexer	4
	F572	Quad 2-1 multiplexer	11
Latch	F595	R-S latch	4
	F601	D-latch	3
	F602	D-latch (with reset)	4
J–K flip-flop	F771	J-K F/F	9
	F774	J-K F/F with set–reset	11
Counter	F961	4-bit sync. binary counter with $\overline{\text{reset}}$	50
	F962	4-bit sync. binary counter with $\overline{\text{reset}}$	34
Comparator	F985	4-bit magnitude comparator	46

[a] Number of cells required per function is given, each cell being two transistor-pairs as shown in Fig. 4.17(b).

Source: NEC Electronics Inc. [31].

having to be used where complex logic relationships have to be realized. However, in addition to simple NAND and NOR gates, a number of other more comprehensive logic gates can be proposed. These are characterized by the fact that their inputs are not always symmetrical, that is, they realize different functions depending upon which order the input variables are connected, unlike simple Boolean logic gates which realize the same output relationship irrespective of any permutation of the gate input connections. The specification of such universal primitives can be derived from function classification theory, and have the capability of realizing any Boolean function of n variables depending upon the input connections, where n is 2, 3, etc. [33–35]. Some of the possibilities for $n = 2$ are shown in Figure 4.19, from which it is evident that each primitive is more complex than a simple NAND or NOR gate. However, these complex primitives need not be made as shown in Figure 4.19 from an assembly of individual Boolean gates, but can be hand crafted in a more compact way to give the same functional relationships. (This may be observed in the design of many commercially available multiplexer circuits, for example.)

In comparing the use of NAND and NOR gates with these more comprehensive primitives, the following conflicting factors arise:

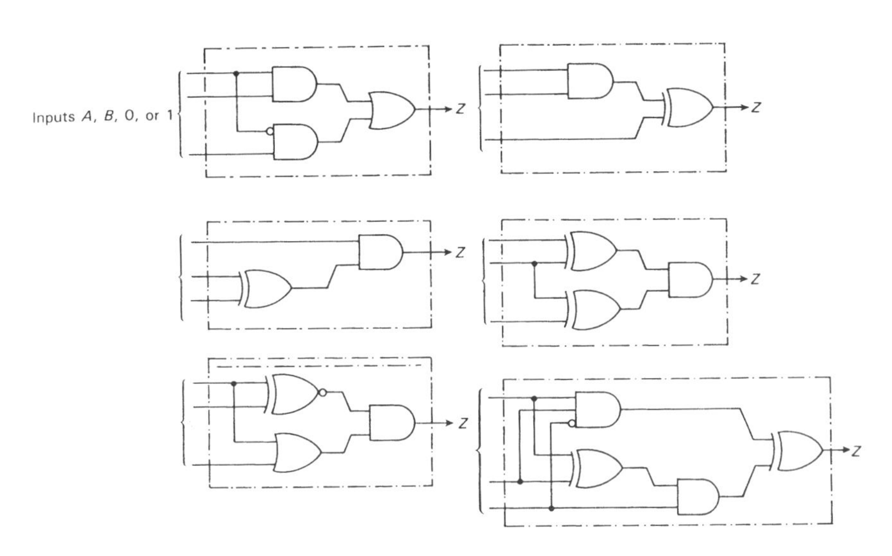

Fig. 4.19 Some single-output universal logic primitives which are capable of realizing any function of two variables A, B, depending upon the input connections, the inputs being A, B, logic 0, and logic 1 arranged as necessary. The first circuit will be recognized as a multiplexer. More complex universal logic primitives capable of realizing any function of three (or more) variables have also been proposed.

NAND and NOR gates:

- simple cell design, with small silicon area per cell
- large number of cells required to realize most logic macros, with cumulative propagation delays
- high local wiring density between cells for cell interconnections, thus requiring wide wiring channels in the array
- the use of NANDs and NORs is, however, familiar to most logic designers

Universal logic primitives:

- possibly complex cell design with a higher silicon area than a simple NAND or NOR cell
- fewer cells required than with NAND or NOR cells, giving potentially higher speed
- lower wiring density between cells than with NAND or NOR cells, thus allowing narrower wiring channels in the array
- the use of such primitives in logic design is, however, initially unfamiliar to most logic designers

Although studies have shown that there is an overall saving in silicon area in adopting a more comprehensive primitive for gate-array use than NANDs or NORs, the latter's more familiar functions remain the usual choice of a gate array vendor. The multiplexer cell (see the first circuit of Figure 4.19) has, however, been used by some vendors, particularly for high-speed applications, and some other primitives may be found in the products of specialized vendors such as Algotronix [36]. (Algotronix is now a subsidary of Xilinx, Inc.) The apparent difficulties of using comprehensive primitives in a gate array should not arise if the vendor provides a comprehensive library of interconnections for a range of standard logic duties, as is done for CMOS component-level arrays and other uncommitted products.

Figure 4.20 illustrates some of the Boolean NAND and NOR functional cells that have been chosen by vendors for their gate-array products. The low power Schottky TTL circuit shown in Figure 4.20(a) was intended to be compatible with standard off-the-shelf STTL circuits, and was therefore relevant for use in an all-TTL product environment. The ISL primitive shown in Figure 4.20(b) is similar to the circuit shown in Figure 2.12, but with a fixed resistor to provide the constant bias current I_B. Its low power dissipation lent itself to large, reasonably fast bipolar arrays, as high as about 500 gates per chip, but there always was a severe designer reaction in using such logic gates with a fixed fan-out and a fan-in of one. Both the Schottky TTL and the ISL gate-array primitives have now been overtaken by CMOS, or perhaps BiCMOS.

Finally, BiCMOS gate arrays, with a cell as shown in Figure 4.20(c), have

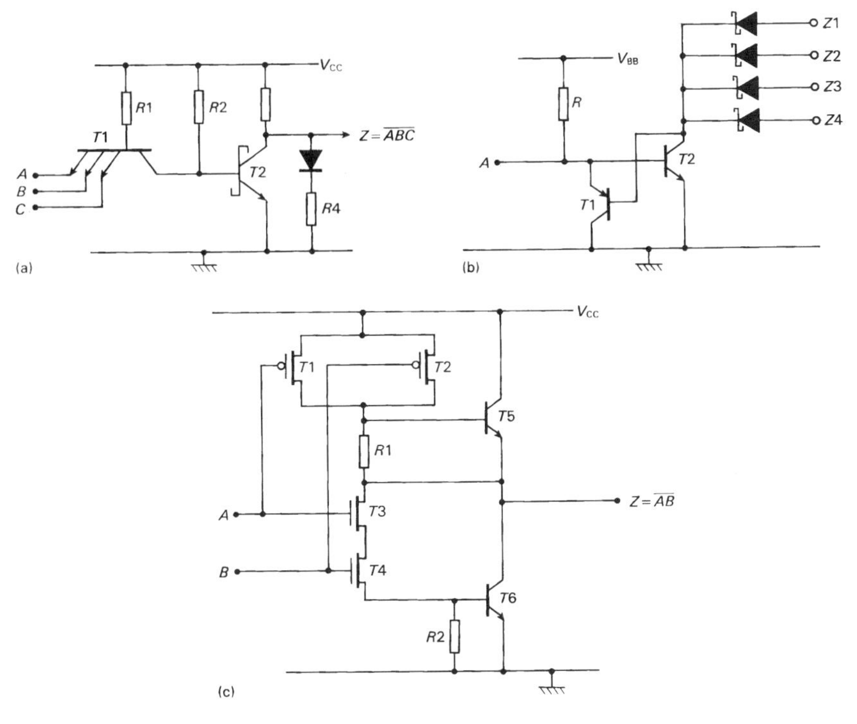

Fig. 4.20 Functional logic primitives that have been used in commercial logic arrays: (a) low-power Schottky TTL 3-input NAND; (b) ISL four-output NOR; (c) high-speed BiCMOS 2-input NAND.

been marketed. The basic cell is a 2-input NAND with bipolar output drive, but the particular distinction of these products is that a CMOS static RAM macro is also included on the same uncommited chip as the BiCMOS array. Subnanosecond gate propagation times are claimed, which is faster than CMOS alone can provide when driving output loads, and the word × bit length of the memory is configurable by the custom metallization [37,38]. These and all other functional-cell USICs rely heavily upon the vendors' data to provide library interconnection details and commitment software.

4.3.2 Routing Considerations

The principal design activity required with all types of custom ICs which use some form of predesigned basis—cell designs in the case of standard-cell USICs and prefabricated wafers in the case of gate-array USICs—is the routing required to complete the final custom chip. Placement of the required functional parts on the chip layout is a necessary prerequisite whose function is to facilitate the subsequent routing so as to provide a compact and acceptable interconnection topology. Bad placement can cause impossible-to-route conditions, or failure to meet the performance specification due to excessively long interconnect runs in critical paths of the design.

With gate arrays, the problems of subsequent routing have to be considered by the vendor at the initial chip design stage, in particular how much silicon area shall be provided in the layout for the eventual custom interconnections. With the usual floorplan consisting of rows of cells interspersed with wiring channels, this means an estimation of how many interconnections are likely to be required in each channel, and therefore what shall be the width of the channels. Clearly, too narrow a choice will make the subsequent routing difficult if not impossible, while too wide a choice will waste silicon area and not be economical. Standard-cell USICs do not have this problem, since the wiring channel widths are part of the custom design.

The number of connections required between building blocks in digital systems has been studied. For macros with a very repetitive functionality, for example, the connections required between cells to form a shift register assembly, the problem is straightforward, but for random or 'average' logic requirements the problem is more abstruse.

Consider the partitioning of a digital system into D subdivisions, each of which consists of a number of individual functions or cells. With B cells per division, the complete system S consists of $B \times D$ cells. Furthermore, let each digital cell have an average of K input/output terminals. This is illustrated in Figure 4.21. An analysis of this situation was first made by Landman and Russo [39], who considered a number of real-life digital systems, varying the number of subdivisions from $D = 1$ (all cells contained within one boundary) to $D = S$ (only one cell per subdivision). It was empirically shown that, provided the number of subdivisions of the system was greater than about $D = 5$, then

$$P = K \times B^r$$

where

P = number of interconnections around the perimeter of the subdivision
K = number of terminals per cell
B = cells per subdivision,

and where the index r had a value in the range

$$0.57 \leqq r \leqq 0.75$$

depending upon the structure of the system. This relationship has been termed Rent's rule, with the index r termed Rent's index. (Rent was involved in this

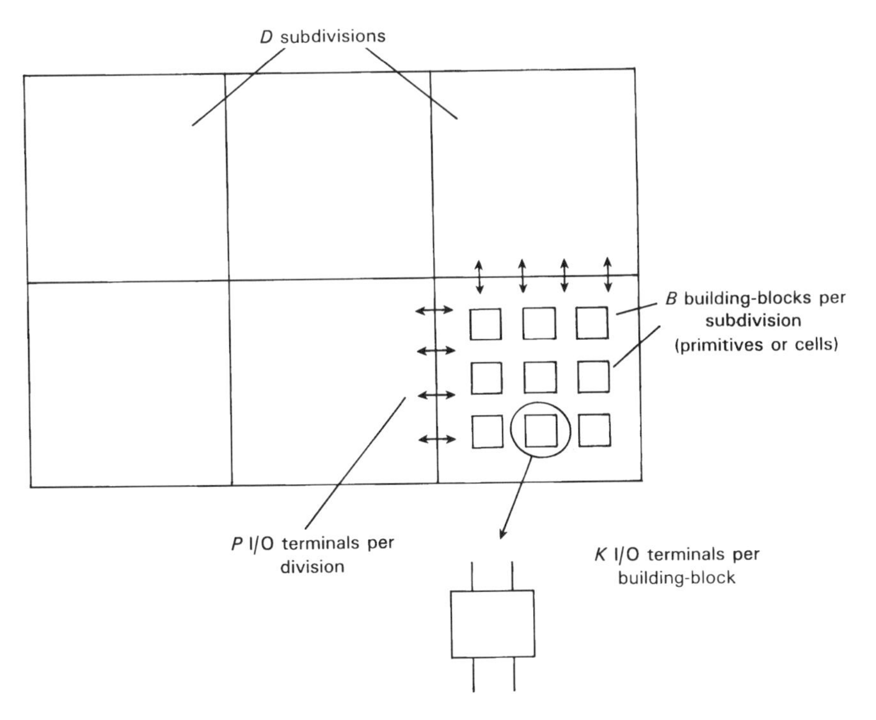

Fig. 4.21 The division of a digital system consisting of S cells of building blocks into D subdivisions, with B cells per division.

subject area at IBM in the 1960s, but does not personally appear to have published his involvement.) Note that for the extreme case of only one cell per subdivision, $P = K(1.0)^r, = K$ which is correct.

The above result was found not to hold as D approached 1; for $D < 5$ some alternative more complex relationships were suggested involving two indices r_b and r_s [33,39], but difficulties in establishing values for these indices are evident. The general equation given above is therefore usually cited. Other statistical studies by Heller, Mikhail and Donath [33,40] and others [41] may also be found, but the work involves a number of parameters whose values are not easy to determine or estimate.

Rent's rule may be applied to channelled architectures as follows. The system partitioning of Figure 4.21 may be applied to a single chip, and the blocks redrawn as a string of cells as shown in Figures 4.22(a) and (b). The connections of the B cells now become the number of interconnections in the wiring channel to and from the block. However, if we assume that there is no local peak interconnection density, which is generally true in gate-array circuits or can be made

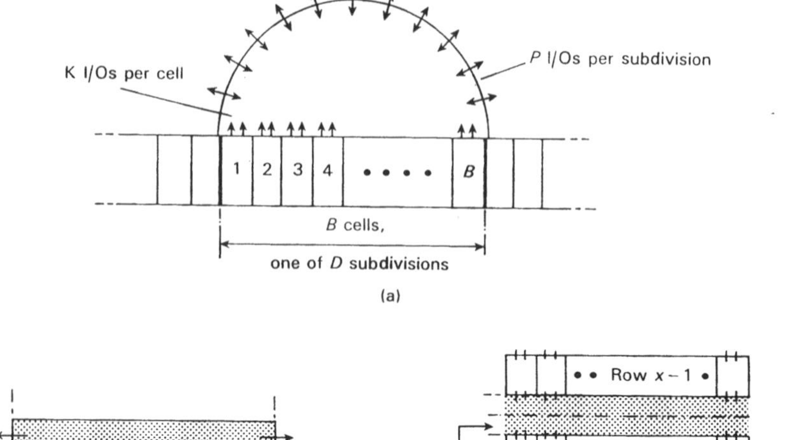

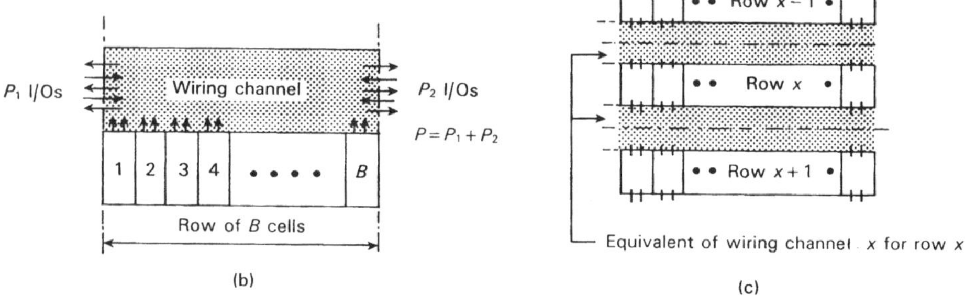

Fig. 4.22 The development from Figure 4.21 into a row and wiring channel topology: (a) B cells in each of D subdivisions; (b) redrawing of (a) into a wiring channel form; (c) alternative possibilities to (b).

true by appropriate placement, then this number of connections will be the same wherever we window our boundary of cells in Figure 4.22(b). If we also assume that half this number of connections crosses the lefthand boundary and the other half crosses the righthand boundary, then the wiring channel interconnect density will be half of the value of P.

As an example, consider a 400-cell gate array arranged in twenty rows of 20 cells per row, each cell having four I/O terminals. The total number of connections associated with a row of twenty cells is then given by

$$\begin{aligned} P &= KB^r \\ &= 4(20)^r \\ &= 22 \text{ taking } r = 0.57 \\ &= 39 \text{ taking } r = 0.75. \end{aligned}$$

Hence, the estimated wiring channel capacity is eleven tracks wide for $r = 0.57$, or nineteen tracks wide for $r = 0.75$.

Should the cells in an array have both-sides access, as most commercial products do, the connections from a row of B cells may be considered to be shared equally between the two wiring channels. This is shown in Figure 4.22(c). However, it will be appreciated that all the wiring channels, except those at the top and bottom of the array, will still have the same total number of connections, and thus there is no change in the required wiring channel capacity.

The problem with Rent's rule is obviously the indeterminacy of the value of r. The more functionality that is contained within a cell the lower the value of r, but with, e.g., a 2-input NAND gate, the value tends to be higher since a greater number of intercell connections is now required. Most USIC vendors now have sufficient company experience to be able to judge what capacity to provide in new products; most gate-array floorplans will be found to have a maximum channel wiring capacity of between ten and twenty interconnections.

The required capacity of the wiring channel around the perimeter of the array is very problematic, since it can vary widely depending upon whether the I/Os have to be allocated in some fixed order to suit a customer's requirements or whether they can be freely allocated to the I/O duties. In general, perimeter wiring channels are given a capacity similar to that of the wiring channels between the rows of cells.

Floorplan layouts other than continuous rows of cells separated by wiring channels have been used for both gate-array and standard-cell USICs. Some of the arrangements are shown in Figure 4.23. Arrangement (b), with wiring access on all four sides of each cell, was the floorplan used for the early bipolar cells

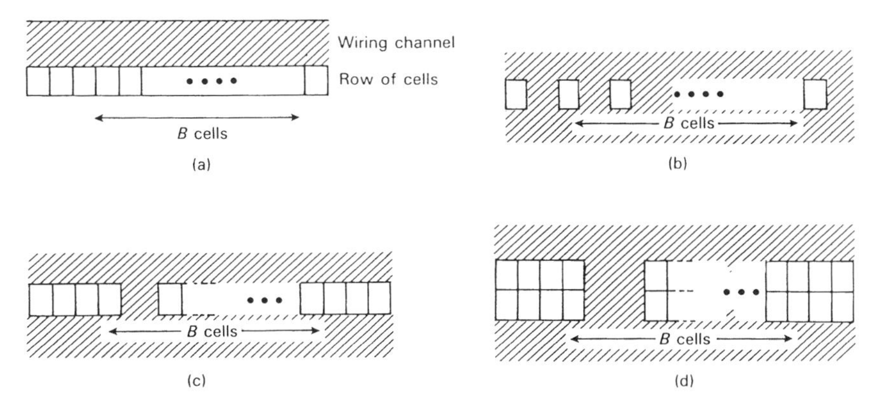

Fig. 4.23 Some alternative floorplan concepts: (a) the channelled array with abutting cells as previously considered; (b) block-cell architecture with four-sides access to each cell; (c) quadruple-cell architecture with top and bottom access per cell; (c) back-to-back eight-cell architecture with single-sided access only to each cell.

shown in Figure 4.16(b). However a theoretical estimation of the wiring channel capacities with these alternative layouts is not very useful, since there are too many degrees of freedom in the routing paths. These alternative floorplans are now all generally obsolete except for analog cells (see Section 4.3.4), but 'channel-less' floorplans have now been introduced as a completely new architecture for uncommitted digital arrays. We will consider this development in the following section.

As well as providing sufficient width to run the required number of interconnections, the basic gate-array design also has to cater for the connections to the cells and means to cross connections under each other to achieve the final routing. With single-layer-metal gate arrays, this is achieved by using the polysilicon level of the fabrication to provide connection paths ('crossunders') at right-angles to the longitudional metal tracks in the wiring channels. Even with two-layer metal polysilicon, connections are still required at this lower level.

Previous illustrations, for example Figures 4.17 and 4.18, show how the polysilicon gate connections of cells together with additional crossunders are provided to the edges of the cells. These connections give both-sides access and also a means of making a connection from one wiring channel to the next one through a cell. However, this does not solve the problem of how to cross over connections within a wiring channel so as to be able to connect to any cell I/O.

The basic solution is to provide a number of polysilicon crossunders or 'fingers,' sometimes known as 'tunnels,' across the wiring channels. These may be extensions of the polysilicon connections from the cells themselves, or completely separate short crossunders spaced at regular intervals along each wiring channel. However, the choice is complicated by the following factors:

- Excessive use of polysilicon rather than metal in interconnections will degrade the circuit performance due to the higher resistivity of polysilicon compared with metal.
- If the contact windows (vias) to the polysilicon crossunders have fixed grid positions, it may not always be easy to route some connections as they may be blocked by previously routed connections.
- On the other hand, if variable via positions to the polysilicon are adopted, then an additional customization mask is necessary as previously noted, but successful completion of routing will be aided by this flexibility.

There is no best solution to this problem, and as a result, many different interconnect topologies have been tried. Further details and discussions may be found elsewhere [33], but in general, variable rather than fixed via positions to the polysilicon level are now used.

Routing considerations for the peripheral wiring channels around the array also dictate the use of polysilicon crossunders, particularly where the logic signal interconnections have to cross the d.c. supply tracks. Again, many and varied arrangements may be found.

The final ease, or otherwise, of routing an uncommitted channelled gate array thus depends upon the original floorplan decisions made by the vendor. The only flexibility available at the customization stage is in the placement (allocation of the functional duties) of the cells in the array, which must be done so as to avoid bottlenecks in the channel routing. If an originally chosen placement proves impossible to route due to too many tracks required in a wiring channel or the inability to cross over to a required via, then a re-placement procedure, possibly iterative, must be done to solve such wiring difficulties. Human intervention is often the best if not the only way of solving such problems if the CAD fails to complete the required routing 100%.

While automatic CAD routing is essential for all uncommitted circuits containing over, e.g., 1000 logic gates, with the above proviso that a design engineer may have to intervene to solve difficult problems, for small arrays it is still possible for the custom interconnect design to be done by hand. This usually involves the use of large sheets which represent the floorplan of the chip, possibly 100 times full size, on which the OEM designer can finalize the required placement and routing [33]. The design data then have to be converted to manufacturing details by the vendor, which may introduce errors if done by hand. Simulation of the design before manufacture should pick up any errors, but clearly this is not as satisfactory as using a good hierarchical CAD system for all the design phases. Nevertheless, some hand commitment design may still be found, but now largely confined to uncommitted analog chips where human experience is often required. We will consider analog array design further in Section 5.1 of the following chapter.

4.3.3 Channel-less Architectures

One of the difficulties of using gate arrays with fixed wiring channels is that this floorplan is exceedingly inefficient in making regular structures such as ROM and RAM. For such structures, transistors can be packed very tightly together with two axes of interconnection (see Chapter 2), but a gate-array floorplan with its wide wiring channels does not lend itself to this requirement.

A *channel-less architecture* was introduced in 1985 in order to provide a more flexible gate-array topology, and is now available from many vendors. The technology of these arrays is CMOS, and the target market is the very high-speed sophisticated product area. The terminology used for these arrays, however, is not entirely consistent, and includes the following:

- Channel-free™
- Compacted Array™

- CHANNELLESS™
- Channelless Array
- sea-of-gates
- sea-of-cells
- gate forest

and possibly other variants. (Channel-free and Compacted Array are registered trademarks of LSI Logic Corporation, and CHANNELLESS the trademark of California Devices, Inc. The terms 'sea-of-gates' and 'sea-of-cells' have also been used to refer to a floorplan with nonabutting individual cells such as that illustrated in Figure 4.23(b).)

As the name suggests, channel-less architectures contain no predefined wiring channels between cells for the custom metallization. Instead the whole active area of the chip is filled with p-channel and n-channel transistors (see Figure 4.24), with the metal levels of interconnect being routed over unused transistors. This gives a much greater flexibility in the dedication, including the ability to make compact RAM and ROM macros, but at the expense of greater complexity in the CAD routing software. However, this routing is no longer constrained by

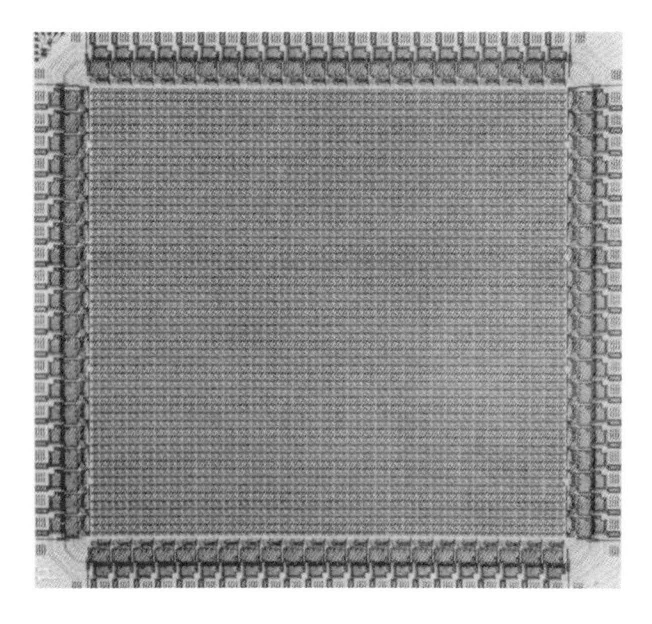

Fig. 4.24 The floorplan of a small CMOS 1.5 μm channel-less array, containing 3808 cells in 56 rows of 68 cells per row, each cell providing three uncommitted transistor-pairs, with 84 I/O pads; V_{DD} supply +3 to +6 V. (Courtesy of Silicon Technology SpA, Italy.)

fixed wiring channels, and two-dimensional freedom to route is available. The percentage of transistors on a chip which are finally used may be as low as 25%, but this is compensated for by the very large number that is packed on the uncommitted circuit. However, for regular RAM and ROM macros, up to 90% transistor utilization may be possible within these areas [42–45].

Like channelled arrays, channel-less arrays are built up from a chosen basic cell which is replicated on the chip, but unlike the former the cells abut both horizontally and vertically, leaving no dedicated wiring channel spacing. Two-layer metalization is employed, and adjacent cells are often mirrored along both the *x*- and *y*-axes so that d.c. supply and common clock lines may be shared between adjacent cells.

The exact arrangement of the p-channel and n-channel transistors provided per cell varies considerably between vendors. However, it is not unusual to have different sizes of transistors in a cell, possibly larger size p-channel transistors than n-channel transistors so as to provide equal drive capability, or 'small' and 'large' transistors for different duties. More n-channel transistors than p-channel may be provided, which is relevant for RAM applications, for example. A further innovation may be the inclusion of a very small p-channel and n-channel transistor-pair whose gates are permanently connected to V_{DD} and V_{SS}, respectively, and which in their permanently cut-off state provide both isolation for the active devices and also tie-down paths to the well and silicon substrate [44].

Figure 4.25(a) illustrates one possible basic cell structure consisting of 'small' size transistor-pairs in the center, and additional 'large' pairs on the outside. If the middle pairs in a cell are not used, up to ten wiring tracks may pass over them, while if the outer transistors are not used up to fourteen tracks may pass over them [46]. With two-level metallization—sometimes three-level with certain products—very compact cell libraries can be provided, the vendor giving the dedication of such cells for a range of duties up to LSI and VLSI complexity. Figure 4.25(b) shows the dedication interconnection pattern for a nest of eight cells to form a scan-path flip-flop (see Chapter 6), this particular cell structure being mirror-imaged in the floorplan on both axes [44]. Notice that interconnections are not constrained to run in straight lines, but are free to 'dog-leg' to complete the routing.

It is clear that channel-less gate arrays cater to a higher level of complexity than do the general range of channelled arrays. Typical specifications are as follows:

- 1.5 μm CMOS technology
- 0.75 ns or faster propagation delay for a 2-input NAND gate driving a fan-out of two
- total gate count per die ranging from thousands to over one hundred thousand, taking one gate as two transistor-pairs
- estimated maximum useable gates 40%

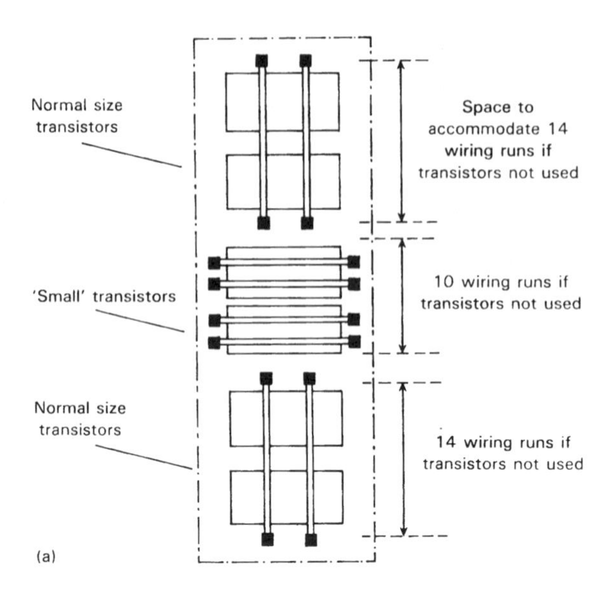
Normal size transistors
Space to accommodate 14 wiring runs if transistors not used
'Small' transistors
10 wiring runs if transistors not used
Normal size transistors
14 wiring runs if transistors not used
(a)

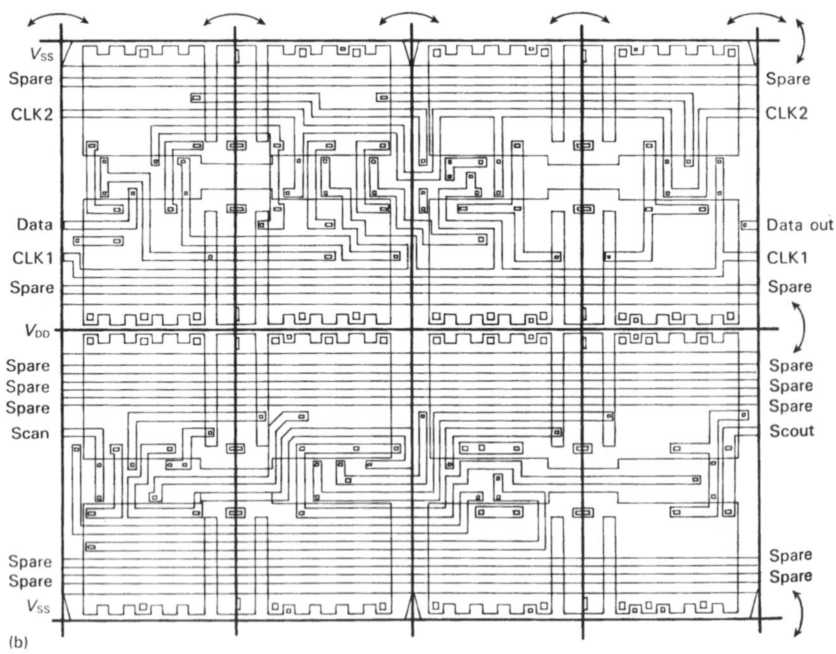
V_{SS}
Spare
CLK2
Data
CLK1
Spare
V_{DD}
Spare
Spare
Spare
Scan
Spare
Spare
V_{SS}
Spare
CLK2
Data out
CLK1
Spare
Spare
Spare
Spare
Scout
Spare
Spare
(b)

- up to 350 or more I/Os per IC
- TTL/CMOS I/O compatability

The total market potential for channel-less arrays is still debatable. It is claimed that they provide the same performance and system complexity as standard-cell USICs, but with a quicker and cheaper design time and fabrication cost due to the reduced mask set required for customization. Better performance than standard cells at any given time has been claimed, since it may be possible to bring improved fabrication capabilities faster to the market place by redesigning channel-less arrays than by redesigning a complete standard-cell library to incorporate such changes, but this claim may not be entirely valid if good CAD is available which can automatically update standard-cell designs to keep pace with technology improvements. It should be noted that five or more masks are required for customization, which is double that necessary for simple channelled arrays, and hence the channel-less USIC is unlikely to challenge the use of the latter unless the particular attributes of channel-less USICs are required [47]. The more complex placement and routing will be discussed in the following chapter.

4.3.4 Analog and Mixed Analog/Digital Arrays

The terminology 'gate array' may imply that the USIC contains an array of digital logic gates only, and would be a misnomer if the array is designed for analog or mixed analog/digital applications, or if it contains separate components rather than logic gates. However the term 'gate array' is very often loosely used to refer to any type of uncommitted IC, including those which have analog capability. Other terms such as uncommitted component array (UCA), uncommitted gate array (UGA) and uncommitted logic array (ULA™, see Section 4.3.1) have been used to refer to either digital or analog products, but the ambiguous terminology 'gate array' still seems to be widely used for any type of uncommitted array product. We will therefore continue to use this term here where generally appropriate. As far as is known, the term 'uncommitted analog array' has never been used.

Fig. 4.25 CMOS channel-less gate array architectures: (a) a possible basic cell containing normal size and 'small' size transistor-pairs, omitting any supply rails for clarity; (b) an alternative cell design showing the committment of a floorplan of eight cells to realize a scan-path flip-flop macro. The mirror image floorplan has been indicated. Note that the wiring runs marked 'spare' are unused through-space only and not interconnection metal, and that the actual interconnections can take random routing paths. (*Sources*: (a) based on Ref. 46; (b) based on Ref. 44.)

In contrast to the very large number of transistors present in channel-less arrays, uncommitted arrays for purely analog (linear) purposes usually contain a much less impressive number of devices. They may, however, contain high-voltage or high-power transistors, and will certainly dissipate much more power in their linear mode of operation.

Early analog arrays consisted of a number of individual resistors, capacitors and bipolar transistors on the uncommitted chip, with no specific floorplan topology, the component count possibly being tens of npn and pnp transistors, a few higher-power output transistors, and a few capacitors together with perhaps a hundred or more resistors of various values. Since these components were scattered all over the floorplan layout, custom routing was invariably done by hand. This lack of a formal grid layout meant that every custom design was done from scratch, with no vendor's library of interconnections normally being available. However, one accessory was introduced, namely off-the-shelf SSI packages, sometimes termed 'kit parts,' each SSI IC connecting and giving access to a small number of the components provided on the uncommitted chip, from which a PCB breadboard prototype could be physically assembled and tested. Because the analog frequencies originally involved were not high, usually under 1 MHz, the performance of the discrete component breadboard was very close to that of the committed IC.

However, the new generation of analog arrays employs a much more structured floorplan. This consists of *tiles*, sometimes termed *mosaics*, each tile being the analog equivalent of the uncommitted digital cell but consisting of a fixed structure of uncommitted transistors, resistors and possibly capacitors per cell. These tiles are replicated on a fixed grid pattern across the chip layout [21,48]. Wiring space is available between components and between cells, but wide wiring channels are not usually provided since the interconnection routing of an analog array is much more localized and less distributed than in digital arrays [49–54].

The problem which is still present when formulating a tile specification is the choice of components to provide per tile so that each may be of maximum use. This dilemma facing vendors is illustrated by the well-known representation shown in Figure 4.26. Among the solutions which have been marketed are the following:

- the use of two or more different standard tile specifications per uncommitted IC to cover general purpose duties
- the inclusion of additional separate resistors and capacitors and sometimes other components such as zener diodes around the chip
- the inclusion of additional separate transistors for high-power or high-voltage duties

Figure 4.27 shows one product which contains two types of tile, termed a 'standard' tile and a 'power' tile. Component details are as follows:

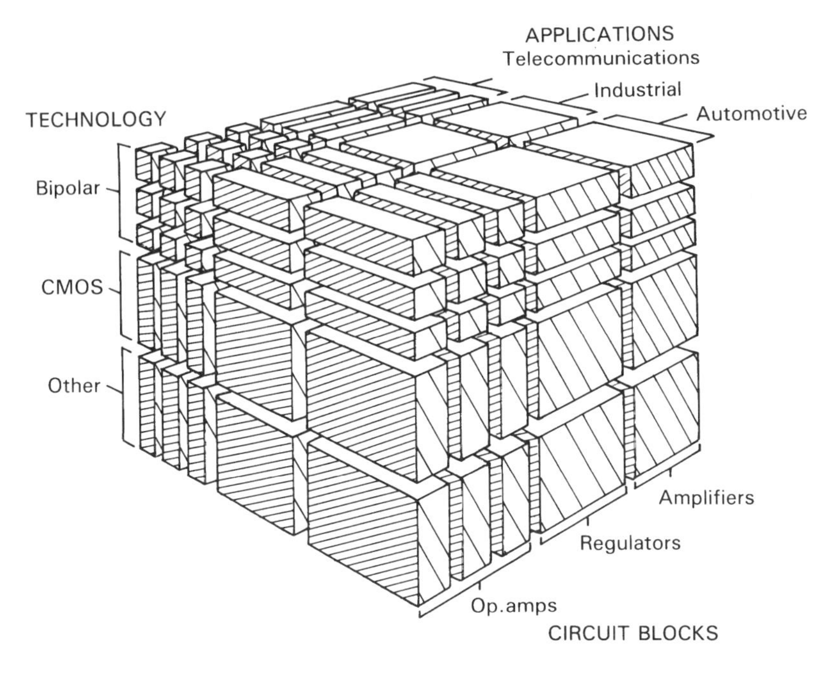

Fig. 4.26 The classic representation of the dilemma in finding a truly universal solution to the specification for an uncommitted analog IC. (*Source*: based on Ref. 48.)

- npn transistors, various size, 8 per standard tile, 6 per power tile
- pnp transistors, various size, 8 per standard tile, 6 per power tile
- resistors, various values, 50 per standard tile, 30 per power tile
- capacitors, 1 per standard tile

The small npn transistors have a typical f_T of 350 MHz and the corresponding pnp transistors an f_T of 300 MHz, each with a collector-to-emitter reverse breakdown voltage of 33 V minimum. All internal connection points are on a rigidly defined grid, which allows the OEM to carry out hand-routing on a master layout sheet if required, aided by macro interconnection library details supplied by the vendor [53].

An alternative design in several ways is shown in Figure 4.28. Here again, there are two types of tile, but now each type consists of small-geometry npn and pnp transistors only, varying beween the two tiles in their placement detail. Additional large transistors, capacitors and zener diodes are provided in columns and around the perimeter of the chip. Another different feature is that the space between all tiles and other components is given a thin-film nichrome cover from which required resistor values can be customized by etching. The resistor values

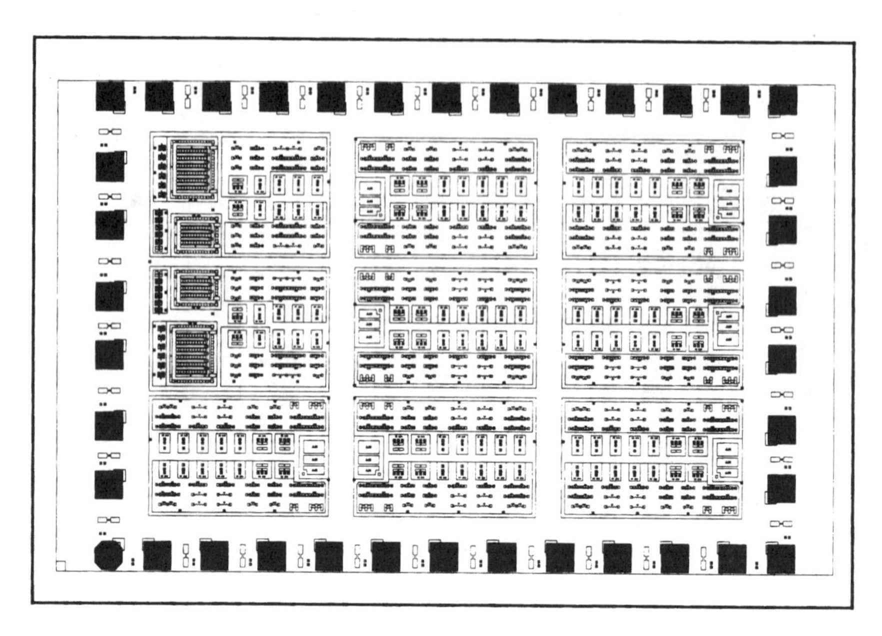

Fig. 4.27 A tile-based bipolar uncommitted analog (linear) array containing seven standard tiles and two power tiles, with 36 I/Os. (Courtesy of AT&T Technologies, PA.)

are given by $R = \rho(L/W)$, where ρ is the sheet resistivity of the nichrome, L the effective length and w the effective resistor width. This is claimed to give good matching ability, very low parasitic capacitance, and the ability, if required, to laser-trim resistor values to high absolute accuracy. As with the product previously considered, this linear array can be routed by hand, with the help of the vendor's macro library.

Both of these products are typical in using bipolar technology, with silicon dioxide isolation between components so that they are effectively in SiO_2 islands. This gives extremely good isolation properties with high voltage breakdown. Very high frequency performance is available from many products, but because the transistor sizes are fixed, system performance cannot match that of a full-custom design where freedom to size all the transistors and resistors so as to achieve optimum performance is available.

One serious drawback of uncommitted analog arrays, such as illustrated in Figure 4.28, is that the components provided in each tile cannot conveniently be used should any digital logic also be required, for example, analog-to-digital and digital-to-analog conversion. A component utilization of possibly less than 20% is achieved if analog tiles are used for logic functions. Early generations of CDI technology uncommitted logic arrays as detailed in Figures 2.14 and 4.16 and

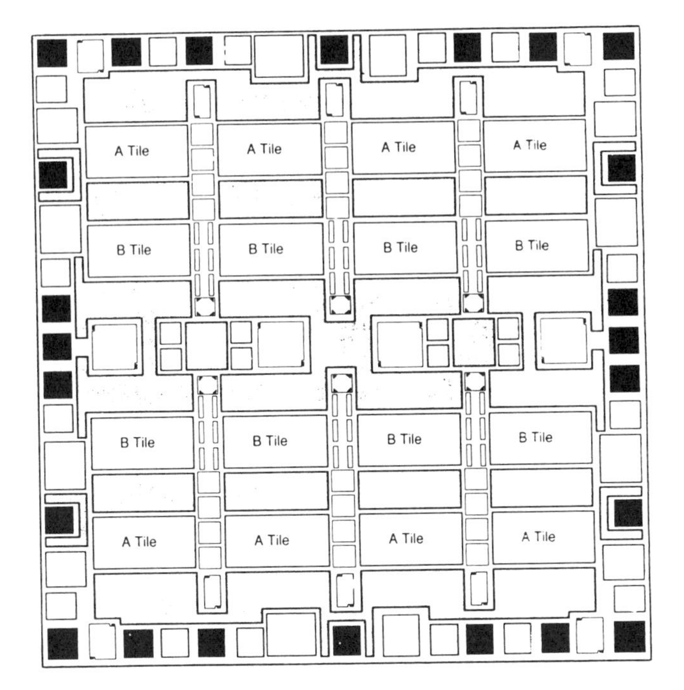

Fig. 4.28 Another tile-based uncommitted analog array. The plain areas between the cells and other fixed components are nichrome areas which may be etched to form precision resistors. (Courtesy of DataLinear, CA.)

marketed by Ferranti, UK, and Interdesign, CA, offered much more satisfactory mixed capabilities; I^2L technology, marketed particularly by Cherry Semiconductor Corporation, RI, also provided good mixed analog/digital capabilities, but present trends are to have separate analog and digital parts on the same chip in order to optimize each duty.

An example of this evolution may be illustrated by a 1990s version of the bipolar ULA, which has a border of uncommitted analog tiles around a core of uncommitted digital cells. This combination gives up to 100 MHz analog capability, with 1 ns gate propagation times in the digital cells. An alternative floorplan architecture uses a channel-less CDI sea-of-gates layout as the digital core. Both analog and digital macro library data are available for custom dedication.

Other recent products, however, combine bipolar and MOS technology for mixed duty arrays, such as illustrated in Figure 4.29. This particular product features a CMOS channelled array of forty-eight cells, each cell consisting of five transistor-pairs, surrounded by eight main analog tiles plus other discrete

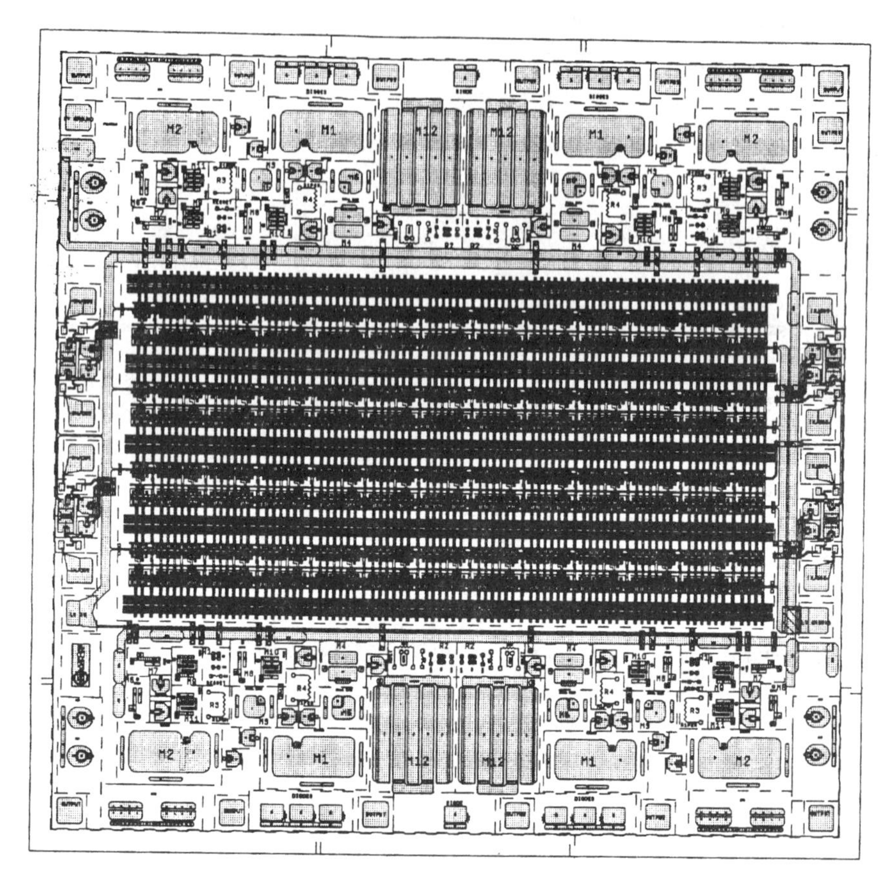

Fig. 4.29 A mixed bipolar/MOS technology array, with CMOS transistor-pairs as the central core and analog components around the periphery. Chip size is 5000 μm × 4830 μm, with 24 chip I/Os. (Courtesy of AT&T Technologies, PA.)

components. This product is aimed at robotics and similar duties, and provides high voltage ratings, greater than 250 V in some of the analog components.

It is apparent that the many and varied analog requirements necessitate a wide range of different array specifications. Hence the OEM designer must always carry out a wide vendor search when considering the possible use of such USICs in new company products. Newer products may be MOS technology only rather than mixed bipolar/MOS, improvements to the analog performance of MOS circuits being such that it is now adequate for many OEM requirements.

4.4 MASKLESS FABRICATION TECHNIQUES

The normal means of producing the required geometric areas on a wafer during fabrication is by photolithography, using individual masks to define each separate level of fabrication. At each mask stage the wafer is first given a silicon dioxide protective layer followed by a thin layer of photoresist, which is subsequently illuminated through a mask with ultraviolet light. If *positive photoresist* is used, then the areas receiving the UV illumination can be etched away, and the areas which have not been illuminated will remain; if *negative photoresist* is used, then the areas which have not been illuminated are removed. (For VLSI fabrication, positive photoresist is preferable because it does not swell as much during development as negative photoresist, and hence reproduces very small dimensions more accurately.) Silicon processing then continues through the windows etched in the photoresist.

Masks are made of special glass with a chromium surface film which is patterned by E-beam lithography. This involves coating the chromium with a resist which is sensitive to an electron beam, rather than photoresist which is sensitive to UV light, and then 'writing' the required pattern on the resist by means of an E-beam machine which focuses, deflects and switches the beam of electrons on and off. Where illuminated by the E-beam, the chemical bonds of the resist are weakened and can subsequently be washed away leaving windows through which the underlying chromium can then be chemically removed, thus producing the optically transparent pattern on the mask. However, because the E-beam can only be deflected by one or two millimeters without unacceptable distortion, the table holding the mask must be moved ('stepped') very accurately under computer control in order that a complete wafer-sized mask may be written.

There are, however, three distinct methods of using the masks in the subsequent photolithographic processes, namely:

- contact exposure, where the mask physically touches the photoresist on the wafer during exposure
- proximity exposure, where the mask is held a few microns above the surface during exposure
- projection printing, where the mask is held some distance above the wafer and an optical sysem is used to project the UV illumination on the surface

This is illustrated in Figure 4.30 [55]. With the latter method, the mask no longer defines the whole wafer patterning but only a small area, possibly one die only, in order to maintain high resolution. The mask is now called a *reticle*, and may be ten times the actual die size, the optical system projecting this enlarged mask pattern down to the required dimensions. A complete wafer is now exposed by

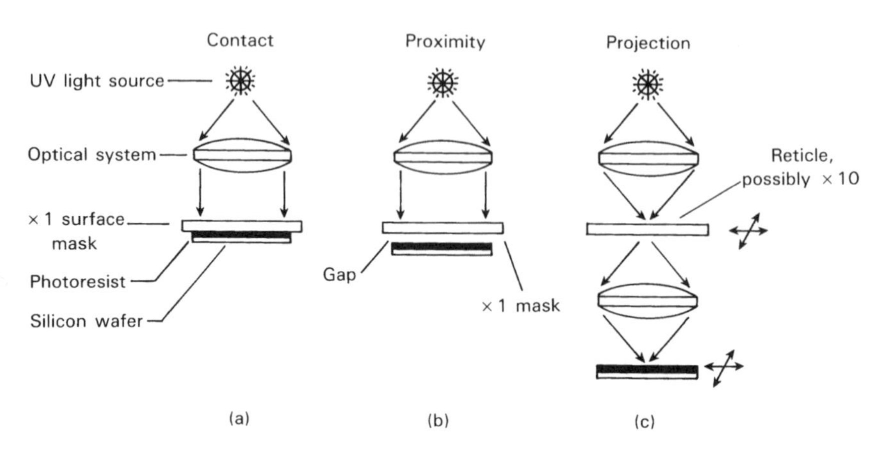

Fig. 4.30 Optical lithographic methods: (a) contact; (b) proximity; (c) projection with computer-controlled stepping of reticle and wafer.

a step-and-repeat process using a precision machine called a *wafer stepper*. (A ×5 or a ×10 reticle may also be used to make a full-size ×1 mask by an optical step-and-repeat procedure, the reticle being held as the master pattern data.) The resolution obtainable with these methods is better than 1 μm.

4.4.1 E-Beam Direct-Write-On-Wafer

It is evident that the production of masks for wafer fabrication is both expensive and time consuming. One mask for a 150 mm diameter wafer may cost several thousand US$, and be manufactured by a specialist company separate from the actual IC vendor. For custom microelectronics where small-quantity production may be involved, means to eliminate this cost and the potential bottleneck in the wafer production are attractive [56].

E-beam direct-write-on-wafer (E-beam lithography) is one commercial possibility. Here an E-beam machine such as that shown in Figure 4.31 is driven by the same design data on disk or magnetic tape as would be used to make the more conventional glass/chromium optical masks, but now the E-beam is used to pattern resist on the wafers directly. With careful control of the E-beam energy, the resist can be patterned without causing any radiation damage to the underlying doped silicon areas. Greater care to avoid damage may have to be taken with MOS technologies than with bipolar.

The advantages of direct-write-on-wafer include the following:

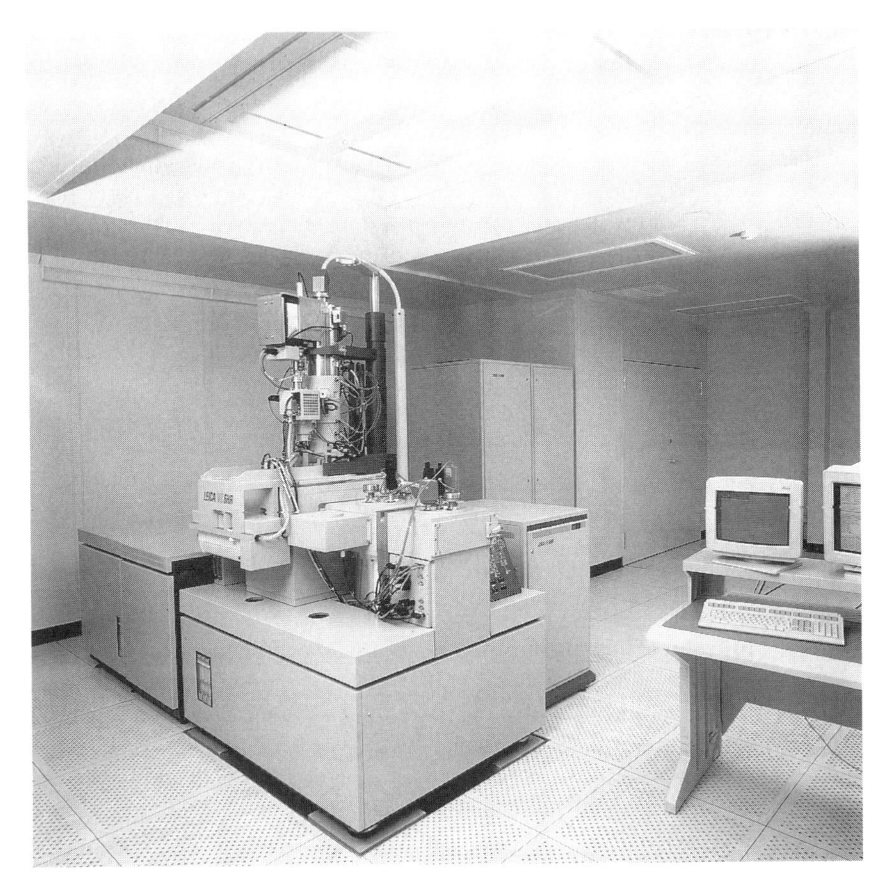

Fig. 4.31 A typical commercial E-beam machine, with line-width capabilities down to possibly less than 0.1 μm and workpiece position controlled by laser interferometry. (Courtesy of Leica Lithography Systems Ltd., UK.)

- elimination of the cost in making optical masks
- possibly improved delivery time for small quantities due to not having to wait for the delivery of optical masks
- compatability of the CAD software output which defines the required geometry with the E-beam machine's computer control
- the ease with which design errors can be corrected
- the ability to mix different customers' designs on one wafer (*multi-project wafers*) in order to reduce individual design costs

Against these real or potential advantages must be weighed the very high capital cost of E-beam machines, possibly $4M or more per machine, and the slow writing speed, requiring possibly more than ten or fifteen minutes to write just one masking pattern.

Because of the latter factor, it is not usually considered realistic to consider E-beam direct-write-on-wafer for volume production purposes, particularly where ten, twelve or more mask patterns per wafer are involved. However, direct-write-on-wafer becomes a very attractive way to produce ten or twenty prototype circuits for verification purposes before full scale production is commenced [9,10]. Hence, there is here something of an anomaly: E-beam direct-write was initially considered as a means of producing custom ICs in very small quantities for OEMs without the need to go through the process of making optical masks, but this has not generally proved to be commercially viable; instead, E-beam direct-write has proved to be very valuable to confirm the functionality of a new VLSI design intended for mass production as a standard part, where time to market is critical and cost is of secondary concern compared with ensuring a fault-free commercial product.

Earlier problems with E-beam direct-write-on-wafer included the following:

- difficulty of alignment of the E-beam with underlying areas of the silicon fabrication
- possible problems with cumulative tolerances and distortion across the full surface of the wafer, sometimes termed 'run-out'
- wafer surfaces not being perfectly flat, particularly with multilayer metalization

However, the technique must now be regarded as a very satisfactory means of producing small quantities of ICs of VLSI complexity relatively quickly once the computer-aided design has been completed, provided the cost of use of the E-beam machine and the time scales are relevant.

4.4.2 Other Technologies

Laser technology is another technology that has been applied in the manufacture of microelectronic circuits. Lasers are extensively used in sophisticated production equipment for accurate positioning and stepping using helium-neon inferometry, and in inspection and marking [55], but they can also be used for part of the on-chip processing [57,58]. The latter area includes the following:

- laser photolithography
- laser cutting and trimming
- laser welding
- laser deposition

The use of a laser source to pattern photoresist is receiving increasing attention as geometries shrink to submicron values. An excimer-type laser is normally used as the light source, giving high-intensity UV radiation exposures as short as 20 ns with 0.1 μm resolution. While this is used for the production of masks and reticles in place of previous UV light sources, direct-write-on-wafer can also be done as an alternative to E-beam direct-write. X-ray lithography is also being considered for submicron lithography, with wavelengths in the 4–50 Å range [55,59].

One particular commercial application which illustrates the use of laser direct-write-on-wafer is the LASARRAY gate array system [60,61]. Here, a helium-cadmium (He-Cd) laser was used to expose the photoresist, following which an etching procedure removed the unwanted metal of an uncommitted wafer. A specific feature of the LASARRAY system was that it could be supplied to OEMs to undertake their own dedication procedure. The system was designed as three self-contained and interlocked clean-room chambers, which provided (i) wafer customization, (ii) wafer slicing, and (iii) testing and packaging, the input being the standard uncommitted wafers and the output being the packaged chips. The uncommitted wafers, however, were specific to the LASARRAY process and not from any other gate array vendor, since they had to contain positioning information for the several stages of automatic processing. The overall floor space of the assembly was 9.03 m × 7.05 m, and could be installed in any reasonably clean area without the need for a super-clean environment. The whole commitment process was under computer control, and typical processing time for a complete 100 mm diameter wafer was about 2–3 hours. However the necessity to use the special LASARRAY uncommitted wafers and the relatively high capital cost of a complete system, on the order of US$ 1M, and the difficulties of upgrading to keep pace with technology improvements, was a severe disadvantage to its widespread adoption.

Laser cutting and laser trimming, however, are both well-established practices, particularly in the field of hybrid circuits where passive components need to be trimmed to close absolute values. Laser cutting has also been employed to customize arrays which contain all possible interconnection paths on the uncommitted chips, the unwanted connections being cut by a computer-controlled laser beam. Further details of laser cutting and laser trimming may be found in the extensive bibliography listed in Ref. 57.

In all the fabrication processes previously considered, the patterning of the top interconnection layers has been a *subtractive process*, unwanted metal being removed leaving only the wanted interconnect; however, with a process known as 'laser pantography' metal may be deposited in conducting strips using a scanning laser. This is therefore an *additive process*.

The principle of direct-writing of conducting strips depends upon a local reaction caused by the energy of a focused laser beam in a reactive atmosphere, this reaction being produced either by the temperature rise induced by the laser spot (*pryolysis*), or directly by proton energy (*photolysis*). Many different combinations of materials and laser sources have been studied to deposit conductors such as copper (Cu), aluminum (Al) and tungsten (W) (see Table 4.2), but one introduced into commercial use uses an argon-ion laser to deposit tungsten or nickel (Ni) for the interconnection tracks.

This particular example is LASA Industries' QT-GA (quick-turn gate-array) process, a process which the company has termed *lasography*, and which provides one-layer-metal or two-layer-metal gate arrays of several thousand 2-input NAND-NOR gate capacity [9,51,58,60–63]. The complete deposition system is shown in Figure 4.32. Preprocessed wafers which have been fabricated up to the I/O bonding pads and the contact windows (vias) are the starting point, and the initial procedural steps are as follows:

- wafer probe test for correct fabrication parameters
- wafer slicing
- packaging of individual die
- wire-bonding of all I/Os
- visual inspection
- insertion of packaged but uncommitted circuits into precision carriers for loading into the deposition machine

Within the machine the uncommitted circuit is positioned in a process chamber and surrounded with an appropriate mixture of gases which decompose to deposit metal under the laser spot. The complete commitment procedure is as follows:

- auto-positioning of the packaged circuit
- first layer of tungsten (or nickel) tracks deposited
- SiO_2 insulating layer applied
- laser etching of contact windows through the SiO_2
- second layer of tungsten tracks deposited
- final insulating layer applied
- lid attached to the package

Table 4.2 Examples of Conductor Materials Deposited by Laser Pantography

Possible reactive atmosphere	Possible type of laser	Conductor deposited
$Al_2(CH_3)_6$	Ar^+(257 nm), Kr^+(476 – 647 nm)	Al
$Au(CH_3)_2(AcAc)_2$	Ar^+(515 nm)	Au
$CuSO_4$	Ar^+(515 nm)	Cu
$[Pd(m - O_2(CH_3)_2]_3$	Ar^+(515 nm)	Pd
$W(CO)_6$	Ar^+(257 nm), Ar^+(350–360 nm)	W

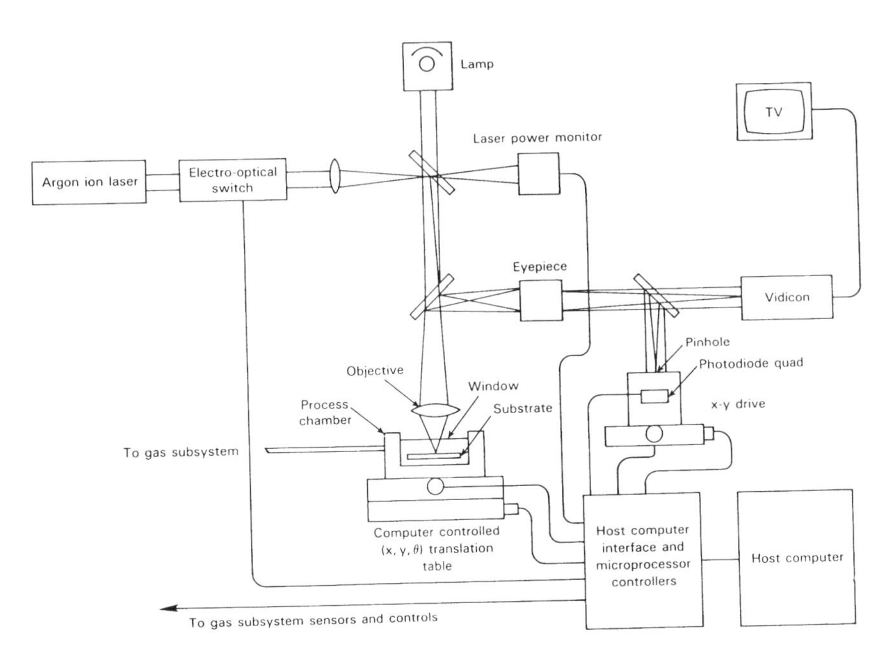

Fig. 4.32 The schematic arrangement of a laser pantography system to deposit conducting paths on an uncommitted die. The die is held in a reactive atmosphere and written with a finely focused laser beam. (*Source*: based on Ref. 57.)

The whole commitment operation is under dedicated computer control once the uncommitted ICs have been initially loaded.

A particular feature of this process is that because the individual ICs are processed and not complete wafers and the metal deposition only takes place at a 1 μm spot under the laser beam, the internal size of the initial process chamber can be made very small, only about 2 m^3. The subsequent two chambers to and from which robot arms move the IC are about 25 × 25 × 12 cm for the SiO_2 processing, and 5 × 5 × 5 cm for the final packaging chamber. As a result the complete processing unit with its chemical and computer control is entirely self-contained. Performance of the system is such that a writing speed of 1–2 cm s^{-1} is available, which allows a 1000-gate two-layer-metal array to be processed in about 5 minutes, or a more complex 6000-gate array in about 120 minutes. Line widths down to around 1 μm with spacings of 1 μm can be produced, with a location accuracy of 0.2 μm.

As with the LASARRAY system and other similar company ventures [60,64], it will be noted that this LASA system also requires specially prepared uncommitted wafers, which therefore locks an OEM into one supplier and/or product. The cost of this system is also appreciable, possibly US$ 2–3 M. Hence, all these maskless fabrication possibilities have not been seen to be viable for ordinary OEM use, but may have a niche market for very large companies that require a very fast (same day) production for research or other urgent applications where the latest state-of-the-art complexity or performance is not required. In general, these interesting and innovative possibilities have been overtaken by the large user-programmable products such as those considered in Section 3.6 of the previous chapter.

4.5 SUMMARY AND TECHNICAL COMPARISONS

Having completed this coverage of custom microelectronic techiques, we may now summarize their principal attributes. We will leave the question of detailed design costs, usually known as nonrecurring engineering (NRE) costs, until Chapter 7, where the choice of design style for a particular project will be considered in detail. This consideration will also involve the cost of computer-aided-design (CAD) resources.

Since the custom mask costs involved in the use of uncommitted arrays is less than that of the full mask set required for standard-cell or full-custom ICs, initial production costs should be lower for arrays, but the smaller die size of the latter, which gives more circuits per wafer, may make subsequent volume production costs cheaper. The overall economics of this are illustrated in Figure 4.33 [33,65,66]. For very small production quantities, the use of any form of custom IC cannot usually be justified on purely financial terms due to the cost of the custom design activities; for extremely large volume requirements, the smallest and most efficient design, that is, full-custom design, may be the best,

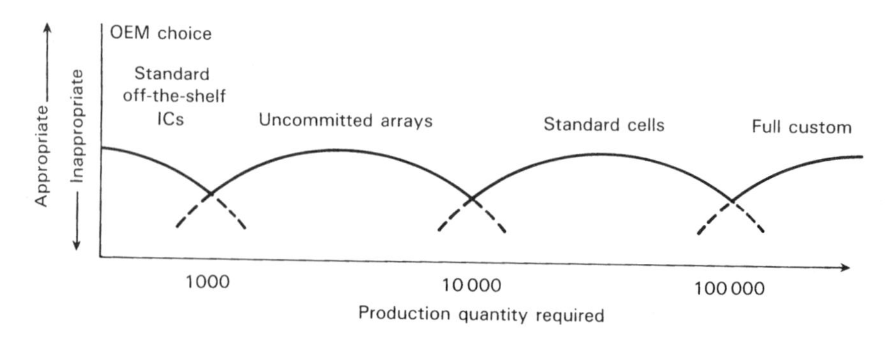

Fig. 4.33 The general range of viable production volumes for different types of digital microelectronic realization.

but in between these two widely spaced extremes uncommitted-array and standard-cell designs offer their respective advantages.

This broad generalization remains reasonably true when detailed design costs are taken into consideration, but the increasing capabilities and use of large programmable logic devices and channel-less arrays are now blurring this previous picture. Programmable logic devices may capture the lefthand half of the picture shown in Figure 4.33, with channel-less products invading the righthand half for complex custom requirements. These considerations will be pursued further in Chapter 7.

Turning to purely technical details, more precise comparisons and summaries can be made. In practice, it may be that technical performance, for example minimum power consumption, dictates the type of product and technology used rather than financial considerations, but in most engineering situations more than one design solution usually needs to be considered. (Reference may be made to Chapter 2, Tables 2.3 and 2.4, for technology comparison details.)

A summary of the custom technologies which are technically most appropriate for different duties is as follows:

- analog-only applications: bipolar
- digital-only applications: MOS or bipolar
- mixed analog/digital applications: no best technology
- very high speed operation: bipolar or GaAs
- very large number of gates: MOS, specifically CMOS
- high operating voltage: bipolar or special technology MOS
- very low operating voltage: MOS, specifically CMOS
- low power dissipation: CMOS
- memory: MOS, specifically nMOS

In general, very high speed digital operation requires high power dissipation, as was indicated in Figure 2.36. This, in turn, can limit the number of gates which can be provided on a single chip. Figure 4.34 illustrates this trade-off.

However, both gate delay times and power dissipation in CMOS circuits are heavily dependent upon operating conditions. The delay time of a simple Inverter gate vs. output load capacitance is shown in Figures 4.35(a) and (b). All these values are, of course, dependent upon the physical size of the FETs. The CMOS power dissipation is largely dependent upon the number of gate on/off cycles per second, since each gate has to charge and discharge an output capacitance C, giving a resultant power dissipation of $P = CV^2f$ watts, where V is the output voltage swing ($\simeq V_{DD}$) and f is the frequency of the on/off cycles per second. For a typical internal gate in a 1 μm gate-array or standard-cell IC, this power may be about 2–5 $\mu W\ MHz^{-1}$, while output I/Os with a 50 pF loading may dissipate around 500 μW at 1 MHz. This a.c. power loss is considerably higher than any d.c. leakage power loss through the p-channel and n-channel

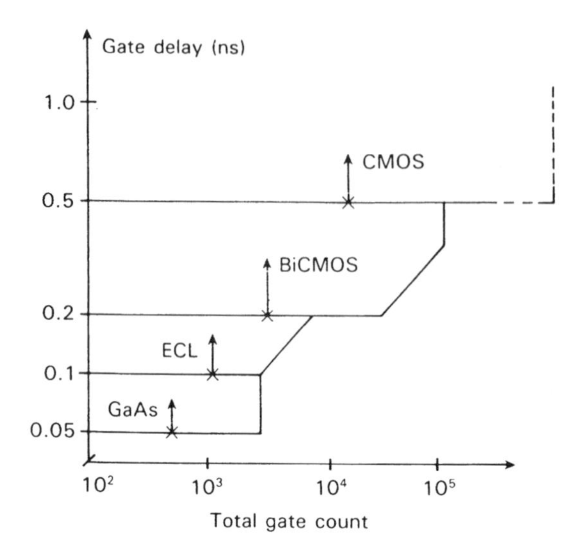

Fig. 4.34 Typical technology limits of minimum gate delay vs. total gate count for custom ICs.

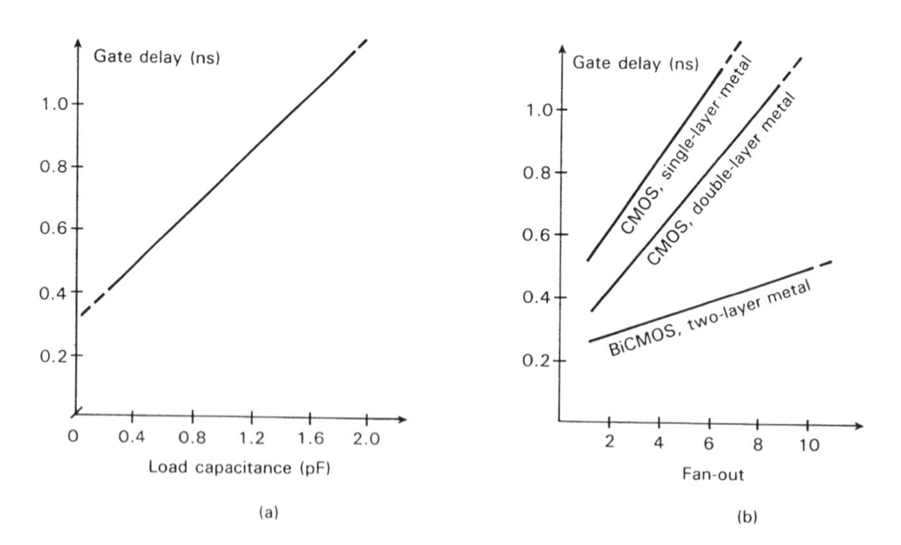

Fig. 4.35 Typical technology characteristics for custom ICs: (a) the effect of output load capacitance on the delay of a 2 μm CMOS gate; (b) the effect of fan-cut on CMOS and BiCMOS gate delay.

transistor-pairs when in the transition periods between on and off, and hence is usually considered to be the power dissipation of the whole circuit. This may typically be 500 mW for a 6000-gate USIC operating at 20 MHz, but depends, of course, upon the number of gates which are switching at any one time.

Full-custom design has been shown to be the most efficient way to tailor a circuit to a specific customer's requirements, but with the disadvantage of cost and time to design. A comparison of full custom with the two principal custom alternatives is given in Table 4.3. It will be noted the cheaper and faster the custom route, the less efficient the custom IC tends to be, but this does not mean an unacceptable performance for the vast majority of OEM requirements.

Perhaps the greatest disadvantage of the full-custom and standard-cell design style is that in the event of a design change after the first prototype stage, a completely new mask set and fabrication run has to be undertaken. For commercial applications where the exact product specification is not absolutely fixed at the initial design stage, this is a serious and potentially expensive problem. One published solution to this difficulty is to include an uncommitted gate-array section on a standard-cell design, the former containing the random logic and other peripheral duties which may require changes or new variants in the light of development [67]. Any such variations after the first prototype has been made can be done by a redesign of the final metallization (interconnection) masks only, not requiring a revised full mask set. This technique, however, has not been actively promoted by normal custom IC vendors.

Table 4.3 General Comparison Between the Three Principal Categories of Custom Circuits[a]

Attribute	Full-custom USIC	Standard-cell USIC	Uncommitted array USIC
Design flexibility	100	50[b]	30[c]
Design and prototype costs	100	25	20
Design and prototype times	100	20	15
Risk factor (probability of prototype circuits not working correctly)	100[d]	50	50
Cost and time for prototype redesign	100	50	50
Die size area	100	150	200
Maximum system speed	100	75	50
Power consumption	100	120	150
Volume production costs	100	200	300

[a] Full custom taken as 100 in each case. Exact comparison depends on many factors, including commercial expertise and competition.
[b] Limited by available standard-cell libraries.
[c] Limited by gate-array specifications.
[d] Highest risk due to human errors and omissions.
Source: based on Ref. 64.

The overall exact place of channel-less arrays is difficult to define. Table 4.4 lists their capabilities in comparison with the other types of uncommitted array architectures, from which their place at the sophisticated end of the market will be noted. Their particular attribute in being able to make memory and similar logic structures efficiently puts them in direct competition with standard-cell architectures, but requiring only the reduced mask set for their dedication. They are, of course, confined (at present) to CMOS technology, and have no forseeable analog capability.

To summarize the analog custom area, uncommitted analog (linear) arrays are now well established, usually bipolar with a floorplan of standard tiles on a rigid grid pattern plus other separate components. Vendors' libraries of interconnections to realize standard analog building blocks are available. Routing may still be done by hand if overall chip complexity is not too great. Analog standard-cell products are also commercially available, but more for mixed analog/digital custom circuits than purely analog [9,68]. CMOS and BiCMOS technologies are receiving more attention than bipolar for these mixed duties, since it is the digital macros rather than the analog macros that tend to dominate the mix. Figure 4.36 attempts to overview the whole range of custom and semicustom solutions which we now have for both the analog and the digital application areas.

In conclusion, we find the following:

- uncommitted gate arrays, both single-layer-metal and multilayer-metal, are widely established
- standard-cell USICs are also very widely established
- CMOS is the dominant technology for digital applications
- ECL and GaAs cater to the very high speed specialist market
- channel-less arrays and mixed analog/digital standard-cell architectures are both newer entries to the custom IC market and are receiving continuing attention
- fast prototyping using some form of direct-write-on-wafer is of great

Table 4.4 Typical Capabilities of Five Types of Uncommitted Array[a]

Type of uncommitted circuit	No. of 2-input gate equivalents per chip	Typical gate delay (ns)	Typical number of usable I/Os
Silicon-gate CMOS (channelled)	150–20,000	1–5	20–300
Channel-less CMOS	3500–230,000[b]	0.6–0.8	60–256
Bipolar CDI	150–10,000	0.5–5.0	20–200
Bipolar ECL	100–15,000	0.1–0.5	20–300
GaAs	2000–4000	0.1	20–120

[a] Exact capabilities depend on vendor expertise and target markets and are continuously evolving.
[b] Total per die; not all usable, e.g., 40% maximum utilization.

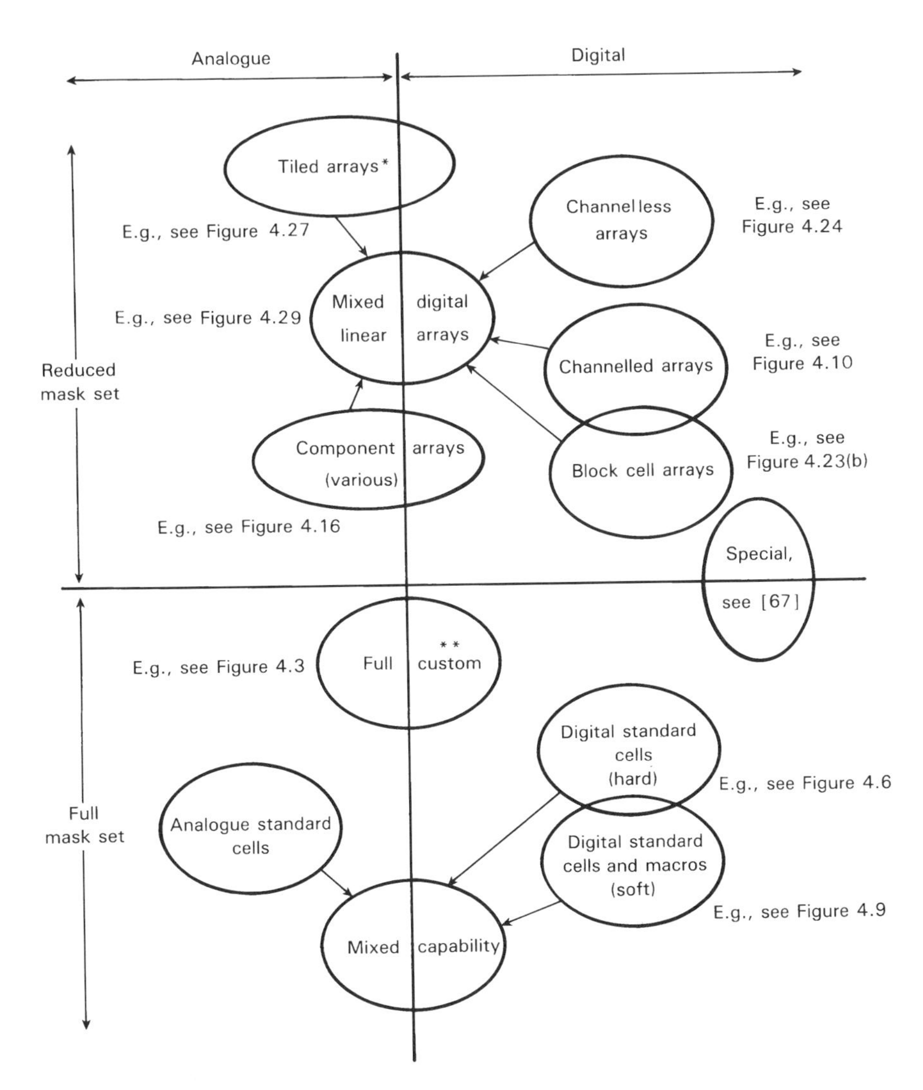

Fig. 4.36 A final summary of the principal types of custom microelectronics, with considerable blurring of the boundaries between types. Architectures which span the reduced mask set and the full mask set are internal to certain companies and not commercially available. Complex programmable logic devices, however, increasingly challenge the whole digital area.

importance where very rapid production of a few prototype ICs is necessary, usually where time rather than cost is the main criterion

All the above custom techniques, however, depend upon good CAD resources for design, simulation and manufacture. This will therefore be the subject of Chapter 5.

REFERENCES

1. Mead, C. and Conway, L., *Introduction to VLSI Systems*, Addison-Wesley, MA, 1980
2. Pucknell, D.A. and Eshraghian, K., *Basic VLSI Design: Systems and Circuits*, Prentice Hall, NJ, 1988
3. Conway, L., Bell, A. and Martin, E.N., A large-scale demonstration of a new way to create systems in silicon, *Lambda*, Vol. 1, No. 2, 1980, pp. 10–19
4. Lyon, R.F., Simplified design rules for VLSI, Lambda, Vol. 2, No. 1, 1981, pp. 54–59
5. Griswold, T.W., Portable design rules for bulk CMOS, Lambda, Vol. 8, September 1987, pp. 62–67
6. Lipman, J., A CMOS implementation of an Introductory VLSI course, Lambda, Vol. 2, September 1987 No. 4, pp. 56–58
7. Weste, N.H.E. and Eshraghian, K., *Principles of CMOS VLSI Design*, Addison-Welsey, MA, 1985
8. Haskard, M.R. and May, I.C., *Analogue VLSI Design: nMOS and CMOS*, Prentice Hall, NJ, 1988
9. Integrated Circuit Corporation, *ASIC Outlook 1990: an Application Specific IC Report and Directory*, Integrated Circuit Corporation, Scottsdale, AZ, 1990
10. B.E.P. Data Services, *Semicustom IC Yearbook*, Elsevier Scientific Publications, Amsterdam, 1988
11. Steinkerchner, S. and Rowson, J.A., A 75000 gate graphics chip designed in 10 weeks, *Electronic Product Design*, Vol. 10, October 1989, pp. 37–47
12. Lin, Y.L.S., Gajeski, D.D. and Toga, H., A flexible-cell approach for module generation, *Proc. IEEE Custom Integrated Circuits Conf.*, 1987, pp. 9–12
13. Rossbach, P.C., Linderman, R.W. and Gallager, D.M., An optimizing XROM silicon compiler, *Proc. IEEE Custom Integrated Circuits Conf.*, pp. 13–16
14. Dawson, R.H., Witkosky, B.J. and Kotcherlakota, S., ASIC design using VITAL, *Proc. IEEE Custom Integrated Circuits Conf.*, 29–32
15. Watkins, D., Rasmussen, R. and Chang, Y., Megafunctions and megacells for ASIC designs: a Comparison, *Proc. IEEE Custom Integrated Circuits Conf.*, pp. 375–378
16. Saucier, G., Read, E. and Trilhe, J. (Eds.), *Fast Prototyping of VLSI*, North Holland, Amsterdam, 1987
17. Lau, P.K., Eden, R.E. and Lee, E.S., GaAs standard cell family designed with current limiting capacitor diode FET logic approach, *Proc. IEEE Custom Integrated Circuits Conf.*, 1988, pp. 101–104

18. Gayakwad, R.A., *Op-Amps and Linear Integrated Circuits*, Prentice Hall, NJ, 1988
19. Rips, E.M., *Discrete and Integrated Circuits*, Prentice Hall, NJ, 1986
20. Barna, A. and Porat, D.I., *Operational Amplifiers*, Wiley, NY, 1989
21. Allen, P.E., Computer aided design of analogue integrated circuits, *J. Semicustom ICs*, Vol. 4, December 1986, pp. 23–32
22. Dedic, I.J., King, M.J., Vogt, A.W. and Mallinson, N., High performance converters on CMOS, *J. Semicustom ICs*, Vol. 7, September 1989, pp. 40–44
23. Pletersek, T., Trontejl, J., Trontejl, L., Jones, I. and Shenton, G., High-performance designs with CMOS analog standard cells, *IEEE J. Solid-State Circuits*, Vol. SC.21, 1986, pp. 215–222
24. Olmstead, J.A. and Vulih, S., Noise problems in mixed analog-digital integrated circuits, *Proc. IEEE Custom Integrated Circuits Conf.*, 1987, pp. 659–662
25. Dugan, T.D., GaAs gate arrays, *High Performance Systems*, February 1989, pp. 64–74
26. Singh, H.P., A comparative study of GaAs logic families using universal shift registers and self-designed gate technology, *Proc. IEEE GaAs IC Symp.*, 1986, pp. 11–14
27. Andrews, W., ECL process drops power and picks up speed and density, *Computer Design*, Vol. 28, July 1989, pp. 37–41
28. Ruyat, S., CMOS power and density in ECL arrays, *Electronic Product Design*, Vol. 8, October 1987, pp. 77–81
29. Texas Instruments, *HCMOS Gate Array Design Manual*, Texas Instruments, Inc., Dallas, TX, 1985
30. LSI Logic, *Data Book and Design Manual*, LSI Logic Corporation, Milpitas, CA, 1986
31. NEC Electronics, *CMOS/CMOS 4A Gate Arrays*, NEC Electronics, Inc., Mountain View, CA, 1987
32. Read, J.W., *Gate Arrays: Design and Applications*, Collins, UK, 1985
33. Hurst, S.L., *Custom-Specific Integrated Circuits: Design and Fabrication*, Marcel Dekker, NY, 1985
34. Edwards, C.R., A special class of universal logic gate and their evaluation under the Walsh transform, *Int. J. Electronics*, Vol. 44, January 1978, pp. 49–59
35. New, A.M., Statistical efficiency of universal logic elements in the realisation of logic functions, *Proc. IEE*, Vol. 129, May 1982, pp. 93–98
36. Algotronix, *Configurable Array Logic Users' Manual*, Algotronix Ltd., Edinburgh, Scotland, 1992
37. Bennet, P.S., Dixon, R.P. and Ormerod, F., High performance BiCMOS gate arrays with embedded configurable static memory, *Proc. IEEE Custom Integrated Circuits Conf.*, 1987, pp. 195–197
38. Nishio, Y. *et al.*, 0.45 ns 7 K Hi-BiCMOS gate array with configurable 3-port 4.6 K SRAM, *Proc. IEEE Custom Integrated Circuits Conf.*, pp. 203–206
39. Landman B.S. and Russo, R.L., On a pin versus block relationship for the partition of logic graphs, *Trans. IEEE*, Vol. C.20, 1971, pp. 1469–1479
40. Heller, W.R., Mikhail, W.F. and Donath, W.E., Prediction of wiring space requirements for LSI, *Proc. IEEE 14th. Design Automation Conf.*, 1977, pp. 32–42
41. King, H., A statistical approach to route estimation, *Electronic Product Design*, Vol. 9, May 1988, pp. 61–65

42. Berry, J. and Grassick, E., Channel-less arrays increase flexibility, *Electronic Product Design*, Vol. 8, October 1987, pp. 43–45
43. Takahashi, H., A 240 K transistor CMOS array with flexible allocation of memory and channels, *ISSCC Digest*, February 1985, pp. 124–125
44. Anderson, F. and Ford, J., A 0.5 micron 150 K channelless gate array, *Proc. IEEE Custom Integrated Circuits Conf.*, 1987, pp. 35–38
45. Beunder, M., Kernhof, J. and Hoefflinger, B., Effective implementation of complex and dynamic CMOS logic in a gate forest environment, *Proc. IEEE Custom Integrated Circuits Conf.*, pp. 44–47
46. Kubosawa, H. *et al.*, Layout approach to high density channelless masterslice, *Proc. IEEE Custom Integrated Circuits Conf.*, pp. 48–51
47. Andrews, W., Small CMOS arrays flourish as channelless types emerge, *Computer Design*, Vol. 28, January 1989, pp. 47–59
48. Bray, D. and Irissou, P., A new gridded bipolar linear semicustom array family with CAD support, *J. Semicustom ICs*, Vol. 3, June 1986, pp. 13–20
49. Sparks, R.G. and Gross, W., Recent developments and trends in bipolar analog arrays, *Proc. IEEE*, Vol. 75, 1987, pp. 807–815
50. Tektronix, *Quickchip 2 Designer's Guide*, Tektronix, Inc., Beaverton, OR, 1990
51. Meza, P.J. and Gross, W.J., Standard tiles: a new design method for high-performance analog ICs, *Proc. IEEE Custom Integrated Circuits Conf.*, 1987, pp. 639–643
52. Crolla, P., A family of high-density tile-based bipolar semicustom arrays for the implementation of analogue integrated circuits, *J. Semicustom ICs*, Vol. 5, December 1987, pp. 23–29
53. AT & T, *Semi-custom Linear Array Brochure*, AT & T Technologies, Allentown, PA
54. Raytheon, *RLA Series Linear Array Design Manual*, Raytheon Corporation, Mountain View, CA
55. Sze, S.M. (Ed.), *VLSI Technology*, McGraw-Hill, NY, 1985
56. Carter, R.C., Hardy, C.J., Jones, P.L. and Lawes, R.A., Customisation of high-performance uncommitted logic arrays using electron beam direct write, *J. Semicustom ICs*, Vol. 5, March 1988, pp. 23–32
57. Leppavouri, S., Laser processing in ASIC fabrication, *J. Semicustom ICs*, Vol. 6, December 1989, pp. 5–11
58. Elsea, A.R. and Draper, O.L., Advances in laser assisted semi-conductor processing, *Semiconductor International*, Vol. 10, April 1987, pp. 443–449
59. Santo, B., X-ray lithography: the best to come, *IEEE Spectrum*, Vol. 26, February 1989, pp. 48–49
60. Cole, B.C., Gate arrays' big problem: they take too long to build, *Electronics*, November 1987, pp. 69–71
61. LASA Industries, *LASARRAY Technical Brochure*, Lasa Industries, San Jose, CA, 1986
62. Petach, P., Lasography allows ASIC gate array prototyping in hours, *J. Semicustom ICs*, Vol. 6, March 1989, pp. 11–15
63. Cole, B.C., LASA's mobile ASIC fab may do the job in minutes, *Electronics*, November 1989, pp. 72–74

64. Or-Bach, Z., Pierce, K. and Nance, S., High density laser programmable gate array family, *Proc. IEEE Custom Integrated Circuits Conf.*, 1987, pp. 526–528
65. The Open University, *Microelectronic Matters*, Microelectronics for Industry Publication Ref. PT 504 MM, The Open University, UK, 1987
66. The Open University, *Microelectronic Decisions*, Microelectronics for Industry Publication Ref. PT 505 MED, The Open University, UK, 1988
67. Hornung, R., Bonneau, M. and Waymel, B., A versatile VLSI design system for combining gate arrays and standard circuits on the same chip, *Proc. IEEE Custom Integrated Circuits Conf.*, 1987, pp. 245–247
68. Design Staff Survey, Survey of analog semicustom ICs, *VLSI System Design*, Vol. 8, May 1987, pp. 89–106

5

Computer-Aided Design

From the preceding chapters it will be evident that computer-aided design (CAD) plays a key part in the field of custom microelectronics, being used in the following three principal activities:

- in the circuit design (synthesis) process
- in simulation before any manufacturing is commenced
- in the silicon fabrication process

The CAD, or more strictly the CAE or CAM, used in the fabrication activities is normally the province of the vendor, and will not concern the OEM in any depth. We shall have no need to consider this area in any detail; instead, we will concentrate here on the design and simulation aspects of the custom circuits.

The necessity for CAD to be available is largely due to the volume of data involved rather than the difficulties of design, plus the need to ensure that the design is as far as possible error-free before incurring any unrecoverable manufacturing expense. Time and money are therefore the fundamental driving factors behind the need for good CAD resources, this being especially necessary for custom design activities where design cost and time to market are critical.

As discussed in Chapter 1, early CAD tools for IC design were only concerned with the geometric layout and subsequent mask-making procedures, and were entirely in the hands of the IC vendors or very large manufacturing companies undertaking full-custom designs. These tools provided complex geometrical manipulation, with design rule checking (DRC) to ensure that line widths and spacings and other dimensional details were all valid. Little or no help was available in optimizing the overall floorplan of the die, and no electrical rule checking

(ERC) or simulation of the final layout (postlayout simulation) was available. Also, there was no standardization of software and data formats between companies, each vendor tending to write its own in-house software programs.

Since then there has been an increasing development of CAD for circuit design and simulation, the ideal being a hierarchy of resources where the data at each level of activity are able to be passed on to the next lower level without any manual interface processing. Equally, any problem experienced at a lower level should be capable of being transmitted back to a higher level for appropriate correction. This hierarchical structure is illustrated in Figure 5.1.

There is some blurring in the distinction of the levels shown in Figure 5.1. In particular, the exact demarcation lines between the following phases are difficult to define:

(i) the behavioral-level design, where the overall behavior of the required system and flow of information is being considered
(ii) the architectural-level design, where the behavior of each large block or partition and its timing requirements are being considered
(iii) the functional-level design, where each architectural block is further partitioned into functional boxes such as counters, adders, multiplexers and so on

Also, because human creative ability and judgement are involved in these areas, particularly towards the top of this hierarchy, CAD support at these levels is difficult to provide—such CAD support should employ formal languages and be knowledge-based to be of true assistance to a designer in this creative design phase.

Another important factor faces the OEM designer at the initial stages of any new microelectronic-based product design. This is the choice of design style in which to complete the required circuits. The factors involved in making this choice are complex, involving both technical and financial considerations, as will be discussed in Chapter 7, but ideally one universal CAD suite should be available whatever the chosen design route. Figure 5.2 illustrates this further aspect facing the OEM. At present most CAD systems are not sufficiently flexible to cater to all the divergencies of Figure 5.2; a fully hierarchical suite may be available for, e.g., a programmable logic device solution, but it is unlikely that this resource would be of much use for the detailed design of a standard-cell IC.

The overall IC design process for any chosen design style with ideal CAD support is shown in Figure 5.3. This may be compared with the design activities given in Figure 4.7 and 4.11. The most well-established areas of this hierarchy (see also Figure 5.1) are the back-end activities where there is little creative work but which mostly involves the mechanical translation of the (ideally proven) design into silicon. The least well-established CAD areas are the architectural and behavioral levels. Hence, up to now many software suites start at about the func-

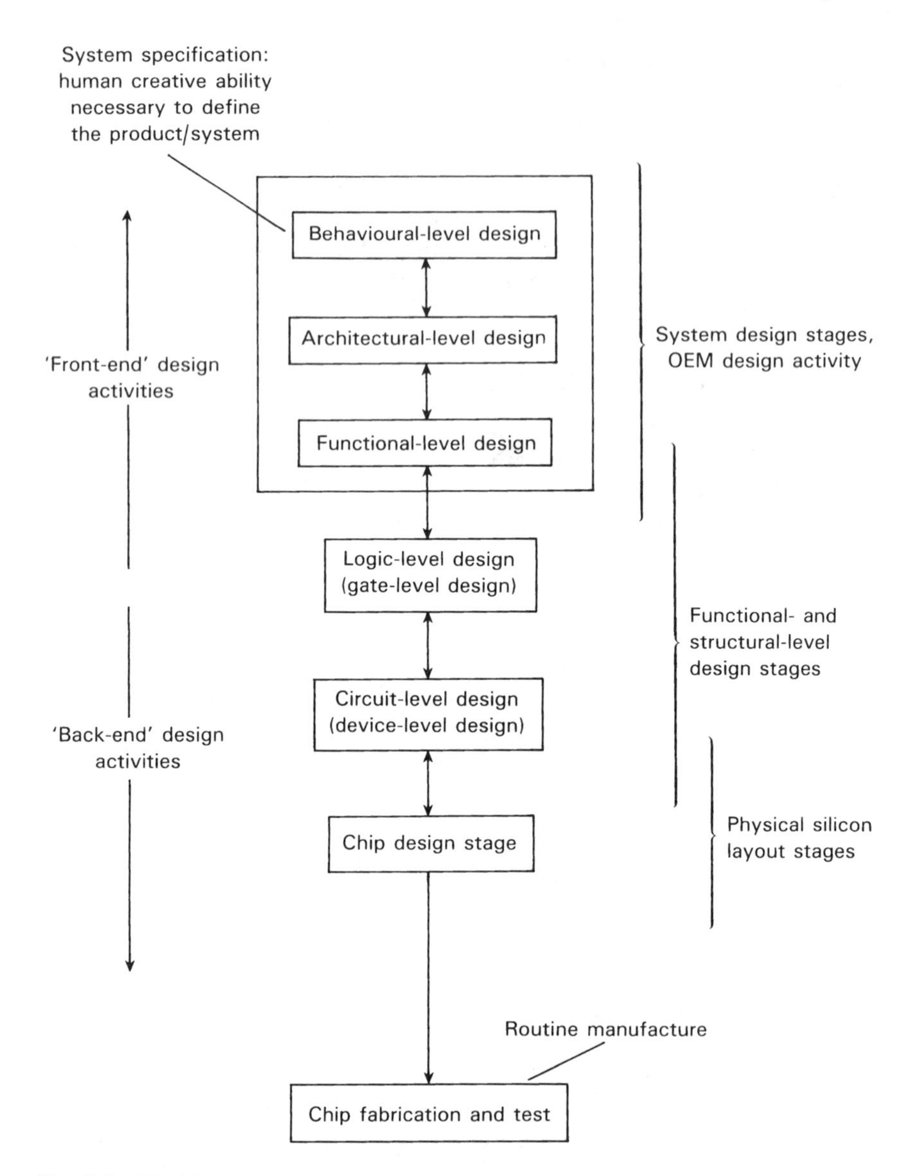

Fig. 5.1 The ideal hierarchy of design stages for which CAD resources should be available with the software programs at each level able to transmit data to and receive data from adjacent levels. Simulation activities must also be available alongside these design activities in order to ensure correct first-time designs as far as possible.

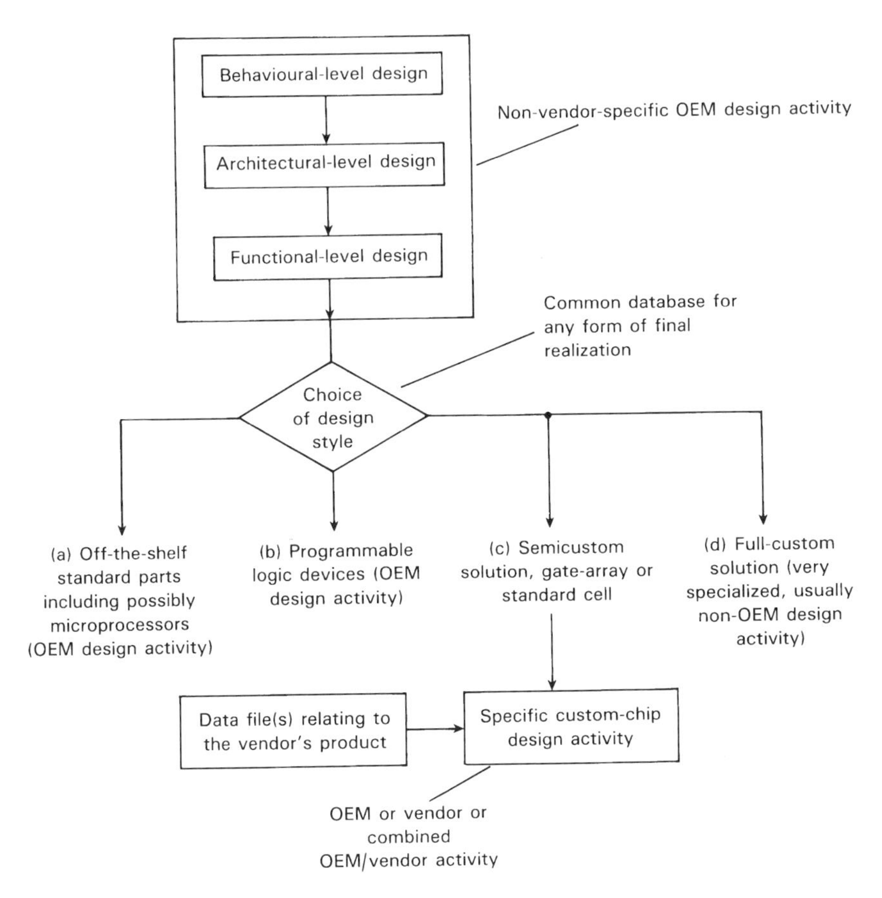

Fig. 5.2 The design style choices facing the OEM, where in an ideal situation the hierarchical CAD resources of Figure 5.1 would allow the choice of design style from some common database at the functional level.

tional level, or even at the gate level in more simple packages, which is adequate for many OEMs who are not concerned with designs involving complex data flows. Further general and detailed information may be found in the literature [1–8].

All the preceding comments refer largely to digital logic design and not to analog. However, in the following sections, which look more closely at the CAD tools, we will bring into the picture both analog and mixed analog/digital requirements as well as digital.

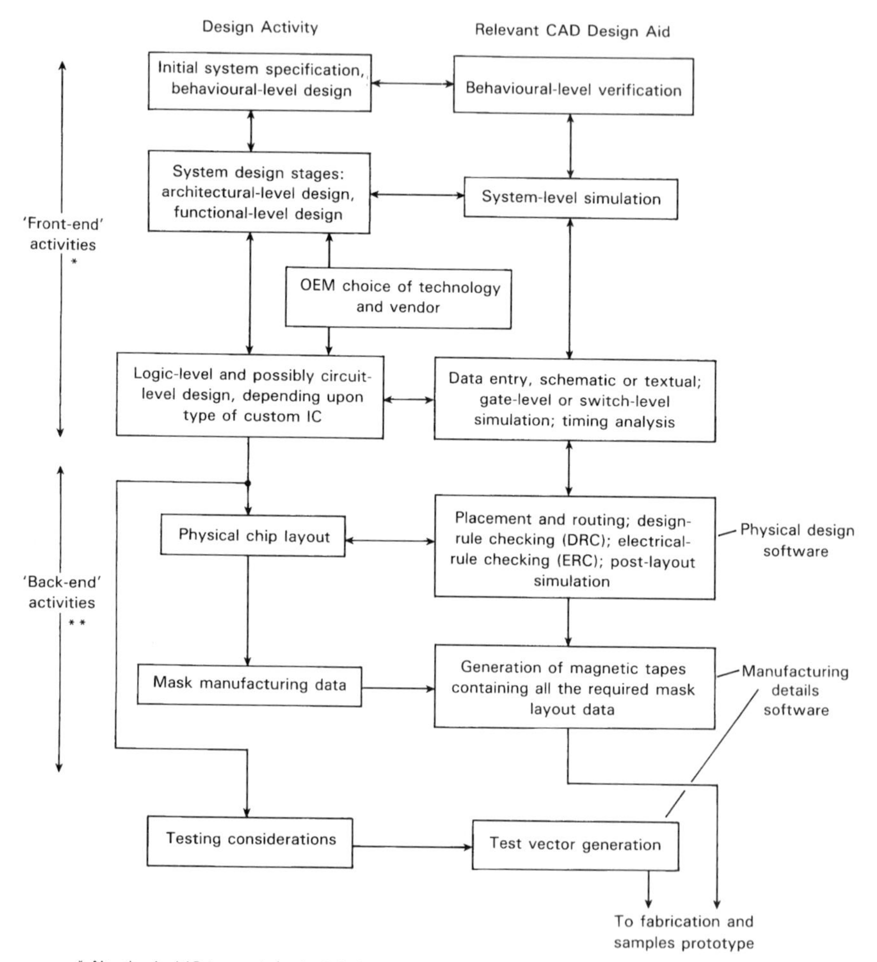

Fig. 5.3 The overall IC design process for digital ICs and the corresponding CAD software needs. For all custom microelectronic design some predesigned library details are available from vendors so that detailed geometric design at the silicon level is not required.

5.1 IC DESIGN SOFTWARE

From the preceding discussions, an ideal CAD system would allow a top-down design procedure, starting at the behavioral and architectural levels and moving down to the final silicon level. At each level of activity, corresponding simulation resources should be available so that design effort does not proceed too far without confirmation of correctness; these essential simulation activities will be considered in Section 5.2.

Although top-down system design is always initially necessary, there are also certain bottom-up design constraints in all forms of custom microelectronics, since the final silicon realization will invariably employ predesigned elements from a vendor's library or other source, rather than every transistor and other components being designed in a new and unique manner. Technology limits of performance, etc., will also be present which will limit the freedom of design at the silicon level.

For digital systems we may consider design software under the following four main headings:

(i) behavioral- and architectural-level synthesis
(ii) functional- or gate-level synthesis
(iii) physical design software
(iv) test requirements software

Category (i) above should largely be independent of the way the circuit is finally made; category (ii) includes the design entry and schematic capture procedures which generate the data for the subsequent placement and routing of the required circuit building blocks in the silicon layout; category (iii) includes the placement and routing of the required building blocks in the silicon floorplan, and (iv) includes the generation of the appropriate data with which the IC may subsequently be tested. Analog design software does not readily follow this hierarchy, and will be considered separately in Section 5.1.5. Software for programmable devices, being very vendor-dependent, will also be treated separately (see Section 5.1.6).

5.1.1 Behavioral and Architectural Level Synthesis

To cater to the growing complexity of VLSI circuits, CAD tools are increasingly sought to handle top-down design starting at the behavioral and architectural levels. Such tools should link with, or encompass, the design duties at the structural decomposition level where the gates or other library parts are introduced. This abstraction above the structural level is illustrated in Figure 5.4.

The purpose of behavioral/architectural level CAD is to allow the required system design to be expressed in such a way as to allow the following:

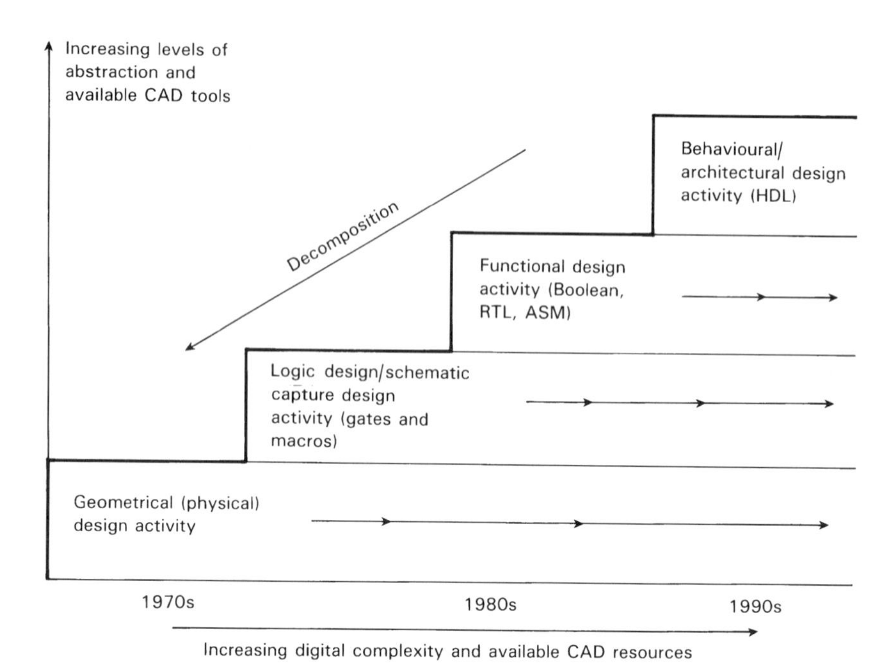

Fig. 5.4 The levels of abstraction in the digital system design process, where the behavioral and architectural level CAD is ideally independent of the final form of realization. RTL = register transfer language, ASM = algorithmic state machine.

- manipulation and optimization of the system design at this abstract level, with ready means to try different ways of composing the required system (the system architecture), for example, serial vs. parallel data processing
- verification of this abstract design in order to ensure correct behavioral response

Verification is therefore an integral part of this level of CAD. However, this level cannot include detailed timing simulation, since this can only be incorporated at a lower level when gate or silicon layout details are known. Ideally, this high-level CAD should be independent of any subsequent choice of implementation and be technology independent, thus allowing a system designer to investigate a number of alternatives before becoming involved in gate-level design and other details.

These high-level tools do not, however, relieve the designer from having to undertake the *initial creative activity* of specifying what the required system has to do in terms of recognizable events and responses. CAD is only an aid to verification of the designer's first ideas, clarifying any key features and interface requirements, and generally optimizing a system structure before more detailed design work begins.

To model and simulate a high-level description of a digital circuit or system requires a behavioral-level computer language. The particular languages used in integrated circuit CAD are termed *hardware description languages* (HDLs). These are often derived from high-level programming languages, and hence HDL models often resemble a program written in a high-level language such as Pascal or C. Hardware description languages should contain many of the procedural constraints found in high-level languages [19,20], such as arithmetic operators, logical operators, if-then-else, while and case statements. These constraints allow the behavior of a digital system to be modeled in an unambiguous English-language-like manner, and allow computer simulation of the behavioral response of the circuit model to be done. Time relationships based upon clock periods (cycles) can be built in. Hence, HDLs can support the optimization and verification of the architectural details of a proposed design, but do not overcome the necessity of a good architectural proposal in the first place or the designer's skill in suggesting alternatives; nor do they guarantee that the design can finally be made within the limits of a chosen technology.

A number of hardware description languages have been developed. The one most widely quoted is VHDL, which was originally developed from the US Very High Speed Integrated Circuit (VHSIC) government-funded research program, and is now an IEEE and US Department of Defense standard (IEEE Standard 1076-1987). All custom designs for the Department of Defense are required to be described in VHDL. However, a difficulty with the original VHDL was that it was developed as a means of documenting a vendor's IC design so that its behavior and structure could be checked, communicated and filed, and was not originally intended as a design tool for use during the actual design phase. For documentation purposes it proved to be satisfactory, but when attempts were made to use it also for design, certain shortcomings were encountered. The result of this has been that several commercial variants of VHDL have now been marketed which provide design as well as documentation capability; prominent among these is the Verilog HDL from Gateway Design Automation. Several others such as HELIX™ from Silvar-Lisco, M-HDL from Silicon Compiler Systems, System VHDL™ from Viewlogic and ASYL-PlusVHDL from Innovative Synthesis Technology have also been marketed. The problem which is currently present is that while VHDL is the *ipso facto* standard HDL language, the commercial variants do not exacly conform, and as a result translators are necessary to convert between commercial HDL design data and the standard VHDL documentation format.

While HDL can be used to define elements as small as individual gates using Boolean-type expressions, for example:

SIGNAL, x, y, cin, cout, sum: BIT

sum <= x XOR y XOR cin

cout <= (x AND y) OR (x AND cin) OR (y AND CIN)

which describe a full-adder with no timing clause (default to 0 ns), the essential power of a hardware description language is to allow both:

(a) the texual description of a system at the abstract level using control flow information
(b) the textual description of a system at the architectural level, using data flow descriptions of the architectural partitions

Once the general functionality of a system has been verified at the behavioral level, then the functionality of the partitioned system can be simulated and compared with the former for agreement. This hierarchical partitioning may be extended to smaller and more detailed partitioning when appropriate. The ability to use (*instantiate*) a circuit building block (a design *entity*) many times in an overall architectural description and to give generic data to allow families of devices to be constructed, plus assertions which must be true at all times, are further features of HDLs, together with necessary commands such as WAIT, GUARD and AFTER. Figure 5.5 shows a short VHDL description.

The general procedure using a hardware description language is given in Figure 5.6. A textual describtion of the behavior of the required system is first entered, from which the HDL software compiles and checks a behavioral description file. Details of the system input signals are also entered to build up the HDL stimulus file. Simulation of the behavioral or architectural system can then be made, with the output being printed out or plotted for checking and verification. If both behavioral-level and architectural-level models are entered, the HDL system may automatically output any functional discrepancies between the two.

Details of VHDL, which like all other description languages is a complex programming language, and its application and commercial variants may be found published [1,7,9,21–29]. A useful worked example of a circuit which implements a parallel sorting algorithm is given in Ref. 26, which emphasizes that estimated timings only can be given at the HDL stage, and that the final implementation may deviate significantly from these high-level estimates. In-depth details of some of the commercial variants are not so widely published in this literature as the VHDL structure.

At some stage in the design process the HDL data have to pass on to the structural level of design, and here there may be a gap between the HDL data and the data format acceptable to the more detailed CAD tools used at these

```
entity sorter is
  port (din:      in vlbit_1d(15 downto 0);    -- data in
        dout:     out vlbit_1d(15 downto 0);   -- data out
        phi1, phi2,                            -- clocks 1 and 2
        reset,                                 -- reset control signal
        read,                                  -- sort dir'n ct1 signal
        double,                                -- double-word ct1 signal
        newword: in vlbit);                    -- word boundary ct1 signal
end;
```

(a)

```
procedure insert  (variable data:   inout   vlbit_2d(0 to
                                            255, 0 to 15);
                   constant size:   in      integer;
                   constant din:    in      vlbit_vector) is
  variable wix:        integer;
begin
  -- find location for new data word, shifting greater words up:
  wix := size - 1;
  while wix /= 0 and data(wix - 1) > din loop
    data(wix) := data(wix - 1);
    wix := wix - 1;
  end loop;
  -- insert new data word in correct location:
  data (wix) := din;
end insert;
```

(b)

```
architecture behavior of sorter is
  constant width:        integer := 16;   -- width of words to sort
  constant length:       integer := 265;  -- number of words to sort
begin
--
--concurrent assertion statement:
--flag attempt to use unimplemented feature after initialization:
  check: assert not (double = '1' and now > 0ns)
                  report "Double-word sorting is not implemented.";
--
--main sorting process:
--insert each new word in order in data bank and emit sorted words:
  main: process
    variable data:        vlbit_2d(0 to length - 1, 0 to width - 1);
                          --data words in process of sort
    variable inwix,
      outwix:integer := 0;-- word indices for input, output
  begin
wait until newword = '1';-- wait for new word boundary
    for clockperiod in 1 to width - 1 loop
      wait until phi1 = '1'; wait until phi2 = '1';
    end loop;                       -- wait for last clock period
    wait until phi1 = '1';          -- wait for start of clock period
    dout <= data (outwix);          -- emit next data word
    wait until phi2 = '1';          -- wait for second phase
    if reset = '1' then
      init: for wix in 0 to length - 1 loop
        data (wix) := X"FFFF";      -- initialize data bank
      end loop;
      inwix := 0;                   -- initialize data indices
      outwix := 0;
    else
      if read = '1' then            -- increment input index and
        inwix := (inwix + 1) rem length;
        insert (data, inwix, din);  -- insert next data word in order
      else                          -- increment output index
        outwix := (outwix + 1) rem length;
      end if;
    end if;
  end process;                      -- return to top of process
end behavior;
```

(c)

Fig. 5.5 VHDL description of a sorting circuit which accepts a stream of up to 256 16-bit positive integer numbers and shifts them out in ascending order of magnitude: (a) the entity declaration which defines the input and output signals of the macro; (b) the behavioral model which defines the required action of the macro; (c) the architectural model which describes one possible design of the macro. Note: An additional procedure may follow to define a sequence of input test signals (logic 0 and logic 1) and determine the output response in order to verify (c). See also Figure 5.21. (*Source*: Viewlogic Systems, MA.)

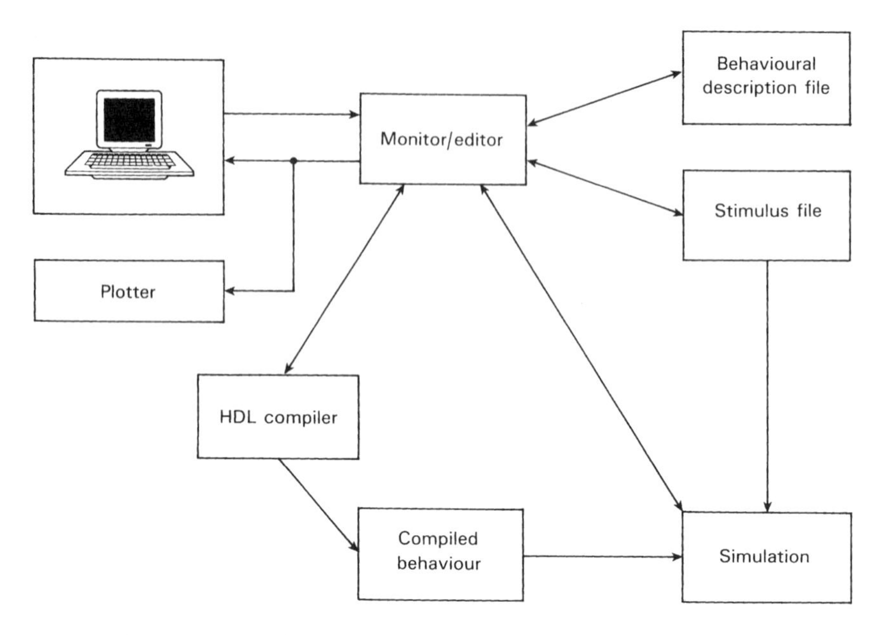

Fig. 5.6 The high-level abstract design and verification structure of HDL software.

later design stages. Figure 5.7 indicates the hierarchical links which are required. Vendors' tools at this lower level of design therefore need to accept the HDL input data.

One problem is that an abstract HDL behavioral-level description implies no particular hardware realization, and hence there has to be an implied or accepted architectural partitioning for the detailed design synthesis to be able to begin. Nevertheless, many CAD/CAE vendors now provide interface links to HDL, for example, the Design Compiler from Synopsis which will accept Verilog HDL and Viewdesign™ from Viewlogic which accepts VHDL. Vendors of programmable logic devices also now provide an interface route. These commercial CAD links may, however, impose some restrictions on the HDL design level, but continuing evolution is very evident in this area.

Although VHDL is an approved US standard, the present concepts of hardware description languages for design purposes are still under attack from certain sources. Some of the published critisisms are as follows:

- While it may be appropriate for system documentation, a computer language is not a suitable medium for system design.
- It is difficult, if not impossible, to describe fully a microprocessor or other very complex flexible macro in HDL.

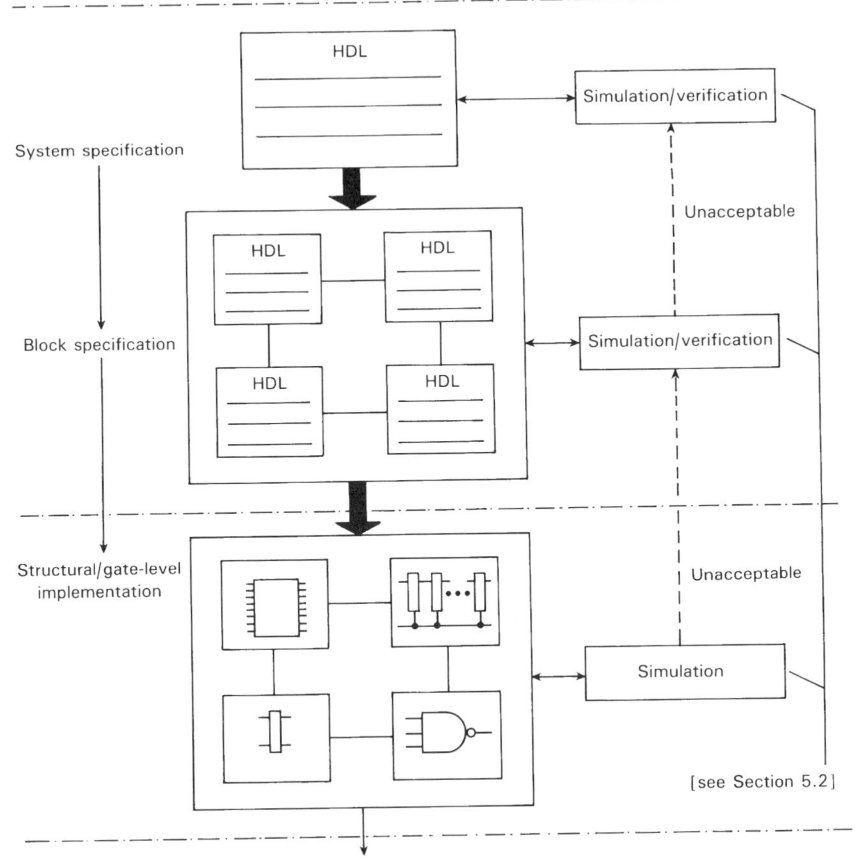

Figure 5.7 The hierarchical levels between HDL synthesis and verification and the lower design levels.

- Diagrams have a superior readability over text for design purposes, and hence some computer means of handling high-level schematic input data rather than complex textual encoding would be preferable.
- HDLs provide a difficult design environment for system designers who are engineers rather than computer scientists, with little overall 'feel' of the design evolution from the textual data.
- High-level synthesis and high-level optimization should be considered as two design activities that are best served by two separate CAD tools specifically designed for the purpose, rather than one tool.

- Since there is still little standardization of data formats that are used in the various levels of IC design, it is premature to try to standardize on a single HDL language for design purposes.

Many of these critisisms have now been addressed, in particular to provide more schematic data to assist the design engineer and also standardization and interchange of data formats. For example, the series of high-level synthesis and simulation tools from Viewlogic Systems, Inc. provides seamless text and graphics software packages, with multiwindow displays such as those illustrated in Figure 5.8. Hence, behavioral and architectural level synthesis using commercial HDLs is now well established, possibly more so for large systems which consist of data flows and fixed logical requirements rather than for more flexible systems which have to cater to a range of activities controlled by RAM, ROM and PLA structures. Very small system designs would not justify the expense of the HDL software unless the design office required it for more complex design activities.

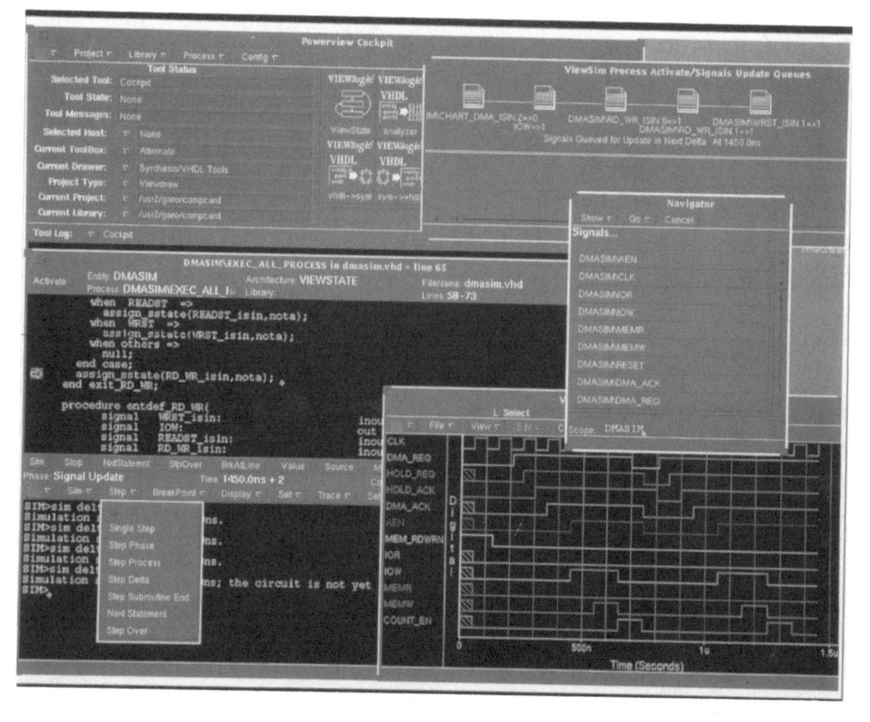

Fig. 5.8 A multiwindow display available with the Viewlogic CAD tools, in this case spanning VHDL textual design descriptions and editing through to simulation and waveform presentation. (Courtesy of Viewlogic Systems Inc., MA.)

5.1.2 Functional and Gate Level Synthesis

In contrast with the above behavioral and architectural levels, the specific structure of the final circuit realization has to be introduced at the functional and gate levels. This involves the types of gates and other macros which are physically available for the final silicon design, rather than the possibly abstract building blocks at the behavioral level.

The most common synthesis procedure at this level for VLSI circuits is to take each block of the architectural level and manually partition it until it is finally described at the available gate or macro level. There is therefore no specific CAD synthesis tool involved. Simulation is, however, essential, and hence the major CAD resources at this design level are for (i) schematic assembly of the circuit gates, etc. (*schematic capture*), and (ii) simulation, rather than synthesis. Most CAD/CAE vendors' software covers this area, since it is the first level at which their library of available cells is formally introduced into the top-down design procedure.

However, it is still possible to describe the required system at this level using a somewhat lower level CAD model than HDL. This is the *register transfer language* (RTL) model, which is a data flow type of model that can describe hardware blocks in more detail than the abstract HDL functional modeling. It may or may not be vendor independent. It is the highest level of modeling which can currently be simulated in a hardware accelerator (see Section 5.2.6). There is clearly an area of overlap between the use of HDLs and RTLs for digital system modeling, with protagonists for each saying that only one should be further developed to cover all hierarchical levels from gate level upwards, but in general RTL is largely hardware specific, whereas HDL is more general and abstract, not being tied to specific vendors' library parts.

The RTL model representations can be used for synthesis purposes to try out design variations or simplifications, but it remains a modeling rather than a true synthesis tool. Details of RTLs may be found in the literature [1,3,22, 30].

There are, however, some true synthesis tools available for functional and gate level design. These tools are possibly most prominent in the proprietary CAD resources made available for programmable logic devices (see Section 5.1.6), since there is often considerable emphasis on optimization and minimization of LSI-size designs in order that they may fit into the smallest, and hence cheapest, PLD. These tools may be considered in two parts, namely:

(i) those which deal with the states of the system and attempt to minimize the state assignment design
(ii) those which deal with the associated combinatorial logic, particularly the random, or 'glue,' logic, and attempt to minimize this digital component.

Logic design is a subject outside the general coverage of this text, but it currently involves binary decision diagrams (BDMs) and algorithmic state machine (ASM) synthesis as well as the more familiar areas of truthtables, Boolean equations, state assignments, minimization and other standard digital logic concepts. Details of all these areas may be found in many standard texts and elsewhere [2,31–36]. However, logic minimization, which is classically taught using Karnaugh maps and the Quine-McCluskey minimization algorithm, is currently best performed for large functions by a series of more efficient CAD programs, particularly ESPRESSO [37,38]. Others such as TAU [39] provide similar fast minimization procedures without an intolerable escalation of the data during minimization, as can occur with the Quine-McCluskey algorithm. These design aids usually form part of a vendor's CAD package, rather than being purchased separately.

Outside the PLD area, the creative synthesis of very complex digital logic networks at the functional or gate level is not well served by available CAD tools, and is still largely dependent upon human ability to do the innovative work. The CAD resources do provide help in optimization and verification once the basic design ideas have been formulated. Further information may be found in published literature [5,8,20,40], but increasingly, the design of the more complex circuits is begun at the higher architectural and behavioral levels considered in the previous section.

CAD tools do, however, come into their own at this functional and gate level stage of design in the following two noninnovative areas, namely:

(i) to cater for 'soft' cells and macros in a vendor's library, where the exact cell specification has to be matched to the circuit designer's requirements
(ii) for schematic capture purposes, where the detailed circuit schematic diagram is drawn by the designer using the CAD graphics capability

The concept of soft cells and macros was introduced in Section 4.2.2 of the preceeding chapter; it caters to the need to be able to make available different capacity circuits such as ROMs, RAMs and PLAs, or to optimize cell performance to meet a given timing specification. The CAD tools which do this are without exception IC vendors' tools, since they are intimately concerned with specific library components, and as a result the software details are usually proprietary. The use of soft macros does, however, slightly blur the interface between this functional and gate level design activity and the following lower-level physical design activity, since in the subsequent placement and routing it may be advantageous to revise the order of inputs or outputs, or otherwise manipulate the actual physical instance of such cells, rather than taking them as fixed entities from the synthesis level. Functionally, however, the soft cells and macros will not change at the placement and routing stage.

On the schematic capture side, where the CAD resources provide a non-creative role to capture the design as the detailed synthesis builds up, CAD tools from both IC and CAD/CAE vendors are readily available which enables the designer to:

- build up the circuit schematic on the CAD screen from a library of symbols
- interconnect the resulting schematic circuit diagram
- replicate completed subsystems such as shift registers, counters, etc.
- change or correct ('undo') the schematic as necessary

In addition, the schematic capture CAD will:

- ensure that no invalid connections, such as a gate output connected to a power supply, are made
- ensure that no input nodes are left unconnected
- ensure that all connections are terminated at both ends
- check that the fan-out from each output node is within the permitted maximum
- compile an inventory of all the cells and macros used in the final schematic
- compile a list of all the connections and the cells to which they are connected

Schematic capture software thus provides (i) electrical rule checking (ERC), and (ii) data in the form of the netlist of interconnections for final prelayout simulation (see Section 5.2.3), and for the following physical design activities.

5.1.3 Physical Design Software

Once schematic capture has been done and simulation completed to verify the design, all original creative work is complete. If subsequent postlayout simulation reveals any timing or other problems, then of course some iteration of the design activity may become necessary. There is therefore always a loop back from a lower to a higher hierarchical level to cater to unforeseen difficulties.

The remaining design procedures at the silicon level are largely complex mechanical procedures and can therefore be handled easily by computer, although human intervention and ingenuity may still be necessary to provide guidance or solve difficulties which the CAD program cannot resolve. As introduced in Chapter 4, the principal design activities at this level are:

(i) *placement*, where the required functions are allocated positions on the chip floorplan
(ii) *routing*, where the detailed geometric pattern of all the required interconnections between functions and I/Os is completed

This is indicated in Figure 5.9. With uncommitted arrays, the floorplan is fixed and the routing is constrained within the available wiring channel area, the principal flexibility here being in the size of the chosen array, which as a general rule should have a total gate count at least 50% higher than the total gate count of the schematic design. (Such gate counts are usually in 2-input NAND gate equivalents.) For cell-library and full-custom designs this constraint is not present, and the usual objective now is to make the smallest final floorplan area (die size) for the required circuit.

Placement

The principal aim of placement is to facilitate and optimize the subsequent geometric routing. Unfortunately, the results of placement cannot be fully tested until the detailed routing is undertaken, and hence most procedures involve a provisional placement design followed by some re-placement if unacceptable or impossible-to-route situations are subsequently encountered.

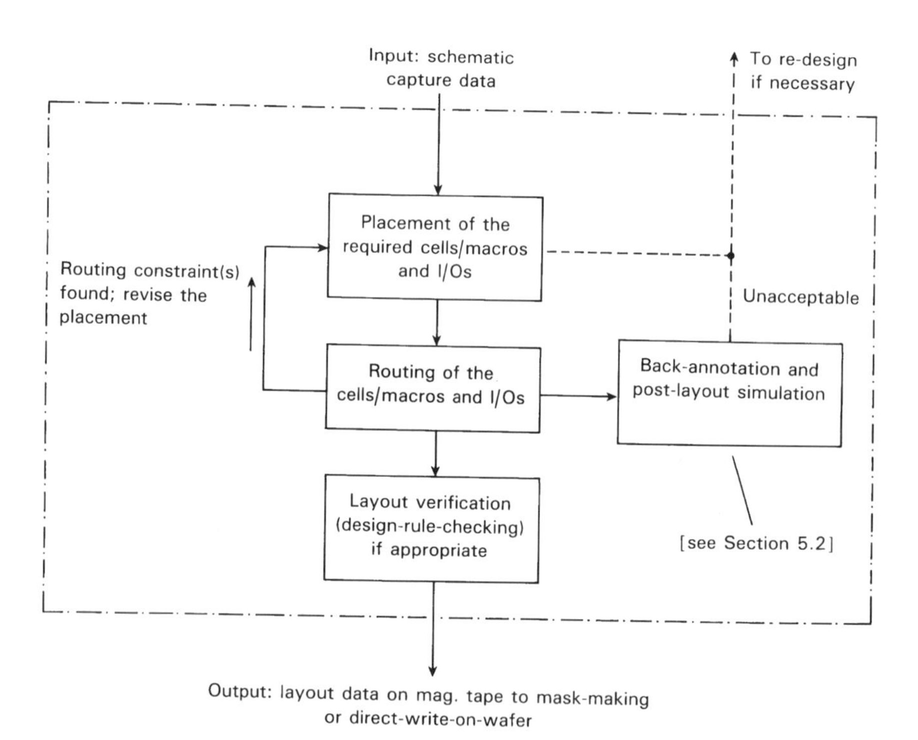

Fig. 5.9 The physical (geometrical) design level, where the circuit layout and interconnection routing are determined, the placement and routing activities being interactive if necessary.

The computation involved in investigating every possible floorplan placement is completely impractical, and hence most placement procedures involve some global placement procedure followed by local iterative improvements. (For m cells there are $m!$ possible adjacency placements, ignoring rotation and mirror-imaging of each individual cell, but as half of these will be global mirror-image placements of the others, there are effectively $\frac{1}{2}\,m!$ possible different floorplan layouts without cell rotation and cell mirror-imaging.) The suggested placements are based upon a global consideration of the wiring requirements rather than individual paths, with a target of minimizing the total interconnect length, or some other chosen wiring criteria. Very critical paths which may have to be kept as short as possible to minimize propagation delays may be given priority in this initial placement. In the case of gate arrays a placement is acceptable once the routing can be accommodated within the fixed wiring channels, provided no unacceptable propagation delays are present.

Placement procedures can be either heuristic, based upon the designer's knowledge of the circuit schematic, or algorithmic. The following three classes of placement algorithms may be found:

(i) constructive initial placement, in which an analysis of the degree of connectivity between cells is first made from which a first placement solution is generated
(ii) branch-and-bound methods, whereby cells are initially grouped into large blocks, each of which is then repeatedly subdivided so as to form a decision tree of associated cell groupings
(iii) an interactive re-placement procedure, in which cells are interchanged, possibly randomly or under some connectivity indicator, the result of each action being observed

In practice, there may be combinations of these classes within a particular place-and-route CAD program.

In many placement procedures a measure of the 'goodness' of the initial placement is the minimization of the total wiring length. However, with channelled architectures there is what is termed a *Manhatten interconnection topology*, with the Manhatten distance M between two points x_1, y_1 and x_2, y_2 being defined by

$$M = \{|x_1 - x_2| + |y_1 - y_2|\}$$

Alternative definitions of length are:

- the summation of distances along the wiring channels only, the 'trunk length,' ignoring individual cell connections across the wiring channels and all perimeter connections
- one half of the perimeter of the smallest rectangle which encompasses all terminal points of a complete net

Note that the latter definition is not necessarily the same as considering the summations of the length of the individual connections which go to make up a complete net, but it is equivalent to the lowest bound tree length in orthogonal routing between all points of the net, provided that the net routing is not blocked by intervening obstacles. In all cases a weighting of individual interconnection distances may be made in order to emphasize critical connections at the expense of less important connections.

Hence, if a chip consists of T total connections, then using, e.g., the Manhatten length, the total on-chip wiring length L_M is given by

$$L_M = \sum_{i=1}^{T} M_i$$

and if each connection is given a weighting factor w_i, then the total Manhatten length is

$$L_{MW} = \sum_{i=1}^{T} w_i M_i$$

Minimization of this parameter may be considered as a goal for optimum placement.

Other suggested 'goodness' parameters include minimization of the total number of interconnect crossovers, or minimization of the total number of 90° turns in the completed routing, assuming north-south/east-west routing directions only in the latter.

The placement algorithm often used for constant-height cells is *linear placement*. In linear placement it is assumed that all cells are first arranged in a single horizontal row. Permutation of cell positions is then undertaken in order to minimize the total Manhatten distance or other distance summation. However, since it is not feasible to consider all possible permutations of cell positions in this minimum summation search, some 'clustering' algorithm is desirable in order to provide a first global placement, following which local perturbation of cells or clusters of cells may be invoked to fine-tune the final placement.

Clustering aims to bring together strongly associated cells at the expense of dispersing those which have little commonality. The connectivity V_{PQ} between two cells P and Q may be defined as

$$V_{PQ} = \left\{ f(P)\frac{C_{PQ}}{D_P} + f(Q)\frac{C_{PQ}}{D_Q} \right\}$$

where

$f(P), f(Q)$ are empirical factors related to the size of the cells P, Q, respectively

C_{PQ} = the number of connections between P and Q
D_P = the number of connections from P not routed to Q
D_Q = the number of connections from Q not routed to P

For example, in Figure 5.10, taking $f(P)$, $f(Q)$, $f(R)$ and $f(S)$ all as unity, we have:

$$V_{PQ} = \left\{\frac{4}{1} + \frac{4}{3}\right\} = 5.33$$

$$V_{QR} = \left\{\frac{1}{6} + \frac{1}{5}\right\} = 0.37$$

$$V_{RS} = \left\{\frac{3}{6} + \frac{3}{0}\right\} = \infty$$

$$V_{PR} = V_{PS} = V_{QS} = 0$$

From this it will be noted that loosely connected cells have a low connectivity value, being zero if no interconnections exist, while any cell which is exclusively connected to another has infinite connectivity value.

In this example it is obvious that cell S should be adjacent to cell R. Equally, there is no requirement for cells P and R, cells P and S, and cells Q and S to be specifically paired with each other. The clustering algorithm therefore proceeds as follows.

(i) compute the connectivity value of all pairs of cells
(ii) combine the two cells with the highest connectivity factor, and call this a new cell
(iii) recalculate all new cell connectivity factors
(iv) repeat (ii) and (iii) above until all cells have been clustered

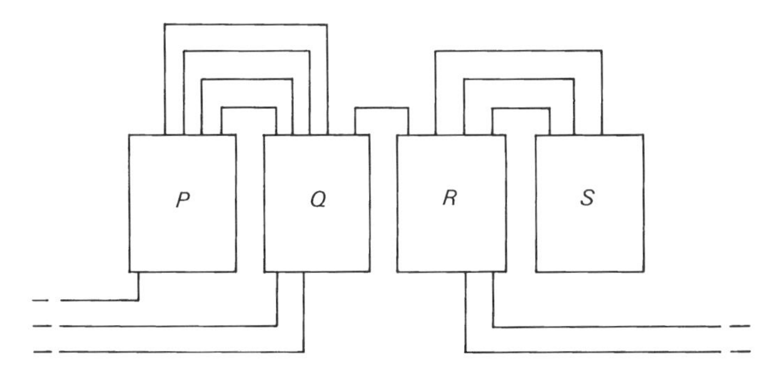

Fig. 5.10 The clustering of cells and connectivity parameters.

This is the outline procedure. Problems arise, however, as 'cells' become large in this clustering procedure, since they tend to dominate the connectivity values. This is where the introduction of the $f(P)$, $f(Q)$, etc., empirical factors becomes necessary in order to counteract the increasingly dominating effects of the large clusters. Other difficulties may arise due to connections running between three or more cells, which may require further empirical adjustments. However, a reasonably sensible clustering of the original cells results, but whether it differs markedly from an initial placement which represents a logical layout of the given circuit schematic is debatable.

Having produced some initial cell clustering, perturbation of the cell clusters can be undertaken if desired in order to improve the total placement. Finally, the single row of placed cells has to be folded into rows to produce a floorplan of the desired shape. It will be noticed that this linear routing technique does not take account of the fact that cells can be adjacent to each other in one row and an adjacent row once the string of cells is folded, and this may well be advantageous in the routing if top-and-bottom access to the cells is available.

More general placement algorithms which are not necessarily restricted to constant-height cells and channelled architectures include min-cut placement, stochastic placement, and simulated annealing placement. These alternative algorithms are appropriate for standard-cell and channel-less sea-of-gates uncommitted-array realizations.

In the min-cut placement algorithm all the cells and macros of the circuit are initially considered as one block, as shown in Figure 5.11(a). A cut $C1$ is applied to divide the assembly into two blocks B_1 and B_2, the interconnect lines crossing this cut being termed 'signal-cuts.' The min-cut placement algorithm then seeks to minimize the number of signal-cuts by swapping cells between B_1 and B_2, which is equivalent to trying some other cut C_1 to divide the two parts. This procedure is then repeated by the introduction of a second cut (see Figure

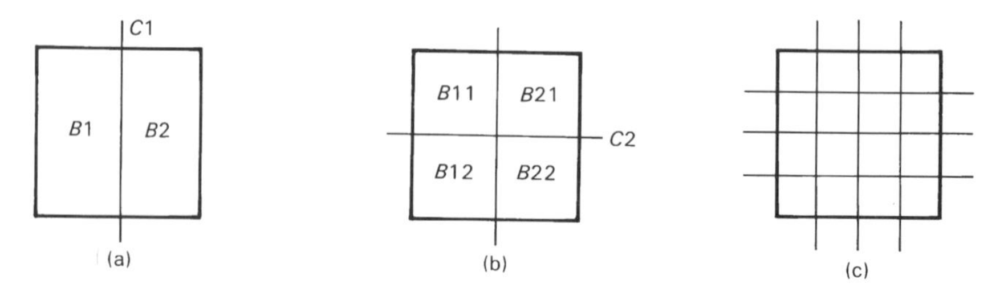

Fig. 5.11 The min-cut placement algorithm: (a) the first cut $C1$; (b) the second cut $C2$; (c) subsequent cuts. Note that the partitions after each cut are not necessarily the same physical size on the layout.

5.11(b)) with the minimization of signal-cuts across C_2 then being sought. Further cuts and repeats of the minimum cut search continue until a satisfactory solution has been reached, which may be at the ultimate single cell per block level or may be terminated at some prior stage.

This min-cut procedure is an application of a previous partitioning algorithm known as the Kernighan-Lin partitioning algorithm [41]. However, applied to the placement problem there is the practical situation that the size of cells being swapped across the min-cut boundaries may not be the same, and hence a final chip layout to implement placement based only on the minimimization of signal-cuts may not be satisfactory.

Stochastic placement requires the construction of a large number of solutions for comparison using random varables to select different block placements, the usual measure of 'goodness' being the total interconnect length. After a range of solutions has been generated, the best positions for the major blocks are accepted, and local iterations are applied to fine-tune the final results. This method uses more computer power than the min-cut algorithm, but does not converge on a solution which may not be globally optimum which the min-cut algorithm can.

In the simulated annealing program, the procedure is modeled upon the physical process which takes place during the cooling of a molten substance such as glass. Here the stresses between molecules are minimized by the random movement of molecules during the cooling down process, such that each molecule tends to find a position at which there are minimum forces acting on it, the amount of movement which each molecule is allowed to make being dependent upon the molton material temperature. Thus, in the placement algorithm, a variable termed the 'temperature' is set to some initial value which defines the amount of movement that each block can have in searching for an improved position where there is minimum 'force' acting on it from other blocks, this force being based on the number and length of the block interconnections. A Monte Carlo technique is used to generate random block displacements from an initial layout, and the 'temperature' of the process is gradually reduced to zero to give a final placement solution.

This procedure requires large computer processing power, but can produce good results, particularly as it does not suffer from becoming locked into locally optimum solutions at the expense of a global solution.

Other algorithmic procedures which may be found, particularly in the iterative improvement phase of placement, include the following:

(i) iterative pairwise interchange, in which pairs of modules are interchanged; if a reduced total interconnect length results, then the interchange is accepted; if not, then the original placement is retained

(ii) relaxation methods, in which each connection is thought of as a stretched elastic string containing a force (f) proportional to the length (l) of the connection; relaxation techniques to minimize the total force $\sum f$ between all cells by re-placement are then applied
(iii) force-directed pairwise exchange, which is a combination of (i) and (ii) above
(iv) the use of 'seed cells,' that is, an arbitrary cell is chosen and all cells closely connected to this seed cell are then drawn into proximity; further seed cells are chosen until all cells are finally grouped

However, in spite of the development of such CAD assistance, placement still requires human involvement, possibly relying upon an initial heuristic placement based upon knowledge of the circuit architecture and functional partitioning, followed by interactive intervention to accept the CAD solutions or suggest alternative placements. The subsequent routing and postlayout simulation may also reflect back on this placement, and require revisions to the floorplan if problems are encountered later.

For further theoretical information on placement, see reports by Rubin [10], Goto and Matsuda [14], Hanan and Kurtzberg [42], and Schwartz [43]. Further information on particular algorithms and practice may be found in the literature [1,14,44–54].

Routing

Routing is the final phase of the physical design process before any manufacturing costs are involved. It features prominently in all vendors' suites of CAD software, and is still under continuing improvement. (Both placement and routing go back to pre-LSI days, having origins in the design of printed-circuit boards containing discrete components or SSI/MSI integrated circuits. Indeed, placement and routing of PCBs may now be extremely complex due to the use of multilayer boards which may have up to twelve or more layers of interconnect in some cases.)

Between an initial placement procedure and the fully detailed geometric routing there may be a global routing (or 'loose' routing) procedure, the objective of which is to produce a routing plan in which each net is defined in one or more segments, allocated to an appropriate interconnect level, and given a category of priority for the subsequent routing [54,55]. In the case of simple single-layer-metal gate arrays, this loose routing may not be necessary; on the other hand, with complex circuits it is sometimes regarded as an integral part of the placement procedure since it strongly influences the placement efficiency, particularly as the correlation between the efficiency of placement based upon total interconnect only and the subsequent ease of routing is debatable [56]. Global routing may therefore be inseparable from the placement procedure, and be strongly linked to the type of product which has to be routed.

A common problem in global routing is to find the 'best' way of connecting a net which links more than two nodes, each segment of which is subsequently separately routed by the detailed routing procedure (see later). A net with n nodes may be linked by $n - 1$ segments, but there are n^{n-2} ways in which this may be done, neglecting the possibility of intermediate connecting nodes. The problem of selecting the optimum segments so that the total interconnect length is minimum is the 'travelling salesman' problem, but if intermediate nodes are also allowed then the problem escalates. The latter is called the Steiner tree problem, and like the travelling salesman problem is NP-complete [54,57,58]. (NP-complete–nondeterministic polynomial-problems are those which cannot be solved by any known algorithm; if a correct solution does exist it can, however, be verified [58].)

Consider the four nodes shown in Figure 5.12. The travelling salesman solution is given in Figure 5.12(a), with a minimum spanning tree given in Figure 5.12(b) [59,60]. However, allowing intermediate nodes, the alternative Steiner spanning trees shown in Figures 5.12(c) and (d) become possible [42,54].

The multiple-node routing problem can, however, be tackled during the

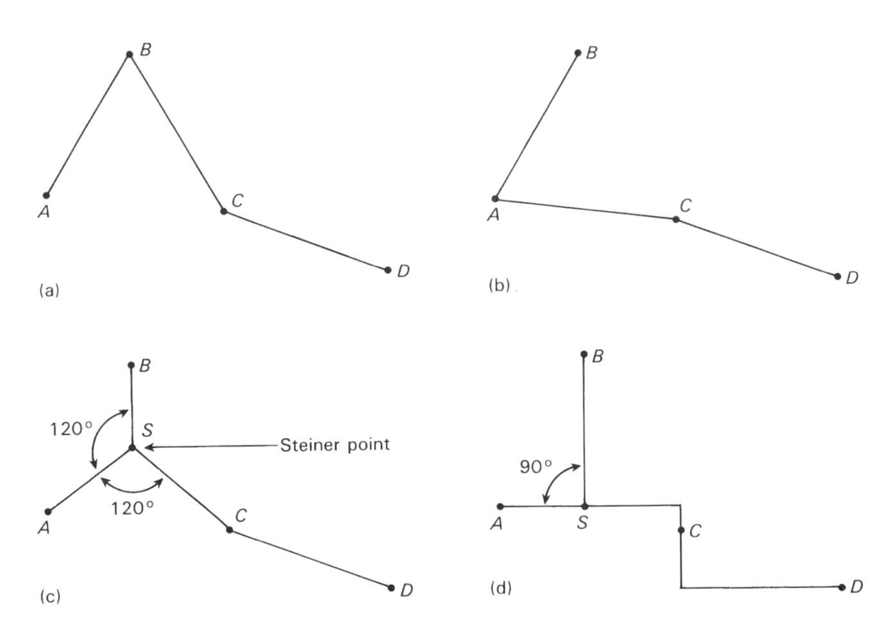

Fig. 5.12 Minimum spanning trees for a 4-node net *ABCD*: (a) the travelling salesman (minimum distance with no backtracking) solution; (b) minimum spanning tree; (c) minimum Steiner tree in nonrectilinear geometry; (d) minimum Steiner tree in rectilinear geometry.

detailed routing procedure by giving the routing algorithm a *source node* and then defining all the other nodes as potential *target nodes*. The first target node reached is accepted to give a completed segment. A second source node is then chosen with the remaining nodes as targets. Routing is completed when all segments have been determined.

The further tasks of global routing, namely allocation of interconnect level and priorities for the detailed routing, are highly dependent upon the particular product being routed. Also, the given routing priorities may have a marked effect on the difficulties of the subsequent routing, since high-priority routes such as clock lines and busses may hamper the routing of remaining nets [42,55,61].

The final fully detailed geometric routing is the last phase of the physical design process. Like placement, it features prominently in all vendors' suites of CAD software. The detailed geometric routing algorithms may be divided into two main classifications, namely:

(i) general routers which deal with the routing between any two points A and B, not confined to specific wiring channels
(ii) channel routers, which as their name suggests are relevant for channelled arrays and similar floorplans

General routers are represented by the following two principal types:

- the wavefront or maze-running type of router, sometimes also known as a 'grid-expansion' router
- the line-search or line-expansion type of router

Channel routers may be divided into three principal types, namely:

- the left-edge router
- the dog-leg router
- the 'greedy' router

(see later), although many commercial routers may contain combined features, plus additional heuristics introduced to cater to specific layout requirements or problem areas.

The most widely known general router is Lee's wavefront router, first introduced in the 1960s [62]. Here, a routed path follows a rigid grid pattern from a source node to a target node, moving from one square in the grid to an immediately horizontal or vertical adjacent one. For example, consider the situation illustrated in Figure 5.13(a), where the white squares indicate space available for interconnection routing and the black squares indicate obstructions due to prior routing or other obstructions. If we require to route from a source square position S to a target square position T, we first number the available white squares as follows:

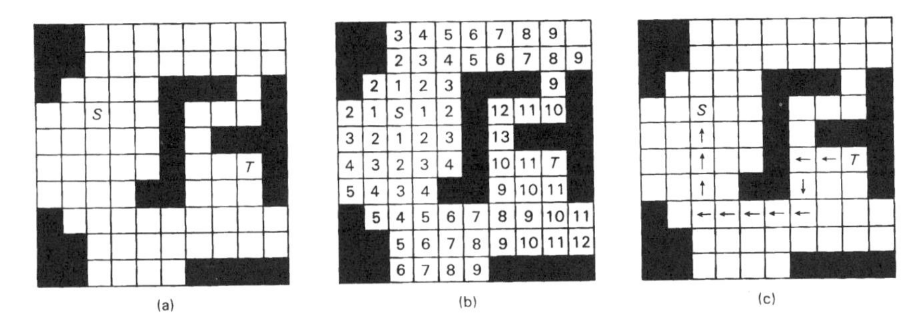

Fig. 5.13 Lees wavefront routing algorithm: (a) grid division of the floorplan, with S the source node and T the target node; (b) grid numbering spreading out from S towards T; (c) possible routing back from T towards S.

(i) all squares immediately surounding S are numbered '1'
(ii) all squares immediately surrounding squares numbered '1' are numbered '2'
(iii) continue numbering squares surrounding every already-numbered square with the next higher integer number until the target square T is reached

This is shown in Figure 5.13(b). Lee's routing algorithm then selects a routing path between S and T by back-tracking from T towards S, always moving from one grid square to a next lower-numbered square, such as shown in Figure 5.13(c).

The advantages of Lee's algorithm are as follows:

- provided the initial square numbering can be completed from S to T, then a routing path between S and T is guaranteed
- the path chosen will always be the shortest possible Manhatten distance between S and T
- it can readily cope with nets involving more than two points S and T

Disadvantages, however, include the following:

- complications arise where there is a choice of squares in the back-tracking, since in theory any cell numbered $(x - 1)$ may be chosen to follow square x
- the choice of a particular route between S and T may preclude subsequent netlist connections from being routed, with manual rip-up and rerouting of previously completed nets becoming necessary to acheive 100% completion
- a considerable and unnecessary area of the available wiring space may

be enumerated by the grid radiating outwards from S, with high computer memory requirements and long CPU running time

Various techniques to refine the basic procedure have been proposed [63,64], in particular to avoid the labeling of grid squares which are generally running in an inappropriate direction from the target square.

The alternative line-search algorithm of Hightower [65] does not require any comprehensive numbering of grid squares radiating from S to T. Instead, north-south and east-west lines starting from S are established, from which a search for the nearest point from which a perpendicular escape route towards T avoiding constraints can be made. This is a new source point S_1. A similar procedure is also started at T, aiming generally at S, giving a new source point T_1. This is illustrated in Figure 5.14(a).

This procedure is repeated from the new source and target points S_1 and T_1 until eventually a line which traces back to S cuts a line which traces back to T. Hence, a Manhatten connection path between S and T is established. This is illustrated in Figure 5.14(b).

The advantages ot this and similar line-search algorithms include the following:

- the algorithm tends to find the path with the minimum number of changes in direction
- very greatly reduced storage and computer times are needed compared with Lee's algorithm
- the final routing path is less likely to block subsequent routing paths compared with Lee's algorithm

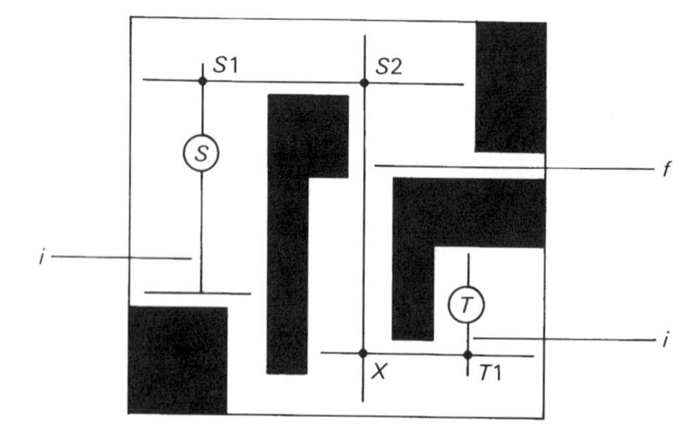

Fig. 5.14 Hightower's line-search algorithm, with i the initial routes from S and from T, and the final routing for connection path $S1$ $S2$ X $T1$, X being the meeting point in the final routing path f.

Against these points the minimum Manhatten distance may not necessarily be found, and unlike Lee's method no guarantee of finding a routing path, if one exists, is present. However, modifications to the algorithm have been reported which guarantee 100% routing if a path exists [66].

In real life, a large number of the required connections are relatively simple straight connections, with a minimum of bends in each route. Hence, many commercial routers include the means to route first the simple (possibly local) connections together with all critical paths which should be kept as short as possible, followed by, e.g., a Lee router to complete the remaining more difficult routes.

Turning to channel routers, rather than completely routing each net in turn over possibly long distances, the wiring requirements along each individual channel are first undertaken. The perimeter wiring around the working area may subsequently be considered as additional wiring-channel routing, or as a final general routing requirement.

With channel routers, we normally assume that the channel may be represented as a rectangular grid with uniform spacing. Assuming that the rows of cells and wiring channels are horizontal, then the cell terminals are spaced at uniform grid positions along the channel, with the connections to be routed running as horizontal paths along the channel connecting with orthogonal polysilicon fingers from the cell terminals.

The best known channel router is the left-edge channel router of Hashimoto and Stevens [67]. Its development is shown in Figure 5.15. It starts by selecting the extreme lefthand cell connection, and routes this at the top of the channel. In the case of two directly facing cell connections, it selects the top one of the two. The next connection routed is the first cell connection to the right of the routing just completed, which can be routed at the same track level in the wiring

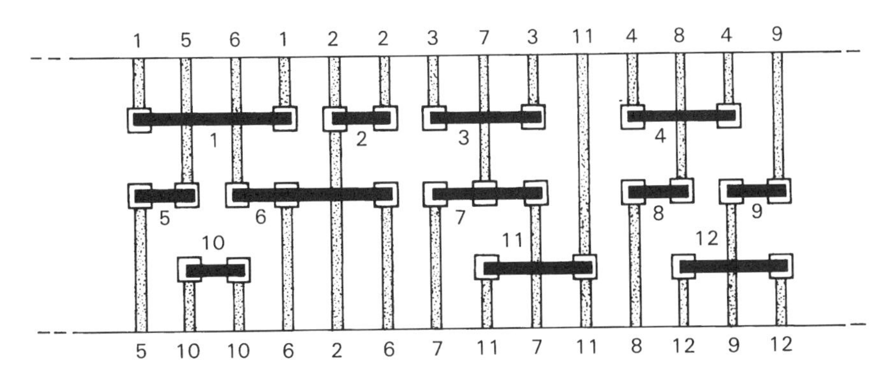

Fig. 5.15 The left-edge channel router, consecutively routing horizontal nets 1, 2, 3, etc.

channel as the first connection. This procedure repeats until no more connections can be made at this initial level, following which it returns to the lefthand edge and selects the next available node for routing using the next-level track position in the channel, thus building up parallel routing along the channel.

It will be observed that this left-edge channel routing algorithm is subject to the following two fundamental constraints:

(i) a horizontal constraint, in that a horizontal overlap of nets on the same track level cannot be allowed
(ii) a vertical constraint, in that a vertical overlap of directly facing connections from cells cannot be allowed

When these two constraints are combined, we may encounter a cyclic constraint which precludes routing a required net. This problem is shown in Figure 5.16, and may be overcome by cell re-placement or by using a second level of interconnect (polysilicon or second-level metal) to circumvent the difficulty [67,68].

The inclusion of 'dog-legs' in the channel router may also overcome cyclic constraints, but at the expense of a more complex routing procedure. Deutsch's dog-leg router [69] allows channel routing to be completed by means such as those shown in Figure 5.17. Branch-and-bound techniques are also available [68], but may involve excessive computer run time if many iterations of the branching are investigated.

Other variants of channel routing algorithms may be found, including Zonal [70], Greedy [71] and Hierarchical [72]. A bench mark for these various algorithms has evolved over the years, this being a series of netlists which the channel routing algorithms are asked to route. The most complex of these netlists is Deutsch's 'Difficult' example, which has a lower bound of nineteen tracks across the wiring channel to route completely; early channel routers did not succeed in

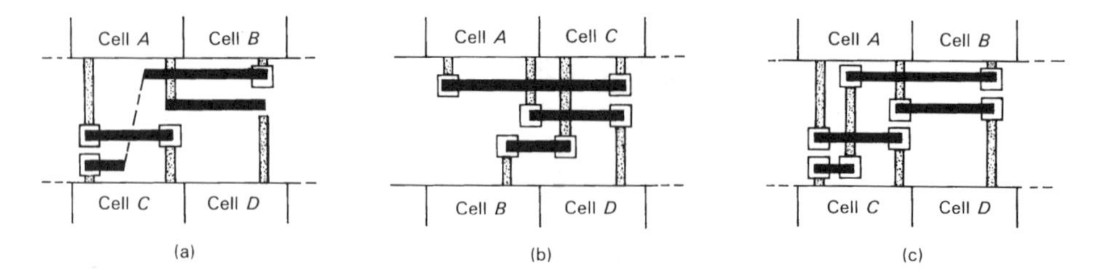

Fig. 5.16 A difficulty encountered by the left-edge channel router: (a) the route between cell *B* and cell *C* cannot be completed with a horizontal segment: (b) re-placement of cells *B* and *C* to achieve a horizontal run; (c) the use of a polysilicon crossunder to achieve the route with the original placement.

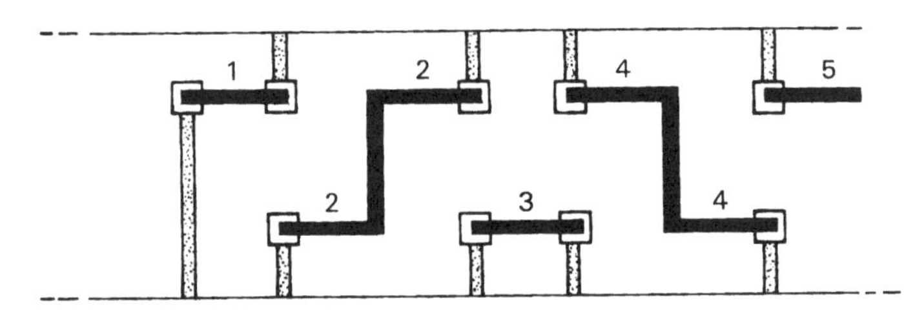

Fig. 5.17 The dog-leg router of Deutsch, which may circumvent cyclic constraints.

achieving this lower bound, but later algorithms such as the hierarchical router do achieve this minimum [72–74].

Channel routers such as Deutsch and others can effectively route to all four sides of a rectangular wiring area, as distinct from longitudional routing only of simple channel routers. They are therefore sometimes referred to as 'block' or 'switchbox' routers, and the areas they route as 'four-sided channels' [10,73,74]. They may be applied to architectures other than channelled provided the area to be routed is a four-sided rectangle; if the area is not rectangular, for example, between large macro blocks of different size, then partitioning into rectangular blocks and switchbox routing of each block may be done. However, in spite of the sophistication of many switchbox and channel routing algorithms, it is often also necessary to provide, e.g., a Lee router to solve particularly difficult interconnection paths which other routers cannot solve. In the case of single-layer-metal channelled floorplans with fixed via positions to the cells, 100% automatic routing can rarely be achieved; human intervention and re-placement becomes necessary to achieve the final routing (see the subject literature [10,14,55,74–78] and the further references which they list, particularly Ref. 78).

As a final point concerning the expediency of routing, it will be evident that as die size increases, CPU times will also increase. Times given in the mid-1980s for typical channel routing and maze routing algorithms were as follows:

- chip complexity 1000 gates, channel routing time = 1 minute, maze routing time = 1 minute
- chip complexity 10,000 gates, channel routing time = 16 minutes, maze routing time = 1.6 hours
- chip complexity 100,000 gates, channel routing time = 2 hours, maze routing time = 144 hours

which gave approximate relationships of $t \propto (\text{No. of gates})^{1.2}$ for channel routing and $t = (\text{No. of gates})^2$ for maze routing. The times quoted above are now likely to be considerably less, but the relationships remain generally valid.

The completion of routing constitutes the end of the physical design activity. However, before the physical design is translated into manufacturing data, it is usual to undertake final checks and postlayout simulation. This is perhaps more important for custom circuits involving a full mask set than for very simple single-layer-metal gate arrays, but some final verification must always be undertaken.

As implied in Figure 5.9, this verification involves some or all of the following:

- layout verification (design rule checking [DRC]) to ensure that all geometric widths and clearances, etc., met the process design rules
- circuit verification to ensure that the layout is a correct implementation of the required circuit
- timing verification to ensure that the circuit response time with the final routing of the circuit is acceptable

The last two checks require the extraction of circuit details from the layout information. This may involve the recreation of the full circuit diagram or the interconnection netlist from the layout data, which can then be compared with the input design data. Timing verification involves resimulation of the circuit using the actual resistance and capacitance values extracted from the layout, instead of the nominal values which had to be used before the physical design details were completed. This will be referred to again in Section 5.2.4.

These activities are usually undertaken by the vendor or independent design house, and the results sent to the OEM for approval. Should the postlayout timing simulation prove unacceptable, then some redesign or re-placement of the circuit in order to achieve the required performance is necessary.

Following design approval, the final physical design activity is the automatic conversion of this layout data into the format required for mask making or direct-write-on-wafer. This vendor activity will use an agreed data format, such as the following:

- Geber format, which is a standard first established for plotters and similar instruments
- Calma GDS II format, developed specifically for LSI and VLSI geometries
- Caltech Intermediate Form (CIF), also developed for LSI and VLSI geometries
- Electronic Design Interchange Format (EDIF), which has now become the international standard for the exchange of electronic design data

The latter has a hierarchical format which is capable of descending to the geometric level to describe floorplan geometries. Further details of these formats may be found in reports by Rubin [10], and will be referred to again in Section 5.5.6.

The vendor or independent design house may also do a final check on the design data by requesting a print of each subsequently manufactured mask layout before it is used, perhaps 100 times full size, which can be used as a visual check that nothing has gone grossly wrong during the transfer of this data, for example, the use of an incomplete or incorrect magnetic tape. These prints may be known as 'color-keys' or 'blowbacks,' and may be transparent to enable several to be viewed on top of each other for a rapid check on alignment, etc.

Further details of IC layout and mask making may be found in the published literature [1,10,79].

5.1.4 Test Program Formulation

An essential part of the front-end design activities must be a consideration of how the final IC will be tested. This becomes critically important as the size and complexity of the circuit increases, and above, e.g., 5000 gates special means to ease the task of testing becomes increasingly necessary.

The problem of testing is complex because of the following two factors:

(i) only the input and output pins of an IC are available for test purposes, probing of the interior nodes of the circuit not being possible under normal test conditions
(ii) the time to undertake an acceptable production test procedure is limited

If time were not a constraint it would be possible to test any circuit exhaustively through all its possible modes of action, and check for correct functionality under all conditions. However, consider even a simple 16-bit accumulator IC which can add or subtract a 16-bit input number from any 16-bit number already in the accumulator. In total there may be 19 input signals (16 bits plus 3 control signals), plus the 16 internal latches. Thus, the number of different input combinations which can be applied is 2^{19}, and the number of different input numbers which may be in store in the accumulator is 2^{16}. This gives a total of $2^{19} \times 2^{16} = 2^{35} = 34{,}359{,}738{,}368$ possible different logic conditions for this simple circuit. Hence, to test this circuit exhaustively would theoretically require over 34×10^9 input test vectors, which at an applied rate of one million per second would take almost ten hours to complete the test. In general, any n-input circuit with s internal storage elements (latches or flip-flops) theoretically requires 2^{n+s} test vectors for exhaustive testing.

Clearly, this problem must be addressed at the design stage in order that:

- some appropriate nonexhaustive test sequence shall be defined which checks the correct functionality within certain limits
- the circuit design shall incorporate special circuit details which facilitate

testing by providing a test mode whereby the circuit is partitioned or otherwise modified under test conditions.

The latter technique is termed *design-for-test* (or *design-for-testability*), and the techniques involved are referred to as DFT techniques.

Because of the great importance of test, it will be the sole subject matter of Chapter 6. CAD resources are available to aid the designer in developing an appropriate test strategy and to determine appropriate test vectors. However, this is additional to the hierarchical design activities which we have considered in the preceding pages, and does not alter the concepts already covered in any way.

5.1.5 Analog and Mixed Analog/Digital Synthesis

The subject of analog system synthesis spans the design of the basic individual building blocks such as amplifiers, oscillators and converters which go to make up a complete system, plus the system design which is the interconnection of these blocks to meet any required system specification.

Basic analog design is a skill which must be learned, and many excellent texts are available [80–88]. However, few CAD resources are available to help in the initial creative work, but CAD assistance is extensively used—and necessary—in simulation (see Section 5.2.5). Analog design is further complicated by the difference between designing using discrete components, where accurate value, close tolerance passive components are readily available, and designing monolithic circuits where these attributes are not present, plus the diversity of requirements which may be needed (see Figure 4.26).

Most OEM designers will therefore use some form of predesigned analog circuit blocks, since he or she cannot afford to become an expert designer at the silicon level. Assuming that off-the-shelf ICs and other components are not used, the custom choices are as covered in Chapter 4, namely a vendor's library of proven circuits which can be made from uncommitted component-level arrays, or a vendor's library of analog cells for a standard-cell solution.

The following CAD help which is available in the design phase is largely graphical, as follows:

- symbol generation
- capability to draw on the VDU screen the circuit as it is being built up
- schematic capture and automatic formatting of the design for subsequent simulation

The essential part, however, is the subsequent simulation, the results of which may be presented as appropriate graphical plots [17,89–92]. This will be illustrated further in Section 5.2.5.

Considerable CAD help is available in one particular design area, namely in filter design activities, although even here simulation forms the essential background. In this area, we have the nearest analog equivalent to the silicon compiler used for digital synthesis purposes (see Section 5.3). With filter synthesis programs it becomes possible to enter the required filter characteristics into the CAD system, and the program will attempt to determine the appropriate component values [83,84,93–95]. Nevertheless, it still involves designer expertise to suggest the appropriate type of filter to meet the specification, and to accept or reject the results of the synthesis.

Mixed analog/digital design is not well served by CAD resources. A common practice is to design the two parts separately, utilizing human expertise to handle the interface. No single simulation resource is currently available which can handle the detailed device-level simulation necessary for the analog sections and the gate or higher hierarchical level of simulation necessary for the more populous digital parts. Much research work continues to be done in this area, but in the absence of a single comprehensive design theory which covers the wide range of functions used in analog systems, it remains difficult to formulate CAD tools to cover this widely diverse area. Human expertise and creative talents backed up by good CAD simulation and verification are likely to remain the general means of analog and mixed analog/digital circuit design [83,89,96,97].

5.1.6 Programmable Logic Device Software

In contrast to the relative sparsity of CAD tools for analog synthesis, CAD tools are widely available for all the design activities associated with PLDs, being part of the resources marketed by PLD vendors and distributors. Digital devices only are involved at present.

The CAD support usually includes the following:

- design entry, usually incorporating vendor's library data
- design consistency checks
- simulation, either before or after the following steps or both
- automatic placement and routing of the design on a particular PLD—the selection of the optimum size of PLD from the vendor's range of devices may be part of the program
- down-loading of this design data into an appropriate programming unit for the commitment (dedication) of the PLD
- design documentation

At the design entry phase, one or more of the following levels of design may be used:

- high-level logic statements
- state machine statements, specifying states and conditional branching and also inputs and outputs of the state machine
- Boolean equations

- truthtables
- schematic capture, possibly using a vendor's menu of primitives and macros
- netlist data defining the required system

This is indicated in Figure 5.18.

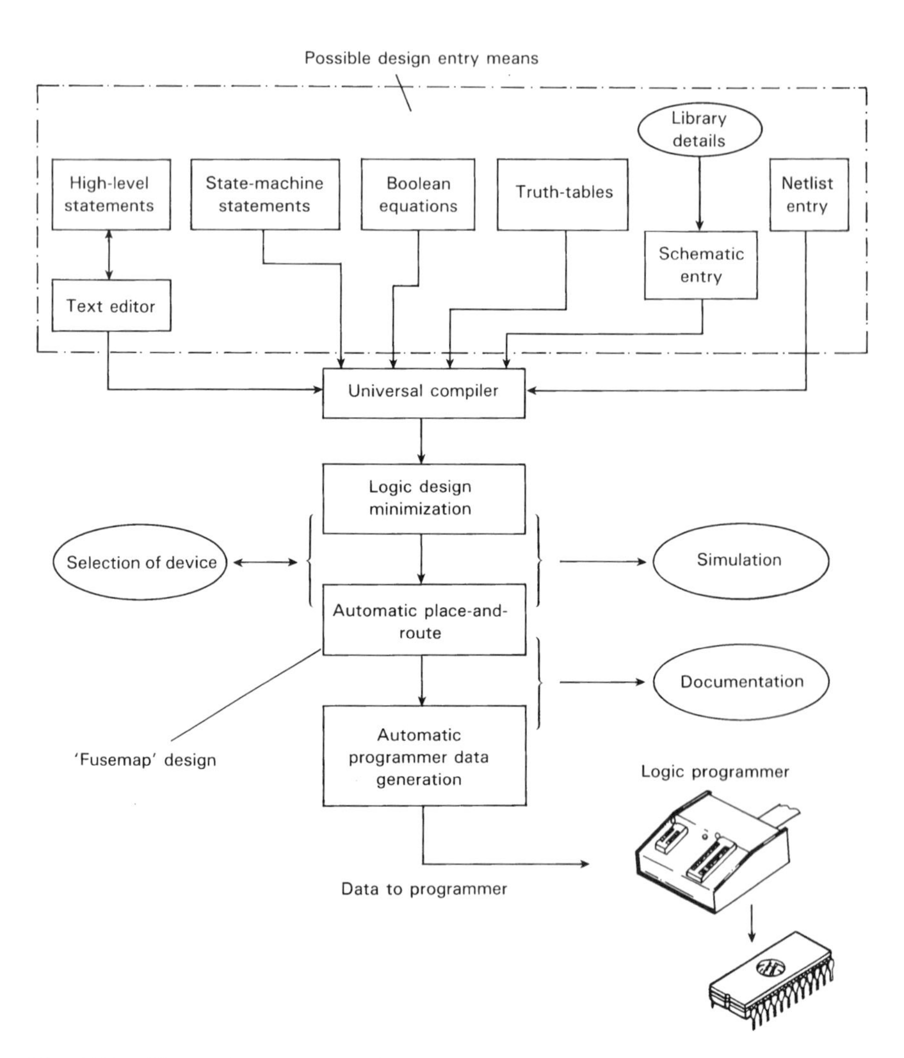

Fig. 5.18 The CAD resources available for programmable logic devices. Not all the possible entry means may be used for the less sophisticated range of PLDs. Place-and-route in this context means the mapping of the required function on the chosen PLD plus the determination of the programmable interconnect pattern.

The high-level logic statements which may be used in the design of the larger and logically more powerful types of PLD are usually written in a language such as Pascal or C, but are not necessarily as comprehensive as in the general-purpose description languages (HDLs) introduced in Section 5.1.1. This is because functional simulation is not always done at the architectural level, but instead the emphasis is more upon a top-down design route targeted upon specific PLD architectures, with simulation at the PLD level. An example is Altera's high-level language AHDL, which is a VHDL derivative with enhancements to aid the initial design entry using library macros which can be the equivalent of the 74** series ICs, and where the target PLD architecture determines the hierarchical decomposition [22,98]. The supporting graphics editor allows multiwindow representation of the synthesis procedure, as shown in Figure 5.19.

There are, however, many CAD suites available which support a range of vendors' products. Among them are ABEL, PALASM, CUPL and PLPL; details of these and others may be found in vendors' literature and elsewhere [99–106]. The syntax of these software tools is, unfortunately, not completely standardized, for example, logical AND and OR in CUPL is & and #, respectively, while in PLPL it is * and +, but translators to convert from one format to another have been produced.

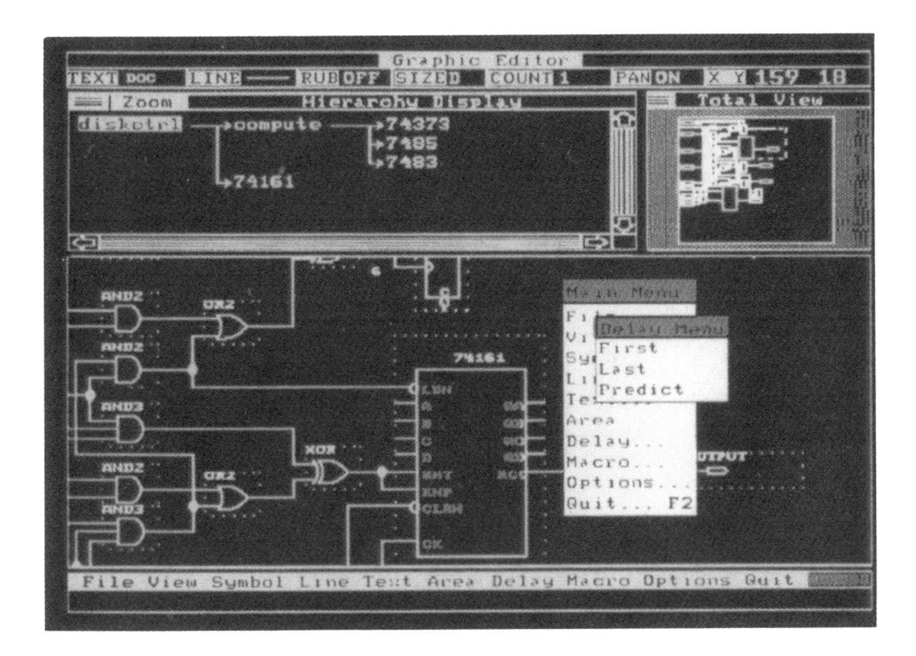

Fig. 5.19 The MAX-PLUS hierarchal graphics editor which provides either top-down or bottom-up design facilities. (Courtesy of Altera Corporation, CA.)

Currently, one of the largest and most complex user-programmable off-the-shelf CPLD products available is the logic cell array (LCA) (see Chapter 3, Section 3.6). The vendor's CAD tools to support LCA design also support design entry interfaces from several other schematic entry formats, including ABEL, PALASM and others, as indicated in Figure 5.20, which allows designs already undertaken for other types of PLD to be translated into a LCA realization [104].

For the more simple types of PLD which do not have such dense logic as the more complex products, simple schematic capture or Boolean equations may be entirely adequate. An essential part of all PLD design software is Boolean minimization and factorization in order that an efficient and complete design tool is available for OEM use to fit the architecture of the device [99]. Such CAD resources specifically targeted on the final IC realization thus provide the most complete means available to an OEM to progress a design from concept to a final IC realization, without recourse to any outside design help or vendor participation. These hardware and software resources will be mentioned further in Sections 5.3 and 5.4.

5.2 IC SIMULATION SOFTWARE

It will be apparent from the preceding pages that simulation is the essential element of microelectronic CAD. While there has been much talk of 'correct-by-construction' in higher levels of design, verification is still needed to ensure that all design details are, as far as possible, correct before manufacture is commenced. The importance of simulation is reflected in the relative times which have been quoted for undertaking the various design activities in digital system design [40], as follows:

- Man-hours: 35% on system specification and design, 30% on logic and circuit design, 25% on initiating and accepting simulation activities, 5% on physical layout activities, and 5% on layout and test pattern generation; total, 100%.
- CPU hours: 5% during system specification and design, 10% during logic and circuit design, 40% on different simulation activities, 15% on physical layout, 30% on layout and test pattern generation; total, 100%.

From these figures it will be seen that at the initial creative stage human activities are prominent, falling off as the more mechanical aspects begin to dominate. However, the required CPU activity increases as the human activity decreases, being particularly heavy for simulation and verification.

The different levels of simulation which are encountered in digital IC design are as follows:

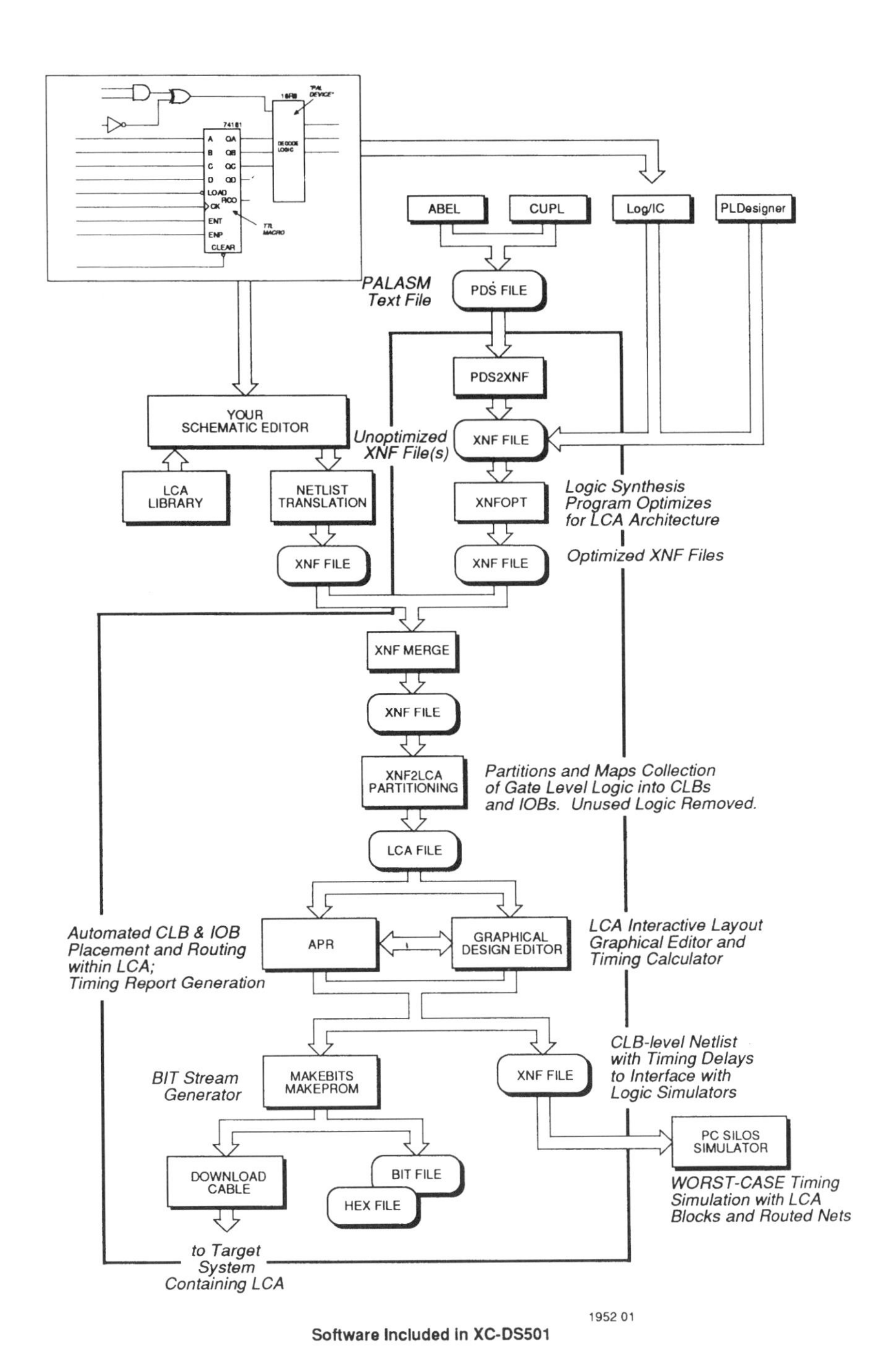

Fig. 5.20 A Xilinx CAD system which supports inputs from ABEL, CUPL, LOG/IC and PLDesigner, plus schematic netlists and text files.

- behavioral level
- architectural level
- functional level
- gate level
- switch level
- circuit and device level
- the final silicon fabrication level

Behavioral and architectural levels are concerned with the system design before it is partitioned at the functional level into specific macros and logic functions; gate level is concerned with the individual gates which make up the functional level, while switch level is an alternative to the gate level where the logic gates are modeled as on-off switches. The latter is particularly relevant for CMOS circuits. Circuit and device level is concerned with the transistor level of design, while the silicon fabrication level models the physics of the actual silicon fabrication process. We shall have little need to consider the process level [11] in this text. Analog circuit simulation does not involve this full hierarchical structure, and the usual simulation level is at the circuit and device levels only.

The lower the level of simulation, the more accurate are the simulation results, but the amount of CPU time increases rapidly. On the other hand, at the top hierarchical level, either broad confirmation that a complex system proposal is correct or easy means to evaluate alternative architectures is all that is required, not detailed timing or other verification which can only be done when the design has progressed to a more detailed level. For more simple circuits, e.g., with fewer than 5000 gates, functional-level simulation may be the highest level of simulation necessary, since this may be done with an acceptable amount of computing.

All levels of simulation involve the computer application of input signals to the model of the circuit, and checking of the computed output response. This output response may be converted into graphical waveform representations, which are much easier to intrerpret than textual print-out of logic values. As well as functional verification, the detection of spikes and hazards at the lower levels of simulation are a necessary part of this activity.

5.2.1 Behavioral, Architectural and Functional Level Modeling

Since it is difficult to draw a clear line of distinction between these three levels of digital simulation, we will consider them together. All involve a programming language which describes the system and its appropriate degree of partitioning, with varying degrees of timing information from zero delays to estimated nominal or average values. The terminologies *mixed-level simulation*, *multilevel simulation* and *mixed-signal simulation* may be found to cover procedures which span several categories [6,12,30,107–110], although this must be distinguished from simulation systems which attempt to cover both analog and digital circuits and which may be referred to as mixed-mode or multimode simulators [83,109]. Unfortunately, the use of these adjectives is not always clear, with 'mixed-mode' sometimes being used to refer to mixed-level digital simulation.

Unlike analog simulation, which requires the accurate modeling of continuously varying voltages and currents, digital simulation can use models with the binary values 0 and 1, with time as a further parameter. At the more detailed lower levels of simulation, other concepts such as impedance must be incorporated, giving rise to unknown or undefined logic states or further distinctions such as low or high impedance source and drive conditions, etc., but this degree of detail is not required at the behavioral, architectural or functional levels.

Simulation at these high levels requires the choice of an appropriate hardware description language such as considered in the preceding sections. VHDL is possible, but more widely promoted are the HDLs whose structure is directly relevant for simulation duties, such as Verilog HDL, Silvar-Lisco HHDL and others, or even those specifically designed for simulation rather than synthesis, such as HILO-GHDL and SIMULA [21]. The OEM is largely dependent on what is commercially available from chosen vendors, with cell library data and good graphics being key features. Unfortunately, rapid obsolescence is present in many commercial software products, which makes the OEM choice of CAD software a short-term solution in many cases.

The VHDL listing to verify the functionality of the sorter circuit given in Fig. 5.5 is shown in Fig. 5.21, where PHI_PW defines a 100 ns positive clock pulse for clock lines phi1 and phi2, the clock pulses being equally spaced within a cycle period P of 400 ns. No propagation delay timing is present in this particular VHDL description, as may be present in other proprietary HDLs.

System designers always prefer graphical rather than textual input and output data, and hence most commercial HDL systems provide VDU displays such as previously illustrated in Figure 5.8. Windowing is a powerful advantage in this area, as it allows the designer to give priority to the different aspects of the design and simulation activities as required.

5.2.2 Gate and Switch Level Modeling

Some of the high-level HDLs which may be applied to the architectural and functional level also descend to the gate level for simulation purposes, for example, GenRad's HILO and others. These general-purpose tools have the capability of incorporating many different signal conditions, for example, logic 0, logic 1 and X (unknown), each at either strong, weak or high impedance. This gives a total of nine different possible digital signal conditions. Others have five (or more) different strengths available, giving fifteen (or more) possible conditions.

```
architecture action of driver is
  constant PHI_PW:              time := 100ns;
  constant P:                   time := 400ns;
  constant PHASE:               time := P/2;
  constant NEWWORD_DLY:         time := (P-SKEW-
                                PHI_PW)/3;
  constant NEWWORD_PW:          time := 30ns;
  constant NBITS:               integer := 16;
  constant NWORDS;              integer := 256;
begin
--
--
  timing: block
    begin
    phi1      <=  '0'   after PHI_PW when phi1 = '1' else
                  '1'   after P-PHI_PW when phi1 = '0' else
                  '1'   after P-SKEW-PHI_PW;
    phi2      <=  '0'   after SKEW when phi1 = '0' else
                  '1'   after SKEW when phi1 = '1' else
                  '0';
    newword   <=  '0'   after NEWWORD_PW when newword = '1' else
                  '1'   after NBITS * P-NEWWORD_PW
                        when newword = '0' else
                  '1'   after NEWWORD_DLY;
  end block timing;
--
-- setup test
--
  test: process
    begin
--
-- initialization
--
    reset <= '1';
    read <= '1';
    double <= '0';
    din <= b"0000000000000000";
    wait for NBITS * P;
--
-- load initial data
--
    reset <= '0';
    for nword in 1 to NWORDS loop
      nextval(nword, din);
      wait for NBITS * P;
      end loop;
--
-- shift out data
--
    read <= '0';
    for nword in 1 to NWORDS loop
      wait for NBITS * P;
      checkval(nword, dout);
      end loop;
    wait;
  end process init;
end action;
```

Fig. 5.21 A continuation for verification purposes of the VHDL example given in Figure 5.5; here a sequence of input vectors is applied to the architecture and the output response generated. No detailed cell architecture or timing is involved at this stage. (Courtesy of Viewlogic Systems, MA.)

Many of these conditions may be necessary for full-custom design activities where many different transistor sizes may be present, but may not be so relevant for semicustom gate-array and standard-cell design activities.

However, gate-level simulation has been extensively developed in its own right, since it forms the cornerstone of the design verification process. Many of the algorithms and practices developed at gate level are incorporated in higher-level simulators, and it is thus appropriate to consider this and the similar switch-level simulators in some detail.

Gate-level simulation involves the following four interrelated activities:

(i) check of the input data
(ii) reformatting of these input data into an appropriate form for simulation
(iii) the simulation procedure
(iv) output display formatting, either textual or graphical or both

(i) The Input Data

Although the input data may have been graphical as far as the designer was concerned, this has to be converted by the software into textual form with statements defining each of the primitives and their interconnections, plus statements defining propagation delays and other signal information. In this 'flattened' form, all complex macros are partitioned into their constituent primitives.

These data are then checked for possible errors, for example, duplication, omission of interconnections, illegal primitives and so on, but this check cannot, of course, detect wrong design data such as a NOR gate being specified instead of a NAND, or an unwanted Invertor gate being included. Indeed, finding such errors is one of the purposes of the simulation.

(ii) Reformatting

A 'model compiler' generates an appropriate model of the circuit for simulation purposes from the above syntactically correct description, either by generating a *compiled code model* in which the circuit description is organized into a logical sequence which progressively mirrors the data flow of the circuit, or by grouping the circuit description into a set of *interlinked data tables* which are subsequently operated on by the simulation program.

In the compiled code model, the order in which the gates is compiled is significant, since the simulated output from a given gate must be updated before this signal can be used in a following gate input. This process can provide very fast simulation if a single-pass simulation through all gates can be achieved without any revision of the gate ordering.

In the case of the table-driven format, each table contains information relating to the connections from one table to other tables, as well as full information relating to the associated primitives, their input signals and their timing characteristics. By this means a selective trace simulation can be subsequently implemented, allowing the simulation to follow pointers from one table to another as the simulation proceeds. This table-driven format allows more simulation information to be included than in the compiled code model, and also allows changes to be made in a table without having to recompile the whole circuit [110].

(iii) The Simulation Procedure

The basic simulation algorithm performs the following actions:

(i) set all gate outputs at time $T = 0$ to X (unknown)
(ii) apply the first input stimuli to the model
(iii) evaluate the effects of this stimuli, and schedule the changes in gate outputs in an appropriate event queue or timing schedule
(iv) continue to propagate this simulation until no further logic changes take place
(v) repeat with further input stimuli

The effect of gate input changes on gate outputs is given by look-up tables in the software program, which document the effect of all possible gate input stimuli, including timing information. Therefore, no detailed analog computations are involved in gate-level modeling, although the vendor's original generation of this look-up tabular information may have involved such simulation (see Section 5.2.3).

The most important part of this simulation process is determination of the method of scheduling the changes as the simulation run proceeds. This may be in the form of an *events queue*, in which the results of the simulation at time t are placed in the events queue at time (t + *delay*) for the subsequent simulation activity. However, with dissimilar propagation delays, an event in later time may cause a subsequent event earlier than one previously scheduled. Hence, a continuous update and shuffle of the next-events list of the simulation program is necessary in order to avoid misscheduling, which may involve appreciable CPU housekeeping. On the other hand, no waiting time occurs in moving on from one event in the list to simulation of the next scheduled event.

The alternative to a next-events scheduling list is a *time-mapping* events schedule. Here, every propagation delay is expressed as an integer multiple of a small time delay Δt, the simulation program scanning each Δt period for the next simulation activity required. The advantage of this form of scheduling is that there is no rescheduling of priorities as there is with a next-events list, but against this CPU time is wasted in scanning empty Δt time slots where no event is scheduled.

These principles of scheduling are detailed more fully in the report by Lightner [30] and elsewhere in the literature [43,110].

As circuit size increases, simulation time may be reduced by further techniques. For example, in large networks, only a very small percentage of the gates change state as a result of a preceding change, and hence it is only necessary to simulate the changes in this fan-out area. This 'selective trace' may be incorporated to exploit the latency of a large network. Parallel simulation of parts of the network which do not interact may also be done. All these forms of event scheduling for gate-level simulation are in contrast to most circuit simulators (see later), where all circuit nodes are examined at each time increment, whether they have or have not changed state.

Switch-level simulation is specifically relevant for CMOS circuits, since it overcomes certain problems in trying to model MOS circuits at the gate level. In switch-level simulation, each transistor is treated as a conducting or nonconducting perfect switch having the truthtable as shown in Table 5.1. As in gate-level simulation, X is an unknown (undefined) logic signal, but the additional signal Z introduced here is 'open-circuit,' that is, the switch is nonconducting and hence the output state is not defined by the input node. Also, device threshold voltages are not considered in this simple truthtable.

Further reasons why switch-level simulation is often more appropriate for digital logic simulation than gate-level are as follows:

- since CMOS gates contain only p-channel and n-channel devices and no resistors, the conventional CMOS logic gates may always be regarded as networks of switches (see Figure 5.22(a)).
- many transmission gate circuit configurations which do not have a direct Boolean NAND or NOR representation are found in vendors' libraries (for example, see Figure 5.22(b))

Table 5.1 The Conducting/Nonconducting/Unknown Truthtable for MOS Transistor Switch-level Simulation

Source (input) logic signal	Gate (control) signal	Drain (output) logic signal	
		n-type device	p-type device
0	0	Z	0
1	0	Z	1
X	0	Z	X
0	1	0	Z
1	1	1	Z
X	1	X	Z
0	X	X	X
1	X	X	X
X	X	X	X

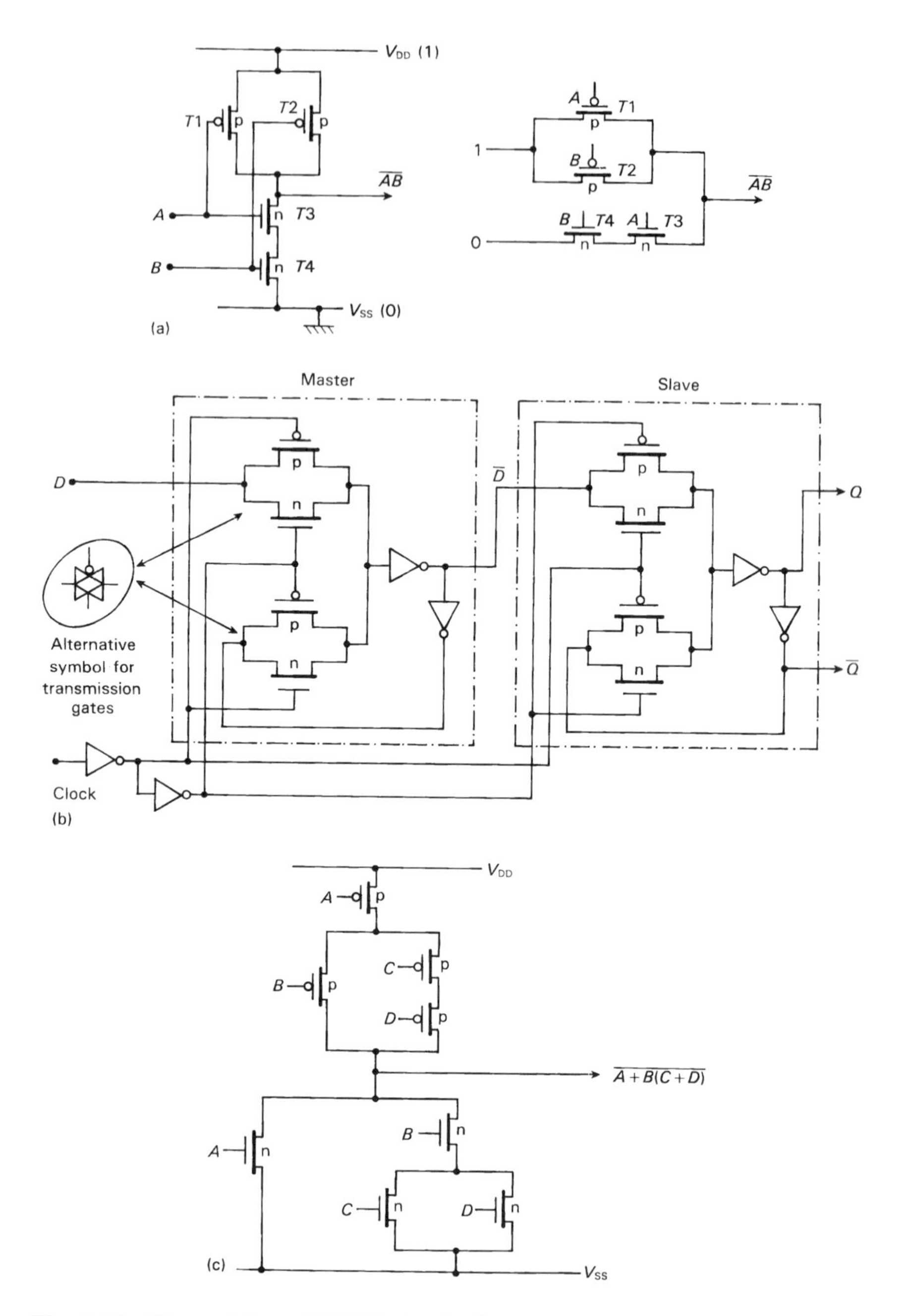

Fig. 5.22 The modeling of MOS logic circuits as switches: (a) conventional CMOS NAND gate remodeled to emphasize the transistors acting as on/off switches; (b) CMOS D-type master-slave circuit using transmission gate configurations; (c) complex CMOS logic gate realizing the function $\overline{A + B(C + D)}$, which may be remodeled as in (a).

- similarly, complex logic functions (for example, see Figure 5.22(c)) do not have a direct NAND/NOR representation
- MOS devices may be used as bidirectional switches so that logic signals flow both ways through them, again something that cannot be easily represented by Boolean gates

Thus, as all CMOS logic circuits contain only these active switches, the switch-level model is an obvious possibility for simulation purposes. This modeling of CMOS gates as switches will also be referred to again in Chapter 6 when considering the fault modeling of logic circuits.

Switch-level simulation of CMOS circuits therefore consists of a network of nodes interconnected by bidirectional switches (see Figure 5.23). This topology can be extracted directly from the circuit layout *without reference to the functionality of the circuit*, and hence forms a powerful further check on the circuit and its physical design.

The actual simulation activity can be a behavioral check on the circuit without any detailed timing considerations. In this case, the effects of an input

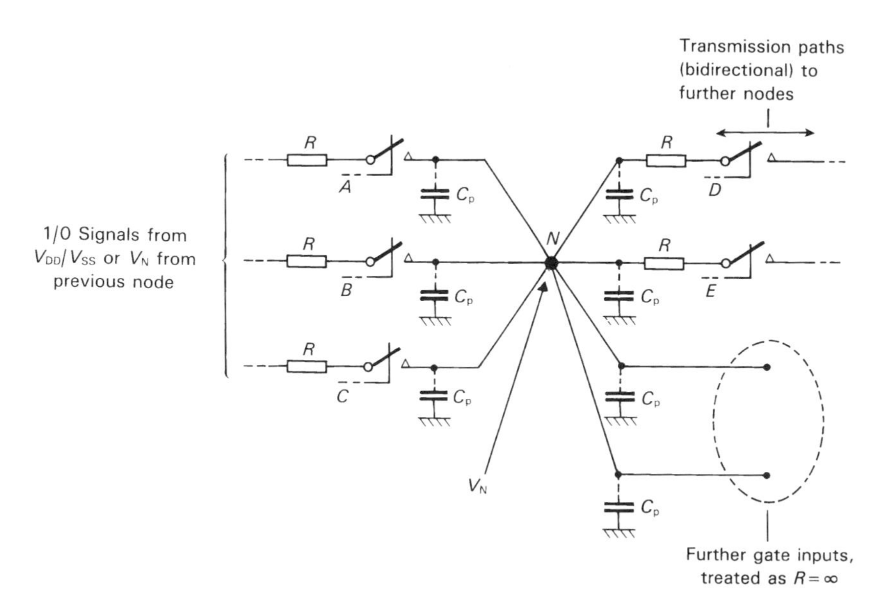

Fig. 5.23 The basic concept of switch-level simulation where perfect switches switch voltages to a summing node N and thence to further nodes. Kirchoff's law must hold at every node. For timing simulation parasitic capacitance values C_P may be introduced. Note that the transmission paths are bidirectional, and also that any closed path from V_{DD} to V_{SS} must be a design error.

simulation is evaluated by tracing the voltage changes through all the perfect switches until all changes have been evaluated, following which the next test input is evaluated. Note that the FET gate inputs may be taken as infinite impedance, and do not therefore affect this simulation. One problem which may be encountered is where an undefined logic value at a node is encountered, as would be the case if all transistor switches connected to this node were open together. The problem with an undefined node is that this undefined state is likely to propagate extensively through the rest of the circuit, and hence the simulator may stop when this condition is first encountered and flag a possible design fault in the circuit being simulated.

Alternatively, the switch-level model may simulate timing by including nominal circuit resistance and capacitance values which provide a measure of the circuit propagation times. Because these parameters are at transistor level, the CR times to charge or discharge nodes can give results without the wide range of signal levels, circuit impedances and timing details that may be necessary in gate-level modeling. However, it is generally impossible to detect race conditions and hazards, but against this, the simplicity of switch-level modeling allows very large CMOS circuits to be rapidly simulated.

In total, switch-level provides an alternative means of simulation to the very detailed circuit-level simulation (see Section 5.2.3) and also to gate-level simulation using vendors' data on logic primitives. The scheduling techniques such as those used in gate-level simulation are also used in switch-level procedures. Unfortunately, both gate-level and switch-level CAD tools are largely vendor-specific, without the ability to be directly integrated into a hierarchical general-purpose simulation package [107]. However, the standardization of interchange formats such as EDIF (see Section 5.5.6) now increasingly allows translation from one CAD tool to another without the difficulties of human intervention and error. For further information see Refs. 110 and 111 and elsewhere [12,13,30].

(iv) Output Display Formatting

The output display format can be textual in the form of logic 0 and 1 lists for each input and output node, being printed out for each simulator time interval. It is very tedious for the human designer to check this data, except possibly for very short duration spikes and glitches which show up as a logic 0 (or 1) in a steady stream of logic 1s (or 0s), and it is therefore common practice for the output of gate-level simulation to be converted into logic waveforms for checking purposes. The overall picture of gate-level simulation is therefore as indicated in Figure 5.24. Further details may be found elsewhere, although the fine details of most vendors' programs are proprietary [1,3,12,13,30,43,110].

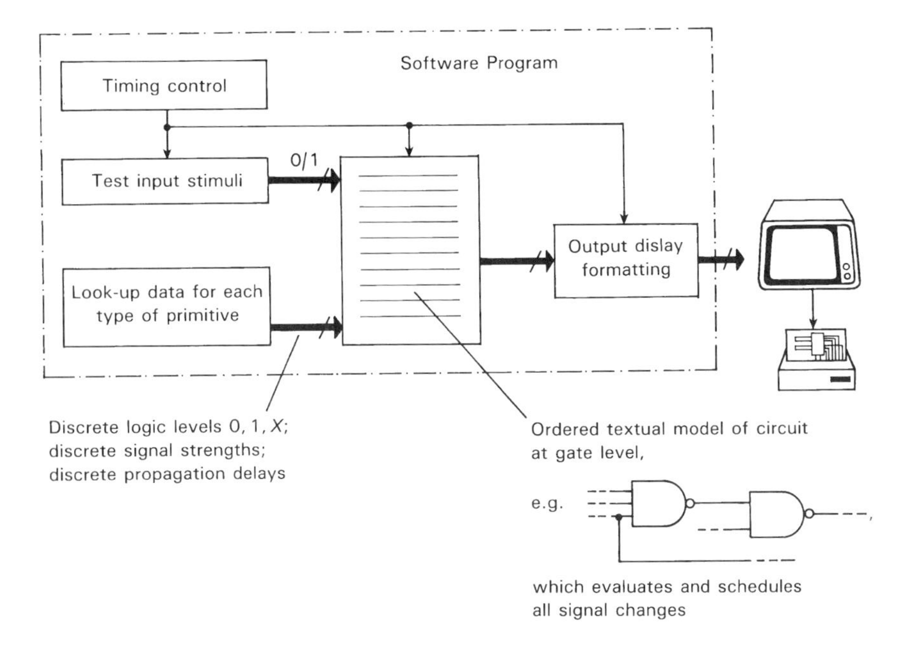

Fig. 5.24 The schematic of gate-level simulation, where the electrical performance of each type of gate is modeled by discrete values with input/output relationships given by look-up tables and not computed.

5.2.3 Silicon Level Modeling

Silicon level, or component level, simulation is the most detailed and accurate level of simulation, but because of the detail involved cannot be realistically applied to circuits consisting of more than a few hundred transistors. It is therefore appropriate for basic building blocks of SSI, MSI and perhaps LSI complexity, but not for VLSI.

The almost universally used program for silicon-level simulation is SPICE ('Simulation Program, Integrated Circuit Emphasis'), and particularly its many commercial derivatives such as PSPICE, HSPICE, etc. (see Section 5.5.2). SPICE was first developed at the University of California at Berkeley in the early 1970s, and is based upon the small-signal a.c. equivalent representations of active and passive electronic devices. These, in turn, are based upon the the physical properties and terminal characteristics of each device, with particular emphasis on the nonlinear characteristics of the active devices (transistors). Basic details of small-signal equivalent circuits are summarized in Appendices C and D. It should be appreciated that as the operating point (current and voltage) in a nonlinear device changes, so the small-signal equivalent circuit values also change, and hence it is necessary for a simulation program such as SPICE to perform a large number of simulation calculations with different operating points as the circuit currents and voltages vary.

SPICE requires the following information to be given:

- the circuit interconnection details
- *R*, *C*, *L* values of all linear devices and perhaps interconnections
- details of all nonlinear devices (transistors, diodes and perhaps monolithic capacitors) and their parameter values at some chosen operating point
- the d.c. supply voltage(s)
- the input signal(s) to the circuit
- which output(s) or other details the program is to calculate and display

Some of the above information will not be required in digital circuits, e.g, discrete inductors or capacitors, or monolithic capacitors, etc., but may be present in analog circuit simulation (see Section 5.2.5).

A particular feature of the SPICE program, and an indication of its power, is that it is not necessary to specify the operating points of each nonlinear device, nor is it necessary to give the small-signal parameter values over a range of operating points. Instead, the program will first calculate the d.c. operating point of each device from the d.c. equations it knows for the device and from the circuit information that it has been given, and then will work out the small-signal parameters at this point from the known values at some other point. In order to perform such calculations, the equations relating the small-signal equivalent values at one operating point to those at other operating points are built into the program, which will perform this revaluation of parameter values many times in a simulation run as circuit currents and voltages change.

The method employed in SPICE for determining an operating point is an iterative procedure, whereby the program first suggests certain current and voltage values and checks to see whether these values are consistent with the circuit topology and the device d.c. equations. For example, for bipolar transistors we have $I_C = h_{FE} I_B$, $I_E = I_B + I_C$, and $I_E = I_S (e^{40V_{BE}})$ (see Appendix C). For MOS transistors we have the d.c. equations given in Appendix D. If the suggested values are not consistent, they are repeatedly disturbed until they converge on consistent values. Unfortunately, in some early versions of SPICE, convergence did not always occur, particularly if zero value had been entered for some of the less important parameters, and as a result the simulation could not proceed. Later versions, particularly the commercial derivatives, do not have such convergence difficulties.

All these procedures are hidden in the SPICE program, and the user does not have to be conversant with the details. Further resources within the program are default values should the user not wish to specify certain parameters. For example, the forward-biased pn junction diode equation of

$$I_D = I_S(e^{40V})$$

(see Appendix C) is given by a further enhancement to cater to particular physical constructions, being

$$I_D = I_S(e^{40V/n})$$

where n is some constant whose value is between 1 and 2 [112,113]. If not specified, the program uses the default value of 1. Similarly, if not specified, the value for I_S is taken as 10^{-14}A for silicon. Table 5.2 shows some of the bipolar and unipolar transistors used in SPICE, together with their default values. Provision

Table 5.2 Some of the Transistor Parameters and their Values Which are Used in SPICE Simulations

Description of parameter	Symbol	SPICE designation	Typical value	SPICE default value
Saturation current	I_{CS}, I_{ES}	is	10^{-15}A	10^{-16}A
Transconductance	g_m	–	$40I_C$ A V^{-1}	$40I_C$ A V^{-1}
Forward-current gain	β	bf	200	100
Reverse-current gain	β_R	br	0.5	1
Base resistance	r_b	rb	100 Ω	0
Series emitter resistance	r_e	re	1 Ω	0
Series collector resistance	r_c	rc	50 Ω	0
Forward transit time	τ_t	tf	4×10^{-10}s	0
Early voltage	V_A	vaf	200 V	infinity
Emitter capacitance, $V_{BE} = 0$	C_{te0}	cje	2 pF	0
Emitter–base contact potential	ψ_{BE}	vje	0.75 V	0.75 V
Emitter–base grading factor	mje	mje	0.33	0.33
Collector capacitance, $V_{CB} = 0$	C_{tc0}	cjc	1 pF	0
Collector–base contact potential	ψ_{BC}	vjc	0.75 V	0.75 V
Collector–base grading factor	mjc	mjc	0.33	0.33

(a) epitaxial npn bipolar transistor

Description of parameter	Symbol	SPICE designation	Typical value	SPICE default value
Body-junction saturation current	I_S	is	10^{-15} A	10^{-14} A
Transconductance parameter	$\varkappa$	kp	3×10^{-5}A V^{-2}	2×10^{-5}A V^{-2}
Body threshold parameter	γ	gamma	0.37	0
Channel length modulation	λ	lambda	0.02	0
Zero bias threshold voltage	V_T	vto	1 V	0
Series drain resistance	r_d	rd	1 Ω	0
Series source resistance	r_s	rs	1 Ω	0
Oxide thickness	t_{ox}	tox	10^{-7} m	10^{-7} m

(b) n = Channel enhancement mode unipolar transistor

for simulating circuit performance at temperatures other than at the nominal value of 25°C is also available within the SPICE program, which involves appropriate modifications to the normally used equations.

To run a SPICE digital simulation program it is necessary to provide full details of each primitive and all interconnections. The internal structure of individual types of gate, for example, a 3-input NAND, may be listed only once and then picked up as a module in the rest of the netlist, or it may be available in the simulator library data. The individual transistor models use library or default values, although for MOS transistors the *W*/*L* ratio may be individually specified—this is particularly necessary for custom designs where the transistor geometries are flexible so as to optimize the circuit performance. Input logic signals may be given zero rise and fall times, or any required ramp from logic 0 to logic 1, and vice versa.

An example of the netlist for a single 2-input TTL NAND gate simulation is given in Figure 5.25(a). The five transistors q1 to q5 are all of the same type, namely 'mod1,' and similarly the diodes are all of the type 'mod2.' Details of these two models have to be given. Input B is being held at 5 V, while input A is a pulse input falling from 5 V 4 ns after time t = 0 with a fall time of 5 ns, dwelling at 0 V for 45 ns, and rising again to 5 V with a rise time of 7 ns, this input being repeated every 90 ns. The output response is to be calculated every 1 ns for 90 ns at the output node 3.

The response of three such NAND gates in cascade may be requested, as shown in Figure 5.25(b). Similar netlists may be compiled to cover more complex digital networks, but clearly the amount of detailed simulation involved at this device level escalates rapidly as the number of gates and transistors increases.

Details for the simulation of a 2-input CMOS NAND gate are given in Figure 5.26, using the same input waveform as in the previous example. The two p-channel transistors m1 and m2 are of type 'mod1' with length and width geometries $L = 2$ μm and $W = 8$ μm.

Because the full electrical performance of the active devices, together with all other circuit parameters, can be built into the SPICE simulation, very detailed simulation results can be obtained, the accuracy depending solely upon the accuracy of the model and the parameter values. Every node of the circuit is evaluated for changing conditions at each simulation interval, which means a great deal of unnecessary calculation occurs in digital circuits which have a high degree of latency, for example, where a long counter is involved with only one or two stages changing state, and hence there is the need for the higher but less detailed gate-level or switch-level simulation for most digital circuits of LSI complexity and above, as we have previously considered.

Further details of SPICE, exactly how it computes the d.c. operating points and the exact details of the transistor models, may be found in the published

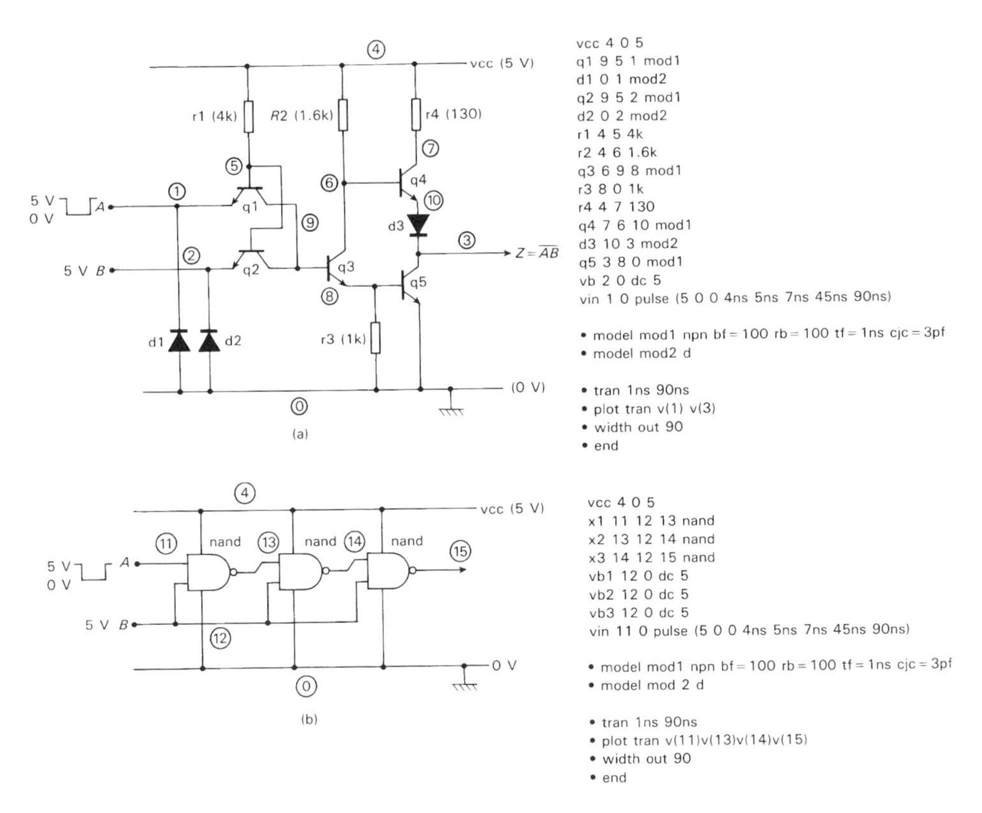

Fig. 5.25 SPICE simulation of a 2-input TTL NAND gate: (a) the gate circuit with all internal nodes identified and the resulting netlist: the diode model 'mod2 d' is the default mode; (b) simulation of these such NAND gates in cascade, with the signals at input node 11 and output nodes 13, 14 and 15 to be printed out: the complete netlist for this simulation must begin with the netlist for nand, which will be the same listing as given in (a). Note, a SPICE netlist may use all lowercase letters as here for component identification or all capital letters, but not both. Also, the precise syntax in some commercial SPICE programs may vary slightly.

literature [112–115]. Since SPICE is a general-purpose simulation covering linear, nonlinear and transient analysis, we will refer to it again in Section 5.2.5.

5.2.4 Back-Annotation and Postlayout Simulation

As has previously been noted, it is not possible to include accurate resistance and capacitance values associated with gate, macro or I/O interconnections in

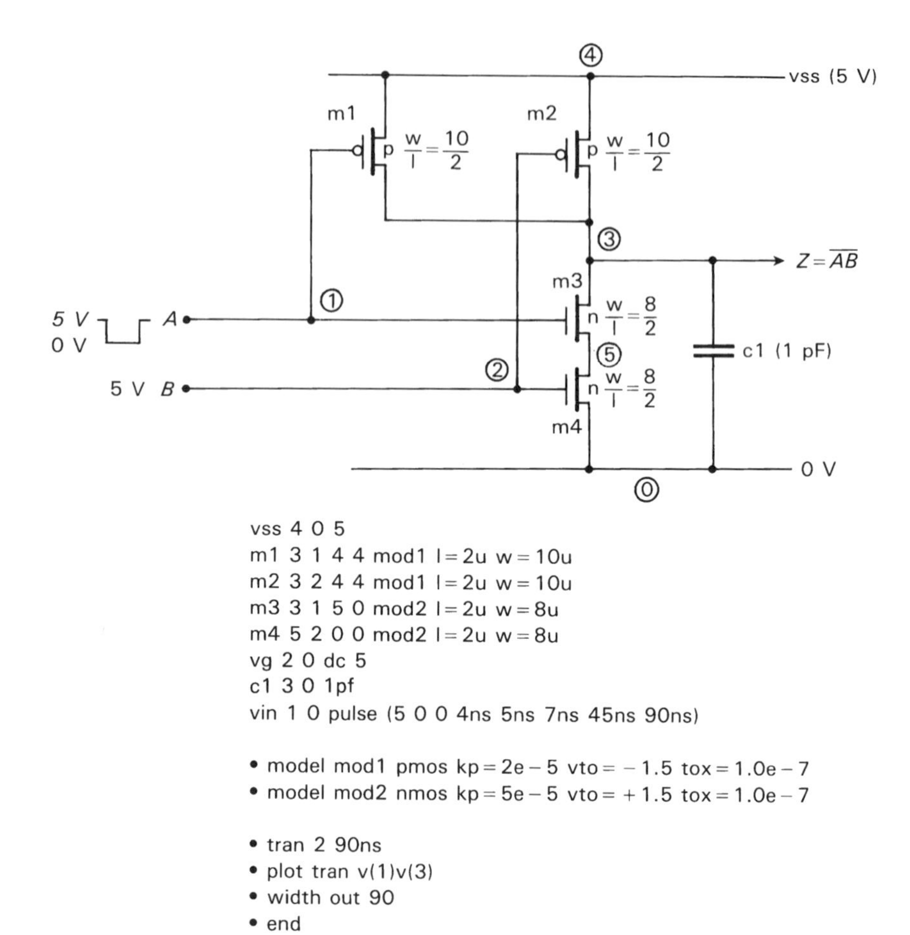

Fig. 5.26 SPICE simulation every 2 ns of a 2-input CMOS NAND gate, where the *W/L* details of each transistor must be specified. Note that the substrate connection of each MOS transistor also has to be specified, = the most positive rail for p-channel devices and the most negative rail for n-channel devices.

any simulation activity until the physical design has been completed. Prior to this, estimated or nominal *R* and *C* values have to be used, for example, to consider each gate output having to drive a capacitance of $n.\ C_G$, where n is the fan-out of the gate and C_G is some nominal value of capacitance associated with one gate input. However, in the actual floorplan there may be certain long interconnect paths and/or the use of a number of polysilicon underpasses which could materially affect the maximum speed of the circuit.

The layout extraction tools providing this detailed information on interconnection paths are invariably vendors' tools. The OEM may use them in some circumstances, but generally it is part of the final activity undertaken by the vendor or independent design house just before manufacture is commenced. The extraction tools provide the following data:

- details of each individual interconnection run, and hence the interconnect R and C values
- identification of long interconnect paths or other critical paths
- the annotation of the netlists previously used for prelayout simulation with this detailed information—hence the term *back-annotation* which indicates this addition to the previously compiled netlists

This is indicated in Figure 5.27 [116–119].

The postlayout simulation using this information is the closest simulation to the final product that can be performed before manufacture. The problems which postlayout simulation may find and which were not evident before in earlier simulations invariably concern long interconnection paths in critical paths of the circuit, so that the circuit speed does not meet the required specification. This could be the fault of the original specification working too close to the limit of performance of the chosen technology or vendors' product, but whatever the reason a redesign is necessary either to:

- revise the placement in order to improve these critical interconnect paths
- revise the circuit design to reduce gate fan-out and hence increase speed
- relax the original specification to meet what has been achieved
- change the technology or product to a higher specification

Any design or technology or product change is, of course, likely to be expensive, and hence conservative figures should always be used wherever possible in the original design activity.

The most powerful vendors' postlayout tools can also create a new circuit netlist from the final placement and routing, and hence a circuit diagram. If this extraction is then simulated instead of back-annotating data into the original netlist, the most comprehensive check of all the design stages, including the physical design activity, is available. But whatever the checking procedure undertaken after placement and routing, the OEM will finally have to give the vendor or design house a formal approval and sign off the design as acceptable before any further work is undertaken.

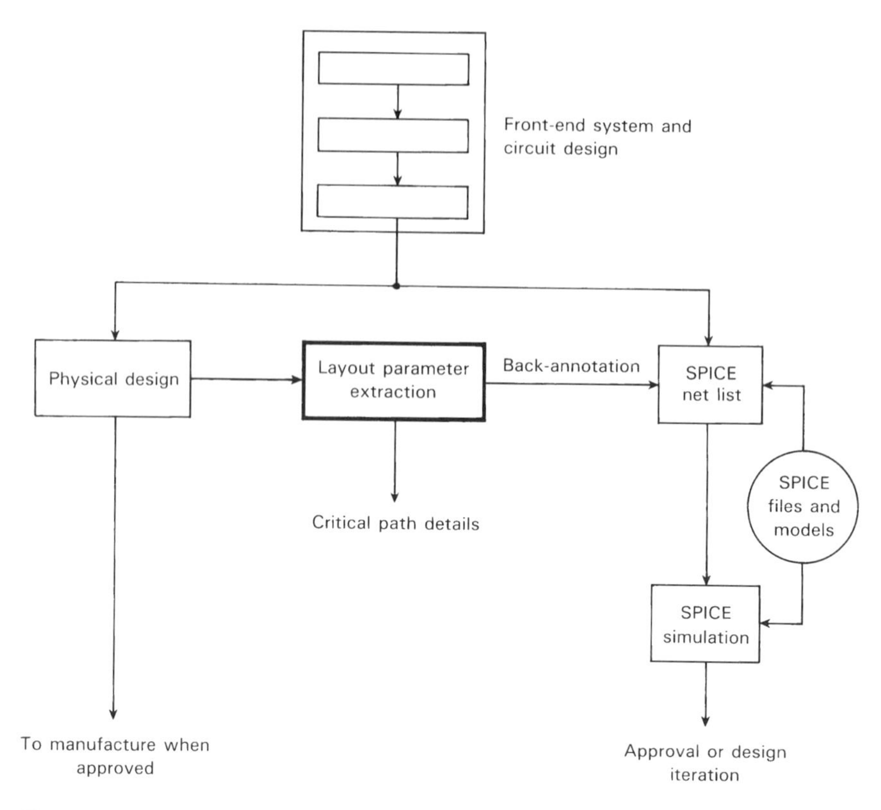

Fig. 5.27 The concept of back-annotation whereby the layout extraction tool feeds back parametric data on the routing to a previous simulation netlist. SPICE simulation is shown here, but equally gate or switch level simulation may be involved, particularly with VLSI layouts.

5.2.5 Analog and Mixed Analog/Digital Simulation

Analog-Only Simulation

For analog-only circuits, the almost universal simulation tool is SPICE and its commercial derivatives. The simulation accuracy of SPICE over a wide range of voltage, current, frequency and temperature makes it supremely relevant. Its principal disadvantage, namely its inability to handle circuits of VLSI complexity due to the detailed calculations involved, does not arise in analog circuits, since it is rarely the case that more than a few hundred transistors at the most are involved.

The use of SPICE for analog simulation is described in many texts [13,80–

83,112,114,120]. Most commercial SPICE-based tools provide a graphical as well as a textual output, with the capacity to display frequency characteristics, etc., in the conventional manner with which a designer is familiar (for example, see Figure 5.28). The ease of changing component values or tolerances in a simulation and observing the results makes analog design a very much faster process than having to construct a physical breadboard layout, but this does not, of course, relieve the designer from having to provide all the creative input.

Mixed Analog/Digital Simulation

For mixed analog/digital simulation it is exceedingly difficult to provide a single comprehensive CAD tool, largely because of the fundamentally different requirements of fine detail but low component count of the analog parts and the coarser detail but high component count of the digital parts. This is sometimes referred to as the 'little-A, big-D' situation. The present solution is either (i) to have separate analog and digital simulators and manually transfer data between them, or (ii) to have one system containing two separate simulators within one simula-

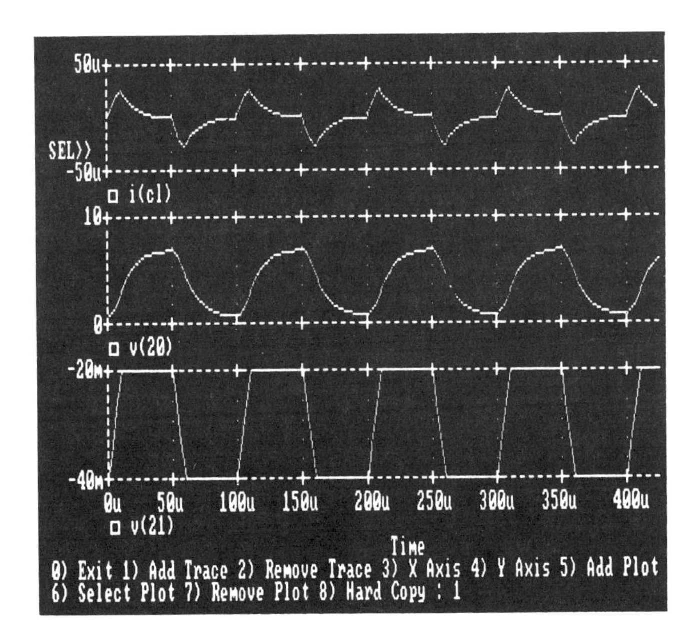

Fig. 5.28 Screen display of an SPICE simulation of an analog circuit with nodal simulations being performed for each 1 μs interval. For obvious reasons such graphical displays are sometimes referred to as 'oscilloscope' outputs. (Courtesy of MicroSim Corporation, CA.)

tion environment, with the user programming which one to apply to the different parts of the circuit. For example, a SPICE derivative may be used for the analog parts, and switch-level or gate-level simulation for the digital parts. However, for 'little-A, little-D' situations, for example uncommitted analog arrays which also have some limited digital capability, SPICE may be relevant for simulating the complete circuit, provided sufficiently comprehensive CAD computing power is available.

In some commercial packages, the ability to move from one simulation tool to the other, for example to initialize circuit conditions before switch-level or gate-level simulation is started, or to analyze individual digital macros in greater detail than is provided by the digital simulator, may be particularly useful, capitalizing upon the feature that the circuit simulator analyzes every individual node in a simulation step, whereas the digital simulator processes only event changes [108].

However, it is the method of organizing the interface between the two simulators that distinguishes different vendors' approaches to this area. This is made more complex if there is feedback between the digital and analog parts, for example in phase-locked loops, than in the case where there is only a one-way transfer of data, for example in D-to-A and A-to-D conversion. The five principal methods which have been adopted to handle mixed simulation are as follows:

- (i) unidirectional coupling of separate analog and digital simulators, running one at a time
- (ii) bidirectional coupling of separate simulators either running in parallel or running alternately
- (iii) unified coupling with the appropriate analog or digital simulator being automatically called for by subroutines
- (iv) single integrated simulator, sometimes termed a 'core' or 'extended' simulator, consisting of a core analog simulator with provision for analog behavioral models of digital cells and macros
- (v) similar to (iv) but with the core simulator being a digital simulator with analog behavioral models for the analog parts

In the first three of the above categories, an appropriate boundary interface must be provided to link the two quite separate types of simulator. This data link tends to be vendor-specific, as are the behavior models used in the last two of the above categories.

Among the difficulties in linking the analog and digital simulations is the dissimilarity of information in the two domains. For example, an analog circuit driving a digital gate requires knowledge of the gate input impedance characteristics before the analog output signal can be accurately determined, but this is not usually available in digital circuit modeling. Conversely, in the digital simulation, discrete information on signal driving strengths (see Section 5.2.2) and on unknown (X) conditions is required, which is foreign to analog simulators.

A single mixed-mode simulator with one unified modeling language which can handle both analog descriptions and digital descriptions, and with hierarchical ability to cover primitive to architectural level, would clearly be advantageous. Considerable research work is continuing in this area, but the fundamental difference between analog and digital simulation will always remain [108,121–124].

The graphics available for mixed analog/digital simulation, however, are already very useful. Figure 5.29 shows a schematic entry of a mixed circuit, which can subsequently be simulated to give the analog and the digital behavior. Top-down schematic capture is available, the analog cells being characterized in

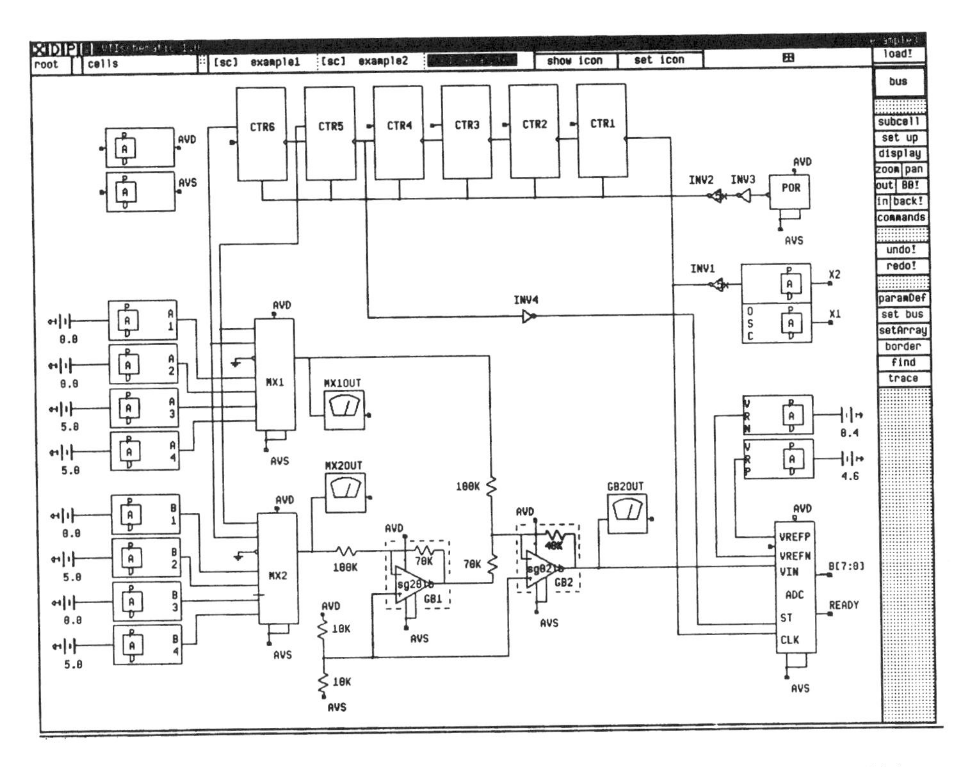

Fig. 5.29 The mixed analog/digital simulator MIXsim of Sierra Semiconductor, which provides schematic entry of mixed analog and digital macros from a fully characterized standard-cell library, with meters added to the schematic to highlight analog nodes of particular importance. Other forms of highlighting may also be used. (Courtesy of Sierra Semiconductor Corporation, CA.)

a standard-cell analog library so that detailed SPICE netlists do not need to be entered [121].

Textual output or graphics output of mixed simulation activities is also available. In particular, the use of windows to display both analog and digital results side-by-side is particularly useful. Figure 5.30 illustrates the type of capability which is available in a mixed graphics-input/graphics-output resource.

5.2.6 Simulation Accelerators and System Simulation

Gate-level and switch-level simulation is appropriate for most digital IC designs, but as the complexity increases to greater than, e.g., about 50,000 gates simulation run times become unacceptably long, and computer memory requirements very great. A quoted gate-level simulation time for a smaller 15,000 gate design running on a 68030-based workstation is over 20 hours, using over 4 Mbyte of

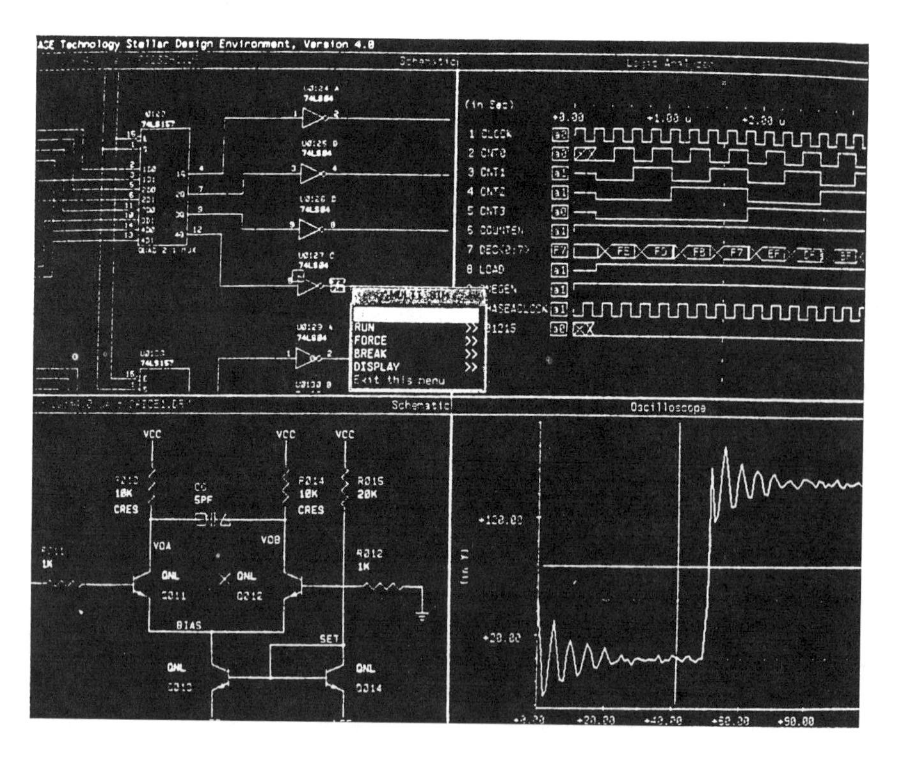

Figure 5.30 The multiple window display availability with MultiSim, which supports both analog and digital simulators working on the same mixed-schematic circuit diagram. (Courtesy of Teradyne. Inc., MA.)

memory for the data, and this time will increase rapidly with an increase in gate count. Such CPU times also depend upon the accuracy of the gate-level model, that is, how many logic states and impedance levels are involved, as well as the computer power running the software.

The use of mixed analog/digital simulation reduces CPU time at the expense of accuracy, and most vendors provide this resource in their digital simulation tools. However, there still remains the difficulty of accurately modeling very complex macros, particularly microprocessors, which due to their complexity and software dedication are difficult to model, and may require many thousands of lines of code. This, in turn, makes the simulation of such modules very time consuming. Architectural-level and behavioral-level simulation, on the other hand, will not be sufficiently accurate for the final confirmation of a design.

A further problem which we have not mentioned in the preceding sections is that a custom IC is often only part of a more complex system, and may have to work in association with other off-the-shelf ICs of LSI or VLSI complexity. The microprocessor may, for example, be a separate IC rather than a macro on the custom chip, and therefore the simulation of the complete system may be particularly important to confirm the details of the custom design.

The following approaches may be found to speed up the simulation of large digital systems:

(i) the use of computers specifically designed for simulation duties
(ii) the use of actual physical circuits (e.g., a working microprocessor) as part of the simulation tool rather than doing all the simulation in software

The former is usually referred to as a *simulation accelerator*, and the latter as a *hardware modeling* system.

Simulation Accelerators

Simulation accelerators support event-driven simulation at gate level and possibly at functional and architectural level, but not at the detailed circuit level which is conventionally handled by SPICE-like simulation software. The improvement in simulation speed is achieved by:

(i) dispensing with a normal general-purpose computer architecture since simulation algorithms do not map conveniently on to general-purpose computers, and designing the accelerator processors to match the simulation algorithms
(ii) by the use of parallel or pipelined processing to increase the throughput

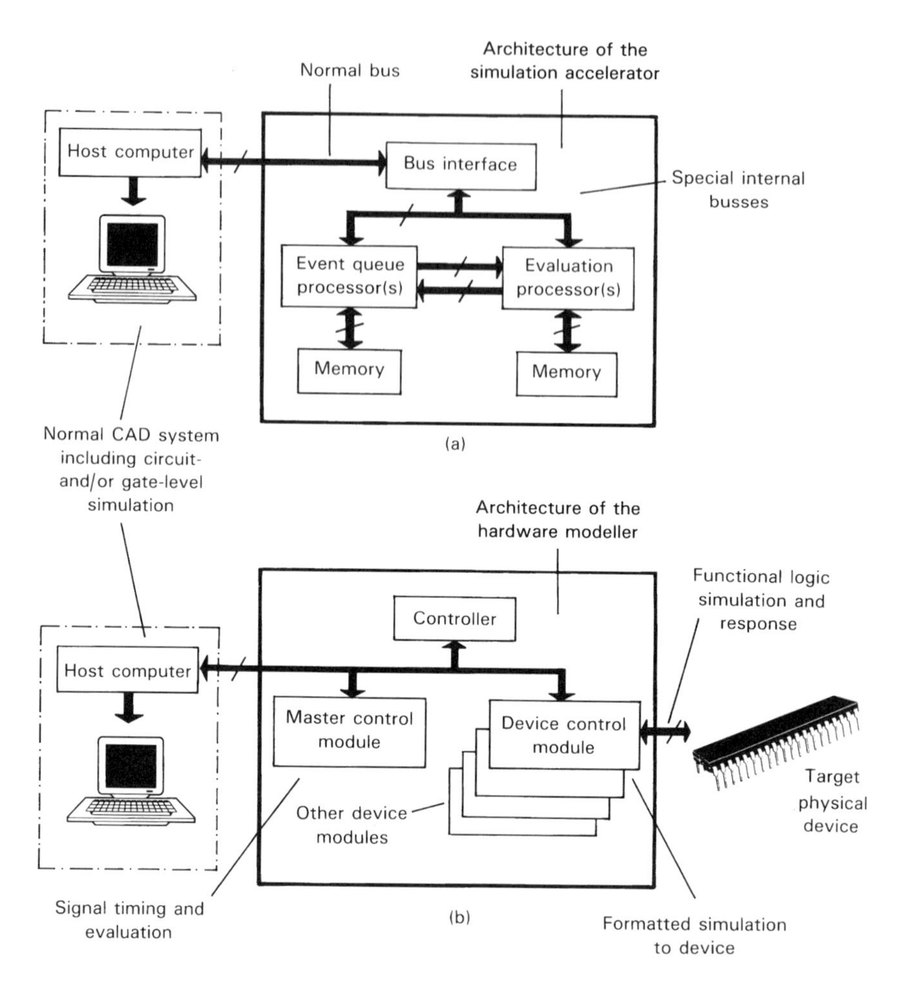

Figure 5.31 Multisimulator multiplatform simulation environments, with design data downloaded from the host computer for fast simulation: (a) the simulation accelerator with high-speed custom-designed processors; (b) the hardware modeling system with it target physical VLSI device used in the simulation loop to provide simulation signals.

A conventional host computer used for the design activities, and perhaps for some circuit-level simulation, is still required to preprocess and postprocess the design simulation data, but does not take part in the accelerated simulation process. This is indicated in Figure 5.31(a). Simulation throughputs of over 1 million events per second with one million logic gates simulation capability are provided in commercial products. Further details may be found in published literature [7, 107,125–127] and in vendors' literature.

Hardware Modeling

Hardware modeling, however, sometimes referred to as *physical modeling*, is unlike any of the previously considered simulation methods in that it combines both software and target hardware in one CAD tool. By using an actual physical device, e.g., a microprocessor or other functional VLSI component, its response to inputs received from the simulator program provides a fully functional response back to the simulator in real time for the continuation of the simulation program for the system design under test.

Like simulation accelerators, a hardware modeling system is a resource additional to the normal CAD tools. The resource comprises circuitry to configure the simulation signals to the physical device and to process the returned signals, and to provide all the other necessary interface requirements for the rest of the CAD system. This is illustrated in Figure 5.31(b). The vendors of such products provide library details to interface to a range of standard off-the-shelf LSI and VLSI devices, so that an OEM can use the system with a working device which acts as an accurate signal source for the rest of the simulation program. Guaranteed working devices may also be supplied by the vendor, but in any case diagnostic test procedures may be built in to confirm continuing fault-free operation of these physical devices [7,107,124,128].

The cost of both simulation accelerators and hardware modeling systems is high, running into many tens of thousands of dollars, and hence such resources would normally be provided for each large design office as a one-off shareable resource, being networked to multiple users on an Ethernet or other local area network (LAN) facility (see later). Additionally, both these types of accelerator may be combined into one system simulator which enables them to communicate directly with each other instead of having always to go through the serving host computer.

These complex simulators are beyond the need of most OEMs, and are likely to remain the tools of major vendors and design offices. They do, however, have a role to play in connection with very large custom circuits, for example, large channel-less arrays containing hundreds of thousands of transistors, and for the final simulation of large systems containing custom ICs in association with other large circuits. A further factor is that the increasing speed of conventional computers and workstations is extending the size of simulation that they can acceptably handle, which makes the need for these more powerful simulation tools somewhat less than was previously considered necessary.

5.3 SILICON COMPILERS

In its fullest sense, a silicon compiler is a CAD tool which could take a behavioral description of a required circuit and from it generate the silicon floorplan design details ready for manufacture. If this is done without human intervention, then the final design should be correct by construction, and in theory should not require any separate simulation and verification procedure. The term 'compiler' is, of course, taken from computer programming, where the software writer details the required computer program in a high-level language which is then automatically compiled down to the final machine code. The final code is correct by construction and any malfunction must be a result of an error or omission at the human level. The final code may not be as compact and efficient as could be produced by a human programmer, but the amount of effort required to program at machine code level is prohibitive except for very small programs of less than, e.g., a hundred or so lines of machine code. A further discussion on the semantics of software compilers vs. silicon compilers may be found in Russell [1].

However, unlike computer programming where for a specific machine and specific compiler there is one final result of the compiler operating on an input program, a high-level digital system design description can theoretically have an infinite number of ways in which it can be made at the silicon level, due to the flexibility of digital logic design. Guidance must therefore be given to a silicon compiler to tell it what choices to make in the hierarchical expansion down to the physical level, or it must be designed and used for a particular restricted type of digital circuit realization. In most cases, a more accurate description may therefore be a 'silicon assembler' rather than compiler if predetermined building blocks are used in this CAD top-down design activity.

The silicon compilers which have been developed fall into the following categories:

- compilers for memory arrays (ROM and RAM), A-to-D and D-to-A converters, arithmetic logic units and other macro functions with known architectures
- compilers for programmable logic devices where the silicon architecture is fixed and the design process consists of mapping a behavioral specification on to the device topology
- compilers for digital signal processing (DSP) applications, for example, digital filters, where the circuit configuration is known or defined
- compilers for telecommunications and control and data path applications, again where the general circuit configuration is known or defined
- certain analog applications, particularly real-time filters, using a menu of standard analog circuit configurations but requiring the computation of component values to meet the given specification
- compilers for digital circuits where the designer first has to decide from an available menu which types of building blocks to use in the design, the compiler then assembling and optimizing them for the particular application

The first category above is more often referred to as a *module generator* rather than a silicon compiler, since it deals with only one block in a larger total system. Module generators usually only optimize the size and silicon layout of a soft macro design to meet the required specification [10,129], but may be associated with a larger compiler dealing with the whole system.

Compilers for PLDs, however, are true silicon compilers in that no human intervention is necessary between the process of entering the system requirements in truthtable or state table or other form, and the design output stating how the target device is to be programmed. This is possible because there is little freedom to innovate with the fixed architecture of a PLD, and because the performance available is also largely predetermined by the device design. Correct-by-construction should therefore always result, providing the correct design requirements are entered by the designer [99–106,130].

Compilers for DSP, telecoms and other applications, including analog, have the advantage that the general circuit configuration is known or is entered by the designer, and that the rules for calculating all component values and other variables for the final design are known. Hence, the CAD tool can be entirely rule-based, not requiring any innovative procedure once the design requirements have been fully specified [83,95,131–133]. One of the advantages claimed for compilers used for analog purposes is that the designer does not need any in-depth knowledge of analog design; this is perhaps too simplistic, since considerable innovative work may be necessary to finalize the specification within the bounds of what is possible, and the CAD may still be unable to complete the synthesis, thus requiring designer intervention and guidance.

However, it is in the field of nonprogrammable digital VLSI circuits that developments in silicon compilation have usually been associated, the objective being to encompass all the tedium of a large digital circuit design activity from behavioral level to silicon level. The first known development in this area was the work carried out at the California Institute of Technology (CALTECH) using their 'bristle-block' approach. Here, the required digital system was designed as a data path architecture, involving registers, arithmetic logic circuits and data-shifting circuits linked by appropriate data highways (busses). These 'core' cells were defined as parameterized soft macros stored in outline but not in a detailed layout form, and the CAD tool called up the required cells and stacked them along several parallel data paths so as to form a roughly rectangular layout. Cell 'stretching' or 'compression' was then undertaken to equalize cell heights along each row of the layout.

This compilation technique is appropriate for data-path types of circuit where signals are passed linearly from one cell to another, but not for random logic requirements which do not map conveniently into this topology. Thus, some additional PLA-type structure was also associated with a bristle-block compilation to cater to the more random requirements. Further details of this early work may be found in the literature [74,134], but overall, this technique had a very restricted application and produced a final silicon layout which could be as much as 50% to 100% greater in area than could otherwise be achieved.

Commercial silicon compilers have been marketed by several vendors, although most are more truly described for the reasons previously suggested as silicon assemblers. Among the products of the 1990s have been Seattle Silicon Corporation's CONCORDE VLSI compiler [135], and Silicon Compiler Systems Corporation's GENESIL compiler [136]. The efficiency of certain compilers in comparison with a hand-crafted full-custom design for the 8-bit 65C02 CMOS microprocessor has been investigated by Evanczuk [137], who showed that there was a still considerable silicon area penalty to be paid for the semiautomatic design solutions produced by the compilers. Further details of the theory and practice of silicon compilation may be found in the literature [134,138–140].

In summary, silicon compilers currently require a design to be described at the functional or macro level, with the software automating most of the design activity from there on, and hence they are only a little more powerful than fully automatic place-and-route tools. Simulation still remains a separate and necessary activity to verify that the design details are correct and that the performance meets the required specification. However, on-going research and development into the use of more powerful languages and expert systems for the logic synthesis, and the incorporation of some level of artificial intelligence, is making the possibility of correct-by-construction design from a high-level description more realistic [139], although a target floorplan and/or specified logic functions to be used will possibly always be necessary.

5.4 CAD HARDWARE AVAILABILITY

The general evolution of CAD hardware and software resources for IC design has been introduced in Chapter 1. Here we will amplify this information and indicate its present maturity.

The use of large mainframe computers is now no longer relevant for the actual design activities, since the evolution of the workstation and personal computer (PC) has enabled sufficient computer power to be brought to each designer's desk for most of the design work. However, a powerful mainframe resource, with its mass storage capacity, will still be required by large design offices as a host computer for the following activities:

- to undertake long simulation runs from data downloaded from individual designer's workstations
- to coordinate and handle the data transfer between different areas of activity and generally take charge of the scheduling and housekeeping activities

- to act as the principal data bank for all the company design and manufacturing details, with restricted access to ensure that no part of a product design can be changed without authority and without appropriate wider consultation

This overall CAE product control may possibly be considered as part of managerial or product engineering duties rather than that of the IC designer, but the latter must be aware of how his or her role fits into the complete picture [141,142].

The general diversification of CAD hardware is therefore as follows:

- mainframe and other large general-purpose digital computers, for example, IBM, DEC, etc., with commercial or proprietary software for centralized duties
- stand-alone workstations, for example, HP, VAX, Sun, etc., with commercial or proprietary software for design purposes
- stand-alone IBM or equivalent PCs, usually with commercial software for relatively more simple design duties

All these resources, however, may be networked together (see Section 5.4.3). We will not consider the first of the above categories any further here, but we will look at the latter two in the following sections.

5.4.1 Workstations

The term *workstation* may broadly be defined as a hardware assembly consisting of the following:

- a central processor, currently a 32-bit processor with present μPs
- a VDU, usually 15″ or 19″, monochrome but increasingly color
- input devices, usually keyboard and mouse
- working (main) memory
- bulk memory, invariably several hundred megabytes minimum
- communications (networking) facility to printers, other workstations, host computer, hardware accelerator, etc.

Although there have been workstations designed by IC software vendors who then marketed a unique packet of matching hardware and software for IC design activities, for example, the pioneering activities of Daisy Corporation, Mentor Corporation and Valid Logic Systems Inc. [74], evolution has seen the adoption of general-purpose workstations by IC CAD vendors, leaving them free to concentrate on the development of increasingly powerful software (see Section 5.5). This separate specialization of hardware and software suppliers is an indication of the intensity of the continuing development of the two areas, wherein one company cannot easily remain an expert in both fields.

The suppliers of workstations currently include Digital Equipment Corporation, Hewlett Packard, Sun Microsystems, Intergraph and others, although a great deal of commonality may be found between all the products on the market, particularly in the central processor used in the product. The latter may be from Intel or Motorola or other source, indicating the yet further breakdown of high-technology specialization. However, the following two developments in the current generation of CAD workstations may be observed:

- the almost universal adoption of the UNIX operating system, originally developed by AT&T
- the increasing use of Reduced Instruction Set Computing (RISC) processors

The RISC/UNIX combination was pioneered by Sun in their Series 4 workstations, and initially employed Fujitsu CMOS gate array ICs in their design. Intergraph Corporation was also a pioneer in this field, using the high-speed CLIPPER chip set originally designed by Fairchild. However, it is not realistic to attempt to quote here current workstation performance figures, since the increase of performance shows no signs of abating; Figure 5.32 shows the general rate of increase in computing power in the 1980s, though it is possible that this almost exponential average increase in power with time cannot now be economi-

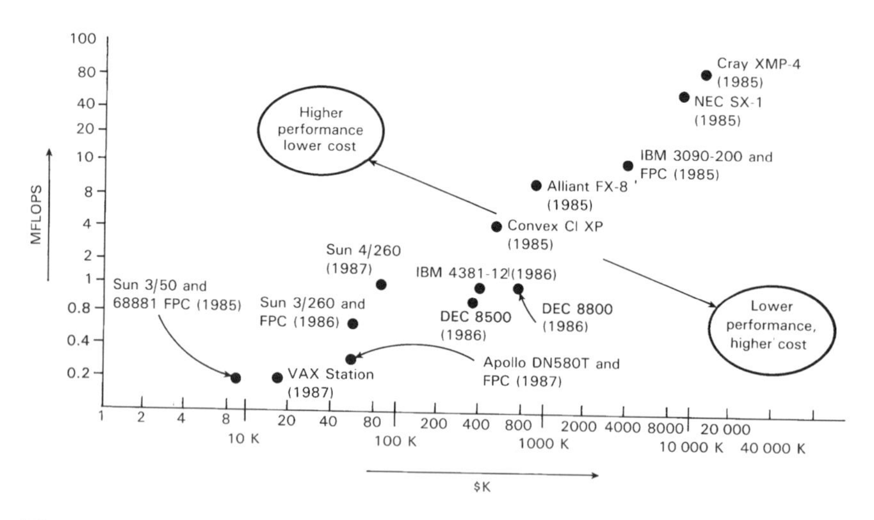

Fig. 5.32 The general capability in MFLOPs vs. cost of hardware computing platforms in the mid-1980s. The trend since this period has been for maximum performance to increase, but performance per dollar to decrease except perhaps at the frontier of performance. FPC = floating point coprocessor. (*Source*: based on Ref. 143.)

cally sustained. Note that this figure has workstation developments at the lower end, and supercomputers at the top end; also, the *y*-axis is in 10^6 floating point operations per second (MFLOPs^{-1}) rather than 10^6 instructions per second, since graphics processing relies heavily upon floating point operations, typically requiring about ten times as many MFLOPs as MIPs to reach a given standard of user response [143]. It is of interest to note that all the hardware platforms shown in Figure 5.30 incurred costs of around \$100,000 per MFLOPs^{-1}, although it is not obvious why this should be so. The end of 1990s workstation should provide a performance of about 25 MFLOPs^{-1} at a lower cost than may be suggested by Figure 5.32.

5.4.2 Personal Computers (PCs)

The personal computer, first introduced by IBM in the 1970s, has increased in power and application so that it is now becoming difficult to draw a clear dividing line between workstations and PCs [143,144]. In terms of operating speed and memory capacity, the PC generally compares with workstations as follows:

- PCs: 2 MIPS, 1 to 2 Mbytes RAM, 10 to 20 Mbytes floppy disk bulk memory
- Early 1990s workstations: 4 MIPS, 4 to 8 Mbytes RAM, 100 or more Mbytes hard disk bulk memory
- Newer RISC/UNIX workstations: 10 MIPS, 8 to 16 Mbytes RAM, several hundred Mbytes bulk memory

These are typical figures only, and should be read as comparative data only between the groups. Table 5.3 gives some further data to illustrate the evolution.

The problem with the first-generation PCs was largely with the MS-DOS operating system, which was good for simple housekeeping and clerical duties but not so relevant for CAD/CAE work since multitasking and windowing cannot be supported by MS-DOS. Additionally, the original MS-DOS could only address 640 K of RAM. However, the strength of the PC lay in its wide global adoption and low cost, with numerous clone versions which encouraged the widespread development of CAD and CAE software from many vendors, and which worked as efficiently as possible within these constraints. The later introduction by IBM of the OS/2 operating system and a new bus structure ('Micro Channel Architecture,' or MCA) was introduced to recover the uniqueness of the IBM product and to prevent cloning, but the subsequent acceptance of UNIX has forced IBM and most workstation vendors to adopt this as the operating system for CAD/CAE purposes—this also forced DEC to provide a UNIX operating system for its range of processors in place of its own and possibly better VMS operating system.

The other original strength of the PC was the ability to plug in additional

Table 5.3 Typical Specifications for a Range of Second-Generation PCs and Workstations, Representative of the Stage When Desk-Top Resources had Become Sufficiently Powerful for Most OEM VLSI Custom Design Activities

Product	CPU	Floating point coprocessor	Clock rate (MHz)	MIPS	Operating system	Mbytes RAM (standard or minimum)	Mbytes hard disk (standard)
IBM PC/AT	80286	None	6	N.A.[d]	MS-DOS	0.64	20
IBM PC/AT	80286	80287	8–10	N.A.	MS-DOS	0.64	20
IBM PS/2 Model 70/80	80386	80387	10–25	N.A.	OS/2	4.0	60–640
IBM System 6000[a]	Own	on-chip	20	29.5	AIX	8.0	120–640
Compaq 386	80386	80287	16	N.A.	MS-DOS	1.0	140
Apollo DN3000	68020	68881	12	N.A.	UNIX	2.0	70
HP Apollo 9000/400 dl	68030	68882	50	12	HP/UX	8.0	200–400
HP Apollo 9000/425t	68040	on-chip	25	20	HP/UX	8.0	200–400
Sun 3/260	68020	68881	25	3	Sun UNIX	8.0	280
Sun SPARKstation SLC[a]	SPARC	SPARC	20	12.5	Sun UNIX	8.0	104–2007
Sun SPARKstation IPC[a]	SPARC	SPARC	25	15.8	Sun UNIX	8.0	104–2007
Sun SPARKstation 470[a]	SPARC	TI 8847	33	22.5	Sun UNIX	32	669–8000
Intergraph 2000 series	Clipper +80386	Clipper	25	12.5	UNIX	16	200
Sony NWS 1500	68030	68882	25	4	UNIX	8.0	240
DEC VAXstation 3100	Own	Own	25	N.A.	UNIX[b]	32	208
DEC VAXstation 3100[a]	Own	Own	25	N.A.	UNIX	32	208
DEC VAXstation 8000[c]	Own	Own	25	N.A.	UNIX[b]	32	208

[a] RISC/UNIX based workstations.
[b] VMS also available.
[c] Enhanced graphics capabilities.
[d] NA = not published or not available.

expansion and processing boards on the PC bus to increase the performance and relevance for specific duties. This has been the principal reason why PCs have been so successful for many industrial design and development purposes, and, conversely, why the later OS/2 operating system with its new MCA bus structure was not welcomed since it was incompatible with the vast investment in enhancement boards that had been developed and marketed.

A further player in the field below the workstation has been the Apple Macintosh computer. The strength of the Apple products was in their intensive adoption of icons and windows, which made them the most favored hardware platform of the 1980s for scientific word processing, desk-top publishing and 3-D pictorial representations. However, the advantages of graphics, icons and windows has now been learned by the PC vendors, and as a result many of the early forecasts that Apple Macintosh hardware would find a place in IC design activities have not happened. This leaves us with the modern general-purpose PC as the most widely used hardware platform now and in the foreseeable future.

With continuing company take-overs, and the blurring of the distinction between workstations and PCs, it is difficult to forecast the future vendors of IC CAD hardware with any great precision. However, it seems probable that there may be only about four major suppliers, for example, IBM, DEC, Hewlett Packard and associates, and Sun. Software vendors (see Section 5.5) will need to ensure that their programs run on all the commercially available platforms, but this should be facilitated by the increasing standardization of formats used in the IC design area.

5.4.3 Networking

Isolated stand-alone PCs and workstations are appropriate for very small design offices, where one designer can do all the required in-house design activity per product, and pass his or her design file(s) over to, e.g., a vendor for further design simulation and fabrication. However, for organizations with more than a small number of machines, it usually becomes necessary or desirable to link resources so that they can exchange data or share expensive common hardware such as large color plotters, etc. This desirability applies equally to PCs and workstations.

The usual means adopted to link distributed in-house resources is a serial bit stream transmitted over a *local area network* (LAN). A number of LAN designs are commercially available, but the two most common ones are:

- the Token Ring, originally produced by IBM
- Ethernet, which is a bus-type interconnect

The Token Ring

The Token Ring is a loop linking all the hardware stations (see Figure 5.33 (a)), the interconnect medium being a fiber-optic cable or a metallic screened twisted

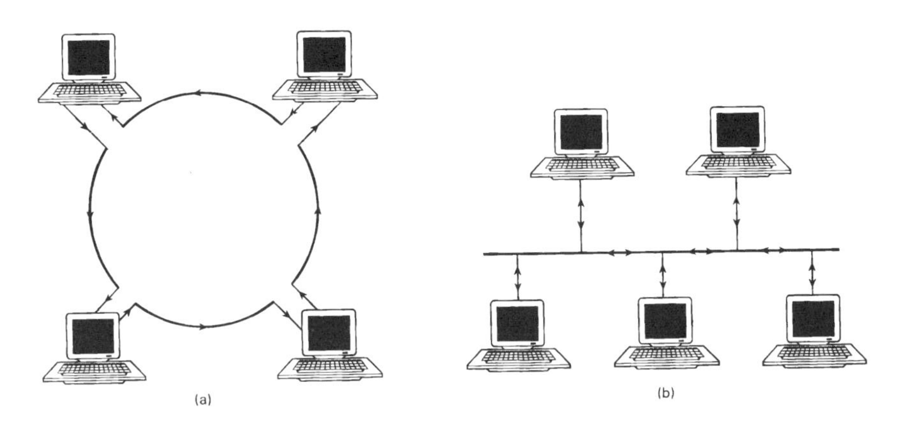

Fig. 5.33 Token Ring (a) and the Ethernet (b) local area networks.

pair. Access to the ring is controlled by an electronic token which is transmitted continuously in one direction around the ring, being set to indicate whether it is alone or accompanied by a destination address plus information for this address. The token is interrogated at each station on the ring, and is then passed on to the next one until it finally reaches its destination address. Here it is accepted, together with the information that accompanies it. In the absence of information the token continues to be circulated.

When a station requires to send a package of information, it checks to see if the received token is alone, and if so adds its information and the required destination address. The Token Ring originally offered a transmission bit rate of 5 Mbits per second, but later interface boards with faster ICs introduced in the late 1980s increased this rate to 16 Mbits per second.

Ethernet

Ethernet differs fundamentally from the Token Ring, and does not have a ring connection pattern between stations. Instead, it is a bus-type interconnect, as shown in Figure 5.33(b), with all stations connected to the single interconnect bus. The stations on an Ethernet contend for access, and if two attempt to transmit simultaneously, then protocols determine the priority. Every station receives on the network, but ignores the data if they do not contain their own station address.

Ethernet can transmit at up to 10 Mbits per second, although the actual speed of data transfer can fall with a high level of traffic due to the priority protocols then involved. Both Ethernet and Token Ring are specified in the IEEE Open Systems Interconnect (OSI) standards, which specify electrical characteristics, protocols, cabling and connection details for local area networks. Ethernet, IEEE Specification 802.3, is more usually used in engineering design and manufacturing environments, with Token Ring, IEEE Specification 802.5, possibly used more in nontechnical business environments.

LANs are typically limited to a total length of about 1 km because of cable losses and interference problems. However, in order to use the higher bandwidth and reduced losses of modern fiber-optic cables and connectors, and to capitalize on their inherent immunity to interference, a new high-speed network topology has now been introduced. This has also become increasingly necessary because of the increase in file sharing, whereby a 'Network File System' allows workstations on the network to share comprehensive data on one set of disks located on the central host computer. To provide this high-speed resource, the *Fiber Distributed-Data Interface* (FDDI) system has been introduced. This consists of a dual-ring interconnection topology as shown in Figure 5.34, and uses a token method of station selection similar to the Token Ring. Both fiber-optic rings may be simultaneously used, giving a total bandwidth of 200 Mbits per second. Further attributes of the FDDI system are that it can span a much wider territory, up to

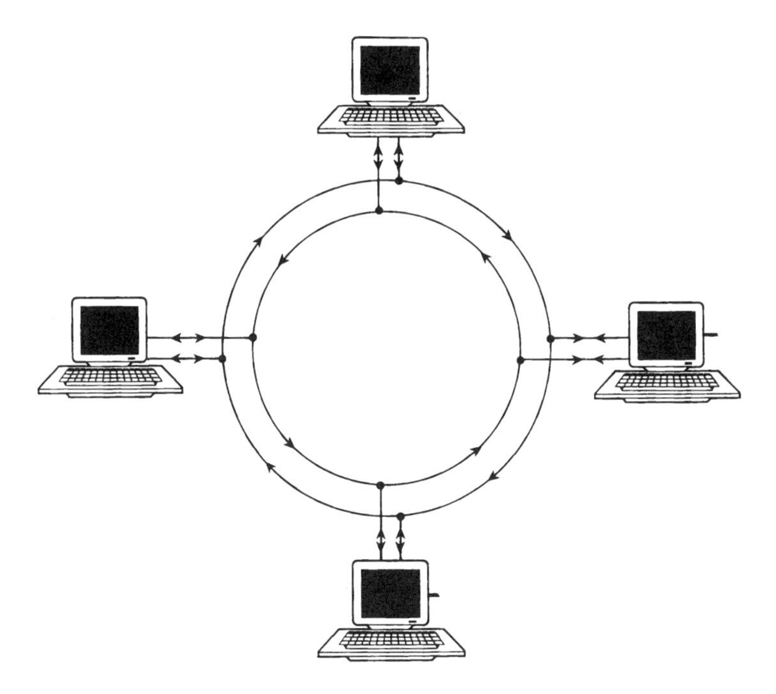

Figure 5.34 The Fiber Distributed-Data Interface (FDDI) local area network, with fiber-optic interconnections.

100 km if necessary, and can tolerate a break in one of the two rings without completely impairing the operation of the system. The FDDI system is therefore more a *metropolitan area network* (MAN) rather than a simple LAN [145].

When data have to be transferred in real time between different CAD resources at the full operating speed of the equipment, then the serial bit transmission of LANs (and MANs) is not appropriate. A local bus interconnect is now required which can link the actual operating busses of the two (or more) equipments so as to provide the required parallel byte interconnections. Among the systems that have been developed are VMEbus, Multibus and Futurebus+, with variants being marketed by individual vendors. These systems can transfer data at rates of from 6 Mbytes per second upwards, and may be combined with LANs for the serial transfer of off-line data. This is illustrated in Figure 5.35.

Bus interconnect systems are normally employed for linking equipments up to a few tens of meters apart, although in theory there is now no theoretical reasons why longer interconnect distances should not be possible using multiple-core fiber-optic cables. However, a big problem is how to link different types of systems, in particular VMEbus and Multibus, which have been extensively used

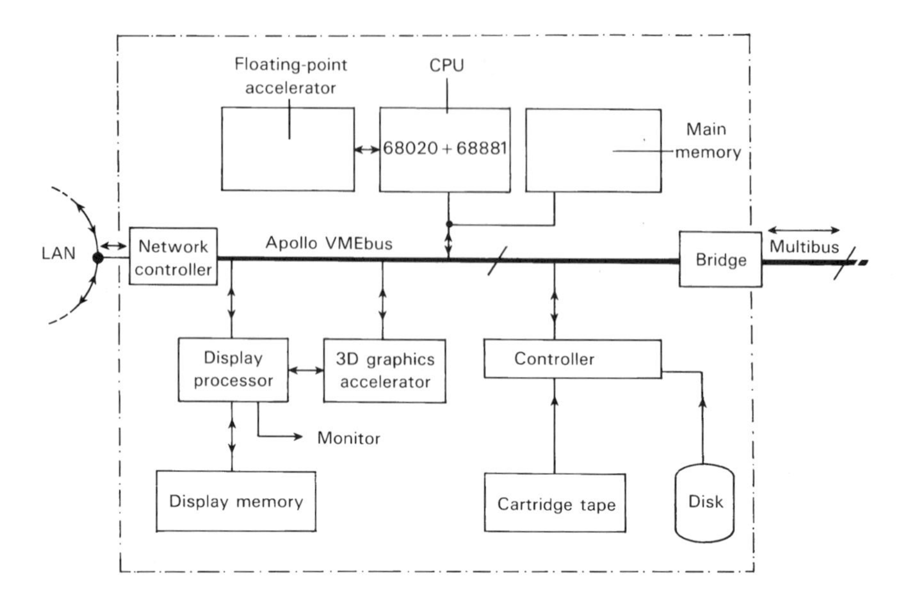

Figure 5.35 The integrated architecture of the HP Apollo Domain workstation environment. With local modules communicating with each other at bus speeds and with LAN communicating to further resources.

in the 1980s, to Futurebus+ which is the 1990s choice [146]. The difficulty here is that these busses utilize completely different event formats, making transfer of data an indirect procedure requiring 'bridge' and 'slave' circuits and registers to provide the parallel data transfer.

5.5 CAD SOFTWARE AVAILABILITY

The choice of CAD hardware and its continuing development is mirrored by the even more diverse and continuously changing range of the software for IC CAD purposes. Managerial decisions on what should be purchased for OEM in-house activities are therefore complex, and will be considered further in Chapter 7.

Another problem that has been an unfortunate feature of CAD software is that data formats have not been standardized. The effects of nonstandardization are twofold, namely:

(i) although individual software programs may be available for all the hierarchical levels of activity shown in Figure 5.3, they may not be able to be easily 'bolted together' to form an integrated suite without encountering interface difficulties—indeed, software is still being developed by academics and industrial vendors which cause great problems when attempts are first made to integrate them with other existing software

(ii) data formats may still differ between vendors, so that even with a simple gate-array choice of design style one vendor's software may not be compatible with, e.g., a PLD vendor's software

These problems largely arise because of the diverse origins of CAD software, compounded by the fact that vendors may have vested interests in not encouraging the free exchange of design data to other vendors' CAD systems. The OEM, however, would clearly prefer vendor independence for all the CAD software programs that he or she may use.

The CAD software which an OEM may use comes from three principal sources, namely (i) the IC vendors themselves, (ii) workstation and other hardware vendors, and (iii) independent software vendors. Software from academic institutions may also be involved, although it usually requires some commercial company to take it over and to make it more robust as well as updating and documenting it—SPICE is the classic example of an academic source of IC design software. These sources of IC software and their relationships to commercial products are illustrated in Figure 5.36.

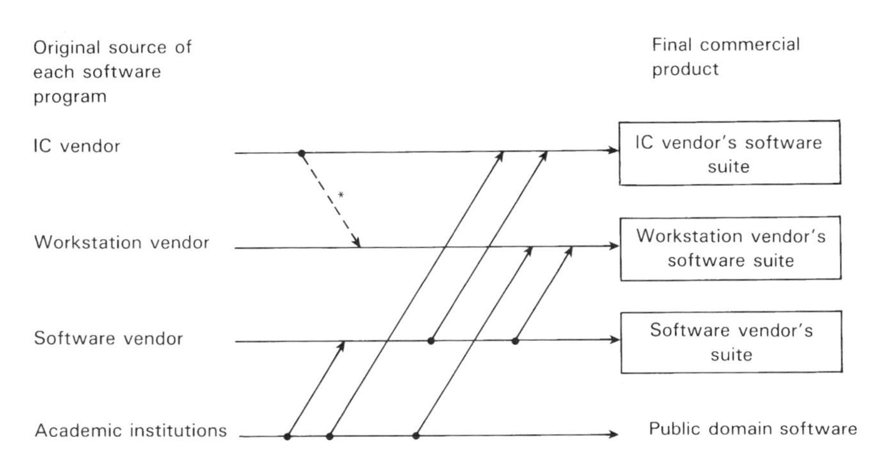

Fig. 5.36 The interleaving sources of IC CAD software. omitting very large OEM companies who may write their own software for internal use only. One software vendor's program may be sold to other software vendors in order to complete a comprehensive suite. (*Source*: based upon Ref. 17.)

The advantages and disadvantages of these various sources of original software are as follows:

- The software from an IC vendor will usually be specific for his own range of product devices, and in general will not cover the design of other IC vendors' products; however, it will cover all the expected OEM design activities, be well proven for its specific duties, and is likely to be inexpensive for the OEM to purchase or lease.
- The software from a workstation vendor may be tailored to the vendor's specific design of the hardware, and may not run on any other hardware platform; it will not be specific in matching the products of any particular IC vendor, and will therefore require further information in order that general design data may be mapped on to a particular vendor's gate-array or standard-cell product. It will usually be an expensive but comprehensive suite of software, suitable for independent design house use as well as by large OEMs.
- The software from independent software companies is likely to run on a range of hardware platforms, but like a workstation vendor's software will not be specific for any particular IC vendor's products. It may be a comprehensive suite of software, or a single software package designed to perform very efficiently one task, for example, simulation.
- Finally, the software from academic institutions is likely to be of an innovative nature when first released, not necessarily completely proven or documented, and therefore possibly requiring further work to turn it into a fully acceptable product.

Because software development costs are becoming so high and hardware platforms are becoming so competitive, increasing rationalization is found in this picture. The growing trend is for highly specialized electronic design automation (EDA) companies to market comprehensive suites of software, which consist of their own programs and also the best software currently available from elsewhere for specific duties, a complete suite of programs being capable of running on a range of standard commercial hardware platforms such as PCs and HP Apollo and Sun workstations. The complex engineering that the EDA company now has to consider is to:

- link the different software packages together to form a comprehensive hierarchical suite (a 'seamless' suite) with one common database, thus enabling the OEM to purchase as much as needed without encountering any incompatibilities
- provide as far as possible different versions of the suites to run on different hardware platforms with different operating systems—UNIX is be-

coming the standard operating system, but even here not all programs written in UNIX are completely compatable
- liaise with all the leading IC vendors in order to provide design data relevant for their products and thus provide the routes to silicon—these are sometimes referred to as 'ASIC Kits'
- provide networking facilities so that different hardware resources may be linked together (see Figure 5.35 for example)
- provide maintenance and update of all these interlinked resources

This is not a task which any OEM would undertake lightly in order to assemble his own company suite of CAD resources.

5.5.1 Design and Layout Software

As discussed in Sections 5.1 and 5.3, the upper levels of IC design are not as fully represented in available software as the tools at the lower design levels, such as schematic capture, placement and routing. The status is generally as follows.

Silicon Compilation Software

This software is available from specialist vendors, and requiring a workstation or equivalent hardware platform.

Schematic Capture Software

This software is very widely available, running on PC or workstation platforms.

Placement and Routing Software

This software is very widely available, running on PCs or workstations for gate arrays but usually requiring the increased memory capacity of workstations for standard-cell designs—this depends upon the complexity of the cell library and the amount of additional memory which can be added to the PC.

Physical Silicon Design Software

Again, this software is widely available, but mainly for vendor or design house use rather than the OEM who has no wish to descend to the silicon level. It is usually workstation based.

5.5.2 Simulation Software

Since it forms a key part in all forms of IC design, simulation software has seen greater emphasis and development than any other associated IC design tool.

High-level digital simulation software running on workstation (or equivalent) hardware platforms and using HDLs (see Section 5.2.1) is increasingly being marketed, but may not yet be accepted and used by the majority of OEM designers. However, the status of the lower levels of simulation is well established, and is generally as follows.

Gate and Switch-Level Simulation

This software is very widely available, previously running on workstations but increasingly on PCs for circuits not exceeding, e.g., 5000 gates.

Circuit (Silicon) Level Simulation

Here, the almost universal simulation tool is based upon SPICE, with virtually every IC CAD vendor providing such a resource. Among the many commercial derivatives from different vendors have been CSPICE, DSPICE, HSPICE, MSPICE, PSPICE, SSPICE, USPICE, VSPICE and ZSPICE, with possibly PSPICE as the most commonly encountered version, plus other versions tailored for very specific applications such as microwave devices [147].

While many have been produced to derive the full d.c., a.c. and transient behavior of any given integrated circuit over a range of temperatures, whether bipolar, MOS or GaAs, many others such as PSPICE from Microsim Corporation are now providing mixed analog/digital simulation capability, the analog simulation being a fully detailed SPICE program with the digital simulation program running separately to handle the (possibly larger) digital part in less detail.

While most SPICE programs require the capacity of a workstation to handle the simulation of circuits containing hundreds of transistors in an acceptable time, an increasing number of vendors are providing PC (or equivalent) based versions which require enhancement of the PC capability. Again, taking SPICE as the example, versions have been marketed for the following hardware configurations:

- Sun, Apollo, DEC and HP workstations, running on a variety of operating systems
- PC/AT or equivalent platforms with 0.64 or 1.44 Mbytes of RAM, type 8087, 80287 or 80387 floating point coprocessor, up to 16 Mbytes hard disk, and MS/DOS operating system
- 80286 and 80386 based PCs with OS/2 operating systems
- Mackintosh platforms with 6881 floating point coprocessor

The first of the above is the most common, with the last now possibly no longer in OEM use.

However, the latest PCs with 150 MHz or more Pentium processor, 16 Mbytes of RAM, 2.5 or more Gbits hard disk and graphics capability are now sufficiently powerful to take over all the simulation duties which the average

OEM may require. Current software will increasingly provide as standard windowing and graphics-in/graphics-out capabilities, together with built-in processing such as Fourier analysis and distortion analysis of the simulated outputs of a.c. circuits.

5.5.3 PLD Design Software

Software for all the OEM design and programming activities associated with programmable logic devices is commercially available from both vendors and distributors of such devices. The most widely marketed tools include the following

- ABEL
- CUPL
- PALASM
- PLDesigner
- PLPL

and others, which are used by many PLD vendors to build up with other software a complete CAD package to cover the design and dedication of their devices. Enhancements to the basic CUPL and other tools may also be incorporated by some vendors. For the larger size logic cell arrays (LCAs), Xilinx Inc. provides its own development system, XACT, which, as shown in Figure 5.20, has provision for accepting input data from other PLD tools [104]. Actel Corporation similarly provides a comprehensive package for its large FPGAs with its Designer Advantage™ design system working in the Cadence™ or Mentor Graphics® or OrCAD™ or other design environment [148].

The hardware platforms for PLD design include PC-based systems and workstation-based systems. The larger PLD devices require considerable enhancement to PC platforms for the execution of the software; for example, Xilinx requires at least a fully compatible 386/486 PC running under version 5.0 (minimum) MS-DOS, with 16 Mbytes of RAM and 60 Mbytes minimum hard disk for OrCAD-based systems [104]. Required workstation requirements are similarly comprehensive; for example, Actel specify Sun SPARC or SPARC2 workstation running under Sun OS version 4.1.3 (or later) operating system, with 32 Mbytes minimum RAM and 60 Mbytes minimum hard disk, or other similar workstation specifications [148].

All the above figures should be read as being illustrative only of the widely available PLD software support [98–106], with many new and improved packages being continuously announced.

5.5.4 Printed Circuit Board (PCB) Design Software

Like PLD design resources, CAD tools for PCB design are readily available from a wide range of specialist vendors as stand-alone tools. However, most CAE vendors who package a complete hierarchical suite of programs with a common database for custom IC design activities also include a PCB tool in their suite of programs, or an interface to one or more of these tools from the specialist companies. As well as PCB layout design, these tools can often also design the layout for wire-wrapped assemblies.

PCB design predates custom microelectronics, being originally developed for PCBs containing discrete components and SSI/MSI integrated circuits. Available tools now range from the relatively simple, catering to up to, e.g., 100 ICs on single-sided or double-sided boards, to extremely complex programs for multilayer boards with perhaps fifteen to twenty layers and containing several hundred IC packages. The latter may require the power of a mainframe or supercomputer to solve the placement and routing difficulties. However, the increase of complexity that can now be put on a single chip is tending to minimize any further increase in the complexity of assembly and routing of PCBs, the complexity being accommodated within the IC packages rather than in the assembly of many IC packages.

Among the accepted industry standard PCB layout tools are P-CAD, Omni-CAD, CADDS, Scicards, and others. For the majority of OEM and custom microelectronic needs, the lower end of the above spectrum of capability is usually appropriate. The hardware platform on which the lower of these design tools will run is again the PC, although increasingly, 80386 minimum based machines are being specified. At least 20 Mbytes hard disk is also usually needed [149–152].

5.5.5 Housekeeping and Product Costing Software

Documentation of complex system designs is a critical factor in the design, manufacture and update aspects of the product, and increasingly, a common company database of all aspects of the product is necesary. In the absence of good housekeeping, there is a danger of a company having individual stand-alone islands of CAD/CAM/CAE which do not communicate electronically with each other, and which then require considerable human effort to exchange data and maintain overall consistency.

Figure 5.37 indicates the commonality which should exist between design activities and documentation. The precise complexity will depend upon company activities, and could range from very simple to an extremely complex nework of resources. Typically, some or all of the following will be involved:

- workstation- or PC-based engineering design stations for IC design activities
- similar design stations for PCB and other electrical and mechanical design activities
- computer-controlled test facilities

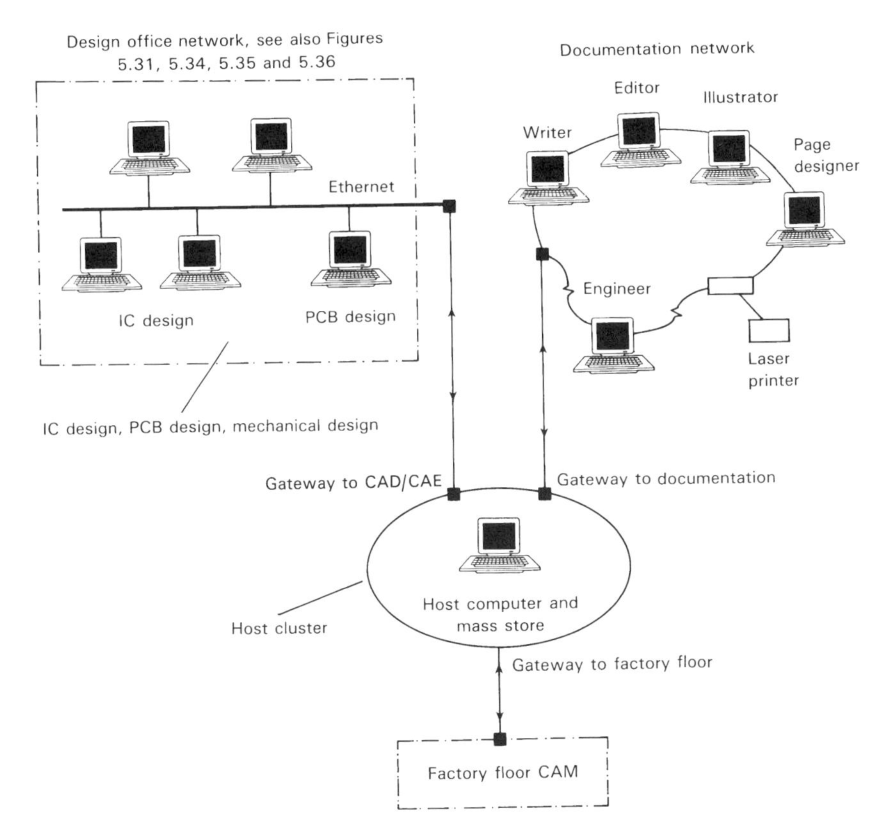

Fig. 5.37 The linking of CAD, CAE and CAM into an integrated system with one common database sometimes referred to as CIM (computer integrated manufacturing), although this designation is more usually reserved for the factory floor postdesign activities. The particular arrangement of Ethernet or equivalent bus to connect closely coupled resources and LAN or equivalent networking for dispersed resources will, of course, vary for each installation.

- laser printers and desk-top publishing resources for documentation, illustrations, maintenance publications, etc.
- overall stock control
- LAN networking and central mainframe computer with mass storage for central data files

This overall company organization is beyond our consideration here. Software to provide all these housekeeping facilities is available, with advice from specialized consultants on how to build up a company structure, but it remains a managerial decision as to the scope of, and necessity for, in-house computer-based housekeeping resources [141,142,153,154]. The software rather than the hardware invariably causes the greatest problems, with interfacing and noncompatibility of data formats a major cause of many difficulties.

5.5.6 Data Formats and Standards

From all the preceding considerations it is clear that there is an increasing need for software tools to be able to communicate with each other, without the time-consuming and error-prone need of human intervention and re-entering of data from one tool to another. The following conditions are therefore desired:

- standardization of data formats in CAD/CAM software for specific duties from different vendors, e.g., the formats for mask-making, formats for automatic test machines, etc.
- interchange formats to allow data interchange between tools doing different parts of the design and manufacturing activity

Unfortunately, it is not possible to acheive global standardization of data formats for all machines doing different manufacturing activities, due largely to the dissimilar duties involved. Instead, several standards have emerged from specialized vendors for particular activities, which are accepted by other companies or machines which use this particular area of activity. For *interchange* of data, translation of data to some common data standard such as EDIF (see below) is therefore required. This is illustrated in Figure 5.38.

Data formats which have received acceptance through widespread use for particular duties include the following:

- *Gerber format*: developed by the Gerber Scientific Instrument Company for plotters, particularly appropriate for the graphics artwork of PCB and wire-wrapped layouts
- *EBES (Electronic Beam Exposure System) format*: developed by Bell Laboratories and used for the control of E-beam machines for direct-write-on-wafer
- *CADDIF format*: developed by Schlumberger-Factron, and used as a standard for testers but not extensively adopted (many commercial IC testers can accept directly the design and simulation data generated during the normal design activity)
- *GDS II format*: developed by the Calma Corporation, it is the most widely adopted standard to describe the geometric details of an IC design for mask-making
- *CIF (Caltech Intermediate Form) format*: developed at the California

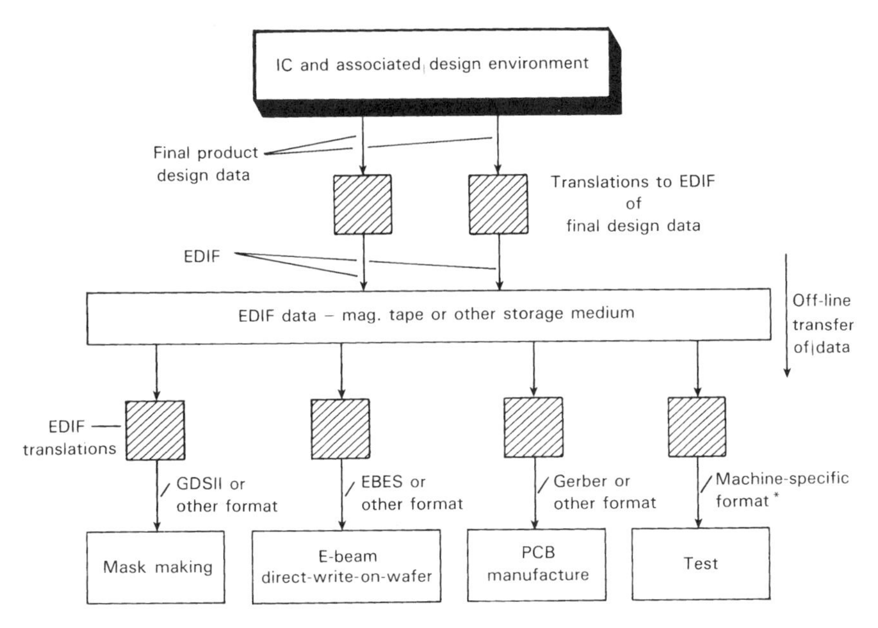

Fig. 5.38 The principle of the translation of non-EDIF design data into standard EDIF format for transfer to other in-house or vendor activities. Note that translation to EDIF may not always be necessary to provide off-line transfer of data, but it provides a common format when multiple-sourcing or vendor- or machine-independent interchange of data is required.

Institute of Technology, and which is an alternative to GDS II for mask-making

The first two of the above are machine-specific, and can therefore be fed directly into the appropriate machines (see Figure 5.38). The others are more general, and need to be converted into the actual data format required by a particular make machine. Further details of these *ipso facto* standards may be found in the literature [10,155–157].

Standard interchange formats, which can act as a link between individual tools, each with their own individual formats, currently include the folowing:

- *IGES (International Graphics Exchange Specification)*: a standard originally developed in the late 1970s for the interchange of data between mechanical CAE systems, particularly relevant for numerically controlled machines and production equipment

- *EDIF (Electronic Design Interchange Format)*: an internationally recognized standard covering all activities in the IC design area, and which is still being consolidated
- *SDIF (Stimulus Data Interchange Format)*: aimed at providing a standard format for all IC testing activities, but possibly being subsumed by EDIF

EDIF has therefore become the increasingly accepted standard in the IC design field, with many vendors now providing EDIF interfaces to their design tools. EDIF adopts a textual format, and is therefore readable by design engineers. It resembles a LISP programming language and has a hierarchical structure enabling any form of information on circuit topology, geometric details and behavior to be incorporated. An EDIF file contains a set of libraries, each of which contains a set of cells. Each cell can be described by 'views,' with the following seven types of views currently being available:

(i) netlist, for circuit topology details
(ii) schematic, for logic symbol structures
(iii) symbolic, for more abstract representations
(iv) mask layout, for layout geometries
(v) behavioral, for functional descriptions
(vi) document, for general textual information
(vii) stranger, for any information which does not fit into the above six areas

Examples of these and the general hierarchy may be found in Rubin [10].

To some degree, EDIF overlaps both IGES and SDIF. It can be used for PCB layout data as well as IC layout data, and can therefore take over a duty which is covered by IGES. At the other extreme, the SDIF interchange format has been developed to address the difficult data problems involved in the testing of ICs [158,159], having a similar syntax to EDIF but being more specific for the definition of testing requirements. Developments to EDIF may therefore incorporate more of the attributes provided in SDIF, making the need for SDIF less secure. Additionally, the new generation of commercial testers, ideally with the capability of testing both analog and digital ICs, may promote the development of an enhanced tester format, which can be linked more easily to a standard data interchange format than the present SDIF format.

EDIF is therefore the now recognized standard for IC data exchange purposes, with continuing developments to widen its scope and universality. As well as the links to testing, work to link to VHDL and other high-level design languages is being developed. Further details may be found in the published literature [10,107,160,161].

5.6 CAD COSTS

It is not possible to put hard and fast costs against the wide range of CAD hardware and software which is available for electronic design purposes for the following reasons:

- the market is highly competitive, with new or improved products being continuously announced
- take-overs and mergers between CAD/CAE vendors is a frequent occurrence, which modifies the sales basis for many products
- nonrealistic prices may be charged in order to penetrate new markets or maintain the existing market share for certain products
- the introduction of new microprocessor chip sets modifies the cost/performance ratio of hardware platforms
- software may be provided by IC vendors at noncommercial rates if this is a means of obtaining custom IC orders from an OEM

What is generally true, however, is that CAD hardware costs have fallen dramatically in the past two decades, first with workstations in competition with mainframe and mini computers, and later with PCs with additional enhancements challenging an appreciable part of the market captured by workstations.

It must also be appreciated that the actual hardware and software costs are only part of the expenditure which an OEM may incur in the design of a microelectronic-based product. Other nonrecurring engineering (NRE) costs, such as mask-making costs, liaison costs and staff time are involved—a consideration of these important NRE costs will form part of the managerial discussions in Chapter 7 [162,163]. The following paragraphs attempt to provide a typical overview of CAD costs; however, the figures quoted must not be taken as current but instead only as indicative of relative or historical costs in the area.

The first introduction of workstations by Daisy, Mentor and Valid [74] involved costs of around \$100,000 per seat. (The term 'seat' is frequently used when quoting CAD/CAE costs; it is the cost of the resources available and used by one design engineer, and may be a single stand-alone workstation plus software or the cost of a network of stations divided by the number of stations.) This figure of \$100 k was noticeably lower than prices previously charged for large Applicon or Calma graphics stations or mainframe or mini computers used for design purposes. By the latter part of the 1990s these costs had fallen to the figures shown in Table 5.4; subsequent reductions in cost have now brought costs down to less than the \$10 k mark per seat for most design work.

The cost of PCs has also fallen since their first introduction, but the need for enhancement for many CAD activities has meant that the basic PC cost usually has to be at least doubled to make a useful IC design resource. Total costs

Table 5.4 1987 Costs per Seat of Workstations, on the Basis of a Network of Four Stations per Design Office and Central Data Bank[a]

Apollo		DEC		Sun	
Workstation resource					
DN3000-C		VAXstation 2000		3/110LC	
68020 CPU		MicroVAX II CPU		68020 CPU	
68881 FPC		78132 FPC		68881 FPC	
4-Mbyte memory		4-Mbyte memory		4-Mbyte memory	
15-in color VDU		15-in color VDU		15-in color VDU	
Apollo Token Ring or Ethernet		Ethernet		Ethernet	
Domain IX and AEGIS		LAV or NFS		NFS	
		VMS or Ultrix		SunOS	
	$8900		$7900		$15 900
	×4 = $35 600		×4 = $31 600		×4 = $63 600
Central resource					
DSP4000		VAX 8600		3/260S	
68020		with 4-Mbyte memory		68020	
68881				68881	
8-Mbyte memory		4-Mbyte memory add-on		8-Mbyte memory	
348-Mbyte disk		456-Mbyte disk		280-Mbyte disk	
60-Mbyte cart. tape		95-Mbyte cart. tape		60 Mbyte cart. tape	
	$29 400		$470 300		$49 500
Total:	$65 000		$501 900		$113 100
	Price per seat: $16 250		**Price per seat: $125 475**		**Price per seat: $28 275**

[a] These figures are for illustrative purposes only to show how per-seat costs may be calculated; current figures will show a reduction in these prices.
Source: based upon Ref. 139.

of, e.g., $5 to $10 k per seat may therefore be involved, which again is blurring the distinction between the workstation and the PC [143].

The cost of software for IC design purposes varies from as low as a few hundred dollars for individual programs running on PCs, for example, SPICE-based simulation programs for very basic transistor circuits, to more than $100,000 for the outright purchase of a fully integrated hierarchical suite of programs from a vendor, covering all the design activities from behavioral level downwards (see Figure 5.3). To a large extent, the OEM purchases as much or as little as required, and thus costs have to be individually determined. The usual company experience is that total software costs are usually at least equal to the total hardware costs.

CAD for specific duties such as PLD and PCB design may be PC-based and available as a complete hardware/software package from specialist distributors as well as vendors. A PLD package may be on the order of $10 k complete; PCB design packages may be as low as around $2 to $5 k for a simple system, rising to three or four times this cost for more sophisticated resources [104,149]. For extremely complex many-layer PCBs, which are way beyond the needs of most OEMs, the total cost of non-PC–based hardware can rise to $100 k or more.

Finally, it must be noted that both hardware and software incur annual maintenance charges. Software maintenance and updating is particularly significant, since except on very simple programs continuous updating is done by the supplier to correct errors or ambiguities, or to improve details of performance. The per-annum costs of CAD/CAE maintenance is usually found to be 10%–20% of the purchase price, and hence over a five-year life is likely to cost at least a half of the original purchase price of the resource.

These financial matters will be looked at again in Chapter 7, but it must be stressed again that actual costs cited here are illustrative only, and should not be taken as precise for the future. The general future trend, however, may be for costs to reduce still further in real terms, or alternatively more power to be available for the same equivalent costs [163,164].

5.7 SUMMARY

The viability of custom microelectronics depends upon the proven availability of silicon fabrication methods and upon good CAD resources to enable custom ICs to be accurately and expeditiously designed. The three distinct levels of design are broadly summarized in Figure 5.39, with the 'front-end' design activities involving creative work not yet being so well served as the 'back-end' standard manufacturing activities. Simulation forms an essential part throughout the design process, ideally preventing the design activity proceeding from one level to the next without first proving the correctness of the design details so far.

CAD Hardware

As we have seen, the CAD hardware has developed into the two principal resources, namely, the workstation and the PC, with the dividing line in usefulness

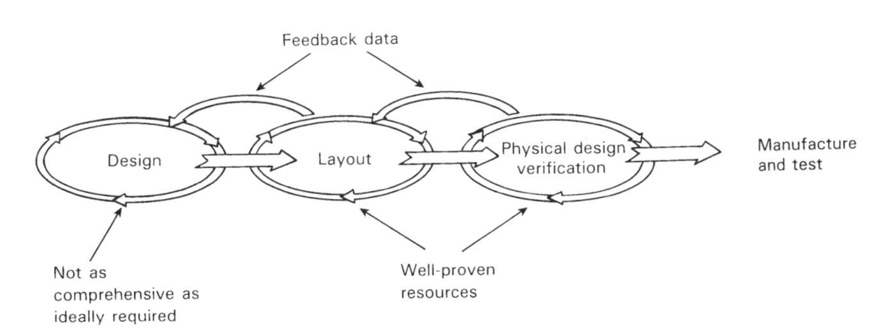

Fig. 5.39 A broad summary of the custom IC design activities which need to be supported by appropriate CAD/CAE resources, with simulation being a key element at each stage.

between them increasingly difficult to define. Future developments are heavily dependent upon new processor chip sets produced by the leading IC manufacturing companies as off-the-shelf standard parts, such new products being used both by the workstation manufacturers and by the PC manufacturers to upgrade the performance of their products [164,165].

IC Design Software

IC design software remains the greatest problem area for the OEM in deciding what and how much is necessary in terms of the required in-house design activities. Certain software packages such as

- the GARDS layout system from Silvar-Lisco for the placement and routing of gate arrays
- the CADAT simulation program from HHB Systems for switch, gate and behavioral level simulation
- HILO from GenRad for high-level and macro simulation and for fault simulation
- the many commercial derivatives of SPICE for transistor-level simulation

may be considered as representative of the specialized developments in these different areas of IC design over the past two decades. However, as we have discussed, it is difficult for an OEM to purchase individual software programs from different sources, and then weld them together to form an integrated software package due to dissimilar and/or incompatible data formats; instead, the evolution is towards specialist vendors assembling and marketing comprehensive suites of programs, which provide the OEM with a hierarchical package of tools to undertake whatever range of design activity is desired. The essential feature here is that the vendor has undertaken all the difficult compatibility work in compiling the suite of programs, and has provided a common database for all the design tools and data.

The development of hardware description languages (HDLs), and in particular VHDL and its commercial variants, is a development which has not yet reached full maturity and acceptance. The use of such high-level languages for design purposes is clearly attractive if it can relieve the OEM designer of detailed work, or if it can hasten a correct-by-construction design style, although it may be considered to be unnecessary for the design of the more simple range of custom circuits, particularly those which do not involve any digital signal processing (DSP). However, considerable further development work in this area is likely.

Data Interchange Standards

We have seen that one of the greatest needs in the IC design area has been the need for some common standard data format so that design and simulation data can be transferred without human intervention between resources. The introduction of the EDIF data interchange standard has been the most successful and internationally accepted means of handling this interchange requirement, and therefore most IC CAD vendors are now providing an EDIF interface on their tools. Continuing development of this standard is anticipated.

Testing

Finally, every IC design has to be tested after fabrication. This also forms part of the design activity, since how-shall-we-test-it must be considered during the design stage.

Testing represents the greatest challenge to the IC designer as the complexity on a single chip increases. Because of this difficulty and its importance, we have not considered it in the preceding pages, but will devote all the following chapter to a complete study of both the problems involved and the various means which have been developed to try to solve the very difficult testing requirements. The following chapter, therefore, will be seen as a continuation of the computer-aided design activities which we have already covered, to make in total a comprehensive procedure with which all IC designers must be familiar in order to finally produce working and tested end products.

REFERENCES

1. Russell, G. (Ed.), *Computer Aided Tools for VLSI Design*, IEE Peter Peregrinus, UK, 1987
2. Hicks, P.J. (Ed.), *Semi-Custom IC Design and VLSI*, IEE Peter Peregrinus, UK, 1983
3. Russell, G., Kinniment, D.J., Chester, E.G. and McLauchlan, M.H. (Eds.), *CAD for VLSI*, Van Nostrand Reinhold, UK, 1985
4. Rohrer, D.A., Evolution of the electronic design automation industry, *IEEE Design and Test*, Vol. 5, December 1988, pp. 8–13
5. Newton, A.R. and Sangiovanni-Vincentelli, A.L., CAD tools for ASIC design, *Proc. IEEE*, Vol. 75, 1987, pp. 765–776
6. Johnson, P., Mixed-level design tools enhance top-down design, *Computer Design*, Vol. 28, January 1989, pp. 85–88
7. Harding, W., System simulation ensures chips play together, *Computer Design*, Vol. 28, August 1989, pp. 70–83
8. Dettmer, R., Logic synthesis: a faster route to silicon, *IEE Review*, Vol. 35, 1989, pp. 427–430
9. VLSI System Design, *User's Guide to Design Automation*, CMP Publications, NY, 1988
10. Rubin, S.M., *Computer Aids for VLSI Modeling*, Addison-Wesley, MA, 1987

11. Engl, W.L. (Ed.), *Process and Device Modeling*, Advances in CAD for VLSI Book Series, Vol. 1, North-Holland, Amsterdam, 1986
12. Hörbst, E. (Ed.), *Logic Design and Simulation*, Advances in CAD for VLSI Book Series, Vol. 2, North-Holland, Amsterdam, 1986
13. Ruehli, A.E. (Ed.), *Circuit Analysis, Simulation and Design*, ibid., Vol. 3 (two parts) 1986
14. Ohtsuki, T. (Ed.), *Layout Design and Verification*, Vol. 4, North-Holland, Amsterdam, 1986
15. Goto, S. (Ed.), *Design Methodologies*, Advances in CAD for VLSI Book Series, Vol. 6, North-Holland, Amsterdam, 1986
16. Hartenstein, R.W. (Ed.), *Hardware Description Languages*, Advances in CAD for VLSI Book Series, Vol. 7, North-Holland, Amsterdam, 1986
17. The Open University, *CAD for Custom Chips*, Microelectronics for Industry, Publication PT 505CAD, The Open University, UK, 1988
18. The Open University, *Computers in the Design Process*, Microelectronics for Industry, Publication PT506COM, The Open University, UK, 1989
19. Marcus, C., *Prolog Programming*, Addison-Wesley, MA, 1987
20. Kelley, A. and Pohl, I., *C by Dissection: the Essentials of C Programming*, Addison-Wesley, MA, 1987
21. Meyer, E., VHDL strives to cover both synthesis and modeling, *Computer Design*, Vol. 28, October 1989, pp. 42–45
22. Harding, W., Logic synthesis forces rethinking of design tools, *Computer Design*, Vol. 28, December 1989, pp. 51–57
23. IEEE Publication, *VHDL Language Reference Manual, Draft Standard 1076/B*, IEEE Computer Society, Publication Department, CA, 1987
24. Shahdad, M., Lipsett, R., Marshuer, E., Sheenhan, K., Cohen, H., Waxman, R. and Ackley, D., VHSIC hardware description language, *IEEE Computer*, Vol. 18, February 1985, pp. 94–103
25. Hookway, R.J., System simulation using VHDL, *J. Semicustom ICs*, Vol. 5, September 1987, pp. 23–29
26. Sullavan, R. and Asher, L.R., VHDL for ASIC design and verification, *Semicustom Design Guide*, CMP Publications, NY, 1988
27. Armstrong, J.R. and Gray, F.G., *Structured Logic Design with VHDL*, Prentice Hall, NJ, 1993
28. Sauceir, G. and Mignotte, A. (Eds.), *Logic Architecture Synthesis: State of the Art and Novel Approaches*, Chapman and Hall, UK, 1995
29. Hunter, R.D.M. and Johnson, T.T., *Introduction to VHDL*, Chapman and Hall, UK, 1996
30. Lightner, M.R., Modeling and simulation of VLSI digital systems, *Proc. IEEE*, Vol. 75, 1987, pp. 786–796
31. Clare, C., *Designing Logic Systems Using State Machines*, McGraw-Hill, NY, 1972
32. Green, D., *Modern Logic Design*, Addison-Wesley, UK, 1986
33. Hayes, J.P., *Introduction to Digital Logic Design*, Addison-Wesley, NY, 1994
34. Roth, C.H., *Fundamentals of Logic Design (Fourth Edition)*, West Publishing, MN, 1992

35. Muroga, S., *Logic Design and Switching Theory*, Wiley, NY, 1979
36. McCluskey, E., *Logic Design Principles*, Prentice Hall, NJ, 1986
37. Brayton, R., Hachtel, G., McMullen, C. and Sangiovanni-Vincentelli, A., *Logic Minimization Algorithms for VLSI Synthesis*, Kluwer Academic Publishers, MA, 1984
38. Rudell, R., *EXPRESSO IIC Users' Manual*, University of California at Berkeley, Department of EECS, 1986
39. Poretta, A., Santomauro, M. and Somenzi, F., TAU, a fast heuristic logic minimizer, *Proc. Int. Conf. on CAD*, November 1984, pp. 206–208
40. Bouhasin, G., EDA pushes towards logic synthesis, *User's Guide to Design Automation*, CMC Publications, NY, 1988, pp. 10–16
41. Kernighan, B.S. and Lin, W., An efficient heuristic procedure for partitioning graphs, *Bell System Technical Journal*, Vol. 49, 1970, pp. 291–307
42. Breuer, M.A. (Ed.), *Design and Automation of Digital Systems*, Prentice Hall, NJ, 1972
43. Schwartz, A.F., *Computer Aided Design for Microelectronic Circuits and Systems*, Vol. 1 (736 pp.) and Vol. 2 (772 pp.), Academic Press, UK, 1987
44. Breuer, M.A., Min-cut placement, *J. Design Automation and Fault Tolerant Computing*, Vol. 1, 1977, pp. 343–362
45. Lauther, U., A min-cut placement algorithm for general cell assembly, *J. Design Automation and Fault Tolerant Computing*, Vol. 4, 1980, pp. 21–34
46. Sechen, C. and Lee, K.-W., An improved simulated annealing algorithm for general cell assemblies, *Proc. IEEE Conf. on Computer-Aided Design*, 1987, pp. 478–481
47. Veechi, M.P. and Kirkpatrick, S., Global wiring by simulated annealing, *IEEE Trans. CAD*, Vol. CAD.2, 1987, pp. 165–172
48. Rutman, R.A., An algorithm for placement based upon minimum wire length, *Proc. AFIPS Conf. SJCC*, 1964, pp. 477–491
49. Wilson, D.C. and Smith, R.J., An experimental comparison of force directed placement techniques, *Proc. IEEE Design Automation Workshop*, 1974, pp. 194–199
50. Chyan, D.-Y. and Breuer, M.A., A placement algorithm for array processors, *Proc. IEEE Design Automation Conf.*, 1983, pp. 182–188
51. Tsay, R.-S., Kuh, E.S. and Hsu, C.-P., PROUD: a sea-of-gates placement algorithm, *IEEE Design and Test of Computers*, Vol. 5, December 1988, pp. 44–56
52. Macaluso, E., Graphical floorplan design of cell-based ICs, in *User's Guide to Design Automation*, CMC Publications, NY, 1988
53. Praes, B.T. and van Cleemput, W.M., Placement algorithms for arbitrary shaped blocks, *Proc. IEEE Design Automation Conf.*, 1979, pp. 474–480
54. Kuh, E.S. and Marek-Sadowski, M., Global routing, in Ohtsuki, T. (Ed.), *Layout Design and Verification*, Advances in CAD for VLSI Book Series, Vol. 4, North-Holland, Amsterdam, 1986
55. Massara, R.E. (Ed.), *Design and Test Techniques for VLSI and WLSI*, IEE Peter Peregrinus, UK, 1989
56. van Cleemput, W.M., Mathematical models for the circuit layout problem, *Trans. IEEE*, Vol. CAS.23, 1976, pp. 759–767
57. Chen, N.P., New algorithms for Steiner trees, *Proc. IEEE ISCAS*, 1983, pp. 1217–1219

58. Garey, M.R. and Johnson, D.S., *Computers and intractability: a guide to the theory of NP-completeness*, Freeman, CA, 1979
59. Loberman, H. and Weinberger, A., Formal procedures for connecting terminals with a minimum total wire length, *J. ACM*, Vol. 4, 1957, pp. 428–437
60. Krusa, J.B., On the shortest spanning subtree of a graph and the travelling salesman problem, *Proc. American Math Society*, Vol. 7, 1956, pp. 48–50
61. Abel, LC., On the ordering of routes for automatic wiring routing, *IEEE Trans.*, Vol. C.21, 1972, pp. 1227–1233
62. Lee, C.Y., An algorithm for path connection and its application, *IEEE Trans. Computers*, Vol. EC.10, 1961, pp. 346–365
63. Akers, S.B., A modification of Lee's path connection algorithm, *IEEE Trans. Computers*, Vol. EC.16, 1967, pp. 97–98
64. Hoel, J.H., Some variations on Lee's algorithm, *IEEE Trans. Computers*, Vol. C.25, 1976, pp. 19–24
65. Hightower, D.W., A solution to line-routing problems on the continuous plane, *Proc. IEEE Design Automation Conf.*, 1969, pp. 1–24
66. Heyns, W., Sansen, W. and Beke, H., A line expansion algorithm for the general routing problem with a guaranteed solution, *Proc. IEEE Design Automation Conf.*, 1980, pp. 243–249
67. Hashimoto, A. and Stevens, J., Wire routing by optimizing channel assignment with large apertures, *Proc. IEEE Design Automation Conf.*, 1971, pp. 155–169
68. Kernigham, S., Schweikert, D. and Persky, G., An optimum channel routing algorithm for polycell layouts of integrated circuits, *Proc. IEEE Design Automation Conf.*, 1973, pp. 50–59
69. Deutsch, D.N., A dog-leg channel router, *Proc. IEEE Design Automation Conf.*, 1976, pp. 425–433
70. Yoshimura, T. and Kuh, E.S., Efficient algorithm for channel routing, *IEEE Trans. Computer Aided Design*, Vol. CAD.1, 1982, pp. 25–35
71. Rivest, R.L. and Fiduccia, C.M., A "greedy" channel router, *Proc. IEEE Design Automation Conf.*, 1982, pp. 418–424
72. Burstein, M. and Pelavia, R., Hierarchical channel router, *Proc. IEEE Design Automation Conf.*, 1983, pp. 591–597
73. Burstein, M., Channel routing, in Ohtsuki, T. (Ed.), *Layout Design and Verification*, Advances in CAD for VLSI Book Series, Vol. 4, North-Holland, Amsterdam, 1986
74. Hurst, S.L., *Custom Specific Integrated Circuits: Design and Fabrication*, Marcel Dekker, NY, 1983
75. Soukup, J., Circuit layout, *Proc. IEEE*, Vol. 69, 1981, pp. 1281–1304
76. Antognetti, P., Pederson, D.O. and de Man, H. (Eds.), *Computer Design Aids for VLSI Circuits*, Martin Nijhoff, MA, 1984
77. Rothermel, H.-J. and Mlynski, D.A., Routing method for VLSI design using irregular cells, *Proc. IEEE Design Automation Conf.*, 1983, pp. 257–262
78. Damnjanovic, M.S. and Litorski, V.B., A survey of routing algorithms in custom IC design, *J. Semicustom ICs*, Vol. 7, December 1989, pp. 10–19
79. Yoshida, K., Layout verification, in Ohtsuki, T. (Ed.), *Layout Design and Verification*, Advances in CAD for VLSI Book Series, Vol. 4, North-Holland, Amsterdam, 1986

80. Haskard, M.R. and May, I.C., *Analog VLSI Design: nMOS and CMOS*, Prentice Hall, NJ, 1988
81. Allen, P. and Holberg, D., *CMOS Analog Circuit Design*, Holt, Reinhart and Winston, NY, 1987
82. Nordholt, E.H., *Design of High-Performance Negative Feedback Amplifiers*, Elsevier, Amsterdam, 1983
83. Trontelj, J., Trontelj, L. and Shenton, G., *Analog Digital ASIC Design*, McGraw-Hill, UK, 1989
84. Gregorian, R. and Temes, G.C., *Analog MOS Integrated Circuits for Signal Processing*, Wiley, NY, 1986
85. Grebane, A.B., *Bipolar and MOS Analog Integrated Circuit Design*, Wiley, NY, 1984
86. Gayakwad, R.A., *Op-Amps and Linear Integrated Circuits*, Prentice Hall, NJ, 1988
87. Gray, P.R. and Meyer, R.G., *Analysis and Design of Analog Integrated Circuits*, Wiley, NY, 1984
88. Tsividis, Y. and Antognetti, Y., *Design of MOS VLSI Circuits for Telecommunications*, Prentice Hall, NJ, 1985
89. Bray, D. and Irissou, P., A new gridded bipolar linear semiconductor array family with CAD support, *J. Semiconductor ICs*, Vol. 4, June 1986, pp. 13–20
90. Crolla, P., A family of high-density, tile-based semicustom arrays for the implementation of analogue integrated circuits, *J. Semiconductor ICs*, December 1987, pp. 23–29
91. Allen, P.E., Computer-aided design of analogue integrated circuits, *J. Semiconductor ICs*, December 1986, pp. 22–31
92. Spence, R. and Soin, R.S., *Tolerance Design of Electronic Circuits*, Addison-Wesley, MA, 1988
93. Trontelj, L, Trontelj, J. *et al*, Analogue silicon compiler for switched-capacitor circuits, *Proc IEEE ICCAD*, 1987, pp. 506–509
94. Assael, J., A switched-capacitor filter silicon compiler, *IEEE J. Solid-State Circuits*, Vol. 23, 1988, pp. 166–174
95. Weder, U. and Möschwitzer, A., SCF, a gate array switched-capacitor filter design tool, *J. Semicustom ICs*, Vol. 8, March 1990, pp. 15–21
96. Davidse, J. and Nordholt, E.H., Basic considerations concerning the application of semicustom techniques for the processing of analogue signals, *J. Semicustom ICs*, Vol. 5, December 1987, pp. 5–11
97. Habekekotté, E., Hofefflinger, B., Klein, H.-W. and Beunder, M.A., State of the art in analog CMOS circuit design, *Proc. IEEE*, Vol. 75, 1987, pp. 816–828
98. Tong, C.S., A new MAX-EPLD architecture which provides logic density, speed and flexibility, *J. Semicustom ICs*, Vol. 7, September 1989, pp. 12–19
99. Bolton, M.J.P., *Digital System Design with Programmable Logic*, Addison-Wesley, MA, 1990
100. Bostock, G., *Programmable Logic Handbook*, Blackwell, UK, 1987
101. Advanced Micro Devices, *PAL® Device Data Book and Design Guide*, Advanced Micro Devices, CA, 1996
102. Texas Instruments, *FPGA Data Manual and Applications Handbook*, Texas Instruments, Inc., TX, 1995

103. Amtel, *Configurable Logic Design and Application Book*, Amtel Corporation, CA, 1996
104. Xilinx, *The Programmable Logic Data Book*, Xilinx Corporation, CA, 1996
105. Altera, *The MAX Data Book*, Altera Corporation, CA, 1995
106. Osann, R., A designer's guide to programmable logic, Parts 1, 2 and 3, *EDN*, January/February, 1985
107. Stump, H., A designer's guide to simulation models, *Computer Design*, Vol. 29, January 1990, pp. 91–98
108. Meyer, E., Mixed-signal simulators take divergent paths, *Computer Design*, Vol. 29, January 1990, pp. 49–56
109. The Open University, *System and Logic Simulation*, Microelectronics for Industry, Publication PT505SIM, The Open University, UK, 1988
110. Russell, G. and Sayers, I.L., *Advanced Simulation and Test Methodologies for VLSI Design*, Van Nostrand Reinhold, UK, 1989
111. *IEEE Design and Test*, Vol. 4, August 1987, special issue *Modeling and Switch Level Testing*
112. Sedra, A.D. and Smith, K.C., *Microelectronic Circuits*, Holt, Reinhart and Winston, NY, 1987
113. Hodges, D.A. and Jackson, H.G., *Analysis and Design of Digital Integrated Circuits*, McGraw-Hill, NY, 1983
114. The Open University, *Circuit and Device Modeling*, Microelectronics for Industry, Publication PT505MOD, The Open University, UK, 1988
115. Vladimirescu, A., Newton, A.R. and Pederson, D.O., SPICE version 2G.1 users' guide, University of California at Berkeley, Dept. of EECS, 1980
116. Di Giacomo, J. (Ed.), *VLSI Handbook*, McGraw-Hill, NY, 1989
117. Hitchcock, R.B., Timing verification and the timing analysis program, *Proc. IEEE Design Automation Conf.*, 1982, pp. 594–603
118. Tamura, E., Ogawa, K. and Nakano, T., Path delays analysis for hierarchal building block layout system, *Proc. IEEE Design Automation Conf.*, 1983, pp. 403–410
119. Wei, Y.-P., Lyons, C. and Hailey, S., Timing analysis of VLSI circuits, *VLSI System Design*, Vol. 8, August 1987, pp. 52–58
120. Johns, D.A. and Martin, K., *Analog Integrated Circuit Design*, Wiley, NY, 1997
121. Friedman, M., A swifter way to simulate analog-and-digital ICs, *Electronic Products*, Vol. 10, September 1987, pp. 28–33
122. Vucurevich, T., SPECTRUM: a new approach to event-driven analog/digital simulation, *Proc. IEEE Custom Integrated Circuits Conf.*, 1990, pp. 5.1.1–5.1.5
123. Kurker, C.M., Paulos, J.J., Cohen, B.S. and Conley, E.S., Development of an analog hardware language, *Proc. IEEE Custom Integrated Circuits Conf.*, pp. 5.4.1–5.4.6
124. Moser, L., Behavioural analog circuit models for multiple simulation environments, *Proc. IEEE Custom Integrated Circuits Conf.*, pp. 5.5.1–5.5.4
125. Blank, T., A survey of hardware accelerators used in computer aided design, *IEEE Design and Test of Computers*, Vol. 1, No. 3, 1984, pp. 21–39
126. Pfister, G.F., The Yorktown simulation engine: an introduction, *Proc. IEEE Design and Automation Conf.*, 1982, pp. 51–54
127. Frank E.H., Exploiting parallelism in a switch-level simulation machine, Proc. IEEE Design and Automation Conf., 1986, pp. 20–26

128. Johnson, P., Software vs hardware models for system simulation, in VLSI System Design, *User's Guide to Design Automation*, CMP Publications, NY, 1988
129. Rossbach, P.C., Linderman, R.W. and Gallacheer, D.M., An optimising XROM silicon compiler, *Proc. IEEE Custom Integrated Circuits Conf.*, 1987, pp. 13–16
130. Kang, S. and van Cleemput, W.M., Automatic PLA synthesis from a DDL-P description, *Proc. IEEE Design Automation Conf.*, 1981, pp. 391–397
131. Leith, J.W., Crystal oscillator compiler, *Proc IEEE Custom Integrated Circuits Conf.*, 1987, pp. 17–19
132. Rabaey, J., Vanhoof, J., Gosens, G., Catthoor, F. and De Man, H., CATHEDRAL II computer aided synthesis of digital signal processing systems, *Proc. IEEE Custom Integrated Circuits Conf.*, 1987, pp. 157–160
133. Helms, W.J. and Byrkett, B.E., Compiler generation of A to D converters, *Proc. IEEE Custom Integrated Circuits Conf.*, 1987, pp. 161–164
134. Trimberger, S., Automating chip layout, *IEEE Spectrum*, Vol. 9, 1982, pp. 38–45
135. Seattle Silicon, *CONCORDE Documentation*, Seattle Silicon Corporation, WA
136. Silicon Computers, *Genesil Documentation*, Silicon Compiler Systems Corporation, CA
137. Evanczuk, S., Results of a silicon compiler challenge, *VLSI System Design*, Vol. 6, July 1985, pp. 46–54
138. Gajski, D.D. (Ed.), *Silicon Compilation*, Addison-Wesley, MA, 1987
139. Parker, A.C. and Hayati, S., Automating the VLSI design process using expert systems and silicon compilation, *Proc. IEEE*, Vol. 75, 1987, pp. 777–785
140. Gajski, D.D., Dutt, N.D. and Prangrle, B.M., Silicon compilation: a tutorial, *Proc. IEEE Custom Integrated Circuits Conf.*, 1986, pp. 453–459
141. Bosworth, M.F., The management of documentation and design bases, *J. Semicustom ICs*, Vol. 4, September 1986, pp. 13–17
142. Burgess, L., PCs for CAE—how powerful are they?, *J. Semicustom ICs*, Vol. 4, December 1986, pp. 44–48
143. Mokhoff, N., Differences blur as PCs take on engineering workstations, *Electronic Design*, Vol. 35, September 1987, pp. 15–31
144. Editorial Staff, Multi-MIPS workstations under $ 15 K, *VLSI System Design*, Vol. 8, August 1987, pp. 78–81
145. Watson, G. and Cunningham, D., FDDI and beyond, *IEEE Review*, Vol. 36, 1990, pp. 131–134
146. Andrews, W., Bridging today's buses to Futurebus, *Computer Design*, Vol. 29, No. 3, 1990, pp. 72–84
147. Bresford, W., Circuit simulators at a glance, *VLSI Systems Design*, Vol. 8, August 1987, pp. 76–77
148. Actel, *FPGA Data Book and Design Guide*, Actel Corporation, CA, 1996
149. Personal CAD Systems, *Answers to the Most Commonly Asked Questions on ICB CAD*, Personal CAD Systems, Inc., CA, 1989
150. Kirkpatrick, J.M., *Electronic Drafting and Printed Circuit Board Design*, Chapman and Hall, NY, 1989
151. Clark, R.H., *Printed Circuit Engineering*, Chapman and Hall, NY, 1989
152. Sloan, J.L., *Designing and Packaging of Electronic Equipment*, Chapman and Hall, NY, 1985

153. O-Reilly, W.P., *Computer Aided Electronic Engineering*, Chapman and Hall, NY, 1986
154. Browne, J., Harhen, J. and Shivan, J., *Product Management Systems: a CIM Perspective*, Addison-Wesley, MA, 1988
155. Geber S.I.C., *Geber Format*, Geber Scientific Instrument Company, Document No. 40101-S00-066A, Beaverton, WA, 1983
156. Factron, *CADDIF Version 2.0 Engineering Specification*, Schlumberger-Factron Corporation, Los Angeles, CA, 1985
157. Calma, *GDS II Stream Fomat*, Calma Corporation, Milipita, CA, 1984
158. Parker, K.P., *Integrated Design and Test: Using CAE Tools for ATE Programming*, IEEE Computer Society Press, Washington, DC, 1987
159. Peiper, C., Stimulus Data Interchange Format (SDIF), *VLSI System Design*, Vol. 7, July 1986, pp. 76–81, and August 1986, pp. 56–60
160. Newton, A.R., Electronic design interchange format: Introduction to EDIF, *Proc. IEEE Custom Integrated Circuits Conf.*, May 1987, pp. 571–575
161. Etherington, E., Interfacing design to text using the Electronic Design Interchange Format (EDIF), *Proc. IEEE Test Conf.*, 1987, pp. 378–383
162. The Open University, *Microelectronic Decisions*, Microelectronics for Industry, Publication PT505 MED, The Open University, UK, 1988
163. McClean, W.J. (Ed.), *ASIC Outlook*, Integrated Circuit Engineering Corporation, AZ, 1990
164. EPD, Towards 2000; Tenth Anniversity Issue, *Electronic Product Design*, Vol. 11, No. 4, April 1990, pp. 47–196
165. Walls, C. and Rosenfield, P., Embedded system development using the 80 $\times$ 86 architecture, *Embedded System Engineering*, Vol. 5, No. 1, January 1997, pp. 30–34

6

Test Pattern Generation and Design-for-Testability

6.1 INTRODUCTION

The need to test custom microelectronic circuits has been noted previously, but due to its significance the details have been left until now in order to consider the whole subject area within one chapter. As has been seen, for example, in Figure 5.3, aspects of testing may be involved in the CAD resources used during the design phase of a custom IC, but in all cases these requirements must always be considered as early as possible in the design of any new circuit.

The testing of any product involving one or more ICs usually involves some verification of the individual ICs before product assembly, plus some-final test of the product before dispatch. The only exception to this may be in very cheap 'give-away' goods, where the cost of individual testing is not justified. Normal testing procedures therefore involve both the OEM and the IC vendor, the former doing the final product test and the vendor and the OEM sharing the individual IC tests in some agreed way.

Figure 6.1 shows the usual prototype and production test procedures. The total testing of a custom IC involves the following:

- tests to ensure that all the fabrication steps have been implemented correctly during routine wafer manufacture (*fabrication checks*)
- tests to ensure that the prototype ICs are functionally satisfactory in all respects to meet the product specification (*design checks*)
- tests to ensure that subsequent production ICs finally used in the product are defect free (*production checks*)

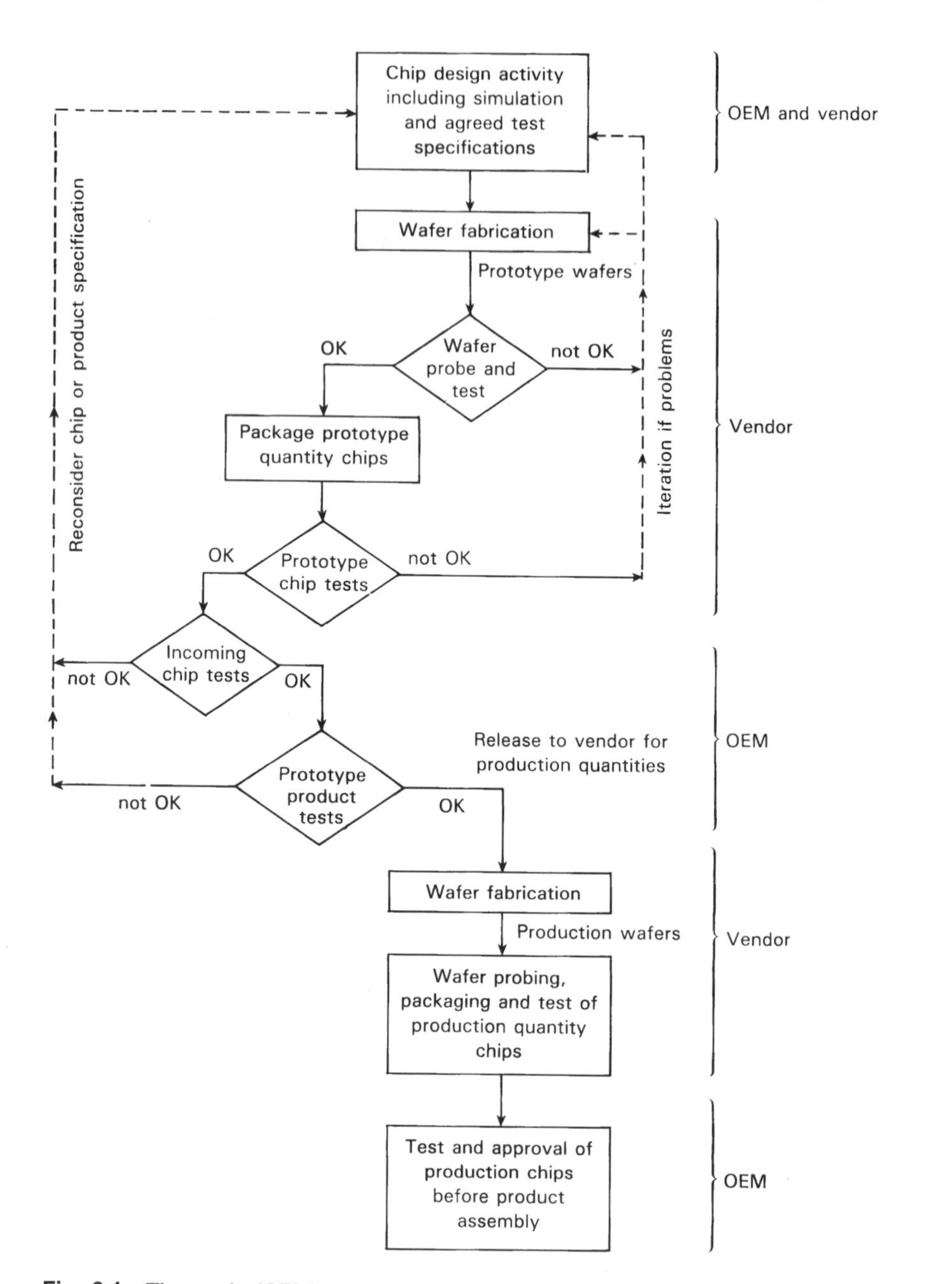

Fig. 6.1 The vendor/OEM test procedures for a custom IC before assembly into the final production equipment. Ideally, the OEM testing should be a 100% fully functional test in a working equipment or equivalent test rig, although this may not always be possible.

The first of these three testing categories is the province of the vendor, and does not involve the OEM in any way. The vendor will ensure that the wafer containing the custom ICs has 'drop-ins' located at scattered points on the wafer, these being small circuits or structures from which the correct resistivity and other parameters of the wafer fabrication can be checked before any more comprehensive tests are begun on the surrounding circuits. This is illustrated in Figure 6.2. Additionally, every mask used in the wafer fabrication will have some identification code

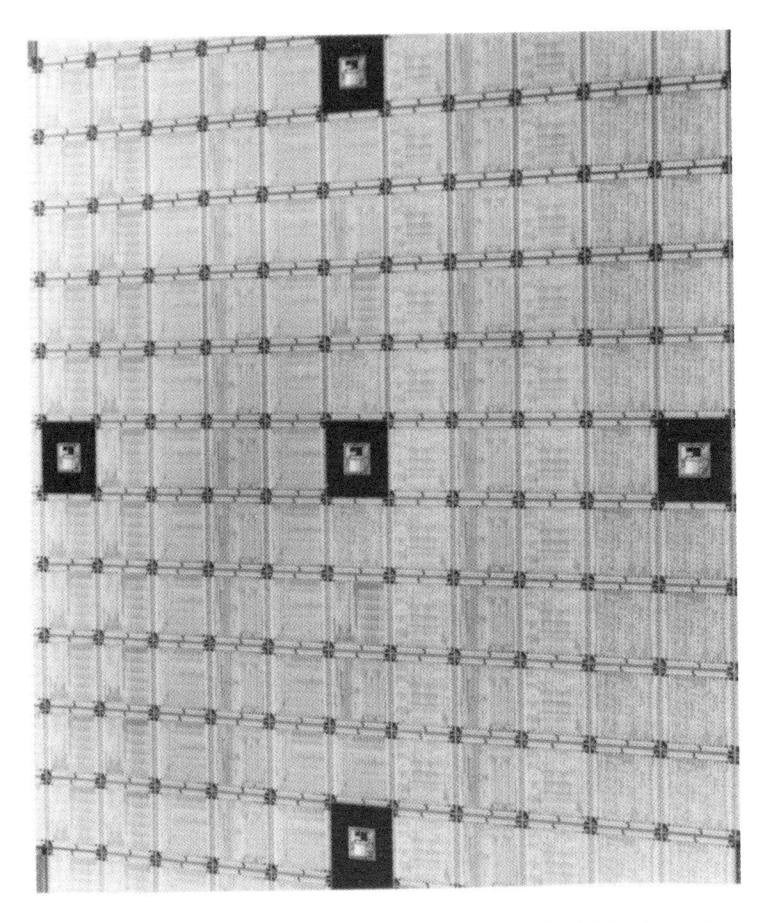

Fig. 6.2 The vendor's means to check on correct wafer fabrication, using small 'drop-in' test circuits at selected points on the wafer. These drop-in circuits may alternatively be known as 'proccess evaluation devices,' 'process monitoring circuits,' 'process device monitors' or similar terms by different authorities. (Courtesy of Micro Circuit Engineering Ltd., UK.)

or symbol on it, which enables the vendor to confirm that all the masks required have been correctly used.

The remaining approval procedures on each individual IC are divided between the vendor and the OEM as indicated in Figure 6.1. These activities involve the following considerations:

- the size and complexity of the custom circuit and how it may be tested
- which tests are to be undertaken by the vendor and which by the OEM
- how much the OEM is prepared to pay the vendor for testing
- how many tests the vendor is willing to apply to each production IC before shipping to the OEM
- what resources the OEM has for both IC and final product testing

All these factors must be considered at the design stage of any custom IC, and cannot be left until the design work has been completed.

In the case of very small USICs containing, e.g., a few hundred logic gates or a very small number of analog circuits, the problem is not very acute. With such small circuits the OEM can usually undertake a fully functional check by plugging each received IC into a test rig or even a final product held for test purposes, and then checking the correct operation of the complete product. (For user-programmable logic devices, where the vendor is not directly involved, this will be the usual situation.) However, if production quantities increase, then there may be pressure to do a restricted series of tests on each IC, which will not exhaustively test its correct functionality. Which restricted tests to do are a matter for discussion, since there is now always a possibility of a faulty circuit being passed on to a final user.

Production testing of the ICs is necessary because it is impossible to guarantee that the fabrication is completely free of defects. If the wafer manufacturing process and the subsequent scribing, bonding and packaging procedures were perfect, then there would be no need to do any testing—every circuit would be fully functional, assuming that the original design had been fully verified. Such perfection cannot be achieved with such complex fabrication processes, and hence some testing of all production ICs is necessary.

However, unless a fully comprehensive and fully exhaustive test of an IC is undertaken, there always remains the possibility that a circuit will pass a given series of tests, but will still not be completely fault-free. The lower the yield of the production process and the less exhaustive the testing, then the greater the probablity will be of not detecting faulty circuits during the test.

This probability may be mathematically analyzed. Suppose the production yield of fault-free circuits is Y, where Y has some value between 0 (all circuits faulty) and 1 (all circuits fault-free). Suppose also that the tests applied to the circuit have a fault coverage (test efficiency) of FC, where FC also has some value between 0 (the tests do not detect any of the possibly faults) and 1 (the tests detect all possible faults). Then the percentage of circuits which will pass the test but will still be faulty in some respect, the *defect level after test*, *DL*, is given by:

$$DL = \{1 - Y^{(1-FC)}\} \times 100\%$$

This equation is illustrated in Figure 6.3.

The significance of this very important probability relationship is as follows. Suppose the production yield was 25% ($Y = 0.25$). Then if no testing at all was done ($FC = 0$), 75% of the ICs would be faulty when they are used. If testing is done and the efficiency of the tests is 90% ($FC = 0.9$), then the percentage of faulty circuits when they are used has now dropped to about 15% (85% fault-free, 15% faulty). This is still about one IC in seven faulty, which is far too high for most manufacturers. Hence, to ensure a very high percentage of fault-free circuits after test, either the manufacturing yield *Y* must be high, or the fault coverage *FC* must be high, or both.

Unfortunately, *Y* is always likely to be low due to the complexities of the

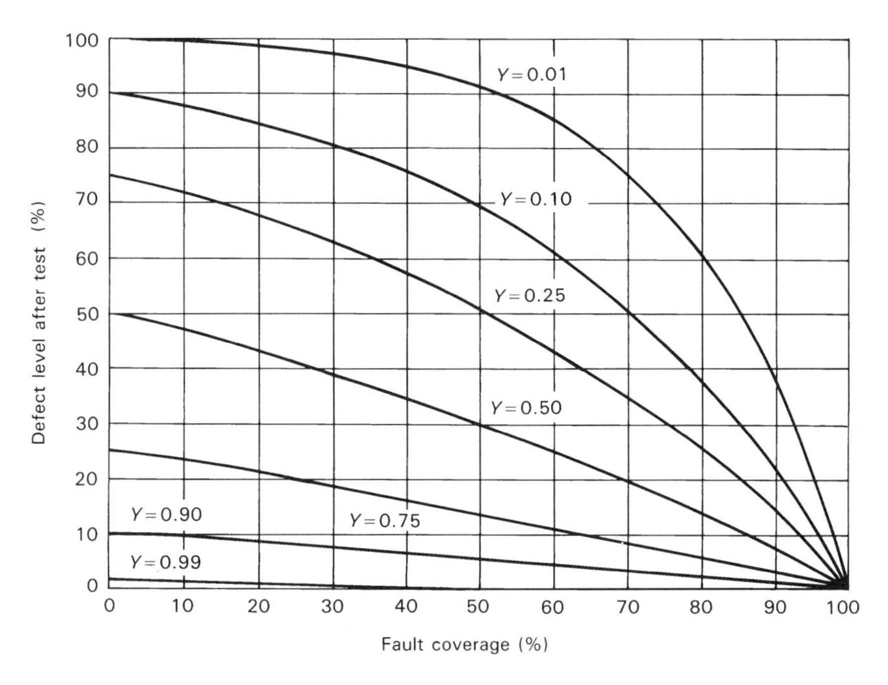

Fig. 6.3 The defect level density *DL* of a circuit after test, with different manufacturing yields *Y* and fault detection coverage *FC*. Note that only $FC = 100\%$ or $Y = 100\%$ (or both) will guarantee 100% fully functional circuits (zero defects) after test.

fabrication. Therefore, to ensure a very low percentage of faulty ICs after test, the testing efficiency *FC* must be high.

This, therefore, is the dillemma of testing complex integrated circuits. If the testing efficiency is low, faulty ICs will escape detection. But acheivement of 100% fault coverage ($FC = 1.0$) may require such extensive testing as to be non–cost effective. Notice that this basic analysis is not unique to integrated circuits, but applies to all manufactured goods from washing machines, automobiles and all other products.

6.1.1 Prototype Circuits

To revert to the problem of acceptance testing of the first prototype USIC: during the design phase, simulation of the circuit will have been undertaken and approved before any silicon fabrication was commenced. This procedure invariably involves the vendor's CAD resources for the final postlayout simulation check, and from this simulation a set of tests for the chip may be automatically generated and down-loaded into the vendor's test equipment. The vendor's tests on prototype circuits may therefore be based on this premanufacture simulation data, and if the circuits pass this means that they conform to the simulation data approved by the OEM.

Unfortunately, prototype custom ICs which pass the vendor's tests often fail to satisfy the OEM's tests when tested under working product conditions. This is not necessarily to say that the ICs are faulty, but rather that they were not designed to provide the exact functionality or performance required in the final product. The original chip specification was somehow incomplete or faulty, perhaps in a very minor way such as a logic 1 input being specified instead of a logic 0, or perhaps because last-minute product changes were not advised to the chip designer. Historically this has been the main reason why custom ICs fail to meet OEM requirements, and it thus demands close managerial control during the design phases to ensure final compatability of product and custom circuit.

6.1.2 Production Circuits

Following the successful acceptance of prototype circuits, in production testing the vendor is primarily interested in being able to test the production USICs as rapidly as possible with the minimum number of test input vectors, normally using very expensive general-purpose computer-controlled test equipment. For these tests a minimum set of test vectors is usually sought, this minimum set being the subject of discussion between the vendor and the OEM. It may, for example, be a set of input conditions suggested by the OEM which checks that the IC performs a certain series of key functions correctly, or it may be a minimum test set based upon fault models (see Section 6.3.2) which will detect a specific set of potential faults in the circuit. However, problems may still arise, such as the following:

- The vendor's general-purpose test equipment may not be capable of applying some of the special input signals met in the final product, e.g., very low or very high voltages or analog signals.
- Similarly, some of the special output conditions which the custom circuit provides may not be precisely measured by the vendor's tester.
- The vendor's test equipment may not be able to test the custom circuit at the normal operating speeds (very slow or very fast) of the final product.

All these aspects, in addition to the problem of nonexhaustive testing, become more serious as the size and performance of the custom circuit increases, and where mixed analog/digital designs are involved.

In the following sections we will consider the fundamentals of testing in more detail, and how design-for-test becomes increasingly necessary for large circuits. However, in all cases detailed discussions between OEM and vendor are always necessary to ensure acceptable testing stratagies. Further general details on testing may be found in the literature [1–9].

6.2 BASIC TESTING CONCEPTS

6.2.1 Digital Circuit Test

Before continuing with a discussion on digital test methods, it may be appropriate to clarify the following three terms.

(i) *Input test vector (or Input vector or Test vector)*: this is a combination of logic 0 and 1 signals applied in parallel to the accessible ('primary') inputs of the circuit under test. For example, if eight inputs are present, then one test vector may be 01101110. A test vector is the same as a word, but the latter term is rarely used in connection with test.

(ii) *Test pattern*: a test pattern is the same as a test vector but with the addition of the fault-free output response of the circuit to the input test vector. For example, if there are four primary outputs, then with the above input test vector the four expected outputs may be 0, 0, 0, 1, respectively.

(iii) *Test set*: a test set is a set of test patterns which in total should determine whether the circuit under test is fault-free or faulty. A test set may be fully exhaustive, or reduced, or minimum (see Section 6.3). Continuing the above example, a test set may begin as shown in Table 6.1.

Table 6.1 An Example Test Set for a Combinational Network with Eight Inputs and Four Outputs

	Test vectors								Test response			
	x_1	x_2	x_3	x_4	x_5	x_6	x_7	x_8	y_1	y_2	y_3	y_4
First test pattern	0	1	1	0	1	1	1	0	0	0	0	1
Next test pattern	0	1	1	0	1	1	1	1	0	0	1	1
Next test pattern	1	0	0	1	1	1	1	1	1	0	0	1
•	•								•			
•	•								•			
•	•								•			

Unfortunately, the terms 'vectors,' 'patterns' and 'sets' are sometimes loosely used, and hence care is necessary when reading literature from different sources.

Consider the test of a simple network containing combinational logic only (no latches or other bistable storage circuits) as shown in Figure 6.4. With n binary inputs 2^n input test vectors are required to test the circuit exhaustively—an exhaustive input test set—with the output response being checked on each of these 2^n input vectors. If $n = 20$, then it is necessary to apply and check the output response to approximately one million input test vectors, which in effect means checking through a one-million-line truthtable. (For CMOS logic gates using pairs of p-channel and n-channel transistors, there is also a question of certain open-circuit faults which may require more than 2^n input vectors for a complete test, but we will come to this in Section 6.3.2.)

The output response check may be undertaken in two distinct ways. It may be possible to check the response of the circuit under test by comparing its output(s) with that of a known good circuit, sometimes termed a 'gold unit,' as shown in Figure 6.5(a). This procedure may be used for the production testing of standard off-the-shelf SSI and MSI circuits, but for more complex circuits and

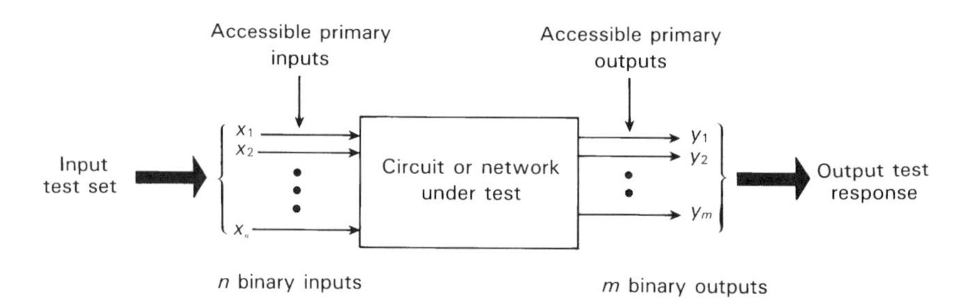

Fig. 6.4 The test of a simple combinational circuit or network containing only combinational logic gates.

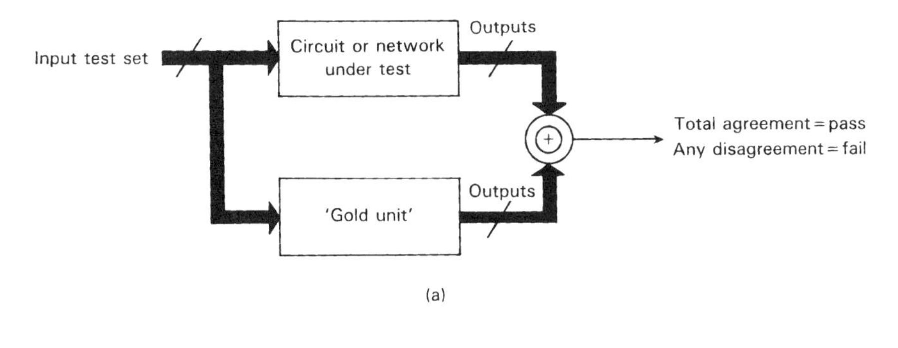

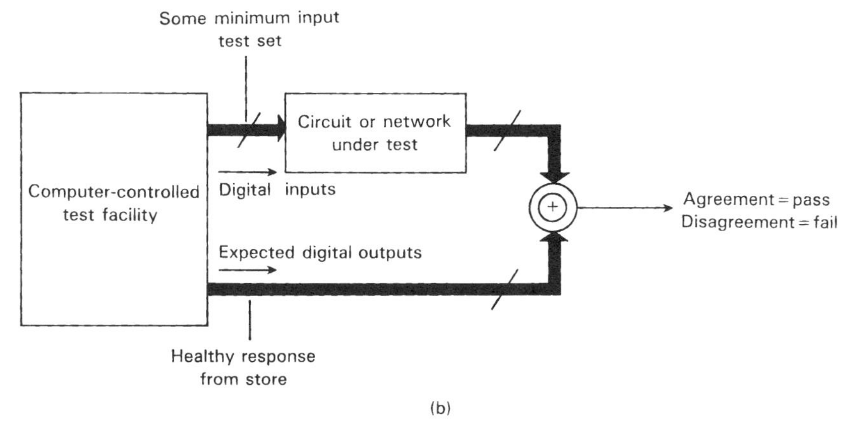

Fig. 6.5 Output response detection: (a) testing against a known good circuit, the input test set usually being fully exhaustive; (b) testing against the stored healthy response of the circuit held in computer memory; the test set may be exhaustive or more usually some reduced (nonexhaustive) test set employed in order to save testing time.

for custom ICs it is more appropriate to use a general-purpose computer-controlled test resource which can be programmed to apply the required test set, monitoring the output response(s) on every input test vector. This is illustrated in Figure 6.5(b).

Because every input test vector and the associated healthy circuit response has to be programmed into a general-purpose tester, this can involve the storage and recall of very large amounts of data, particularly when sequential circuits (see below) are also involved. Hence, there is usually pressure to formulate some reduced or minimum test set which requires less storage and can be applied more quickly, but which will still ensure an acceptable test of the circuit.

By definition, sequential circuits involve bistable (storage) circuits ('latches' and 'flip-flops') to remember present states and to influence the next state of the network. For a network containing s bistable circuits, it is theoretically necessary to test all 2^s possible combination of states in order to provide an exhaustive test. This is clearly necessary if all the s circuits form a single counter, but is less obvious if these circuits are scattered throughout the network in small groups. Nevertheless, there may be bridging or other faults between logic elements which are common to two or more groups, and thus a test of all possible combination of states may still be needed.

Real-life circuits invariably consist of both combinational and sequential elements, and both need to be tested together unless means are provided to separate them under test conditions (see Section 6.6.2). To test a combined combinational/sequential network with n primary inputs and s internal latches under all possible input conditions and internal states now requires $2^n \times 2^s$ input test vectors for fully exhaustive testing. This requirement was previously shown in Section 5.1.4, where it was seen that a simple 16-bit accumulator circuit required over 35 million input test vectors to test the circuit through all its possible logic conditions. The correct 16-bit output response on every one of these test vectors would need to be checked, which is a very tedious and time-consuming procedure for such a simple basic circuit.

There is, however, one very common sequential circuit which can be successfully tested with a method other than 2^s clock pulses to generate all 2^s states. This is the shift register configuration (see Figure 6.6). Since the action of a healthy shift register is that the state of a flip-flop is always passed on to its neighbor on receipt of a clock pulse, it is only necessary to test that logic 0s and 1s can be passed through the register from beginning to end, and that each state can maintain a logic 0 or a logic 1 when its immediate neighbors are in the opposite state. A sequence of tests may thus be as follows:

(i) clear all stages in the shift register to logic 0, and then shift through a single 1 with $s + 1$ clock pulses, thus ensuring that 1 is possible surrounded by near-neighbor 0s

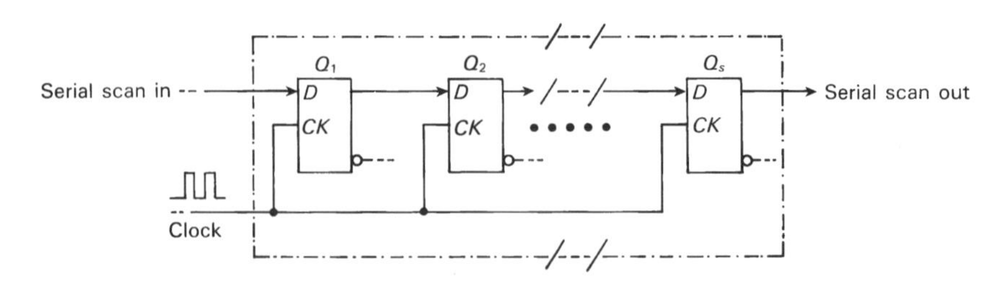

Fig. 6.6 An s-stage shift register assembly which may be tested by scanning in a serial data bit stream at the left and checking that the same data emerges s clock pulses later at the output.

(ii) set all stages in the register to logic 1, and then shift through a single 0 with $s + 1$ clock pulses, thus ensuring that 0 is possible surrounded by near-neighbor 1s

(iii) shift through a pattern of 00110011. . . . , this sequence exercising each stage through the remaining combinations of self and near-neighbor states

For long shift registers it is possible to combine these three types of test by shifting through a sequence such as 0001011100010. . . . , which combines the self and near-neighbor conditions tested by (i), (ii) and (iii). Such tests are sometimes known as *flush tests.*

Hence, it is possible to test a shift register with considerably less than the 2^s clock pulses which are otherwise required by other s-stage networks. As a result the shift register is the most powerful circuit configuration available when testing considerations are involved: it is used in many ‘easily-testable’ and ‘designed-for-testability’ (DFT) IC designs, and will be referred to very many times in the following pages of this chapter.

6.2.2 Analog Circuit Test

The testing of analog (linear) custom ICs is usually less complex than digital IC testing because, in general, analog ICs contain far fewer primitives (amplifiers, etc.) than the large number of gates and other macros involved in digital LSI or VLSI networks. As a result the input/output behavior can be specified more readily, although the parameters involved may be more complex.

Among the test parameters may be the following:

- voltage amplification (gain)
- bandwidth
- crossover and other distortion
- signal-to-noise ratio
- common-mode rejection (CMR)
- offset voltage

and others. A check on all parameters may be essential at the prototype stage, but providing the performance of the prototype circuits is found to be satisfactory then production testing may be relaxed to a subset of the full set, for example, gain and offset measurements, provided that the fabrication process continues to be monitored by means of the drop-in test circuits.

The actual test of analog USICs involves standard test instruments such as waveform generators, signal analyzers, voltmeters, etc., as used in the testing of any analog system. However, it is common for this instrumentation to be in the form of rack-mounted instruments all under the control of a dedicated microcomputer or PC, the whole instrumentation assembly being known as a *rack-and-*

stack resource [1,5]. An alternative to such standard instrumentation may be a specially designed test rig to meet the specific test requirements of a custom circuit. Digital signal processing (DSP) techniques are also being increasingly used in the testing procedures [7,10,11].

It is very necessary for the OEM to discuss the testing of an analog custom IC with the vendor, since the vendor will have no first-hand knowledge of the final product requirements and any critical parameters. In the case of very complex analog circuits, both the design and the testing may be beyond the capabilities of an OEM just starting to use custom ICs, in which case the expertise and guidance of the vendor or specialist design house is essential.

6.2.3 Mixed Analog/Digital Circuit Test

The test of the analog part and the test of the digital part of a combined custom IC each requires its own distinctive form of test, and hence it is usually necessary to have the interface between the two brought out to accessible test points so that the analog and the digital tests may be performed separately. To some extent this mirrors the problem of mixed analog/digital simulation considered in the previous chapter, where it was seen to be difficult to combine these two parts within one simulation resource due to the dissimilar signal characteristics involved.

In the case of a relatively simple mixed analog/digital USIC containing, e.g., an input D-to-A converter, some simple internal signal processing and a D-to-A output converter, it may be possible to define a test schedule without access to the internal interfaces. All such cases must be individually considered, and no hard and fast rules can be subscribed. There are, however, on-going research activities which seek to combine both analog and digital test within one test resource, using, for example, multiple discrete voltage levels to represent the analog signals, or digital signals as voltage-limited analog signals, but no commercial instrumentation using these concepts is known to be currently available. DSP techniques are also of increasing interest for many possible situations [7] and may represent a future test strategy.

However, certain mixed-signal test assemblies, such as illustrated in Figure 6.7, are commercially available. Here separate digital and analog resources are present, being linked by a GP Instrumentation Bus, the whole being under the control of a host computer or PC. If such resources are used, then it is necessary to consider their capabilities during the design phase in order that unforeseen limitations are not encountered at the prototype or production stage.

6.3 DIGITAL TEST PATTERN GENERATION

To summarize the principal difficulties of digital logic testing, the two fundamental problems with VLSI circuits are:

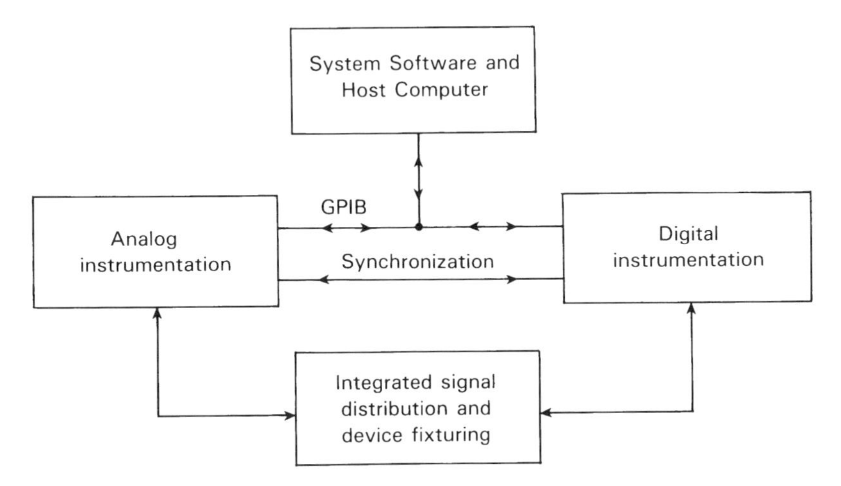

Fig. 6.7 A mixed analog/digital test resource, with the test of the analog and digital test parts synchronized under host computer control. (Courtesy of Integrated Measurement Systems, Inc. OR.)

(i) the restricted access to the circuit, since only the primary input and output pins of the IC are available for test purposes (it is possible for a vendor to probe within an IC if it is necessary to locate some obscure trouble, but this invariably causes scar damage, making the IC no longer usable; some newer techniques using scanning electron microscopy (SEM) may also be used to monitor internal voltage levels without causing damage [12], but neither of these techniques is relevant for normal test purposes)

(ii) the time and cost to test is prohibitive if a fully exhaustive test is attempted

Hence, some reduced test set or other means of reducing test time is needed for large digital networks. As we will discuss later, test pattern generation invariably involves the determination of some nonexhaustive set of input test vectors which will test the circuit to some acceptable confidence level.

6.3.1 Controllability and Observability

Two terms need to be defined before discussing certain aspects of testing, namely, *controllability* and *observability*. The broad concept of controllability is simple: it is a measure of how easily a given node within a circuit can be set to logic 0 or logic 1 by signals applied to the accessible primary inputs. Similarly, the concept of observability is a measure of how easily the state (logic 0 or logic 1) of a given node can be ascertained from the signals available at the accessible primary

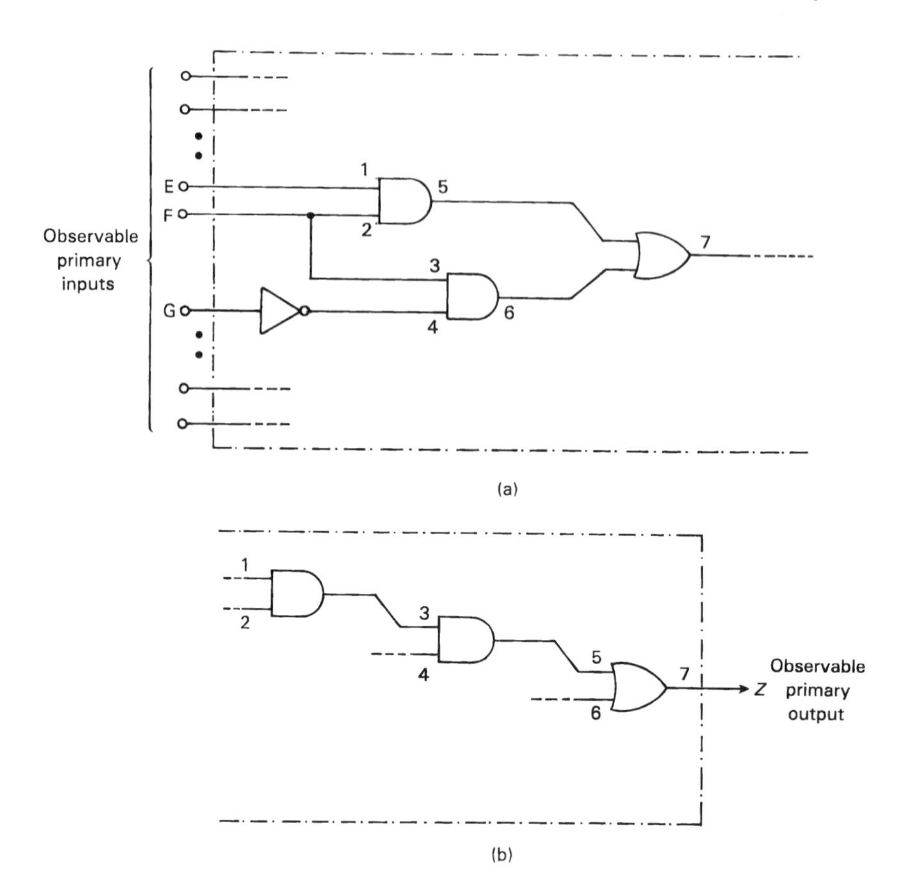

Fig. 6.8 The basic concepts of controllability and observability: (a) the controllability of node 7, see text; (b) the observability of node 1, see text.

outputs. These two terms are central factors when discussing the testability of digital networks.

Consider the small fragment of a circuit shown in Figure 6.8(a). Nodes 1, 2 and 3 are immediately controllable, since they connect directly to the primary inputs. Node 7, on the other hand, is clearly not so readily controllable; it requires either node 5 to be switched between logic 0 and 1 ('toggled') with node 6 held at logic 0, or vice versa, which in turn requires, e.g., input F to be held at 1, input G held at 1, and input E toggled. Hence, the input vectors to control node 7 can be:

Input vector · · · E F G · · ·	Node 7
· · · 0 1 1 · · ·	0
· · · 1 1 1 · · ·	1

Looking at the requirements of observability, consider the small circuit shown in Figure 6.8(b), and suppose it is desired to observe the value of internal node 1. In order to propagate this nodal value to the output, i.e., to make the observable output of 0 or 1 depend upon this node only, it is clear that nodes 2 and 4 must be set to 1 and node 6 set to 0. (This is sometimes termed 'sensitizing' or 'forward-propagating' the path from the required node to an observable output.) Hence, the input signals must be such as to give these logic conditions on nodes 2, 4 and 6 so that the output is dependent upon the value at node 1.

The general features of controllability and observability are shown in Figure 6.9. The further away a node is from a primary input or output, then the ease of controlling or observing this node diminishes. However, provided that there are no redundant nodes in the network, that is, all paths are necessary to produce fault-free outputs, then it is always possible to determine two (or more) input test vectors which will check that a given internal node switches from 0 to 1 and back again correctly. The complexity of determining the smallest set of such vectors for every internal node of a circuit is extremely high, way beyond the bounds of hand computation for all except the smallest of circuits. If sequential circuits are present, then there will be the additional complexity of driving the latches/flip-flops to specific states to give the required nodal observability, which may require a large number of clock pulses to achieve the required states.

With internal circuit redundancy (deliberate or accidental), it will not be possible to control certain internal nodes without incorporating additional circuit details. Consider the monitoring circuit shown in Figure 6.10(a), which has been included to check that the outputs of macros A and B always agree. With a fault-

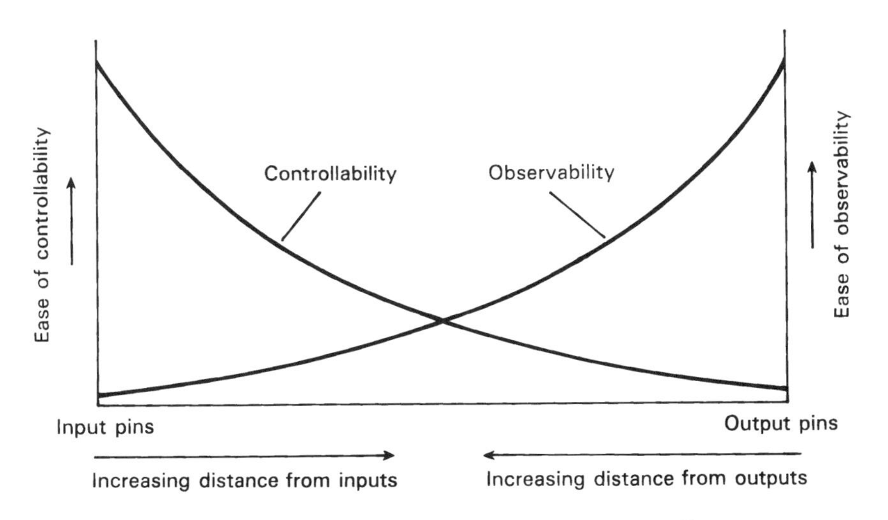

Fig. 6.9 The general features of both controllability and observability.

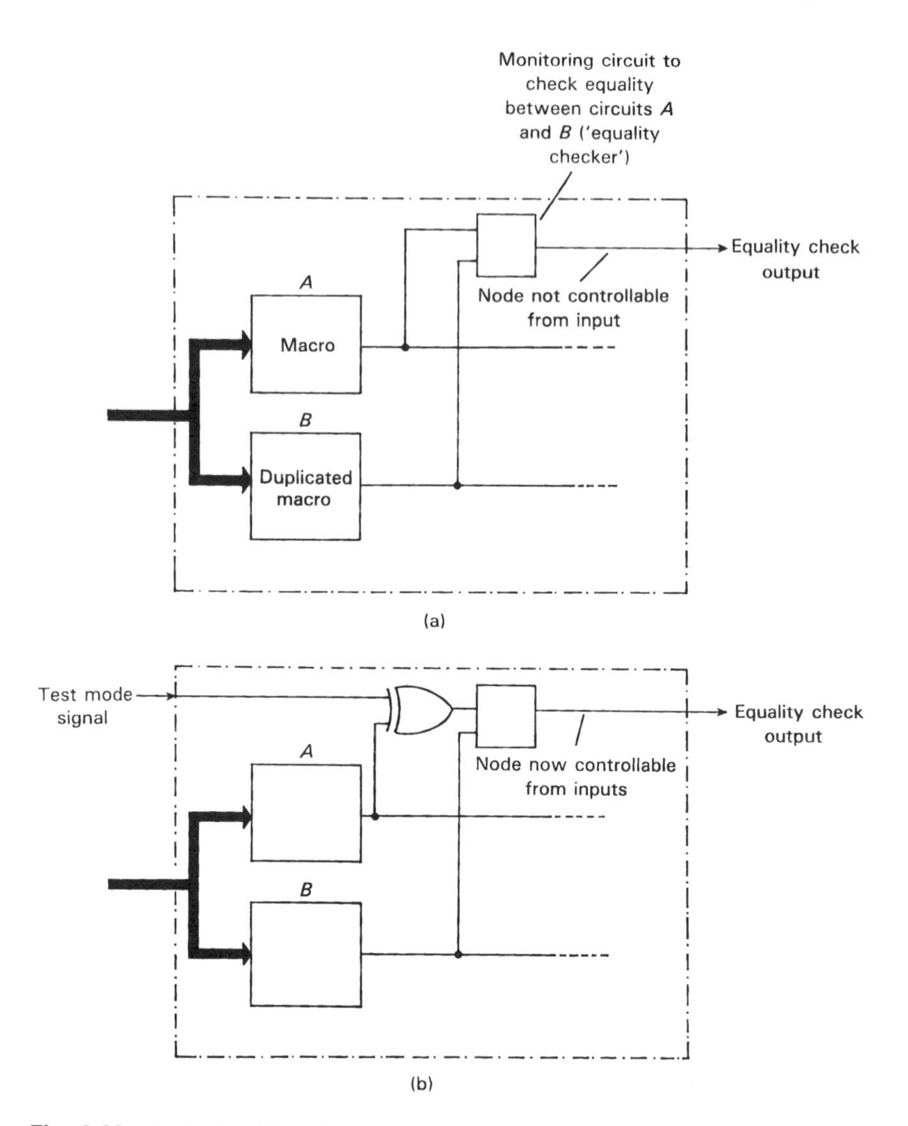

Fig. 6.10 A circuit with an internal node not directly controllably from the primary inputs: (a) a basic circuit with duplicated macros A and B; (b) an addition in order to give controllability of the equality checker circuit.

free circuit, no means exists to influence the output of the monitoring circuit; an addition such as that shown in Figure 6.10(b) is necessary if controllability and observability of this monitoring circuit is required. This is an example of the well-known feature of building in redundancy in any type of system, electrical, electronic or mechanical: unless means are provided for disconnecting or otherwise overriding the redundancy, there is no way in which the correct functioning of this part of a system can be fully checked. On the bonus side, however, if it is ever found that it is impossible to control or observe a node within a logic circuit, then this must be a redundant part of the circuit which the designer has perhaps overlooked

Many attempts have been made to quantify the two parameters of controllability and observability for digital networks. They include the following:

(i) controllability defined as

$$1 - \left| \frac{N(0) - N(1)}{N(0) + N(1)} \right|$$

where $N(0)$ is the number of different input vectors which establish logic 0 on the node in question, and $N(1)$ is the number of different ways of establishing logic 1, and observability defined as

$$\frac{N(P)}{N(P) + N(NP)}$$

where $N(P)$ is defined as the number of input vectors which allows the logic value on the node to propagate to an output, and $N(NP)$ the number of vectors which do not allow this propagation.

(ii) controllability defined by a consideration of the number of nodes which have to be controlled to produce a 0 or 1 at the node in question, and observability defined by a consideration of subsequent nodes which have to be controlled to propagate this signal to an output.

These and other approaches require considerable complexity to accomodate feedback loops, reconvergent fan-out and storage elements. Finally, an average or other amalgamation of the individual controllability and observability values for all the circuit nodes must be made to give some overall measure of the circuit's testability [9].

Testability analysis was widely pursued in the late 1970s and early 1980s. In particular, two commercial software packages may be mentioned, namely HI-TAP (formerly CAMELOT) from GenRad, Inc., based upon (i) above, and SCOAP (Scandia, Inc. Controllability/Observability Analysis Program), based upon (ii). Other commercial packages included TMEAS, COMET, TESTSCREEN, VICTOR and others from software vendors [9]. However, the limitations of testability analysis include the following:

- the analyses do not give an accurate measure of the ease or otherwise of testing
- they are applied after the circuit has been designed, and do not give any help at the initial design stage or guidance on how to improve the ease of test
- they do not give any help in formulating a minimum set of test vectors for the circuit

Hence, although the concepts of controllability and observability are key concepts in the consideration of digital testability, attempted quantification of these parameters is not particularly valuable. Testability analysis should therefore only be regarded as a possible postdesign exercise to identify difficult-to-test nodes, and not as a useful design tool. We will, however, return to consider controllability and observability again in Section 6.3.3.

6.3.2 Fault-Effect Models

The most common way of determining some minimum set of input test vectors to test a digital network involves a consideration of what faults are likely in the circuit. A single test vector can then be determined which would give, e.g., an observable logic 0 at an output when the circuit is fault-free, but a logic 1 if this fault is present on a particular node. This procedure, known as *fault modeling*, can be used to determine a minimum set of input test vectors which will detect the presence or absence of the given type of fault for every internal node of the circuit.

Stuck-At Faults

The most commonly used fault model is the *stuck-at fault model*, in which it is assumed that any physical defect in a digital circuit causes the input or output of a gate to be permanently at logic 0 or 1. Hence, stuck-at faults may be either stuck-at-0 (s-a-0) or stuck-at-1 (s-a-1).

For a three-input NAND gate, for example, there are eight possible stuck-at faults, as shown in Figure 6.11. However, it will be seen that four of these are indistinguishable at the gate output node, all giving the stuck-at-1 output condition. A total of four input test vectors applied to the gate, a *minimum test set*, will therefore detect (but not identify) the presence of these faults, as shown in below:

Input A B C	Expected output	Wrong output	Faults detected
0 1 1	1	0	A s-a-1 or Z s-a-0
1 0 1	1	0	B s-a-1 or Z s-a-0
1 1 0	1	0	C s-a-1 or Z s-a-0
1 1 1	0	1	A or B or C s-a-0 or Z s-a-1

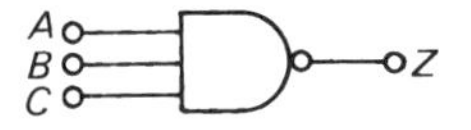

Inputs				Output Z							
A	B	C	Fault-free	A stuck-at-0	A stuck-at-1	B stuck-at-0	B stuck-at-1	C stuck-at-0	C stuck-at-1	Z stuck-at-0	Z stuck-at-1
0	0	0	1	1	1	1	1	1	1	0	1
0	0	1	1	1	1	1	1	1	1	0	1
0	1	0	1	1	1	1	1	1	1	0	1
0	1	1	1	1	0	1	1	1	1	0	1
1	0	0	1	1	1	1	1	1	1	0	1
1	0	1	1	1	1	1	0	1	1	0	1
1	1	0	1	1	1	1	1	1	0	0	1
1	1	1	0	1	0	1	0	1	0	0	1

Fig. 6.11 Stuck at faults on a 3-input NAND gate

Automatic test pattern generation to determine test sets based upon this stuck-at model are widely available. These ATPG programs (see Section 6.3.3) assume that only single stuck-at faults are present at a time, and determine a test to detect this single node stuck at 0 or stuck at 1. The test vector applied to the circuit must be such as to apply appropriate signals to the node being tested (the controllability requirement), and also to propagate the signal on this node to an output (the observability requirement) in order to check for the particular stuck-at condition. This procedure is repeated until tests for all the circuit nodes are covered.

For example, in the simple circuit of Figure 6.12(a), if it is required to check that the output of the Inverter gate *I*1 is not stuck at 0, the input vector shown must be applied. If this node is s-a-0, then the observable output will be 0 instead of the healthy value of 1. Notice that this stuck-at fault is indistinguishable from the Inverter input s-a-1, or the other input of NAND gate *N*1 s-a-0, or its output s-a 1. Hence, in total there is a great deal of commonality in the set of test vectors checking individual nodes s-a-0 and s-a-1, from which a minimum test set can be complied. Figure 6.12(b) gives a larger example, from which it is seen that fifteen input test vectors will detect all the possible s-a-0 and s-a-1 faults.

The possibility of multiple stuck-at faults in a network clearly exists, but the number of possibilities ($3\eta - 2\eta - 1$ compared with 2η single stuck-at faults, where η = the number of nets in the circuit) becomes too great to consider. However, it has been found in practice that the single stuck-at fault model will normally cover multiple stuck-at faults—in other words, it is extremely unlikely that one stuck-at fault will mask another and give a fault-free output from the minimized test set. Note that the location of the stuck-at faults cannot be determined from this test set, but only that the circuit is free from any stuck-at faults or is faulty.

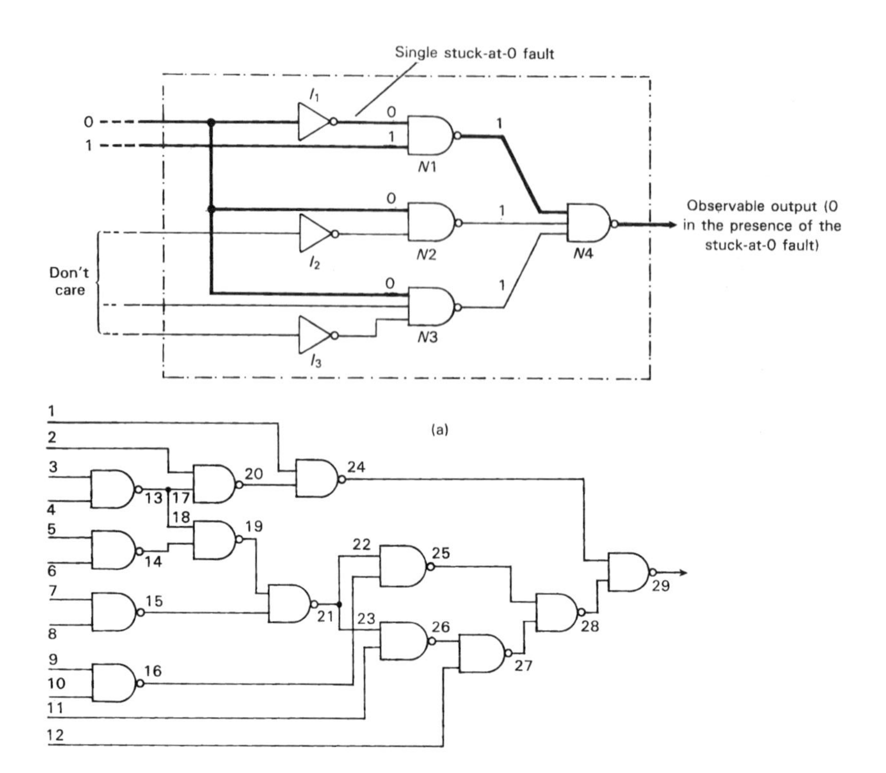

Inputs	Faults covered	Output
10×0×0× × ×01×	1/0 2/1 20/0 24/1 29/0	1
00×0×0× × ×01×	1/1 24/0 29/1	0
11×0×0× × ×01×	2/0 17/0 20/1 24/0 29/1	0
0×11×0×0× ×11	3/0 4/0 13/1 18/1 19/0 21/1 23/1 26/0 27/1 28/0 29/1	0
0101×0×01111	3/1 13/0 18/0 19/1 21/0 23/0 26/1 27/0 28/1 29/0	1
01×001×01111	5/1 14/0 19/1 21/0 23/0 26/1 27/0 28/1 29/0	1
0× × ×11111111	7/0 8/0 15/1 21/0 23/0 26/1 27/0 28/1 29/0	1
0× × ×1101× ×11	7/1 15/0 21/1 23/1 26/0 27/1 28/0 29/1	0
01×0×0× ×011×	9/1 16/0 25/1 28/0 29/1	0
01×0×0× ×101×	10/1 16/0 22/0 25/1 28/0 29/1	0
01×0×0× ×1101	11/1 26/0 27/1 28/0 29/1	0
1111× × ×0× × ×1	17/1 20/0 24/1 29/0	1
0× × ×11×0×0×0	12/1 22/1 25/0 27/0 28/1 29/0	1
011010×01111	4/1 6/1 9/0 10/0 11/0 13/0 14/0 16/1 18/0 19/1 21/0 23/0 25/0 26/1 27/0 28/1 29/0	1
01×01110× ×11	5/0 6/0 8/1 12/0 14/1 15/0 19/0 21/1 23/1 26/0 27/1 28/0 29/1	0

(b)

Fig. 6.12 The propagation of stuck-at faults to an observable output: (a) a simple example showing the test to propagate the output of a gate/$I1$ stuck-at-0 to the output with the three lower primary outputs being "don't care"; (b) a more comprehensive example, showing that 15 input test vectors will detect all the stuck-at-0 and stuck-at-1 faults in the circuit. (Courtesy of Oxford University, UK.)

Bridging Faults

Bridging faults are electrical connections between two (or more) points in a circuit which should be logically separate. The theoretical number of bridging faults possible in a digital network if any interconnection line (net) is assumed to be bridged to any other is extremely high, and thus only 'near-neighbor' bridging faults are normally considered in a bridging-fault model.

However, considerable difficulties are associated with bridging faults, such as the following:

- It is necessary to know the topology of the circuit in order to identify all the 'near-neighbor' connections.
- Does the bridging fault result in the logical AND (wired-AND) or logical OR (wired-OR) of the logic signals on the two lines?
- Does the bridging fault result in the formation of a latch by establishing an erroneous feedback path?

The possibility of forming a latch is illustrated in Figure 6.13, although the exact result of such a bridging fault depends upon the sink or source impedance of the logic signals involved.

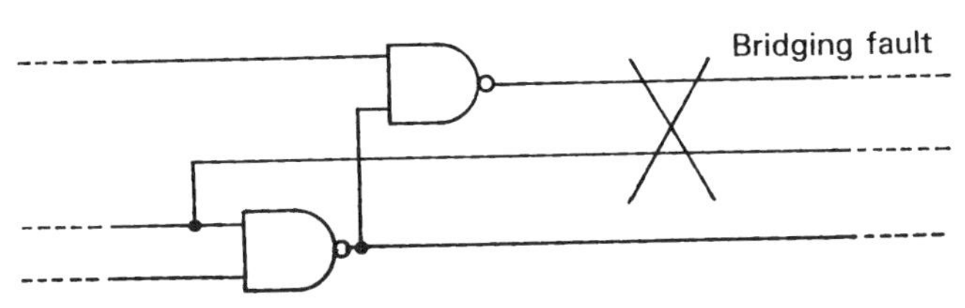

(a) Resultant equivalent circuit of two cross-coupled NAND gates:

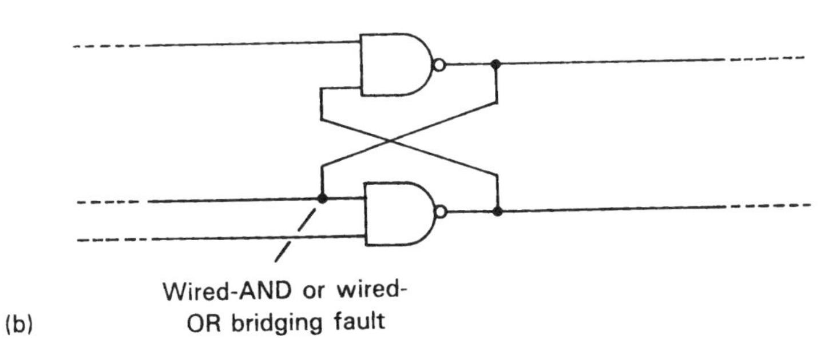

Fig. 6.13 Bridging fault conditions which may produce a feedback path and hence a bistable (latch) configuration: (a) the circuit and the bridging fault; (b) the equivalent circuit with the fault present.

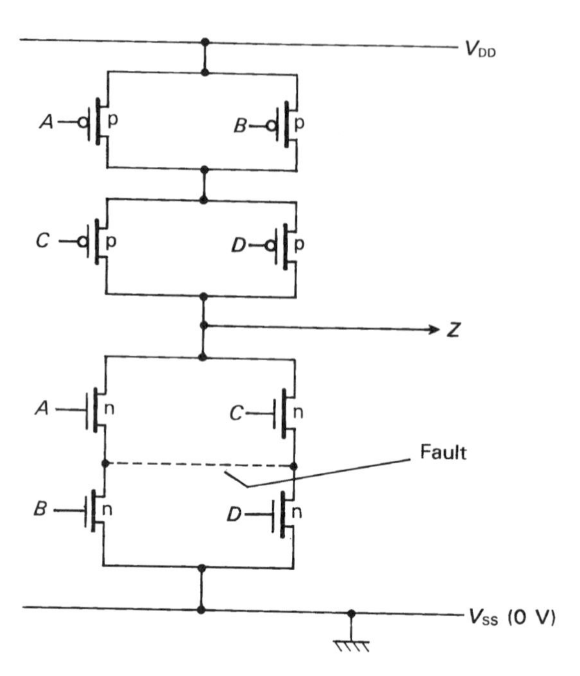

Fig. 6.14 A possible bridging fault shown dotted within a complex CMOS logic gate, the fault-free gate output being $Z = \overline{AB + CD}$. Note that the simple equivalent Boolean gate representation of two AND gates and one NOR gate cannot model this fault.

Bridging faults within a logic gate, particularly in complex CMOS gates, can raise further difficulties, since the logic function may be modified. For example, in Figure 6.14 the presence of the bridging fault, shown dotted, will pull the gate output down towards logic 0 under the input conditions $AB + CD + AD + CB$ instead of the normal conditions of $AB + CD$. However, if the p-channel transistors in this case remain fault-free, considerable power dissipation will result when $AD + CB$ is present. Therefore, such potential faults in CMOS circuits are often monitored by measuring the current taken from the V_{DD} supply line rather than by any functional tests—excessive static current is taken to be a short-circuit condition somewhere within the circuit.

In general, therefore, bridging fault modeling is not employed for the test pattern generation of random layout digital circuits. However, bridging fault modeling is very relevant to consider for PLAs and other regularly structured topologies, and will be specifically considered in Section 6.4 when considering the testing of programmable logic devices.

Open-Circuit Faults

Finally, one must consider open-circuit faults. Unfortunately, this is the fault which is particularly awkward to detect in CMOS logic gates, although less difficult in other technologies where it is usually indistinguishable from a stuck-at fault as far as the observable outputs are concerned.

Consider the problem that can arise in a 2-input CMOS NAND gate (see Figure 6.15(a)). In the fault-free case the four transistors behave as follows:

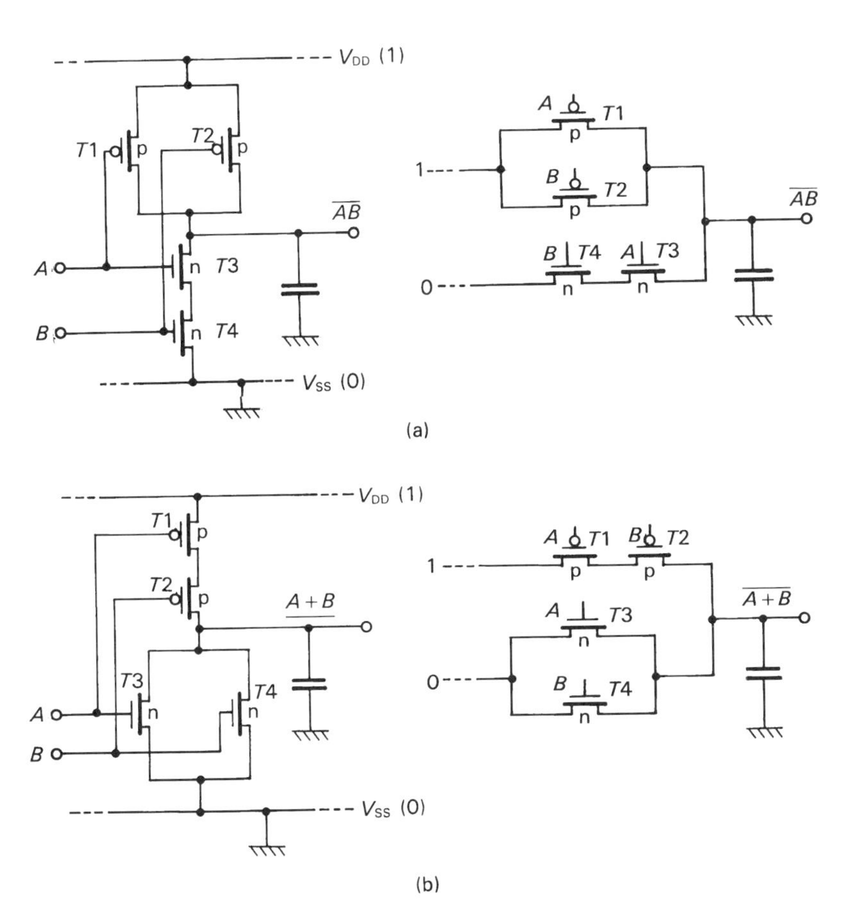

Fig. 6.15 The problem of CMOS open-circuit faults: (a) a CMOS NAND gate drawn conventionally and in an alternative way in order to emphasize the paths from V_{DD} and V_{SS} to the gate output; the capacitance represents the lumped capacitance of all the circuitry connected to the output; (b) CMOS 2-input NOR gate, again drawn conventionally and in an alternative form.

Inputs		State of the transistors				Output
A	B	$T1$	$T2$	$T3$	$T4$	Z
0	0	on	on	off	off	1
0	1	on	off	off	on	1
1	0	off	on	on	off	1
1	1	off	off	on	on	0

Suppose now that the p-channel transistor $T1$ becomes open-circuit so that it is effectively always 'off.' With inputs 00, 10 and 11 the gate output will be unaffected since there is always an 'on' transistor from V_{DD} or V_{SS} to the gate output. With input 01, however, there will be no path to the output from either supply rail, and hence the output will be floating.

If under test conditions the input test vectors are applied in the order shown above, input vector 00 will establish a logic 1 at the gate output through working transistor $T2$, but when the input changes to 01 there will be no immediate change of gate output since the circuit output capacitance will maintain the output voltage at the logic 1 value. Should input 01 be held for a long period, possibly minutes, then the gate output may decay sufficiently to no longer be an effective logic 1 output, but if (as is invariably the case) the input vector is quickly changed, no fault will have been detected in the circuit. The output has 'remembered' the correct output value, and the fault is said to be a 'memory fault.'

However, if in practice the input condition 01 had been preceded by input 11, then the gate output would not have been precharged to logic 1, and on the input 01 it would remain at the incorrect value of logic 0. Hence, to detect this particular open-circuit transistor, it is necessary for the test vector sequence to be 11 followed by 01.

A similar problem can arise with n-channel transistors in parallel (see Figure 6.15(b)), where a lack of a discharge path to V_{SS} (logic 0) may not be detected if the output is already at logic 0. It can be shown that a fully exhaustive test set for any n-input CMOS logic gate requires $n2^n + 1$ input test vectors to detect all possible NAND/NOR failings, rather than the 2^n vectors which is normally considered to be the exhaustive set. The fully exhaustive set for all 2-input CMOS gates is as follows:

- 00
- 01
- 11
- 10
- 00
- 10

- 11
- 01
- 00

Details of how such test sequences may be determined can be found in the literature [7,13,14]. Each overlapping pair of vectors may be regarded as an initialization vector followed by a test vector to check for correct pull-up or pull-down should any change of output from 0 to 1 or vice versa be required. If the precise transistor configuration of all the gates in a n-input network under test is known, then fewer than $n2^n + 1$ vectors may be possible.

The problem with CMOS testing is thus seen to be inherently related to the fact that every CMOS gate consists of two dual parts, the p-channel and the n-channel logic. A logical fault in one half and no fault in the other half, which is perhaps the most likely occurance, gives rise to these problems of testing which are not present in bipolar and nMOS logic. The question of open-circuit faults in CMOS transmission gates and flip-flops such as illustrated in Figure 5.22(b) does not seem to have been specifically considered, or at any rate not known to have been published. Overall, CMOS must be regarded as the most difficult digital technology to test thoroughly because of its dual nature.

Most custom IC vendors do not specifically address this problem of open-circuit memory faults, in spite of CMOS being the dominant logic technology. However, the following points should be appreciated:

- The stuck-at model and the actual circuit failure present may have no relationship with each other; stuck-at tests do not give any reliable diagnostic information on exactly what is wrong in a faulty circuit.
- A collective set of stuck-at tests will not exercise the circuit through all its possible logic states.

Nevertheless, it has been found in practice that if single stuck-at faults are considered at every node (=net) in a circuit, then these tests will collectively cover most of the other possible circuit failures. Hence, the stuck-at fault model is widely used throughout industry to derive test patterns for both digital circuits and printed circuit boards. PLDs, however, may have a different basis for their test vector generation (see Section 6.4). For further details of faults and fault modeling see Refs. 15 and 16 and the extensive references contained therein.

6.3.3 Test Pattern Generation

Test pattern generation is the design process of generating the test patterns required to test a given digital system, whether it is a printed circuit board (PCB) assembly or a custom IC. As previously noted, exhaustive testing is usually prohibitively long, and hence some reduced test set is normally sought. However, as will be seen later, if partitioning of a large circuit into smaller sections is done for test purposes, then it may be possible to test each small partition exhaustively, in which case the test pattern generation problem does not arise.

The generation of a set of test vectors acceptable to both the OEM and the custom IC vendor may be obtained in any one of of three ways, namely:

(i) manual generation
(ii) algorithmic generation
(iii) pseudo-random generation

The formulation of a reduced test set usually requires the internal details of the circuit under test to be known, perhaps from the netlist generated from the gate-level circuit schematic. It may therefore be referred to as 'structural testing,' as distinct from functional testing, which does not need knowledge of the precise internal circuit details.

Manual Test Pattern Generation

Manual test pattern generation for custom ICs is usually undertaken by the OEM who knows what the circuit is intended to do, rather than by the vendor. The test patterns may be compiled in either of the following two methods:

(i) by listing the input vectors which cause every gate output in the circuit to toggle from one output state to the other at least once
(ii) by considering a comprehensive range of working conditions, and listing the input vectors and the expected output responses involved in these operations

Note that the latter is a nonexhaustive functional test rather than a structural test. If a working discrete component prototype circuit had been made, then this may assist the OEM in determining the suggested set of test vectors.

On receipt, the vendor will usually compute the acceptability of the OEM's suggested test set by checking whether this set gives 100% toggling of all nodes in the circuit netlist. If it does not, then the vendor will usually ask the OEM for additional test vectors, or discuss with him the acceptability of the given set. However, it should be noted that a successful 100% toggle test only checks that there are no stuck-at nodes, and does not necessarily confirm the correct functionality of the circuit under all conditions.

Manually generated test sets usually become far too time-consuming to produce when more than a few hundred logic gates and macros are present in a custom IC. Hence, for larger or more complex circuits, algorithmic or pseudo-random test pattern generation is necessary, possibly with formal design-for-testability (DFT) techniques built in at the design stage, as will be considered later.

Algorithmic Test Pattern Generation

Algorithmic test pattern generation, or automatic test pattern generation (ATPG), usually uses a gate-level representation of the circuit, with all gate input and output nodes identified. A fault model is used which assumes that a given type of fault is present on each node of the circuit, the test pattern to detect this fault being determined for every node in the circuit.

The fault model almost exclusively used is the stuck-at fault model introduced above. The following four features may be identified in most ATPG programs:

(i) listing the necessary inputs to a gate which will generate a healthy gate output opposite from that occurring when a chosen stuck-at fault is present on the gate
(ii) determining the primary inputs necessary to establish these gate input conditions (*fault sensitizing*) and to propagate the gate output to a primary output (*path sensitizing*)
(iii) repeating this procedure until all single stuck-at faults have been covered
(iv) combining and sorting this test pattern listing into a minimum set, utilizing the fact that a single input test vector using all the primary inputs and primary outputs may simultaneously test a large number of stuck-at faults in the circuit

The most difficult part in the above is the path sensitizing. The most common methods are based upon Roth's D-algorithm [17], which uses the following five logic values:

- 0 = logic 0
- 1 = logic 1
- D = a fault-sensitive value, D = 1 for fault-free conditions and D = 0 for faulty conditions
- $\overline{D}$ = a fault-sensitive value, $\overline{D}$ = 0 for fault-free conditions and $\overline{D}$ = 1 for faulty conditions
- X = unassigned (don't care) value which can take any value 0, 1, D or $\overline{D}$

From these five logic values the primitive *D*-cubes of failure for logic gates may be defined. See Table 6.2 for the four 3-input Boolean logic gates.

The logic conditions necessary to propagate any D-value through following (fault-free) gates, or to propagate a stuck-at input on a gate to the gate output, are shown in Table 6.3. These conditions are known as the Propagation *D*-cubes. For example, to propagate a stuck-at input condition through a 3-input NAND gate, we can have the D-notation D 1 1 $\overline{D}$ (see the sixth line of Table 6.3).

Table 6.2 The Primitive D-Cubes of Failure for 3-Input Logic Gates with the Input Signals All Healthy

	Inputs			Output	D-notation			
	A	*B*	*C*	*Z* stuck-at	*A*	*B*	*C*	*Z*
AND gate:	1	1	1	s-a-0	1	1	1	D
	0	X	X	s-a-1	0	X	X	$\bar{D}$
	X	0	X	s-a-1	X	0	X	$\bar{D}$
	X	X	0	s-a-1	X	X	0	$\bar{D}$
NAND gate:	1	1	1	s-a-1	1	1	1	$\bar{D}$
	0	X	X	s-a-0	0	X	X	D
	X	0	X	s-a-0	X	0	X	D
	X	X	0	s-a-0	X	X	0	D
OR gate:	0	0	0	s-a-1	0	0	0	$\bar{D}$
	1	X	X	s-a-0	1	X	X	D
	X	1	X	s-a-0	X	1	X	D
	X	X	1	s-a-0	X	X	1	D
NOR gate:	0	0	0	s-a-0	0	0	0	D
	1	X	X	s-a-1	1	1	X	$\bar{D}$
	X	1	X	s-a-1	X	1	X	$\bar{D}$
	X	X	1	s-a-1	X	X	1	$\bar{D}$

The full 5-valued logic relationships for any logic gate can also be defined; for example, for 2-input gates we have the relationships shown in Figure 6.16. The propagation D-cubes are merely particular entries taken from these full 5-valued input/output relationships. An example of this notation applied to a simple circuit is shown in Figure 6.17, from which it will be observed that the fault

Table 6.3 The Propagation of D Values Through Healthy 3-Input Logic Gates

	Inputs			Output
AND gate:	1	1	D	D
	1	D	1	D
	D	1	1	D
NAND gate:	1	1	D	$\bar{D}$
	1	D	1	$\bar{D}$
	D	1	1	$\bar{D}$
OR gate:	0	0	D	D
	0	D	0	D
	D	0	0	D
NOR gate:	0	0	D	$\bar{D}$
	0	D	0	$\bar{D}$
	D	0	0	$\bar{D}$

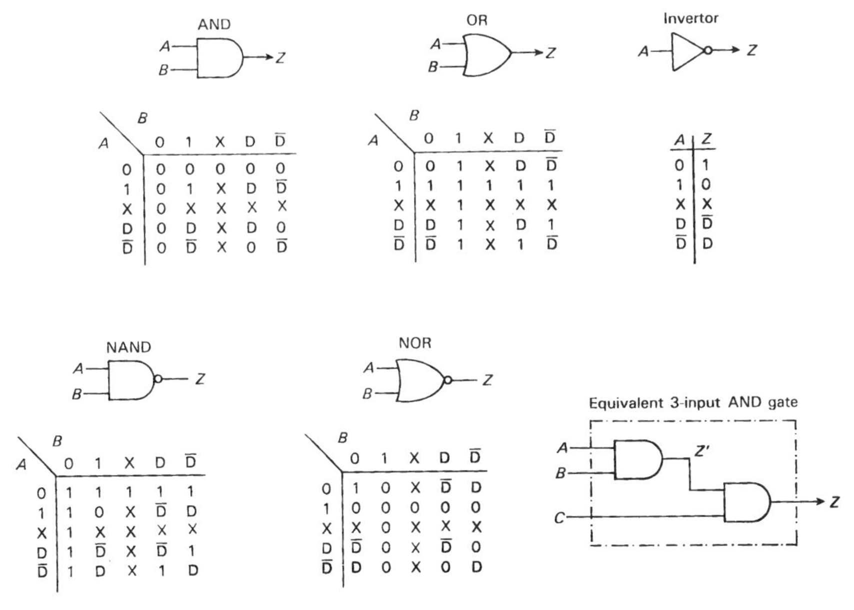

AND

A \ B	0	1	X	D	$\overline{D}$
0	0	0	0	0	0
1	0	1	X	D	$\overline{D}$
X	0	X	X	X	X
D	0	D	X	D	0
$\overline{D}$	0	$\overline{D}$	X	0	$\overline{D}$

OR

A \ B	0	1	X	D	$\overline{D}$
0	0	1	X	D	$\overline{D}$
1	1	1	1	1	1
X	X	1	X	X	X
D	D	1	X	D	1
$\overline{D}$	$\overline{D}$	1	X	1	$\overline{D}$

Invertor

A	Z
0	1
1	0
X	X
D	$\overline{D}$
$\overline{D}$	D

NAND

A \ B	0	1	X	D	$\overline{D}$
0	1	1	1	1	1
1	1	0	X	$\overline{D}$	D
X	1	X	X	X	X
D	1	$\overline{D}$	X	$\overline{D}$	1
$\overline{D}$	1	D	X	1	D

NOR

A \ B	0	1	X	D	$\overline{D}$
0	1	0	X	$\overline{D}$	D
1	0	0	0	0	0
X	X	0	X	X	X
D	$\overline{D}$	0	X	$\overline{D}$	0
$\overline{D}$	D	0	X	0	D

Fig. 6.16 Roth's b-valued logic relationships (D-notation) applied to 2-input logic gates. The relationships for 3-input (or more) gates can easily be derived by considering a cascade of 2-input gates, since normal commutative and associative relationship still hold.

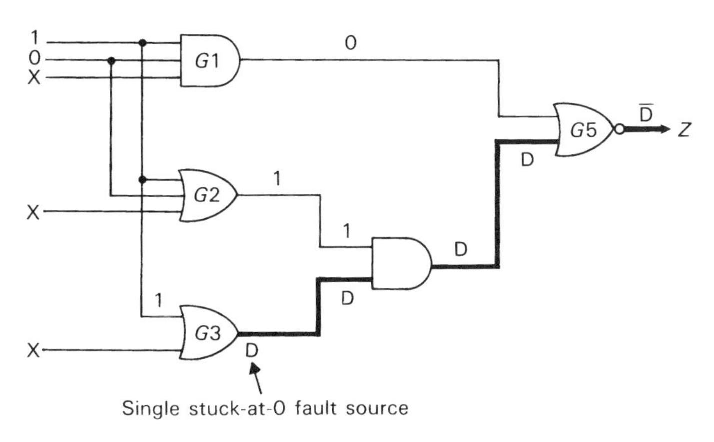

Fig. 6.17 An example of Roth's D-notation where the fault is assumed to be present at the output of gate $G3$ denoted by the symbol D.

source on the output of gate G3 is driven to the observable output by the appropriate signals sensitizing the path from G3 to the output Z. The problem that this poses in a large circuit is, of course, how to establish all the required internal gate input conditions on the forward path to an output without encountering any conflictions, bearing in mind that only the primary inputs are available for controlling the states of all the internal nodes.

Roth's algorithm for test pattern generation therefore consists of the following three main operations:

(i) listing of the primitive D-cubes of failure for each gate and stuck-at condition
(ii) a *forward drive* operation to specify the signals required to propagate each faulty D condition to an observable output
(iii) a *backward trace* (or consistency) operation to establish the required logic 0 and 1 conditions on all gates and primary inputs in order to forward-propagate (ii) above

The algorithm guarantees to find a test for any stuck-at fault if such a test exists (see later in this section), but considerable complexity can arise in the backward trace operation particularly due to (a) reconvergent fan-out, that is, when a fan-out signal from a logic gate travels through distinct paths but reconverges at some further gate before reaching a primary output; and (b) the presence of Exclusive-OR or -NOR gates. For further details of Roth's original developments, see in particular the report by Miczo [18] and other general sources of information [15,16,19].

More recent developments, still based upon the 5-valued fundamentals of the D-algorithm, provide faster and more efficient automatic test pattern generation than Roth's original search mechanisms. In particular, RAPS (Random Path Sensitizing), PODEM-X (Path-Oriented Decision Making), which incorporates RAPS, and FAN (Fan-out oriented test generation) may be mentioned [20–23]. PODEM-X is an ATPG system in which several distinct programs may be used during the test pattern generation procedure. In particular, it encompasses the following:

- an initial global procedure using RAPS, which first sets all nodes in the circuit to X and then arbitrarily selects paths back from each primary output to the primary inputs, determining what fault cover this provides by simulation after each path has been completed [15,21]
- following this procedure, directed tests for each of the nodes which have not been covered by this global procedure are carried out in turn by the main PODEM algorithm, which again uses a backward trace procedure followed by simulation to check how many faults have been covered by each test vector

The fault simulator used in the RAPS and PODEM searches is FFSIM (Fast Fault Simulator), and is a 5-valued zero-delay simulator using the five logic values previously defined [22]. Additionally, PODEM-X includes a controllability analysis of the circuit to define the 'distance' of each node from a primary output; this information is used to guide the search procedures in the choice of a shortest path, which is in contrast to Roth's algorithm which propagates information along all possible paths during a trace procedure. Finally, a compaction program in PODEM attempts to merge as many test patterns as possible in order to produce the final minimized test set.

Worked examples of the RAPS and PODEM algorithms may be found in Ref. 9, and of the PODEM algorithm only in Ref. 16, together with flow charts showing the search procedures.

FAN is similar to but is reputedly a more efficient ATPG program than PODEM. It considers the circuit topology in greater detail, in particular the fan-out points in a network, in order to reduce the amount of computer time wasted in retrying multiple choice paths. The heuristics incorporated in FAN are designed to minimize the number of arbitrary decisions made, which can result in considerable back-tracking (try again) if conflicting signals are subsequently encountered, by keeping account at each phase of the test generation algorithm of the logic values which are still free to be assigned to each node in the circuit. The concept of 'headlines' is also incorporated, a headline being a net in the circuit up to which all gates from the primary inputs have single fan-outs. The advantage of identifying headlines is that once a back trace procedure has reached a headline, then no further back-tracking to the primary inputs is necessary since the inputs can always be set to give the requires logic values at headline points. This is in contast to PODEM, which continues to trace backward until it reaches the primary inputs.

Like PODEM-X, the FAN algorithm can be incorporated into a wider ATPG system, namely FUTURE [24]. This incorporates a global test generator which can be run before FAN is used to complete the test set, but unlike the RAPS procedure in PODEM-X the global test generator in FUTURE applies a series of pseudo-random test vectors, the fault coverage of which is determined by a fault simulator. This is a very fast and simple technique, with the user being able to specify when the pseudo-random tests are to be terminated. It is a characteristic of fault coverage that the first few input test vectors applied to a circuit will normally detect a high percentage of the possible stuck-at faults, but the scoring rate for subsequent vectors drops off as the percentage coverage increases, being asymptotic to 100%. Hence, somewhere between ten and one hundred pseudo-random test vectors are commonly applied before FAN is brought in to complete the test pattern generation. Even then it may take FAN (and the previous PODEM) a considerable and possibly unacceptable time to achieve 100% fault coverage with large and complex digital circuits.

Further details of FUTURE and FAN may be found in Refs. 15, 16, 23, and 24. Further details of fault simulators may be found in Refs. 16 and 25, while an overall discussion of fault models and their problems may be found in Ref. 26.

All the above ATPG techniques guarantee to find a test pattern for any stuck-at fault modeled in a circuit if such a test exists. However, the following should be appreciated:

(i) If a stuck-at 0 fault cannot be detected at a primary output, then this node does not by itself contribute a logic 1 to the following path leading towards a primary output under any input combination.

(ii) If a stuck-at 1 fault cannot be detected at a primary output, then this node does not by itself contribute a logic 0 to the following path leading towards a primary output under any input combination.

(iii) If both (i) or (ii) are present at a given node, then this particular part of the circuit is redundant; all nonredundant nodes must at some time control the value of the logic signal at a primary output.

A situation such as (i) above may be encountered if one of the inputs of an OR or NOR gate is permanently connected to logic 0 due to more gate inputs being present than required; similarly, situation (ii) may be encountered if an extra input on an AND or NAND gate is connected to logic 1. Situation (iii) may be accidental redundancy left in the circuit design, or it may be the result of some deliberate circuit redundancy.

In connection with the last point, there are a number of published examples of circuits which purport to show why a stuck-at fault cannot be detected by fault modeling. Figure 6.18 shows one such published example. What is not always made clear is that these examples may be examples of redundancy in the circuit, which may or may not be appreciated. In the circuit shown in Figure 6.18, the

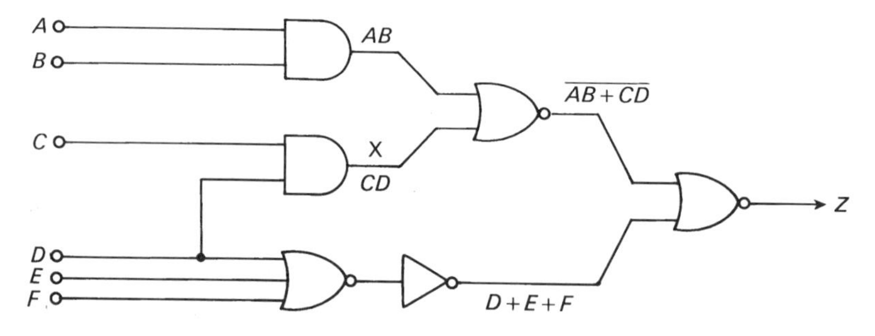

Fig. 6.18 An example circuit where a check for stuck-at-0 faults will show that node X stuck at 0 is undetectable. The circuit illustrated is realizing the function $Z = \{(\overline{AB + CD}) + (D + E + F)\}$, but this may easily be shown to minimize to $Z = A\ B\ \overline{D}\ \overline{E}\ \overline{F}$.

input C is completely redundant, together with the second 2-input AND gate. The 2-input NOR gate may also be replaced by an inverter.

Pseudo-Random Test Pattern Generation

The final technique which may be used for test pattern generation is pseudo-random test pattern generation. This has already been mentioned as the global test generation method in FUTURE to precede the FAN test pattern generation procedure. Theoretically, if a sufficiently long pseudo-random test set is applied, then all possible faults in a circuit will be detected, but such a length is effectively a fully exhaustive test rather than some minimized test set.

The advantage of pseudo-random test pattern generation is that it avoids the difficulties involved in formulating any minimum ATPG program such as previously considered. However, its serious disadvantage if used alone is that if the test set length is short, then the fault coverage (which may be determined by fault simulation) will be inadequate. However, pseudo-random test pattern generation is used extensively in Built-In Self Test (BIST), a particularly powerful form of design-for-testability which we will consider in detail in Section 6.6.3. There we will consider not only the generation of the pseudo-random test patterns, but also how pseudo-random generators may be used to capture the test output responses and provide a 'signature' of the circuit under test. We will, therefore, come back to pseudo-random test sequences later.

The preceding discussions of ATPG have all considered networks which have been flattened to the gate level, with (normally) stuck-at fault conditions being considered at every node. Because the number of individual gates in a complex VLSI circuit is high, consideration has been given to the use of higher levels of functionality in test pattern generation procedures. However, so far this has proved to be a difficult problem, since although there are fewer macros to consider than gates, the difficulties of defining the forward propagation of signals and the backward tracing to establish primary test input signals is considerable. Currently, the effort to tackle this is greater than that involved for ATPG at the gate level, and hence in view of the evolution of design-for-testability which minimizes the TPG problem, there seems little reason to pursue high-level ATPG unless it is an integral part of a high-level HDL design environment. Details of work in this area may be found in Refs. 15, 16, and 27 through 31.

Finally, in these discussions on test pattern generation, it will be appreciated that all the preceding details have implicitly assumed combinatorial networks only, and have not specifically addressed problems of sequential circuit test pattern generation. It is exceedingly difficult to formulate test patterns for sequential networks, since it is necessary to set or reset or clock the circuits into known states before nodal information can be driven to observable outputs. This may involve either a master reset on the circuit, or the running of some 'homing sequence' [7] which will drive the circuits into a known state. The latter technique may not be possible, and in any case a large number of clock pulses may be required before any useful test can begin.

Therefore, in spite of some R. and D. work [18,32] it is generally considered to be impractical to derive effective ATPG programs for circuits containing latches/flip-flops which are deeply buried inside a circuit, that is, where there is poor controllability and observability of these circuit elements. Instead, it is increasingly essential as circuit complexity passes the stage where functional testing becomes impractical for the circuit to be designed in the first place to enable testing of the combinational part of the circuit to be done separately from the testing of the sequential elements; the former can then be tested using ATPG test sets or other means, with the sequential elements tested by their own unique method of test, ideally using a small number of clock pulses for this purpose. This is where partitioning and design-for-testability come into play, as will be covered in Section 6.6. In conclusion, it may be considered that automatic test pattern generation based on fault models has passed its peak of practical usefulness as circuit size has grown, and has been superseded by other functional test methodologies which form part of the design phase of a VLSI circuit.

6.4 TEST PATTERN GENERATION FOR MEMORY AND PROGRAMMABLE LOGIC DEVICES

In the widest sense, programmable logic devices include both RAM and ROM as well as programmable logic arrays (PLAs), complex programmable logic devices (CPLDs) and other variants. All are characterized by internal architectures consisting of regular structures, and hence the most likely internal faults can be readily defined.

The testing of standard off-the-shelf RAM and ROM ICs has received massive attention, and has resulted in established techniques for the test of such circuits (see Sections 6.4.1 and 6.4.2). Should RAM and ROM macros be included within custom ICs, then the same testing techniques are usually applied, the chip design being such as to give access to the memory I/Os. Should this not be done, there arises the very difficult problem of testing the embedded memory through the surrounding logic; this problem has been studied and analyzed mathematically [33–35], but it is a situation which should be avoided if at all possible.

6.4.1 Random Access Memories

The fundamental problem with RAM testing is that every possible memory location in the RAM array may be required to store a logic 0 or a logic 1 during normal operation, but the stored patterns of 0s and 1s are completely fluid during normal system operation. The principal problem, therefore, is to derive a series of tests, based upon the device structure, which is not excessively long but which covers the types of failure most likely to occur. Among these failures are the following:

- one or more bits in the memory array stuck at 0 or 1
- coupling between cells such that a change of state in one cell erroneously causes a change of state in another (usually adjacent) cell—a *pattern sensitive* fault
- bridging between adjacent cells causing the logical AND or OR of their stored outputs
- faults in a decoder which causes it to address additional rows (columns) of cells or entirely wrong rows (columns)
- other faults such as driver circuit faults between bit lines

Additionally, there may be parametric faults such as a dynamic RAM cell failing to hold its charge state for the specified time. These and other possible failings have been extensively studied [15,16,36–39].

The following algorithms have been developed to test RAM circuits. Each involves writing a 1 in each cell and performing a readout and/or writing a 0 in each cell and performing a readout in some chosen sequence of tests. The tests are therefore functional tests, and not tests based upon fault models.

Marching Patterns

In the 'marching-one' pattern test, the memory is first filled with 0s and read out. A single 1 is then written into the first address, and this location output is checked.

The first and second addresses are then written to 1 and checked. This procedure continues until all addresses are full. The converse, the 'marching-zero' pattern test, is then performed, progressively setting the array back to all 0s and checking the 0-valued cells at each step.

This algorithm and its variants are known as MARCHING test procedures.

Walking and Galloping Patterns

The memory array is first filled with 0s and read out. A single 1 is then written into the first address, and all locations except this one are then read out to check that they are still at 0. This is repeated for all other locations individually set to 1 against a background of all 0s, and finally the whole procedure is repeated for a single 0 at each location set against a background of 1s.

Variations of this involve the reading of the single 1 address between the reading of the 0 values, and the reading of the single 0 address between each reading of the 1 values, or other variants, which therefore breaks up the long sequences of the same output value which would otherwise be present.

These tests are variously known as WALKPAT or GALPAT tests.

Diagonal Patterns

Here a single diagonal across the array is filled with 1s against a background of 0s, and every cell is checked. This diagonal is then progressively moved across the array until all cells have at some time been set to 1, checking every cell at each stage. The converse, namely, a diagonal of 0s against a background of 1s, may also be used.

Nearest-Neighbor Patterns

Since large blocks of 0s (or 1s) remote from the single 1 (or 0) are repeatedly checked in the WALKPAT and GALPAT tests, it is much more economical to identify the cells in the immediate surrounding of the single 1 (or 0) cell, and verify the fault-free nature of this cluster of cells at each test step. However, this requires the identification of such cells as part of the test algorithm, but it will provide a considerably reduced total number of read operation in the test procedure.

In connection with the latter considerations, it will be appreciated that the walking and galloping pattern tests require O (N^2) read operations, where N is the total number of cells in the memory array. This clearly is a completely impractical number when memory size exceeds a few thousand bits, even if the tests are performed at hundreds of MHz clock rates, and hence there is considerable pressure to use tests which do not require an order of N^2 operations. In the custom IC area, the vendor must always advise the OEM on the way any memory macros are to be tested, which may reflect back on the most likely fault mechanisms occurring in his particular technology and fabrication process.

Some R. and D. work has, however, involved the provision of a measure of self-test for memory circuits, usually by building in redundancy and incorporating some means of eliminating the effect of faulty cells or rows of cells [7,40]. These considerations are beyond our immediate concern here, but the increase of silicon required to build in such means may have an impact on the yield or failure rate which may be counterproductive in some situations, and in any case does not seem to have become a recognized practice in off-the-shelf memory ICs.

6.4.2 Fixed Memories (ROM and PROM)

Unlike RAMs, read-only memories are, of course, permanently dedicated to give known input–output relationships when in service. In practice, it is not sufficient just to check that the outputs are at logic 1 on the appropriate input vectors, but also that they are not at logic 1 under all other input conditions. This implies that *the full 2^n truthtable of the programmed device must be tested*, where n is the number of inputs.

If programmable read-only memory circuits are in use, then the programmer which the OEM uses to dedicate the PROM to a particular requirement will invariably also have built-in tests which will check the dedicated device functionally and exhaustively. No test set generation by the OEM or other algorithmic complexities are therefore involved. In the case of nonprogrammable ROM macros built into custom ICs, then controllability and observability of the ROM I/Os should be provided so that an exhaustive test can be performed.

The vendor's resources for undertaking both RAM and ROM testing may be very sophisticated, involving computer-controlled test equipment costing several million dollars and operating at clock speeds up to several hundred MHz [41]. These commercial testers may be especially designed for high-speed memory testing, or be large general-purpose testers which can test ordinary logic as well as memory circuits. These are vendor tools suitable for very large size off-the-shelf memory circuits, and are not relevant for the average OEM design and production departments.

6.4.3 Programmable Logic Arrays

Programmable logic arrays are increasingly being used in logic designs in preference to other forms of custom ICs, due to their increasing size and capability. Their testing, therefore, is of growing significance to the OEM.

Off-the-shelf PLAs of small to moderate complexity pose no problems for test, since 100% exhaustive testing can be performed. Indeed, an excellent survey paper by Venkateswaran and Mazumder [42] on PLD design techniques has no direct mention of test. However, for large PLAs there is a need to find some minimum test set and to derive ATPG programs for this purpose; a 30-input PLA, for example, would require over 10^9 input vectors for an exhaustive test at minterm level assuming no sequential elements are present, but internally there may be only, e.g., 50 to 100 product terms generated by the AND matrix. Also, the marching and galloping procedures used for RAM tests are not appropriate for PLAs, since the internal architecture of the two is completely dissimilar (see Figures 2.30 and 3.9).

The particular architecture of the PLA does, however, lead to certain concepts which can ease the testing problem. Basically, the PLA is a two-level AND–OR structure as shown in Figure 6.19(a), and does not have any reconvergent topology such as shown in Figure 6.19(b), which ATPG programs find difficult to handle. However, the suitability of using pseudorandom test patterns for PLA testing is considered to be poor, since to sensitize a particular product term in the AND array requires the exact input literals of the product

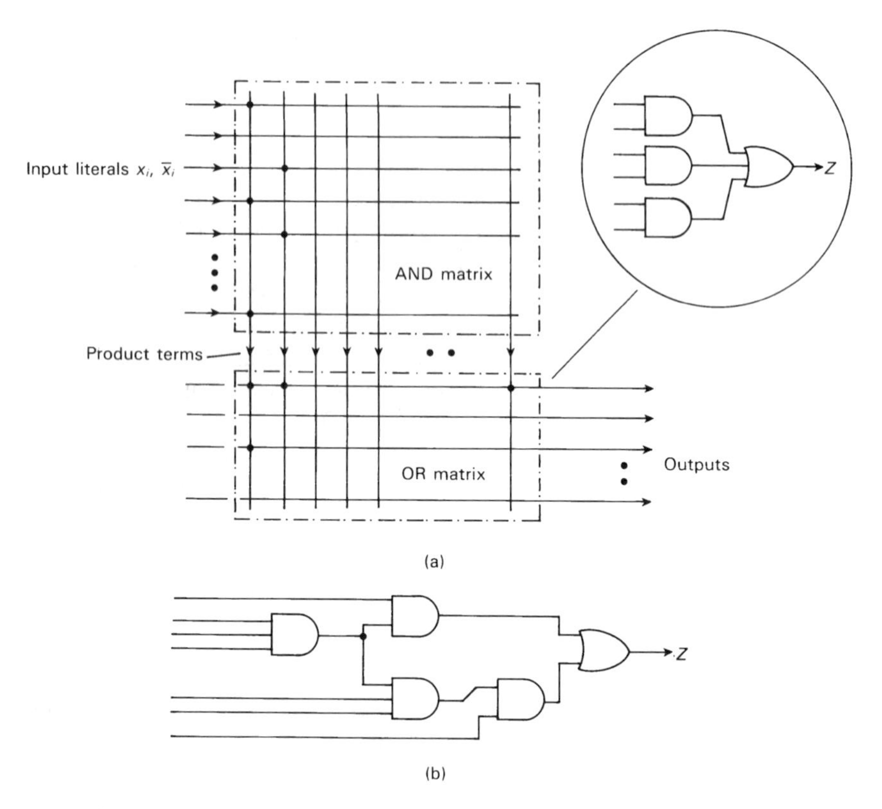

Fig. 6.19 The PLA architecture: (a) the 2-level AND/OR realization produced by the input AND matrix and the output OR matrix—note, in practice it may be a NAND/NAND or a NOR/NOR structure, but this is logically equivalent within complementation; (b) a reconvergent fan-out topology which is never present in a PLA dedication.

term to be applied, and the probability of random input vectors acheiving this is not high.

The PLA outputs are dependent upon the presence or absence of the connections at the crosspoints in the AND or OR arrays. *Crosspoint fault models* can therefore be proposed which represent the presence or absence of these connections. Smith [43] has classified PLA crosspoint faults as follows:

- *Growth Fault*: a growth fault is a missing contact in the AND array, causing a product term to lose a literal and hence have twice as many minterms in the product term as it should
- *Shrinkage Fault*: a shrinkage fault is an erroneous contact in the AND

array, causing a product term to have an unwanted literal and hence lose half of its required minterms, or have none at all if the product term is a single literal and the erroneous contact is the complement of the healthy literal (if the erroneous contact is the complement of an existing literal, this is sometimes referred to as a *vanishing fault* since this input variable disappears from the resulting faulty product term [44])

- *Disappearance Fault*: a disappearance fault is a missing contact in the OR array, causing a product term to be lost in a final output
- *Appearance Fault*: an appearance fault is an erroneous contact in the OR array, causing an additional term to be added to an output

These various crosspoint faults are illustrated in Figure 6.20. Notice that all of them either add to or subtract minterms from the fault-free output function, but

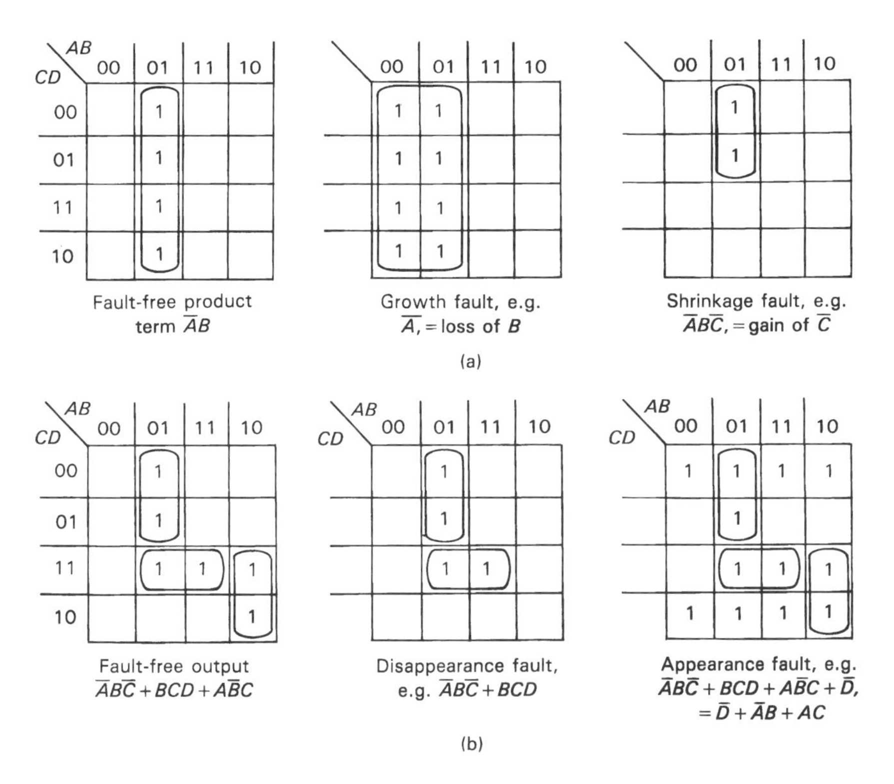

Fig. 6.20 Crosspoint faults in a PLA: (a) growth and shrinkage faults in the AND matrix; (b) disappearance and appearance faults in the OR matrix.

do not cause a complete shift of minterms to another part of the complete input n-space.

A crosspoint fault is also termed a *functional fault* if it alters at least one output of the PLA. This qualification is necessary since an appearance fault, for example, may merely introduce an additional but redundant product term into a final OR output. However, the majority of crosspoint faults will result in functional faults, and in any case a nonfunctional fault is not testable from the primary I/Os.

The process of test pattern generation for PLAs essentially considers single possible crosspoint failures, and determines appropriate input test vectors to show their existance if present. One of the earliest detailed considerations was by Smith [43], followed later by Ostapka and Hong [45], who developed the PLA test pattern generation algorithm TPLA. Subsequent variations have been published, for example, that of Eichelberger and Lindbloom with their PTA/TG algorithm [46]. Reference to these and further work may be found in Russell and Sayers [15], and in Zhu and Breuer [47], who consider and compare a number of published algorithms. Multiple crosspoint faults have been considered by Agrawal [48], who has indicated that coverage of all single crosspoint faults will cover most if not all multiple faults (cf. the single stuck-at model considered earlier). The subject of multiple crosspoint faults remains a topic for academic consideration (see Ref. 49 and the references contained therein), but has little specific industrial emphasis.

However, all these PLA test pattern generation procedures are for standard PLAs which do not have any additional built-in resources for test purposes. Also, the test patterns are entirely dependent upon the functions being realized by the device, and are not independent of the dedication in spite of the fixed topology of the circuit. Therefore, there is attraction in augmenting the basic AND/OR array in some manner so as to produce more easily testable or self-checking circuits, particularly as device size grows to the LSI and VLSI level. Hence, we again see that ATPG based upon fault modeling is being overtaken by other testing philosophies as device complexity increases; we will consider testable PLA designs in Section 6.7 when we deal more specifically with design-for-testability (DFT) techniques.

6.5 MICROPROCESSOR TESTING

The microprocessor is the most complex circuit to test, whether it is a standard off-the-shelf IC or a macro for incorporation within a custom circuit. The OEM can do very little innovation in the test of such circuits due to the following factors:

- the vendor does not usually disclose the full internal details, and certainly not the gate-level equivalent circuit
- general controllability and observability of the circuit is limited to the address and data busses
- the key failure modes of the circuit are not usually revealed

In any case, it is impractical to test at gate level, and hence the only appropriate way is to perform a series of functional tests designed to verify the major components of the architecture, rather than to apply any form of fault modeling.

The evolution of microprocessor testing may be found in the literature [50–54]. Until recently, the most widespread practice has been to undertake functional checks as follows:

- a program counter test
- scratchpad memory test
- stack pointer and index register test
- arithmetic logic unit (ALU) and associated register tests
- further tests on control lines and other peripheral parts

The most difficult of these tests is the ALU, since it is completely impractical to verify exhaustively that it can handle all possible arithmetic operations for all data patterns. Instead, the vendor must propose some minimum series of tests to check that the ALU operates correctly and transfers data for a range of typical operating conditions.

Some theoretical studies of microprocessor testing have been pursued, for example, modeling the microprocessor as a system graph (the *S-graph*), wherein each register in the microprocessor is represented by a node in the graph [50]. Main memory and I/O ports are represented by additional nodes. Another technique has been functional-level fault modeling for bit-sliced processors [52–54], but as bus widths and internal complexity have increased, so the requirement for some form of built-in test and/or design-for-testability has escalated. Hence, all the latest microprocessor designs have built-in means, usually some form of scan test (see Sections 6.6 and 6.8) to ease this difficult problem [55].

The OEM is very largely in the hands of the vendor when it comes to testing microprocessor macros in USIC designs, and full discussions should be held between parties to clarify the test requirements in any custom circuit environment.

6.6 DESIGN-FOR-TESTABILITY (DFT) TECHNIQUES

Because of the difficulties of formulating acceptable tests as digital ICs become larger, and the limits of fault modeling, some test features must be built in at the design stage of the IC, and not as an afterthought once the placement and routing has been completed.

DFT techniques normally fall into the following three principal categories:

(i) *ad-hoc* design methods
(ii) structured design methods
(iii) self-test

The first two of these require the use of some form of comprehensive test facility, but the third minimizes to a large degree the external test resources necessary.

6.6.1 Ad-Hoc Design Methods

Ad-hoc design methods consist largely of a number of recommended or desirable practices which a designer should use when undertaking an IC design. One of the most obvious is to partition a large circuit into functionally identifiable blocks, so that testing of individual blocks can be considered. It is generally accepted that test costs are proportional to the square of the number of logic gates in a circuit, so that halving the size of a block to be tested reduces the testing problems by a factor of four.

During the design procedure, therefore, methods should be adopted with a view to provide good controllability and observability. A common method is to multiplex appropriate primary input and output connections, as indicated in Figure 6.21. Additional I/O pins will be required, which is the penalty to be paid for this increased accessibility. With very large cells in a standard-cell custom IC, such as a processor or a memory array, the vendor will normally have worked out how these macros should be tested. The OEM designer will therefore have no need to decide how to test these large macros, but the chip design must provide the appropriate test access to them.

Long counter assemblies require a large number of clock pulses to test them through all their states. This can be very time-consuming, especially if a given signal from a counter is required before the testing of a further part of the circuit can commence. Long counters must therefore be divided into smaller sections which can be individually tested in order to avoid the necessity for a very large number of clock pulses when under test. This is illustrated in Figure 6.22.

Apart from these general considerations of controllability and observability, the chip designer should also keep in mind the following features in order to facilitate the final test:

- Make a list of the key nodes during the design phase, and check that they can be readily accessed.
- Never build in logical redundancy unless it is specifically required; by definition such redundancy can never be checked from the primary I/Os.

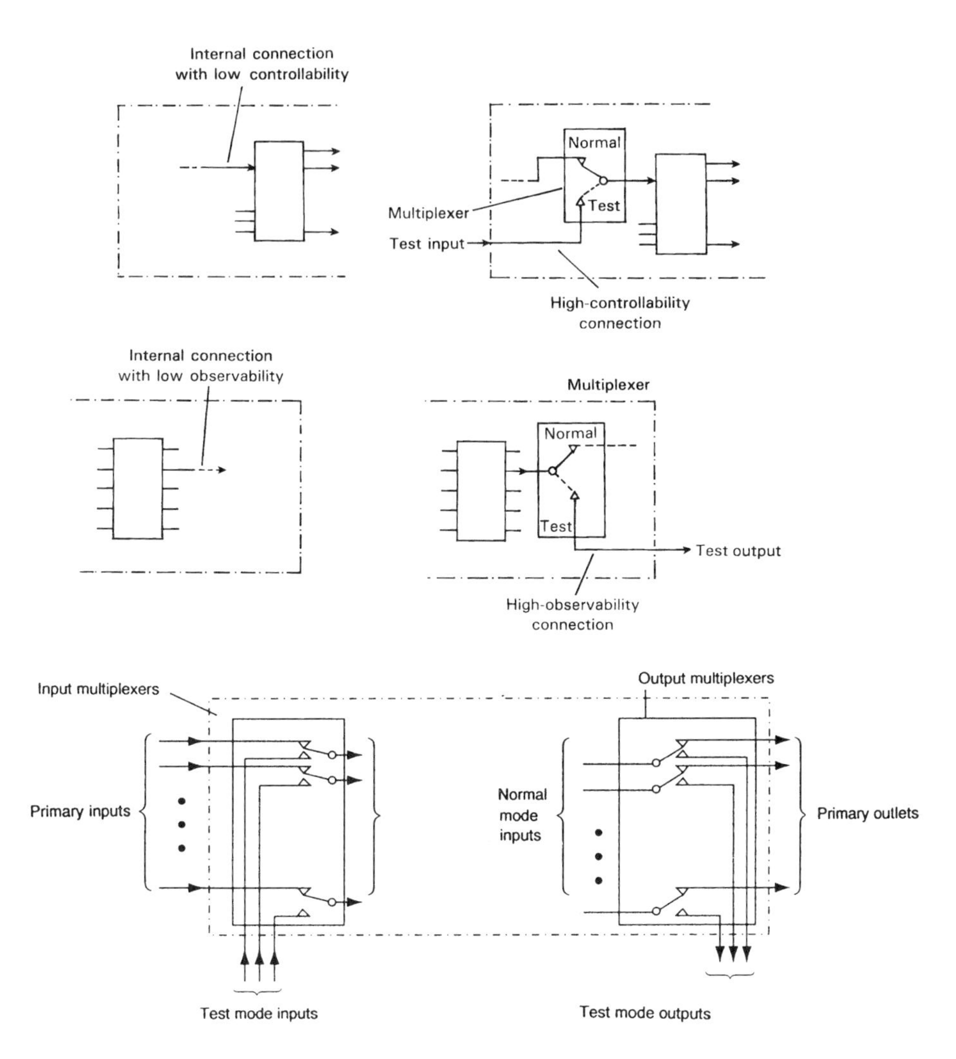

Fig. 6.21 The use of multiplexers to increase internal controllability and observability. The multiplexer paths are switched from normal to test mode under the control of appropriate test mode control signal(s) (not shown).

- Avoid if at all possible asynchronously operating circuits; make all sequential circuits operate synchronously under rigid clock control.
- Use level-sensitive latch (storage) circuits if at all possible rather than edge-triggered circuits (see later).
- At all costs avoid on-chip monostable circuits; these should be off-chip

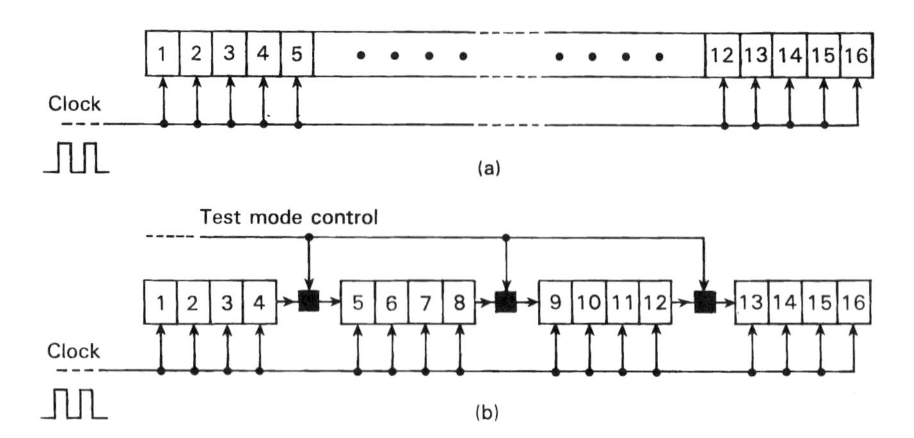

Fig. 6.22 The requirement to divide up long counters during test so as to reduce the testing time. A total of 65,536 clock pulses would be required to test the 16-stage counter (a), but only 16 to test each of the four partitions in (b).

if really required; on chip use some form of synchronously-operating counter/timer.

- Ensure that all bistable storage circuits, not only in counters but also in PLAs and memory and in any local cross-coupled NANDs and NORs which act as local latches, can be easily initialized at the start of test; this usually implies a master reset signal to all such circuits.
- Provide facilities to break all feedback connections from one partition of the chip to another ('global' feedback), so that each partition can be tested independently.
- Keep analog and digital circuits on the same chip physically as far apart as possible, with completely separate on-chip d.c. supply connections.
- Avoid designing 'clever' circuits which perform different duties under different circumstances—one circuit per job is usually easier to test.
- Limit the fan-out per gate as much as possible so as not to degrade performance and to ease the task of manual or automatic test pattern generation.
- It may be useful to consider putting in a separate clock line for test purposes as in Figure 6.23, so that some or all parts of the circuit may be tested more easily, perhaps by one clock pulse at a time.
- Some commercial testers may not cater to unusual input and output signals, such as very slow rise and fall times or nonstandard 0 and 1 voltages; the vendor must be consulted on any unusual logic signals involved.

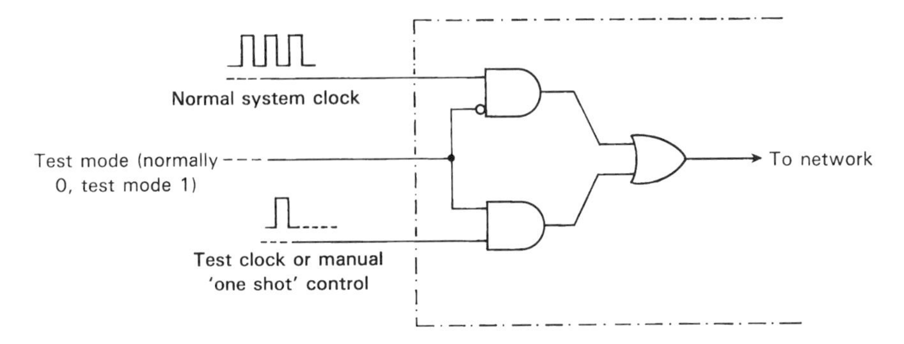

Fig. 6.23 Provision of a separate clock for test purposes, which may be appropriate to step through particular parts of a circuit separately from other parts—this will be found later in several designed-for-testability (DFT) methodologies.

- Remember that a vendor's 100% toggle test on internal nodes is not an absolute guarantee of correct functionality.
- Controllability/observability/testability numbers produced by ATPG programs such as SCOAP, HI-TAP, etc., do not give any indication how to design or redesign a circuit for ease of test; they only indicate possible difficulties which may require further design thought.
- Statistical analysis programs which give an estimate of possible faults covered by a particular test sequence should be treated with care; they may work reasonably well for random combinational logic but not so well for sequential networks or bus-structured networks.
- Finally, never waste any spare I/Os on the custom IC and its packaging—use them if possible for additional test points for internal controllability or observability.

Having produced the USIC design, it remains necessary to formulate the test vectors with which the circuit will be tested. If the circuit is sufficiently small to be tested exhaustively, there is no problem, but for larger circuits there has to be a formulation of some appropriate nonexhaustive set of test vectors. This test pattern generation may be manual or algorithmic (see Section 6.6.3); it is unlikely to be pseudo-random, since this technique is far more commonly associated with self-test methods (see later) than with *ad-hoc* methods of design.

Thus these *ad-hoc* design rules may be summarized by the following four broad statements:

(i) Maximize the controllability and observability of all parts of the circuit by partitioning or other means.

(ii) Do not use asynchronous logic or redundancy unless absolutely essential.
(iii) Provide means whereby all internal latches and flip-flops can be initialized for test purposes.
(iv) Consult the vendor at at stages of the IC and final product design so that mutually acceptable test procedures can be formulated.

6.6.2 Structured DFT Methods

The above good design practices should still be followed even when using more formal design-for-testability techniques, which we will now consider.

Structured DFT methods are formal concepts built in at the design stage. They are principally concerned with sequential logic, and involve some reconfiguration of the sequential parts of the circuit when in test mode to give controllability and observability not only of the storage elements themselves but also of the blocks of combinational logic associated with the storage elements.

Digital logic systems which contain both combinational logic (gates) and sequential logic (latches and flip-flops) can always be redrawn as shown in Figure 6.24(a). This is known as the Huffman model [56]. In most integrated circuit layouts it may be difficult to distinguish the two parts of this represention since they are inextricably mixed, but the two parts will always be present. On the other hand, certain programmable logic devices which have a separate row or block of latches/flip-flops may show this distinction very clearly.

Scan path testing, sometimes known as 'scan test,' is a DFT method which builds upon this general model. It usually provides the following three basic facilities when the circuit is switched from its normal operating mode to its test mode:

(i) reconfiguration of all the sequential storage elements into a continuous shift register configuration, known as the *scan path*
(ii) a 'scan in' procedure, whereby any pattern of 0s and 1s may be serially fed into the shift register
(iii) a 'scan out' procedure, whereby any pattern of 0s and 1s held in the shift register may be serially read out

This is shown in Figure 6.24(b). Procedures (i) and (ii) above may take place simultaneously in certain circumstances.

The test procedure for a complete circuit generally proceeds as follows.

1. The circuit is switched from its normal mode to its scan test mode, which converts the storage elements into the shift register scan path.
2. The correct action of this shift register is then checked by clocking through a pattern of 0s and 1s under the control of the test clock, the

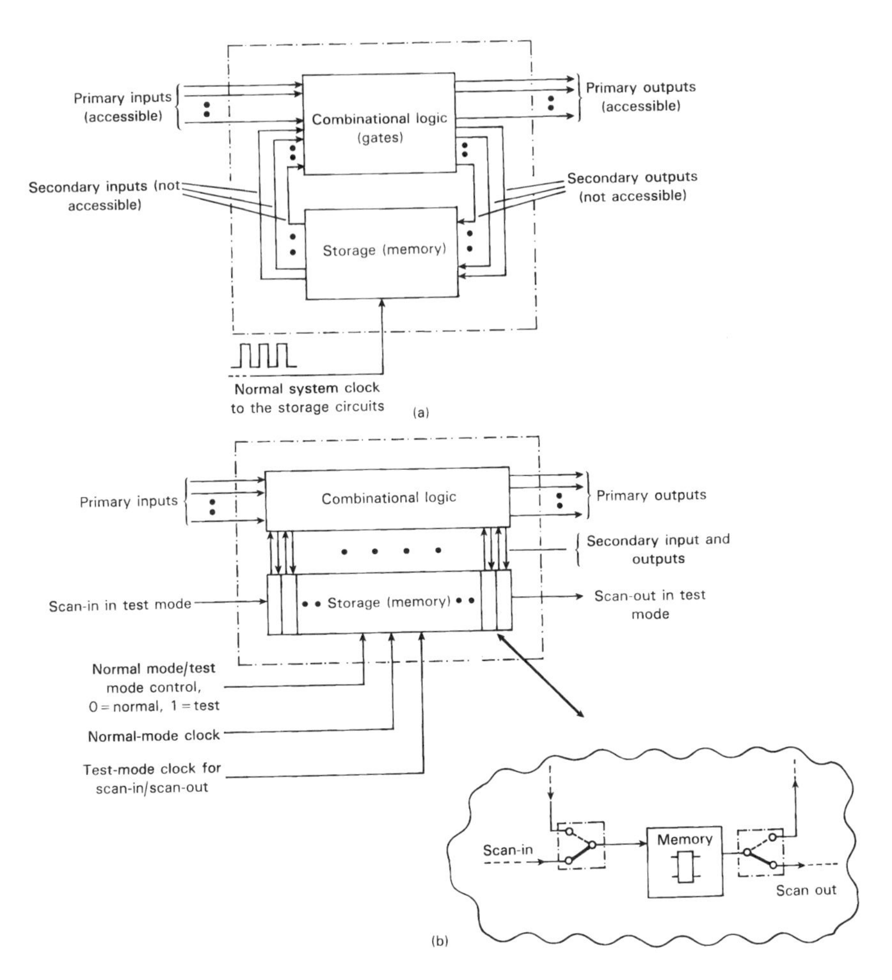

Fig. 6.24 The segregation of any digital network into the combinational logic and the sequential (memory) logic: (a) the theoretical model; (b) the adaption of this model to provide the shift register scan-path means of test.

normal system clock being inoperative. This flush test (see Section 6.2.1) confirms that all the storage elements are functional.

3. A chosen pattern of 0s and 1s is then loaded into the shift register under the control of the test clock, to act as secondary inputs to the combinational logic (see Figure 6.24(b)).
4. The circuit is then switched back to normal mode, and the normal system clock operated once so as to latch the resultant state of the

secondary outputs from the combinatinal logic back into the shift register.

5. With the system switched back again to test mode, the test clock is operated so as to serially feed out these data to the scan out terminal for verification purposes.

Steps 3, 4 and 5 above are then repeated as many times as necessary in order to test fully all the combinational logic network. It will be appreciated that step 3 is applying an input test vector to the combinational logic, with which the primary (accessible) inputs will exercise the combinational logic gates. Step 4 with the primary (accessible) outputs is the response of the combinational logic to this input test data.

The principle behind scan path testing therefore is simple: it provides controllability and observability of internal nodes in a circuit when in test mode, utilizing the facility of a shift register to apply and monitor this internal data without the expense of a large number of additional I/O pins for this purpose. It also capitalizes upon the fact that test vectors for combinatinal logic without feedback can be generated by ATPG programs, whereas test pattern generation for sequential networks is extremely difficult—the need for the latter has, of course, been entirely eliminated by the reconfiguration of the storage elements into a shift register configuration, isolated from the rest of the internal logic gates. It finally provides a means of partitioning the circuit into smaller blocks of combinational logic when required, which may be more easily tested than one large block (see Figure 6.25), although as is seen later in Figure 6.31 that this form of partitioning is most frequently associated with formal built-in self-test methods. Notice that in the circuit of Figure 6.25, it is still necessary to scan in and scan out a lengthy serial bit stream of data even though the individual blocks may be simpler; separate scan paths for each block would reduce the bit stream length, but at the expense of additional scan-in and scan-out I/Os.

The penalties to be paid for this test methodology must, however, be appreciated. They include the following:

- Each of the individual storage elements (primitives) now becomes a much more complicated circuit so that it may be reconfigured (see below).
- The maximum speed performance may be impared slightly because of the extra gates and circuit capacitance in the storage elements.
- Chip size will be larger not only because of the bigger storage circuits but also because of the routing of the additional clock and other interconnections.
- Its adoption imposes a rigorous set of rules on the chip designer in what he or she must do, including careful consideration of the timing and layout of the clock resources.

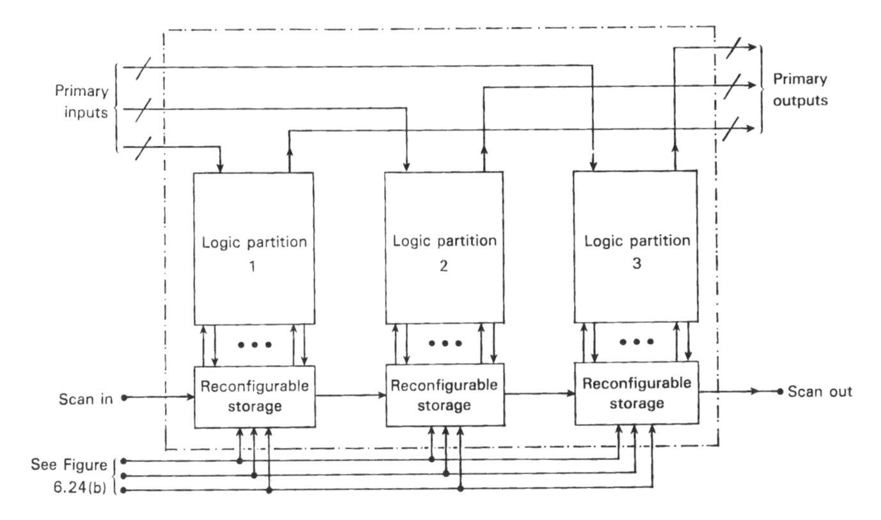

Fig. 6.25 The partitioning of a very large circuit into smaller blocks for test purposes, using scan paths to give controliability and observability of each partition.

- A very considerable amount of housekeeping and data has to be handled by the external test equipment in order to perform the large number of individual load and test operations, and hence the total time to test may be long.

Several variations of scan path testing have been used by different companies, differing mainly in the circuit configurations used in the reconfigurable storage circuits and in whether one, two or more clocks are required. Different company names will also be found. The first and most widely publicized is the *Level-Sensitive Scan Design* (LSSD) system introduced by IBM, which uses the *shift register latch* (SRL) circuit shown in Figure 6.26 [7,57,58]. The terminology 'level-sensitive' refers to the action of all the storage circuits in the LSSD system: a latch (flip-flop) is said to be level-sensitive if (a) all changes to the state of the circuit are controlled by the voltage level of a clock input signal rather than the *dV*/*dt* rise and fall times of the clock edges if the circuit is clocked; and (b) for unclocked latches, the response to changes of logic value on the data input(s) is not dependent upon any gate and interconnect delays within the circuit. What this means is that circuit action is raceless, not being dependent upon any transient (a.c.) performance which may vary or degrade. (The opposite to level-sensitive is edge-triggered: here the change-over of a circuit usually depends upon the propagation times of pairs of gates to control the change of state of the circuit.)

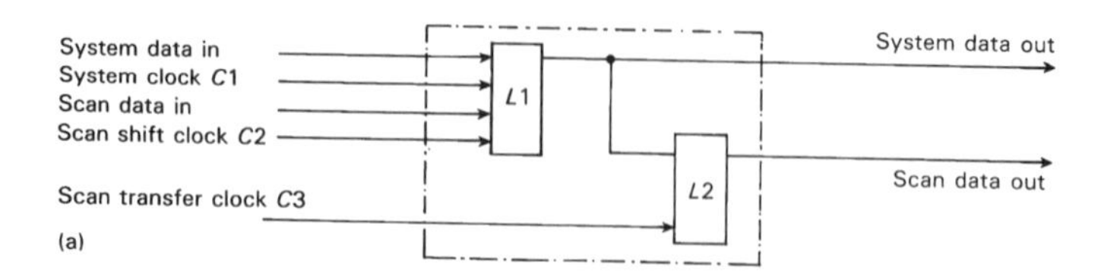

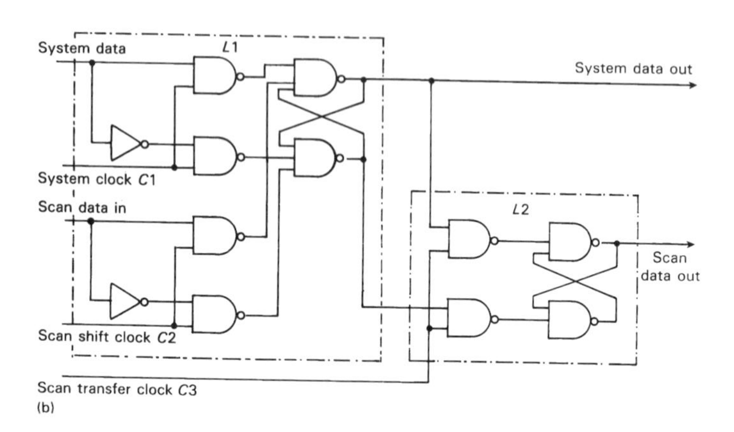

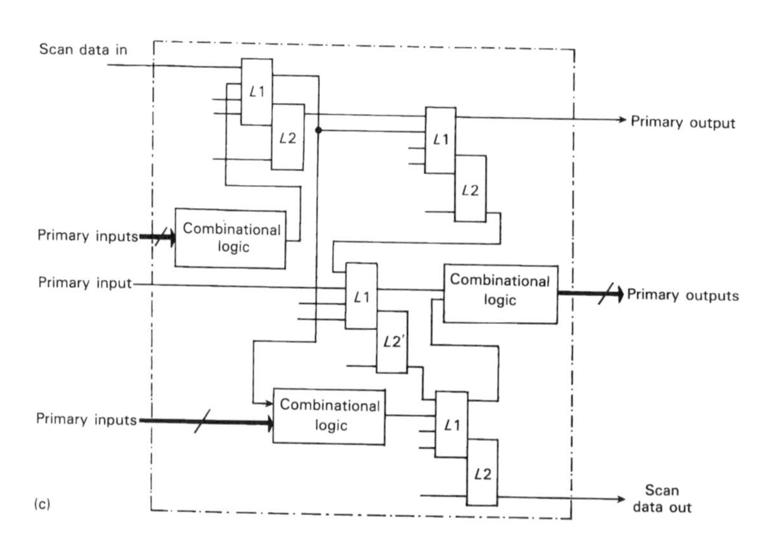

Fig. 6.26 The initial IBM scan-path shift-register-latch (SLR)—note the number of additional I/O pins to give this internal controllability and observability is only four: (a) general schematic with one system clock and two test mode clocks; (b) a NAND implementation; (c) typical system use, with one scan path through all the $L1$ and $L2$ latches. Variations on this SLR circuit are possible (see text).

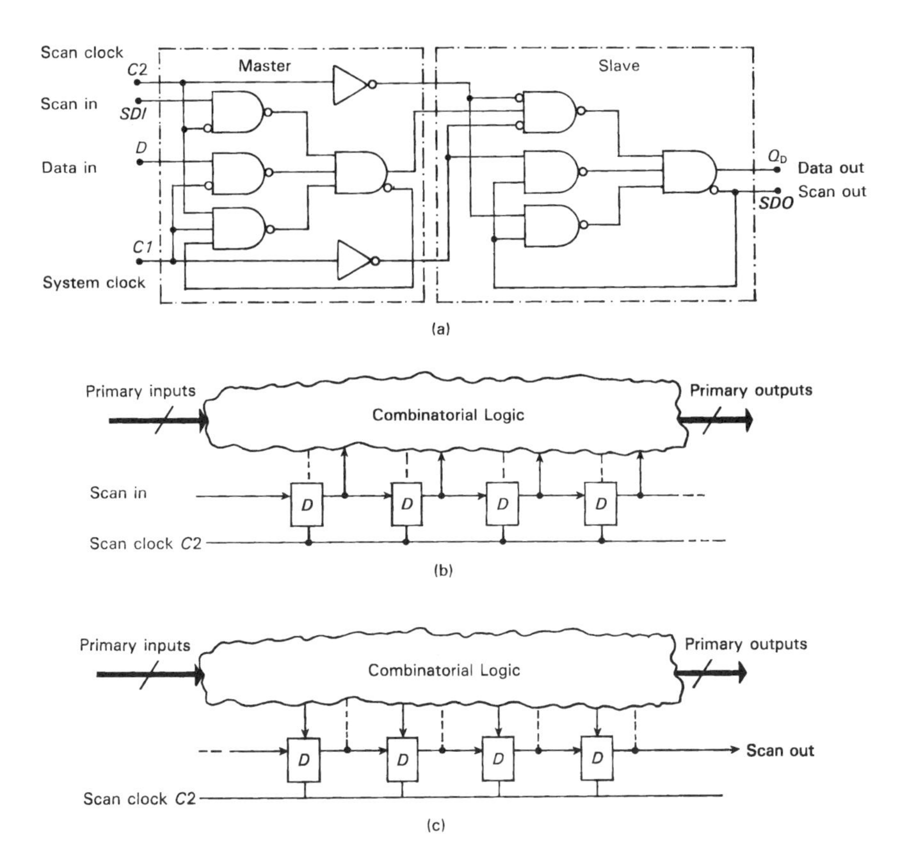

Fig. 6.27 The scan-path method introduced by NEC which employs raceless master-slave D-type flip-flops: (a) the circuit; (b) scan-in; (c) scan-out.

Looking more closely at the circuit of Figure 6.26(b), in normal mode the scan shift and the scan transfer clocks $C2$ and $C3$ are inoperative (logic 0), and only the system clock $C1$ is in use. When $C1$ is high (logic 1), system data are free to enter latch $L1$; when the clock returns to low (logic 0) these data are latched. Latch $L2$ is inoperative during this normal mode. Scan data is latched into $L1$ in place of system data by the scan shift clock $C2$, the system clock $C1$ being inoperative, and finally either the system data or the scan data held in $L1$ may be latched into $L2$ and fed out by means of clock $C3$. Hence, the pattern of clock pulses applied to $C1$, $C2$ and $C3$ which comes from the primary inputs, and which must be nonoverlapping, provides the following three operating modes:

(i) normal operation, output from $L1$
(ii) latch in scan data into $L1$ instead of normal system data
(iii) latch in normal system data or scan data from $L1$ into $L2$, and serially clock out to the scan path output

A variant on this mode of operation is to use both $L1$ and $L2$ in the normal operation of the circuit, clocking the system data first into $L1$ under the control of clock $C1$ and then into $L2$ under the control of clock $C2$. Both the system data and the scan-path data are then taken from the output of $L2$. This operating arrangement is termed the 'double-latch' LSSD system, distinct from the 'single-latch' system where the system data is taken directly from $L1$. Notice that in this double-latch arrangement where both $L1$ and $L2$ participate in normal system operation, in any feedback loop around the combinational logic two clocks and two latches are always operating, and hence no feedback races can ever occur; to ensure this in the previous single-latch arrangement, two separate interleaved nonoverlapping clocks $C1$(A) and $C1$(B) are necessary rather than one common clock $C1$ to all stages, the combinational logic being partitioned such that where any feedback loop is present, then both $C1$(A) and $C1$(B) are involved in the loop. Further details of these clock requirements may be found in the published literature [7,15,16,19,57,58].

These LSSD arrangements involve high silicon overheads—they have two to three times the amount of logic per shift-register-latch compared with a basic D-type circuit. A circuit improvement can be acheived by providing both $L1$ and $L2$ with a system data input and a system clock, thus allowing the output from both $L1$ and $L2$ to be used during normal system operation. This arrangement is known as the $L1/L2*$ shift-register-latch, and clearly reduces the overhead in comparison with the first LSSD arrangement [15,16]. These overhead requirements have been detailed by Williams [16], but care should be taken in reading these and other figures to make sure what is being compared, particularly whether the non-LSSD basis of comparison is a master-slave circuit which by definition contains two latches, the master and the slave, and which is thus very nearly as complex as the LSSD with latches $L1$ and $L2$ shown in Figure 6.26.

Several alternative scan testing methods have been proposed, of which the three most publicized ones are:

(i) the scan-path architecture introduced by NEC [59]
(ii) the scan-set architecture of Sperry-Univac [60]
(iii) the random-access architecture of Fujitsu [61]

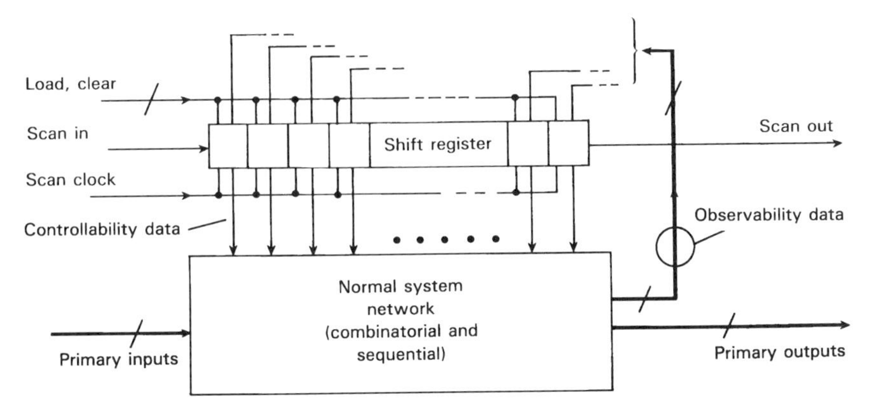

Fig. 6.28 The scan-set methodology introduced by Sperry-Univac, which employs an entirely separate shift register for test access.

Scan-Path

The scan-path approach has a number of similarities to LSSD. It employs what was termed a 'raceless master-slave D-type flip-flop with scan path,' the circuit of which is shown in Figure 6.27. As will be seen, the circuit closely resembles a conventional level-sensitive master-slave D-type flip-flop, but with the addition of a scan-in port and a scan-out port plus a second clock $C2$ to control the circuit when in test mode. The action of the circuit to load test inputs into the combinational logic is indicated in Figure 6.27(b), with the scan out of the resulting logic response being as shown in Figure 6.27(c).

An essential difference between LSSD and scan-path is that the latter uses only one clock for normal mode and one clock for test mode. This means that data pass through the master-slave circuit on one clock cycle, and therefore there is not the absolute security against race conditions happening in feedback loops which there is in the LSSD architecture with its two nonoverlapping clocks. Further details, including the testing strategy, may be found in the literature [16].

Scan-Set

The scan-set test technique is somewhat different from the previous two methods in that it does not employ any form of reconfigurable shift-register-latch in the sequential circuits. Instead, it leaves the working mode storage circuits unchanged, and adds an entirely separate shift register to give controllability to and observability of the internal nodes of the circuit. This is shown in Figure 6.28.

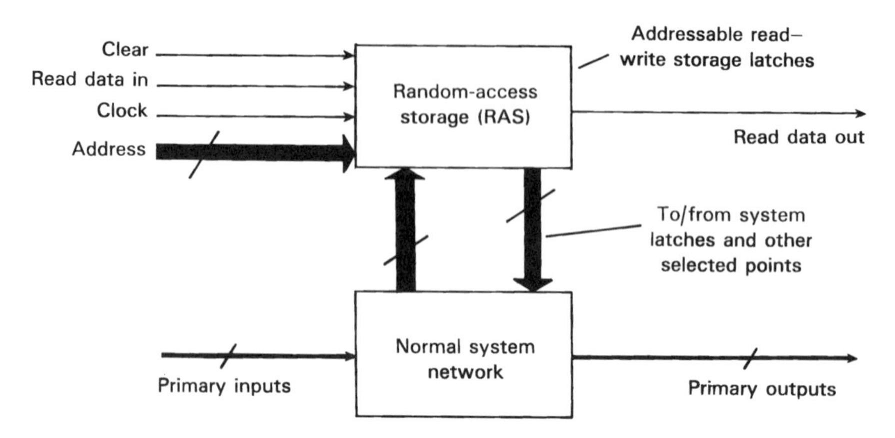

Fig. 6.29 The random-access scan (RAS) methodology introduced by Fujitsu, which employs a form of random-access memory for controllability and observability An optional shift register may be used to read test data in and test results out.

The advantages of scan-set include the following:

- it may be applied to any type of digital network, and does not impose the strict design rules necessary in the LSSD and similar cases
- it does not require more complex latches/flip-flops in the working circuit
- the working system may be monitored under normal running conditions by taking a 'snapshot' of the system data coming into the shift register
- being separate from the working system the scan path testing will involve no additional race hazards and no partitioning problems for the working system

Against these advantages must be weighed the following disadvantages:

- the additional cost of the entirely separate shift register and the possible difficult-to-route interconnections on the chip may be high
- it may be difficult to formulate a good test strategy and decide which nodes of the working circuit shall be controlled and observed when in test mode, bearing in mind that the normal system latches are not automatically disengaged when in test mode

However, from the point of view of the system designer, scan-set may be more immediately acceptable than the rules necessary to incorporate LSSD or scan-path, and possibly can be partially incorporated on some *ad hoc* basis without a great deal of formality.

Random-Access Scan

Finally, there is random-access scan (RAS) [16,61]. Unlike all three preceding methods, RAS does not employ a shift register to give internal controllability and observability. Instead, it employs a separate addressing mechanism which allows each internal latch to be individually selected so that it can be either controlled or observed. This mechanism is very similar to a random-access memory (RAM), consisting of a matrix of addressable latches which may be set either by the test input data or by the normal system latches, the outputs of which may be used to control the system latches or read out for observation. Figure 6.29 gives the general schematic arrangement.

The advantage of the RAS method is its flexibility, since it allows any latch or any node to be addressed. Its disadvantages include a high silicon overhead to provide the RAS matrix and all the system interconnections, and a great deal of time to set test input values into the circuit and to observe the output responses.

Further details of all these techniques may be found in the previously cited references.

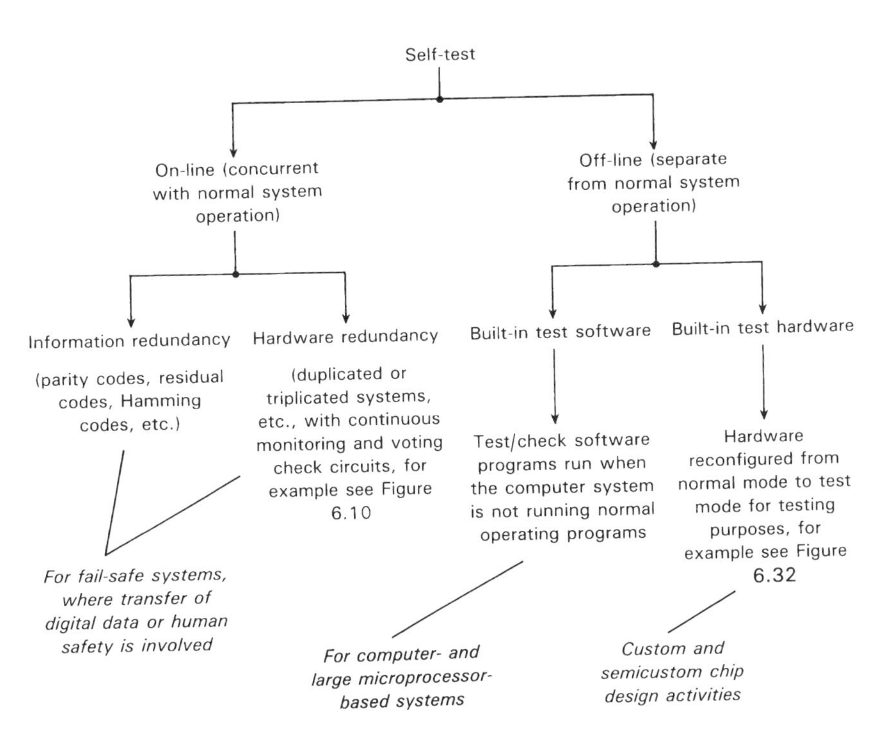

Fig. 6.30 The hierarchy of possible self-test techniques.

While scan test methods and their variants give internal controllability and observability with the minimum of additional I/O pins, and make automatic test pattern generation possible for the combinational blocks of logic, strict adherence to a full scan-based test methodology for large circuits requires CAD resources which implement the design requirements from the basic architectural or functional level. This requirement, together with the considerable silicon overheads and the often long testing times, has meant that such techniques have largely been confined to major product manufacturers such as IBM, where testing requirements dominate. In an attempt to make the basic principles of scan test more widely acceptable, research has been done on *partial scan test strategies*, the objective being to avoid having to incorporate all the latches in the working system in the scan test, thus saving silicon area and reducing the very long length of shift registers which can otherwise accrue.

The problem with the partial scan approach is which latches to make convertable and which to leave as normal system circuits. One approach by Agrawal et al. [62] requires the OEM to supply a set of test vectors using the normal primary I/Os, which caters for as many potential circuit faults as possible (this is not an unusual requirement as we have seen before), following which the faults which are not covered are identified and the latches necessary to be put into a scan mode to access these nodes determined. However, it appears that a high percentage (over 50%) of the latches may need to be in the scan path to achieve the required controllability and observability of these 'difficult to test' nodes, and therefore, it is not clear that the savings in silicon area and testing time are significant. Another technique, RISP (Reduced Intrusion Scan Path), has been proposed by HHB Systems [63], but again no generally acceptable architecture has resulted. Notice that one fundamental limitation will always remain with all these forms of scan test using reconfigurable storage circuits, that is, the need to load in test vectors and verify the result one at a time, thus making it impossible to run the circuit at its normal operating speed during test [7,16,58].

However, we will return to another form of scan test which has wide application later when considering I/O testing in Section 6.8; this is known as *boundary scan* for obvious reasons, and has particular significance in giving controllability and observability of the pins of ICs after they have been assembled into their final PCB or other assembly.

6.6.3 Self-Test DFT Methods

Self-test, or as it is more usually termed in IC design applications, *Built-in Self Test* (BIST), becomes increasingly necessary as chip size increases to VLSI dimensions. As will be seen, BIST has the following advantages:

- It eliminates the need to generate test patterns for the circuit under test.

- It eliminates the need to use extremely expensive test equipment to apply test patterns and monitor the results, and the need for expensive one-off test jigs ('test probe fixtures') to connect a circuit under test to such test equipment.
- The BIST tests are usually performed at the full operating speed of the circuit under test.

Against these advantages must be weighed the fact that the tests are not fully exhaustive under normal working-mode conditions, and therefore there is always a small probability of a faulty circuit being passed as fault-free. However, this is an accepted risk with circuits of VLSI complexity, since 100% functional testing can never be done. (Recall the instances of microprocessor ICs being released on the market with abstruse arithmetic faults which were not apparent from their production testing.)

The general hierarachy of all built-in self-test procedures is shown in Figure 6.30. On-line testing techniques are specialized, and are particularly associated

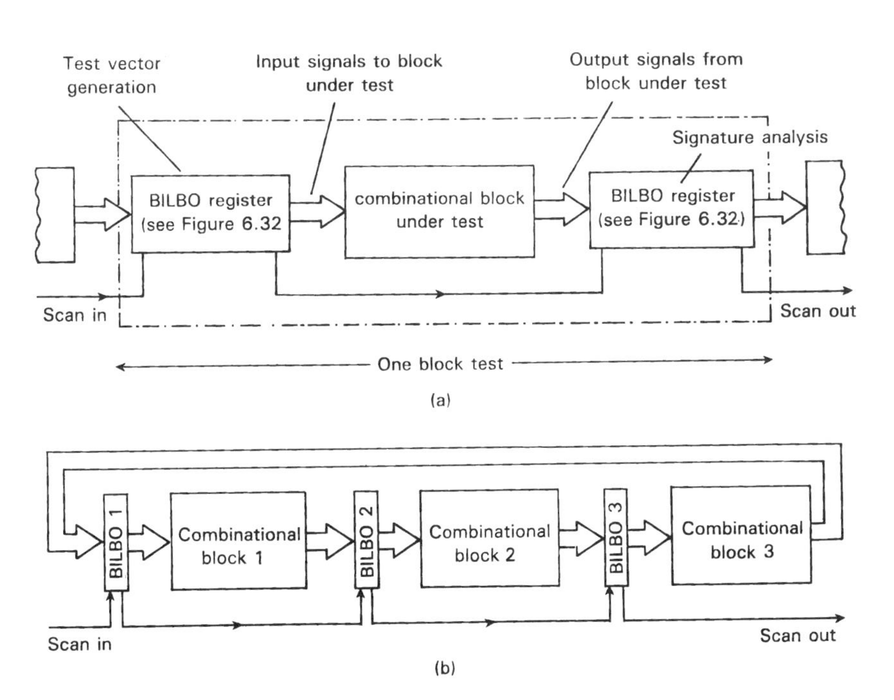

Fig. 6.31 The partitioning of a digital circuit by BILBO registers: the control signals and clocks have been omitted for clarity: (a) the basic concept; (b) example circuit under test partitioned into three blocks when in test mode, each block being individually tested.

with vital communications and control systems [19,64–67]. Off-line testing, particularly with built-in hardware for test purposes, is more relevent for individual VLSI circuits, although we will return to discuss certain on-line redundancy concepts in Section 6.9.

The off-line routine testing of computer systems with test programs is standard computer practice and will not concern us here. Instead, we will concentrate on the extreme righthand path in Figure 6.30, where additional hardware and/or reconfigurable circuits are employed to provide the in situ test procedures. As in all our immediately preceding discussions, we will be dealing with digital-only circuits, and considering some set of test vectors which will provide an acceptable test set for the circuit under consideration.

BILBO

The most widely known BIST technique is *built-in logic block observation* (BILBO). Its general arrangement is shown in Figure 6.31. Each partition of the complete circuit is a block of combinational logic, separated from each other by a set of storage (register) circuits known as BILBOs, which in normal mode form part of the working circuit. However, in test mode, two sets of BILBO registers,

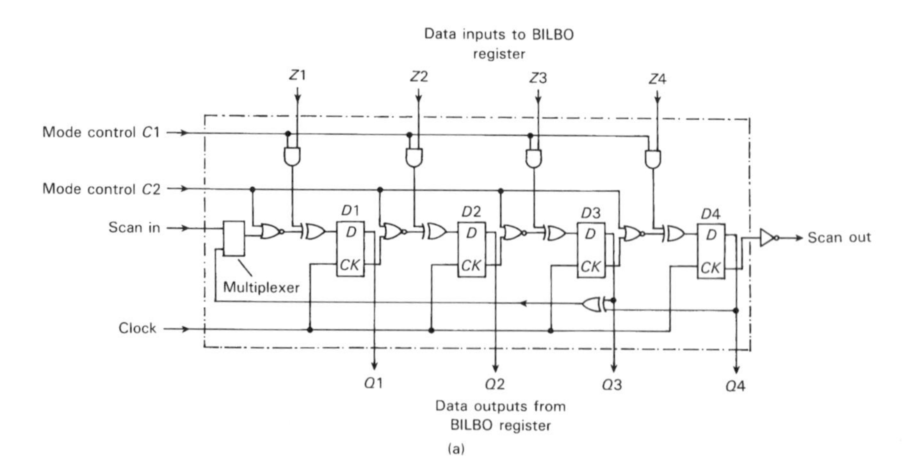

Fig. 6.32 The functional attributes of a BILBO register, showing four stages only for clarity: (a) the overall circuit arrangement; (b) normal mode with $C1 = 1$, $C2 = 1$; (c) scan-path mode with $C1 = 0$, $C2 = 0$; (d) pseudo-random input test vector generation (see Figure 6.33), with $C1 = 1$, $C2 = 0$; (e) output test signature capture (see Figure 6.34), with $C1 = 1$, $C2 = 0$. Note $C1 = 0$, $C2 = 1$ (not shown) resets all stages to zero; the clock signal is applied in all modes of operation. Detailed variations on this circuit arrangement may also be found.

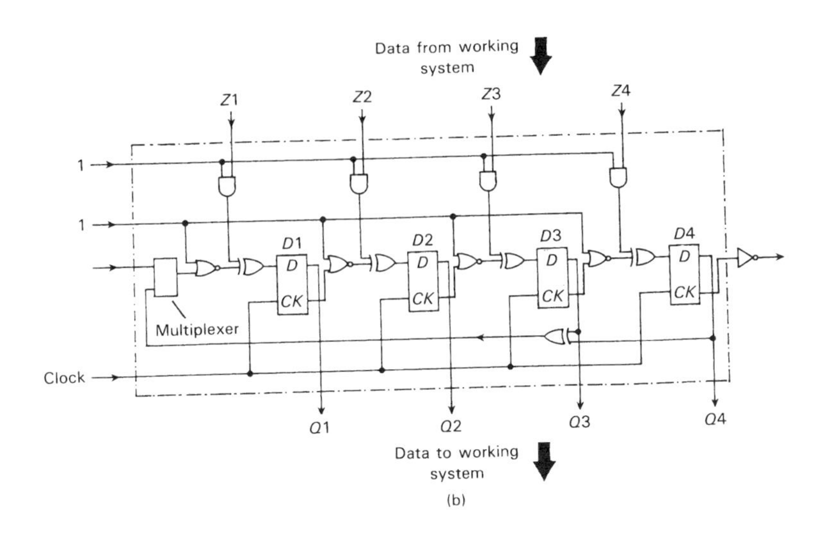

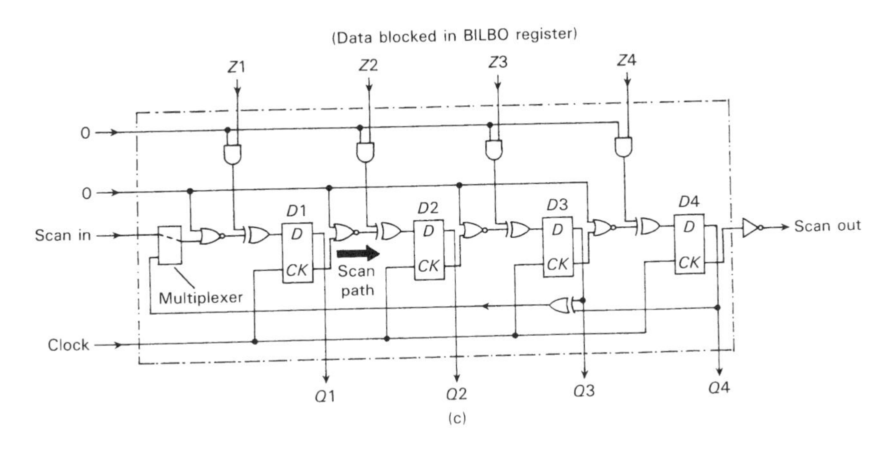

Fig. 6.32 Continued

one preceding and one following each block of logic, are used to test the intervening block, the first to provide input test vectors and the second to provide a test output signature. This signature can be read out in a serial scan-out mode for checking purposes. The first BILBO register can act as the output register for the last block of logic, as shown in Figure 6.31(b).

The BILBO registers are complex circuits, as shown in the developments of Figure 6.32. They operate not only in normal circuit mode and as a scan-in/

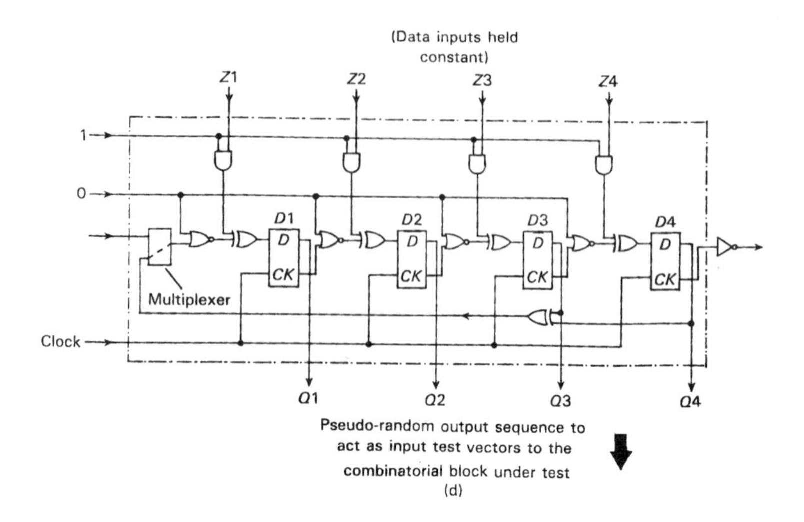

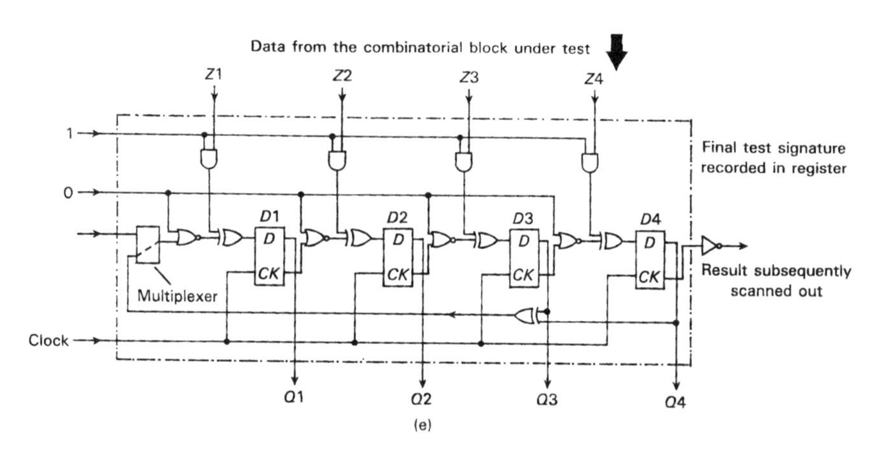

Fig. 6.32 Continued

scan-out shift register, but also as pseudo-random sequence generators for input test vector generation and for output test data capture. The four control signals $C1$ asnd $C2$ operate as follows:

- $C1 = 0$, $C2 = 1$: reset mode, to set all storage circuits to the reset state on receipt of a clock pulse;

- $C1 = 1$, $C2 = 1$: normal operating mode, with each storage circuit operating independently (see Figure 6.32(b))
- $C1 = 0$, $C2 = 0$: scan path mode, with the storage circuits in a shift register configuration (see Figure 6.32(c)); notice that there is an inversion between stages in this particular circuit configuration when in scan mode
- $C1 = 1$, $C2 = 0$: linear feedback shift register (LFSR) mode, which provides either a pseudo-random test vector input sequence to the following block of combinational logic, see Figure 6.32(d), or to give a test signature from the preceding block of logic under test, see Figure 6.32(e).

Each BILBO is controlled by its clock and control signals $C1$ and $C2$. The scan-in and scan-out leads from the blocks may all be connected in series, or kept separate for each block.

Flush testing of all the BILBO registers themselves can be done in the scan-in/scan-out mode as with any scan test system (see Figure 6.25). Loading of individual test vectors serially into an input BILBO register to act as a test input vector for the following combinational logic block can be done, and the resultant test output can be scanned out from the following BILBO register in scan-out mode.

However, the principal reason for the complexity of BILBO is not to use it for logic tests as above, although this is still possible, but to eliminate the need to formulate individual test input vectors and use them one at a time in some lengthy test procedure. Instead, the input BILBO register is normally used in an autonomous LFSR mode so as to apply a pseudo-random set of input test vectors to the following logic block (see Figure 6.32(d)); the output BILBO register is simultaneously also run in an autonomous LFSR mode, but the combinational logic output test response is mixed into this register so as to modify the normal pseudo-random sequence (see Figure 6.32(e)). Notice that each combinational logic output is capable of modifying the sequence of this output BILBO generator and thus the 'test signature' which is finally scanned out.

Considering the circuit principles in more detail, the principle of the linear feedback shift register (LFSR) is shown in Figure 6.33. The normal shift register action which shifts any input pattern of 0s and 1s along the register has taps on its which are Exclusive-ORed together to form the first-stage input data to the register. When these taps are appropriately chosen, the circuit will when clocked count in an autonomous pseudo-random manner through $2^n - 1$ states before the sequence repeats; this is known as a *maximum length pseudo-random sequence*, and contains all the possible states of a n-stage counter except for one 'forbidden' state. When the shift register is assembled by cascading the Q output of each

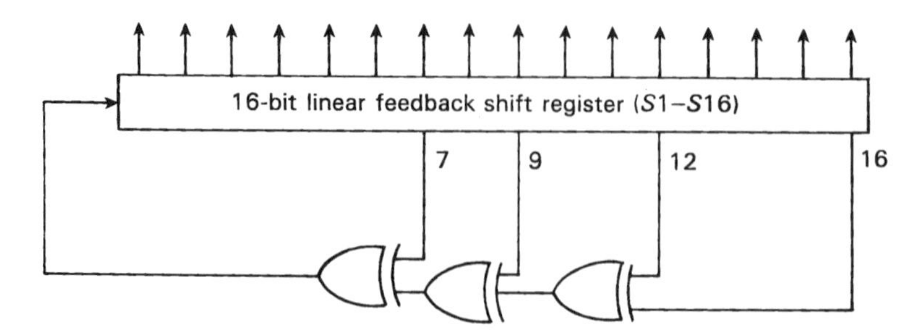

Fig. 6.33 The linear feedback shift register (LFSR) consisting of n D-type flop-flops in cascade with feedback so arranged as to generate an autonomous maximum-length pseudo random sequence when clocked this maximum length being 65 535 ($2^{16} - 1$) in the case of a 16-stage assembly. The clock signal is omitted here for clarity.

stage directly into the D input of the following stage, this forbidden state is the all-zero (00 . . . 00) state; in the case of the configuration shown in Figure 6.32(d) the use of $\overline{Q}$ followed by the NOR inversion gives the same result. The all-zero case may be obtained by resetting the register, but then a seed of 1 . . . . is necessary so that the pseudo-random sequence may commence. (There are ways of making an autonomous pseudo-random generator go through its 'forbidden' state and hence have a maximum length sequence of 2^n, but this involves additional circuitry which destroys its simplicity [7] and is not used in BILBO applications.)

The theory of the generating polynomials which define the taps that are necessary on a LFSR to produce a maximum length sequence may be found in many publications [7,33]. For up to 300 stages it has been shown that never more than four taps are required, one of which must always be the last stage of the register. Lengths of multiples of eight stages always require four taps, but most others require only two (never three) taps. Alternative taps are always possible for $n \geqq 3$. Hence, the LFSR is an extremely simple test pattern generator for automatically generating n-bit test vectors which contain all test vector patterns except one; the major limitation is that it is not possible to order the test vectors other than in the inherent pseudo-random sequence.

If additional 0 and 1 logic signals are fed into an LFSR, then clearly some alternative sequence will result. This is the principal of the test signature, as shown in Figure 6.32(e). The earliest application of this was for the testing of printed circuit boards rather than ICs [68]. Here a hand-held probe was used to sample logic signals at selected nodes of a circuit under test, as shown in Figure 6.34(a). Unless this probe signal is a constant logic 0, it will clearly influence the sequence which the LFSR follows, and will result in a different number being

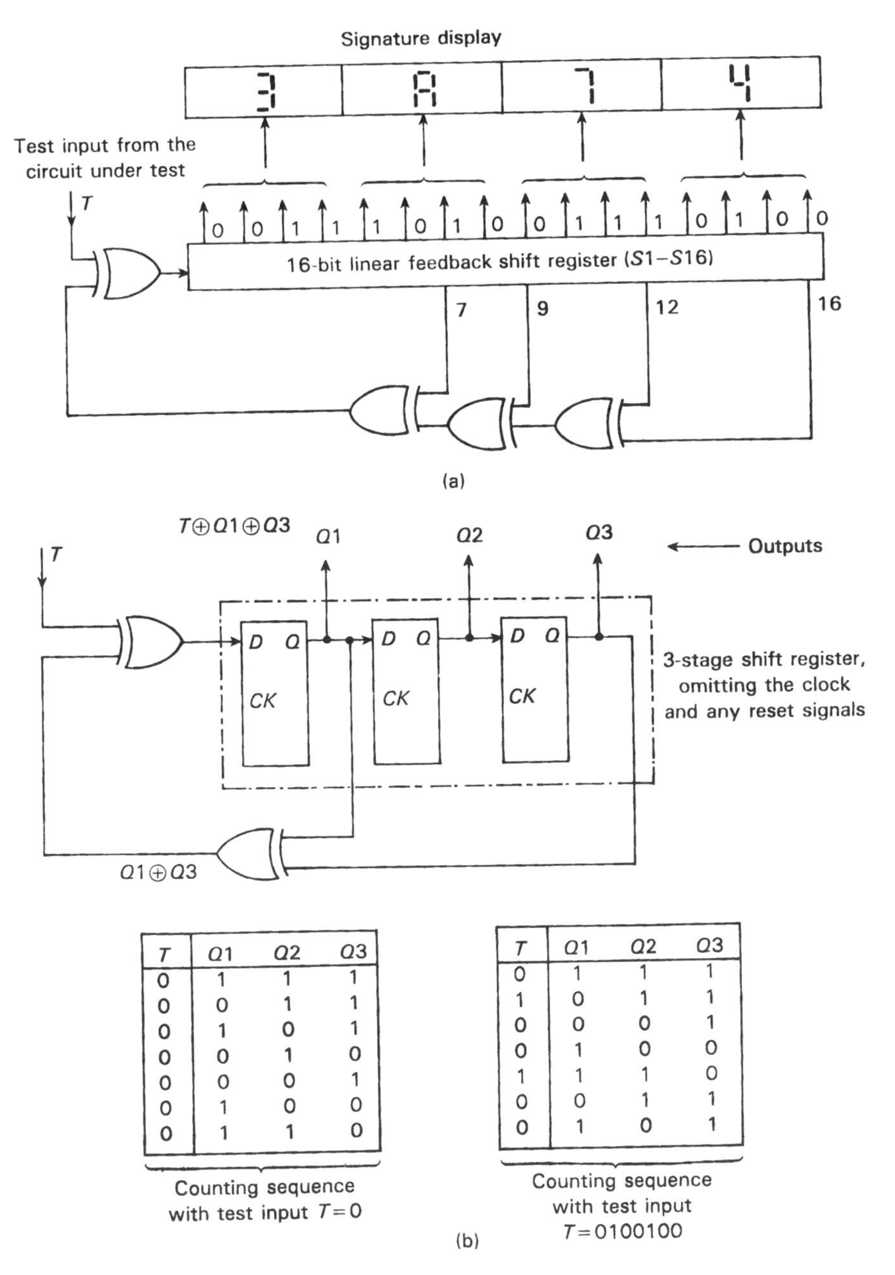

T	Q1	Q2	Q3
0	1	1	1
0	0	1	1
0	1	0	1
0	0	1	0
0	0	0	1
0	1	0	0
0	1	1	0

T	Q1	Q2	Q3
0	1	1	1
1	0	1	1
0	0	0	1
0	1	0	0
1	1	1	0
0	0	1	1
0	1	0	1

Fig. 6.34 The principle of signature analysis using LFSR signature generation: (a) the addition of a single input test input signal to a maximum-length LFSR; (b) example LFSR sequences (signatures) with different test input bit streams a three-stage circuit only being shown for convenience.

stored in the register after a given number of clock pulses have been applied. This is illustrated in Figure 6.34(b). By probing a number of different nodes, this analysis can confirm the correct functionality of the board or identify part(s) which are faulty.

To revert to the BILBO test methodology of Figures 6.31 and 6.32: from Figure 6.32(e) it will be seen that a test signature from the block under test is generated, but unlike the signature analysis method of Figure 6.34, multiple test inputs to the LFSR are now present. Because of these multiple data inputs, the circuit is referred to as a *multiple-input signature generator*, or MISR. The test procedure for each BILBO block of combinational logic is therefore as follows:

(i) initialize the combinational block under test and the preceding and following BILBO registers to known starting states
(ii) apply a given number of clock cycles to both BILBO registers so as to generate the pseudo-random test vector input set and to register the output MISR signature
(iii) scan out the resultant signature for checking

This procedure is repeated for each of the logic partitions.

As with the single-input test signature analysis, there is always a possibility that a faulty circuit may give the same MISR signature as a fault-free circuit under such tests. However, if the number of stages in the LFSRs is reasonably large, then the error masking probability $P(M)$ is theoretically given by:

$$P(M) = \frac{2^{L-1} - 1}{2^{L+n-1} - 1} \approx \frac{1}{2^n}$$

where

n = the number of bits wide in the data, = the number of stages in the LFSRs

and

L = length of the test sequence, $= 2^n - 1$ if maximum length

This compares with the theoretical error-masking probability of the single-input signature analyzer of

$$P(M) = \frac{2^{L-n} - 1}{2^L - 1}$$

which reduces to the same value of $1/2^n$ provided that $L, n \geqq 1$ and $L > n$ [7,33,69]. These analyses are discussed in Ref. 7, and are based upon the assumption of a uniform probability of errors in the test data, an assumption that need not necessarily be true. Nevertheless, it is generally accepted that the error masking probability of LFSR signatures may be taken as $1/2^n$, which gives an acceptably small probability of error when n is, e.g., 16 or more. There are further techniques which have been used to decrease still further this probability of a fault-free signature being given by a faulty circuit, for example, by scanning out the signature at some intermediate point(s) in the test, but we will not pursue these concepts here.

There is, however, one interesting aspect of the mathematics of this fault-masking probability: in the analyses there is no consideration of the structure of the pseudo-random sequences themselves, but only of the length of the test sequences. What this implies is that there is theoretically nothing to be gained as far as fault masking is concerned by using pseudo-random sequences, and that any test vector sequence may be used in the BILBOs, including e.g., a normal binary counter as the test vector generator and a straight-forward shift register for the MISR. The basis for this is that all faults are considered to be equally likely, which is improbable if the faults which can occur are examined in detail. Should the analysis be true, then the only justification for using a LFSR pseudo-random test vector generator is that it is more simple to make than say a binary counter! The mathematics should therefore be treated with caution [7], with the actual fault masking possibly better than these analyses suggest.

BILBO will therefore be seen to offer an attractive method for the test of large digital ICs, enabling them to be tested at their normal clock speeds and not requiring complex test equipments. A n-bit maximum length pseudo-random test sequence is effectively a fully exhaustive test for a n-input combinational logic block, the missing all-zero test vector being possible as an initial or final vector under reset if needed. The principal disadvantages of BILBO, however, are as follows:

- the silicon area overhead to incorporate BILBO is high, possibly as much as a 20–30% increase in chip area
- the inability to apply the test vectors in other than the fixed pseudo-random sequence may be restrictive (see the CMOS open-circuit problem discussed in Section 6.3.2)
- the partitioning of the complete circuit into blocks and the ideal matching of the number of BILBO register stages to the number of combinational logic inputs per block may be difficult—an exact match is not necessary, but there must be some acceptable ratio
- the difficulty of determining the correct fault-free signatures in the MISRs—the 'good machine' signatures—may also be difficult, requiring specific simulation runs at the design stage

If a known-good circuit is available, then good-machine test signatures can be obtained by running the BILBOs, but this is a dangerous procedure to attempt with prototype circuits. On the other hand, full simulation to determine the signatures may be time consuming. Thus, it is possible that some judicious mix of simulation and practical determination may be used.

Variations on the precise BILBO circuit details from that shown in Figure 6.32 have been published [70–76], but the essential principles covered above remain. There has, however, been one development which is an evolution of the architecture shown in Figure 6.32 and which has been used in practice, this being the use of two LFSRs of dissimilar length to act as the test vector generator and as the signature analyzer. The lengths of the two LFSRs per BILBO register are chosen in accordance with a co-prime rule in order to maximize the testing efficiency, giving what is termed a *quasi-exhaustive self-test*, which is reported to be superior to that provided by a single LFSR with the same total number of stages. A fault coverage of over 99.98% has been reported as being achieved. Details of this development may be found in the literature [77]. Some further work using multiple LFSRs working in parallel has also been reported [78].

However, there has been one radically different built-in self-test approach which does not use the autonomous LFSR generator and MISR: the *cellular automaton logic block observation* (CALBO) methodology, which we will now introduce, although it has not yet received the industrial take-up that BILBO has received.

CALBO

The CALBO architecture uses what is termed a *one-dimensional cellular automata* as an alternative to the previous LFSR circuits, this being a one-dimensional row of cells rather than two (or more) dimensional arrays which may be present in some theoretical multidimensional cellular automata [79,80]. The circuit for the CALBO architecture is shown in Figure 6.35, this being a string of four stages only corresponding to the 4-stage BILBO circuit shown in Figure 6.32.

As will be seen from Figure 6.35, the logic to control the next state of any stage Q_k is some logic combination taken from the Q or $\overline{Q}$ output of near-neighbor stages, for example, from stages $k - 1$, k and $k + 1$. The input signal to the data input D of stage k is therefore given by some function $f_k(D)$, where $f_k(D)$ may be defined by an appropriate truthtable:

$Q_k - 1$	Q_k	Q_{k+1}	$f_k(D)$
0	0	0	–
0	0	1	–
0	1	0	–
0	1	1	–
1	0	0	–
1	0	1	–
1	1	0	–
1	1	1	–

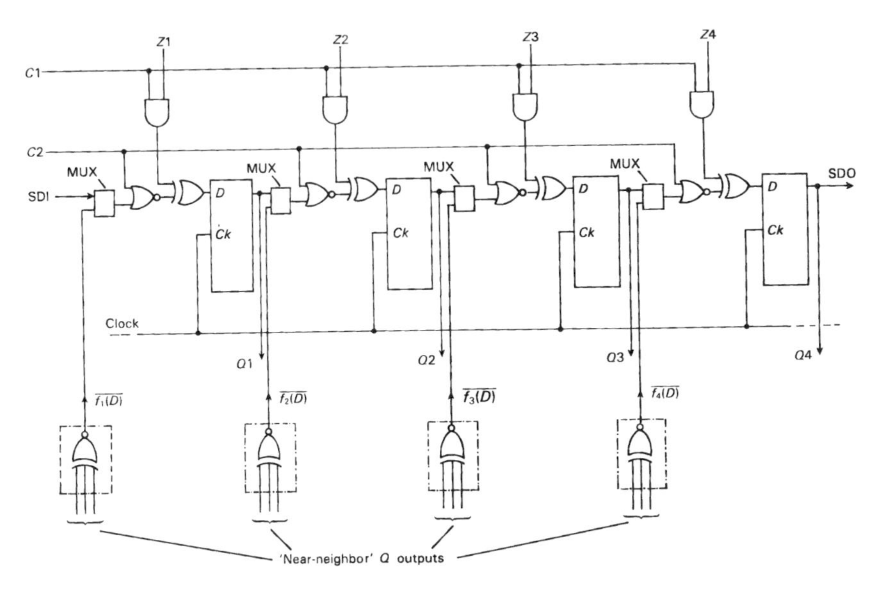

Fig. 6.35 The CALBO circuit details with $C1$ and $C2$ as in the BILBO circuit of Figure 6.32. The feedback to each D input is some locally generated Exclusive-OR function fk (D).

With three inputs there are 256 possible functions $f_k(D)$. However, the majority of these will not be relevant functions for generating useful sequences when the assembly is clocked.

The most useful sequences are generated by some linear (Exclusive-OR) combination taken from near-neighbor stages, since then no information is lost in each resulting function $f_k(D)$. It has been found, and is an intuitive first try, that the two most useful functions are as given in Table 6.4 [80–82]. Reading these two function truthtables as eight-bit numbers with the top entry as the least significant digit, the function $Q_{k-1} \oplus Q_k \oplus Q_{k+1}$ is termed 'function 150,' and $Q_{k-1} \oplus Q_{k+1}$ is termed 'function 90.'

It has been shown that a pseudo-random maximum length sequence of $2^n - 1$ states from a n-stage CALBO register can always be generated by an appropriate string of 90 and 150 circuits, with a logic 0 completing the inputs to $f_1(D)$ and $f_n(D)$. This is shown in Figure 6.36. The 'forbidden' state of all outputs zero exists as in a LFSR, and hence a CALBO circuit also has to be seeded out of this state when used as an autonomous sequence generator. The configurations of 90/150 circuits which produce a maximum length sequence have been documented; like the LFSR generator, there are multiple choices for $n > 2$. It has been shown that given the generating polynomial for any maximum-length LFSR,

Table 6.4 The Truthtables of the Two Useful Near-Neighbor Functions for Generating Autonomous Maximum-Length Pseudo-Random CALBO Sequences

Q_{k-1}	Q_k	Q_{k+1}	$f_k(D) = Q_{k-1} \oplus Q_{k+1}$	$f_k(D) = Q_{k-1} \oplus Q_k \oplus Q_{k+1}$
0	0	0	0	0
0	0	1	1	1
0	1	0	0	1
0	1	1	1	0
1	0	0	1	1
1	0	1	0	0
1	1	0	1	0
1	1	1	0	1

then a corresponding maximum-length CALBO configuration may be determined, but this does not produce the most economical assembly if it is required to minimize the number of the more 'expensive' 150 circuits and maximize the number of the simpler 90 circuits. Indeed, alternative procedures based upon matrix algebra have shown that never more than two 150 circuits are necessary for any n-stage maximum-length CALBO generator. Details of all this algebraic development work may be found in the literature together with tables listing the 90/150 positions for maximum length sequences [7,83–85]

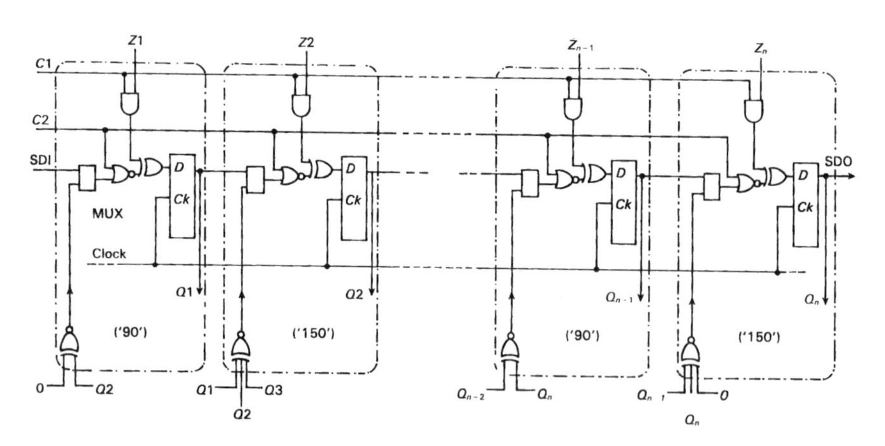

Fig. 6.36 An autonomous maximum-length CALBO pseudo-random test generator composed of type 90 and 150 circuits. The exact sequence of 90 and 150 circuits for maximum length depends upon the number of stages, and will not always be 90, 150, 90, 150, ... Note that a 90 circuit may be made from a 150 circuit by connecting logic 0 to the middle input of the 3-input XOR circuit, but this is very wasteful of silicon area.

The pseudo-random nature of the sequences generated by CALBO are clearly dissimilar to those generated by the previous BILBO LFSRs. However, they have been extensively investigated, and it has been claimed that they possess a more random nature and lower cross-correlation than LFSR sequences. Hence, as test vector generators and signature analyzers for built-in test purposes, CALBO circuits may have preferable testing characteristics, with reduced actual fault-masking compared with BILBO [81,82,84]. The theoretical fault-masking probability assuming all faults are equally likely remains the same as for the LFSR, namely $P(M) \simeq 1 / 2^n$, but as previously noted this result must be treated with caution.

The testing strategy with CALBO will be appreciated as being the same as BILBO. The two control signals $C1$ and $C2$ and the scan-in and scan-out terminals operate in exactly the same way in Figure 6.36 as in the BILBO circuits of Figures 6.31 and 6.32, the CALBO registers being arranged so as to test each combinational logic partition as previously described. Boundary scan (see Section 6.8) using CALBO has also been proposed [86], and further developments in this area may be expected. However, apart from the claimed potential improvement in testing efficiency, another possible advantage of CALBO compared with BILBO is that all the feedback taps to generate a maximum length sequence are local connections, unlike the LFSR, where they have to run the whole length of the shift register. Hence, the length of a CALBO generator may be much more readily increased or decreased by just adding or removing stages and rearranging the local connections; with a LFSR, any modification in length involves a new generating polynomial and hence a change of taps and routing along the register.

This flexibility may be a useful advantage of CALBO compared with BILBO. However, there may be an additional silicon area penalty to pay, since the circuits of Figure 6.36 are marginally more complex than those of a BILBO shift register stage. We will return to consider the silicon overheads involved in built-in self-test in the penultimate section of this chapter.

6.7 PLA DESIGN-FOR-TESTABILITY TECHNIQUES

It is generally accepted that PLAs with an upper limit of eighteen to twenty inputs can be exhaustively tested; a 20 MHz tester clock rate will test such a combinational circuit in about 50 ms. However, for greater than twenty inputs there is an increasing need for alternative test procedures, but in all cases appropriate controllability and observability of the I/Os is necessary if the PLA is a buried macro within a custom VLSI circuit.

As covered in Section 6.4.3, the simple stuck-at fault model is not fully appropriate to PLA testing, and hence the test pattern generation programs that have been developed are not usually based upon the stuck-at model [43–49]. In this section we will introduce some of the concepts that have been proposed for making PLA architectures more easily testable; all involve some addition(s) to the basic AND/OR matrix, and hence the minimization of this additional circuit complexity is of paramount practical importance.

Figure 6.37 shows the span of PLA test methodologies, with the initial division being between on-line (concurrent) test and off-line techniques. Our interests here will not be with all these theoretical divisions, but primarily with those which involve some circuit enhancements to achieve improved testability. As will be seen, the overall picture includes the adoption of BILBO structures and signature analysis [7,16,33,86], as well as code checkers in the case of concurrent testability [15,86–89]. The BILBO and signature analysis techniques may involve the replacement of the normal input decoding circuits which provide x_i and $\overline{x}_i$ from each primary input (see Figures 3.8–3.11) with input latches, which when in test mode are reconfigured into a LFSR so as to provide the input test vectors.

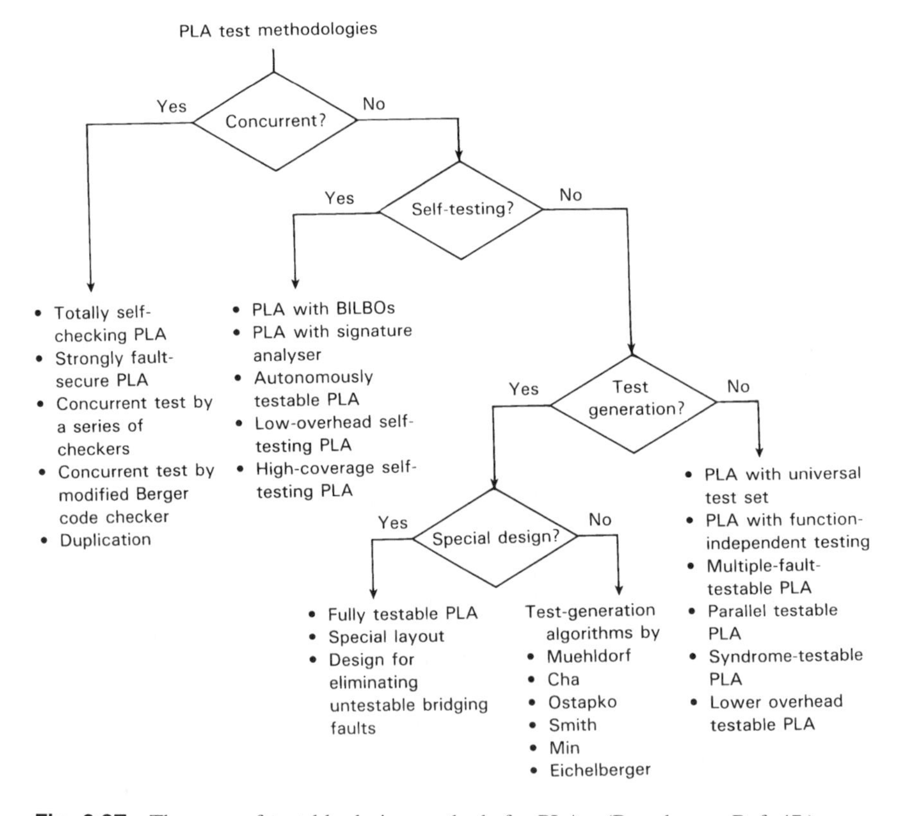

Fig. 6.37 The span of testable design methods for PLAs. (Based upon Ref. 47.)

However, in general, the addition of signature analysis or BILBO adds an unacceptable silicon overhead to the basic array, unless it is necessary because of restricted access to the PLA I/Os.

Concurrent checking is also a topic which we will not here consider in any depth, but we will look at some fundamentals in Section 6.9. However, it is worth noting that complete duplication of PLAs provides an extremely powerful online method of checking (see Figure 6.10), particularly if:

- the primary inputs are arranged dissimilarly between PLA *A* and PLA *B*, each PLA AND array being programmed accordingly
- the primary outputs are likewise dissimilarly allocated
- different product terms are used in *A* and *B* to generate each required output function, or if *B* is designed always to generate the complement of the *A* outputs

Such rearrangements between the two PLAs will hopefully cater to any faults or any weaknesses which may be common to the two devices. However, there remains the problem of verification of the equality (or complementarity) of the two outputs by means of some fail-safe checker, and even if this is done, there still remains a situation where the primary input data from some preceding circuit may be faulty. Hence, excessive checking of just one part of a design should always be considered in the context of the complete system or product. Ideally, the output from any form of checking circuit—electrical, electronic or mechanical—should not be a static condition, e.g., a steady logic 0 or 1, but rather should be some dynamic (continuously switching/oscillating) condition. It is left to the reader to consider the implications of this fail-safe philosophy.

To revert to the other PLA DFT techniques, recall from Figure 3.9 that a PLA has n primary inputs $x_1, \ldots, x_n$, p internal product lines $p_1, \ldots, p_p$, and m primary outputs $f_1(X), \ldots, f_m(X)$. Although referred to as an AND/OR array architecture, the actual arrays may be NAND or NOR structures, but this does not affect our general considerations of the mode of operation. It does, however, need to be taken into account when looking at actual test mode signals, in order to ensure that the logic 0s and 1s are appropriately chosen.

The principal needs when considering PLA design-for-testability techniques are as follows:

- controllability of the 2n x_i and $\bar{x}_i$ input literals, either individually or collectively
- controllability and perhaps observability of each of the p product lines (the AND array outputs)
- controllability of each of the m OR array output lines

Typical ways in which these requirements may be approached are shown in Figure 6.38. As will be seen, there are three principal ideas, namely:

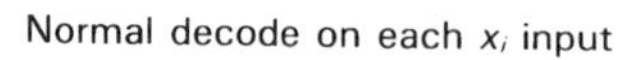

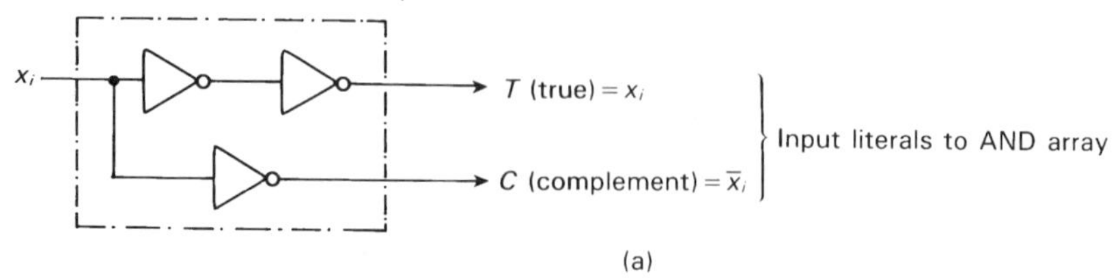

(a)

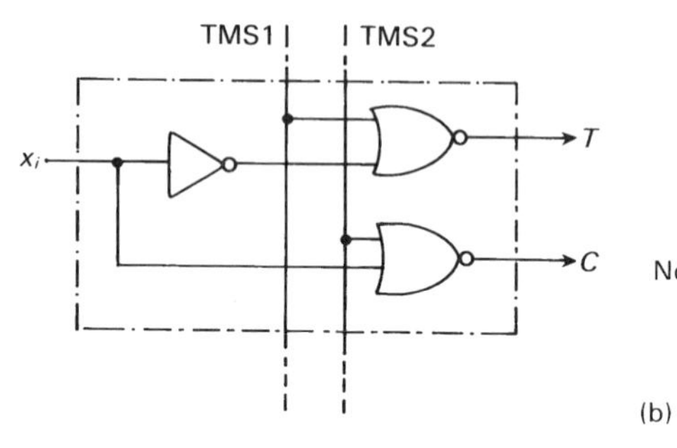

Normal mode

TMS1	TMS2	T	C
0	0	x_i	$\overline{x}_i$
0	1	x_i	0
1	0	0	$\overline{x}_i$
1	1	0	0

(b)

TMS Scan in

x_i

Latch

Latch

$T = x_i$

$C = \overline{x}_i$

Normal mode or scanned-in data

(c)

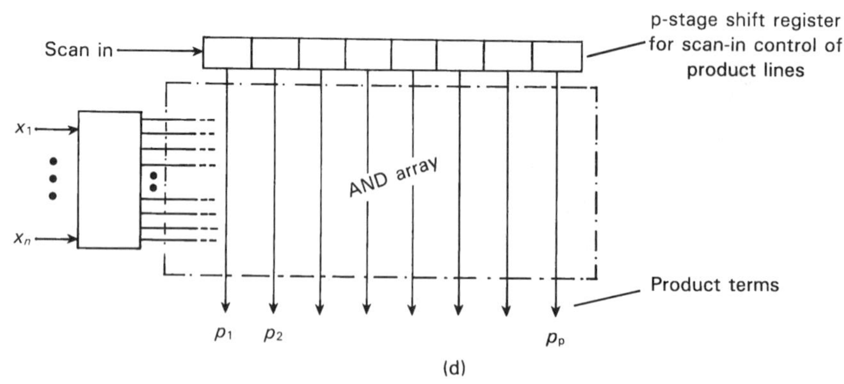

(d)

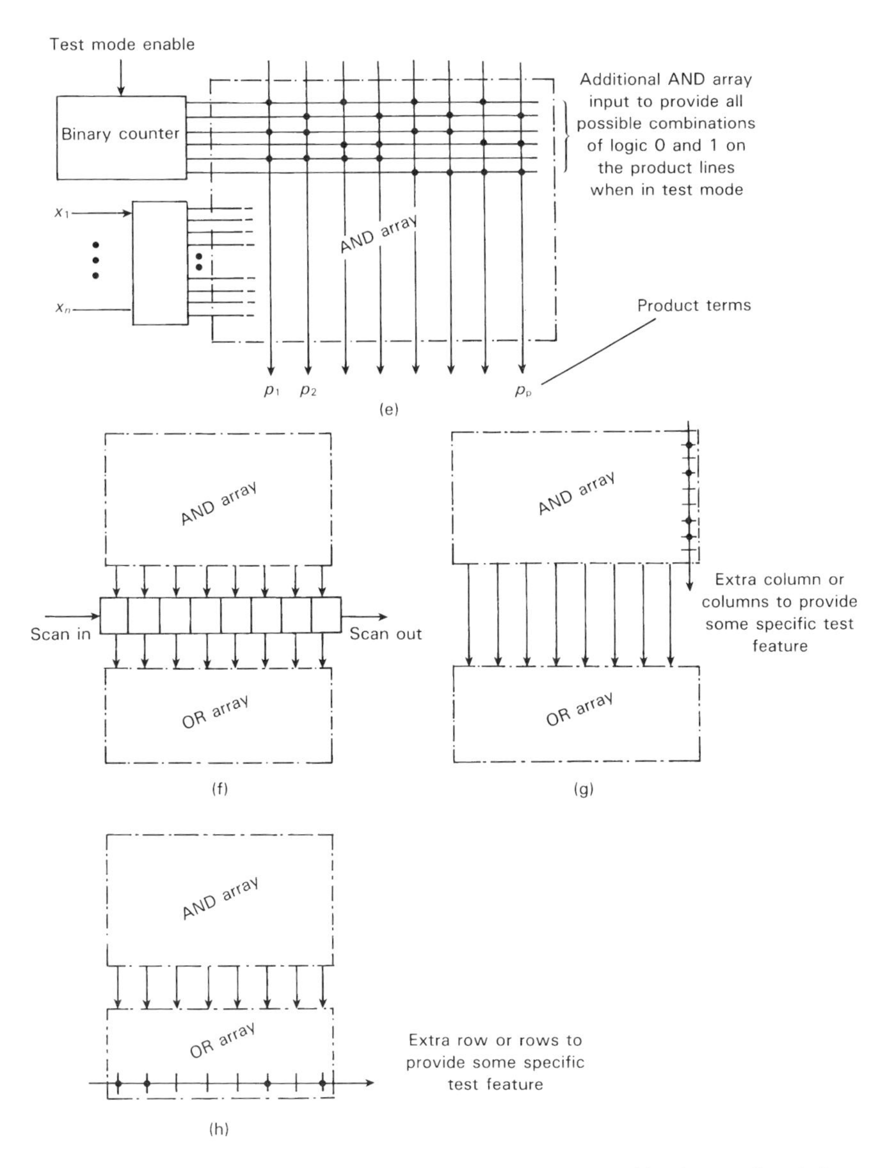

Fig. 6.38 Basic DFT techniques for PLA architectures—combinations and variations on these may be found; TMS = 'test mode signal': (a) the normal primary input decoding; (b) addition to modify the x_i and $\bar{x}_i$ array inputs; (c) scan-path addition to override the normal x_i inputs; (d) scan-path addition to allow removal of product terms from the AND array; (e) an alternative to (d) where the binary counter should have $\log_2 P$ stages so as to be able to supply all possible signal combinations to the P product lines; (f) scan-path addition to provide controllability and observability of the signals between the AND and the OR arrays; (g) and (h) additional column or row lines in the AND and OR arrays, respectively, so as to give some parity or other check feature.

(i) alter the normal x_i input decoders so as to be able to modify the x_i and $\bar{x}_i$ AND array inputs
(ii) add scan path shift registers so as to be able to shift in specific 0s and 1s to the AND and OR arrays, or to scan out product line data for checking
(iii) add additional product columns to the AND array and/or additional rows to the OR array for particular tests

However, there is no one accepted 'best' method for realizing and combining these DFT features.

Examples of DFT techniques which illustrate the above basic concepts are those of Fujiwara et al. [90,91], Hong and Ostapko [92] and Khakbaz [93]. Figure 6.39 shows the FK testing technique of Fujiwara [90], from which it will be seen that the architectural modifications include the following:

- modification of the input decoders to give outputs $x_i + Y_1$ and $\bar{x}_i + Y_2$, thus allowing the true and the complementary input literals to be collectively set to a steady logic 1 value
- the addition of a shift register to the AND array to enable the selective choice of any one product line one at a time—if the AND array is really a NOR matrix, then this requires a logic 1 on all product lines except the one selected, while a NAND matrix would require the opposite logic
- an extra column added to the AND array and an extra column added to the OR array, which can be programmed such that the total number of crosspoint connections in each AND row and in each OR column is odd

The odd-parity check is made by cascaded Exclusive-ORs on the product lines giving a test output $Z1$, and on the sum lines giving a test output $Z2$.

Details of the test patterns required to test for all single faults in the PLA and the additional DFT circuitry may be found in Ref. 90. These tests are independent of the functions provided by the PLA. An extension of this FK scheme is the later FKS proposal [91], which adds a shift register on the AND array input lines so that any row in the AND array may be individually selected. This addition gives complete multiple-fault coverage, but requires a higher number of input test patterns [15,16,91].

The FITPLA test scheme of Hong and Ostapko [92] is similar to the FK and FKS schemes in that a shift register is employed to select individual product lines, with parity checking included in the AND and OR arrays, but the individual line decoding and the parity checking is more complex. Like the FK and FKS methods, the test set is independent of the functions programmed in the PLA; more additional circuitry is involved in the FITPLA method, but the number of test patterns is somewhat less [16]. All these proposals, however, require lengthy

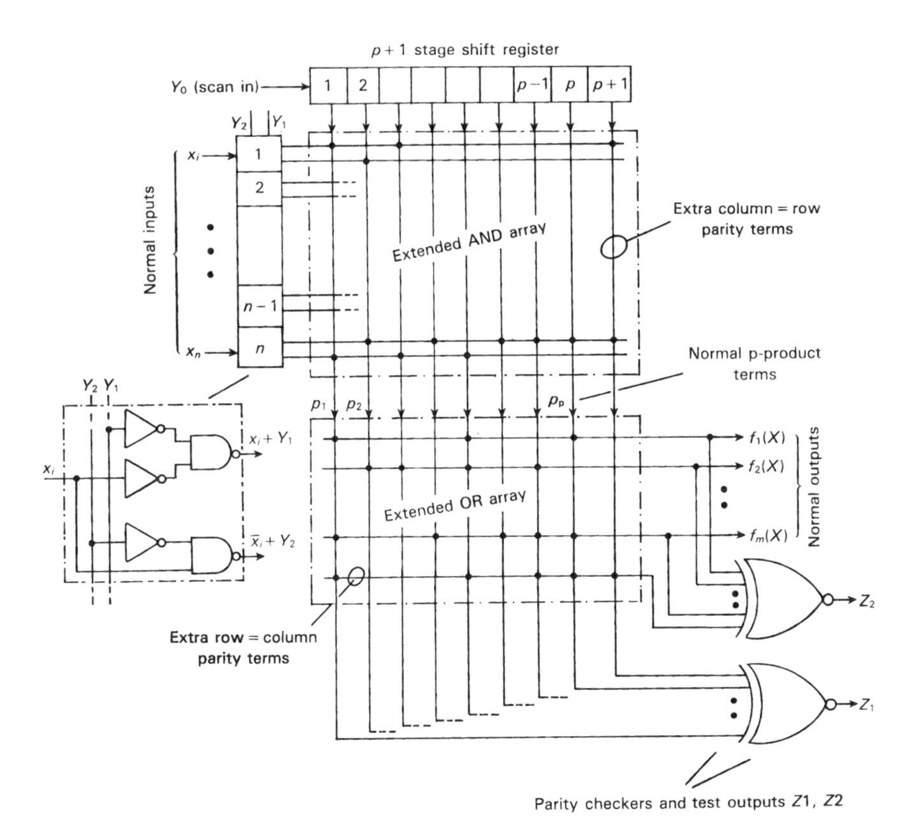

Fig. 6.39 The basic DFT PLA architecture of Fujiawara and Kinoshita, which provides function-independent tests covering all single crosspoint and stuck-at faults. Note the $(p + 1)$ input XOR gate shown here for output $Z1$ would be composed of p 2-input XOR gates, and similarly for output $Z2$ there would be a cascade of m 2-input XOR gates.

cascades of Exclusive-ORs for parity checking, which imply a timing constraint on the test procedures, and a considerable silicon area overhead to implement.

The K-scheme of Khakbaz [93] does not employ any Exclusive-OR parity checking, and thus does not suffer from these silicon area and timing problems. However, the test patterns are function-dependent, and hence the test set has to be determined for each particular PLA dedication. All that is added in the K-scheme is a shift register on the AND array to select individual product lines, plus one additional row in the OR array which is the OR of all the product terms. Knowing the dedication of the PLA, a test set can be formulated which tests (i) the correct operation of the shift register, (ii) the correct presence of each individual product term, and (iii) the correct output from the OR matrix. All single and multiple stuck-at faults and all crosspoint faults which give rise to incorrect outputs can be detected.

Several other variations for PLA testing have been proposed. Attempts have also been made to compare the advantages and disadvantages of these proposals [47], but no single strategy has received universal acceptance. We will come back to some further PLA testing ideas at the research level in our discussions in Section 6.9. As far as buried PLA macros in custom VLSI circuits are concerned, the vendor's views on built-in self-test or on-line concurrent test or other forms of PLA test will influence what the OEM will have to consider in USIC design activities. Current thoughts are not strongly in favor of increasing the complexity of PLAs in order to build in possible test strategies, except possibly where high reliability in service through the use of on-line testing is required.

Further details of and discussions on PLA testing may be found in Refs. 7, 15, 16, and 47–49, and the additional references contained therein.

6.8 I/O TESTING AND BOUNDARY SCAN

The individual I/O circuits at the perimeter of an IC are frequently very complex circuits, since they may be designed to be inputs or outputs as required, and have to interface between various combinations of on-chip and off-chip circuit requirements. Protection against static voltage damage will also be built in in most cases. Further consideration also has to be given to the fact that when the packaged and tested ICs are finally assembled on a printed-circuit board (PCB), final testing of the complete board or system has normally to be undertaken.

6.8.1 I/O Testing

It is particularly important that the input signals provided by the I/Os are correct to drive the subsequent on-chip logic; the output signals provided by the I/Os can be monitored by external test instrumentation. Hence, custom IC vendors may measure (i) the output voltage of each output pin under some worst-case loading or supply voltage conditions, and (ii) the satisfactory performance of each input pin under low input voltage conditions (V_{IL}) and high input voltage conditions (V_{IH}), that is, the threshold voltage design limits of the input circuits. Testing over a range of temperatures may also be performed, particularly for military or avionics applications.

The I/O input tests require additional on-chip logic to check that the input circuits do switch the logic correctly. Figure 6.40 illustrates one internal test circuit whereby each input drives one of a cascade of 2-input NAND gates. To check that each input responds correctly to low and high input signals, $(N + 1)$ tests at V_{IL} (low) followed by the same $(N + 1)$ tests at V_{IH} (high) are applied,

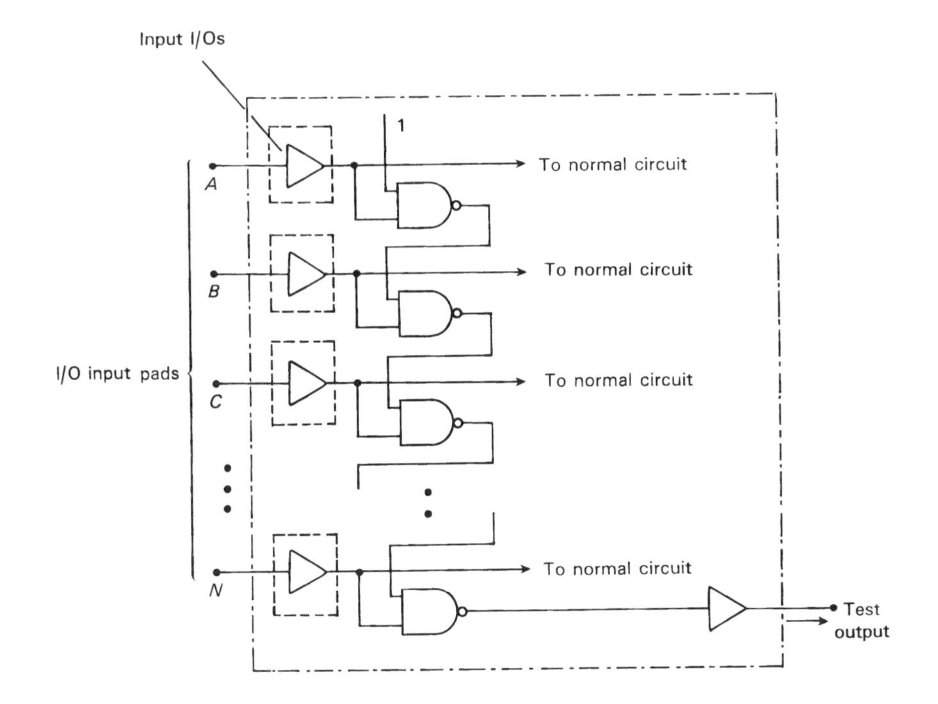

Test input vectors ($N+1$ tests)					Test output	
A	B	C	....	N	N an odd number	N an even number
1	1	1	....	1	0	1
0	1	1	....	1	1	0
0	0	1	....	1	0	1
0	0	0	....	1	1	0
•			•		•	•
•			•		•	•
•			•		•	•
•			•		•	•
0	0	0	....	0	1	1

Fig. 6.40 Vendors' testing of input I/Os with the test input vectors being applied at the design threshold limits of V_{IL} and V_{IH}.

and the resultant signal at the single output pin confirmng the action of each of the inputs is monitored [94].

If the I/Os are bidirectional and have to operate either as inputs or outputs on demand, then an additional test-enable pin has to be provided to disable the output portion of bidirectional circuits when in test mode. The same V_{IL} and V_{IH} patterns may then be applied and the response checked through the NAND cascade.

Tri-stating of I/Os is another feature provided by many vendors. When switched by an appropriate control signal all I/Os take their tri- (third) state, becoming a high impedance and not allowing current to flow in either direction. By this means, the IC on an assembled PCB can be effectively disconnected, allowing tests to be made on the rest of the board or product as though the IC was not present.

However, the most powerful technique for handling the testing problems associated with ICs assembled on PCBs is a scan-path technique, as covered in the following section. Once again, the principle is simple: to provide controllability and observability of I/O signals, this time within a large and possibly complex PCB with difficult test accessiblity, so that either the assembled IC itself can be separately tested, or the board surrounding the IC can be tested, or both.

6.8.2 Boundary Scan and Board Testing

The development of boundary scan originated from design engineers in Europe who were involved in testing very complex PCBs, with increasing use of surface-mount ICs which could not easily be removed after assembly and decreasing spacings between conductors and components making probing more difficult. As a result, the international *Joint Test Action Group* (JTAG) was established, the efforts of which have now resulted in the IEEE Standard 1149.1 for boundary scan architectures and protocols.

The essence of boundary scan is to include a scan latch in every I/O circuit, as shown in Figure 6.41. During normal mode operation, system data pass freely through all the boundary scan cells, but in test mode this transparancy is interrupted and replaced by a number of possible test mode conditions, in total providing the following seven facilities.

(a) normal mode, transparent on-chip operation
(b) test mode, scan in test data to the input boundary cells to act as a test vector
(c) test mode, use data (b) to exercise the on-chip circuits, and capture the resultant IC response in the output boundary cells
(d) test mode, scan out test results (c)
(e) test mode, scan in test data to the output boundary cells

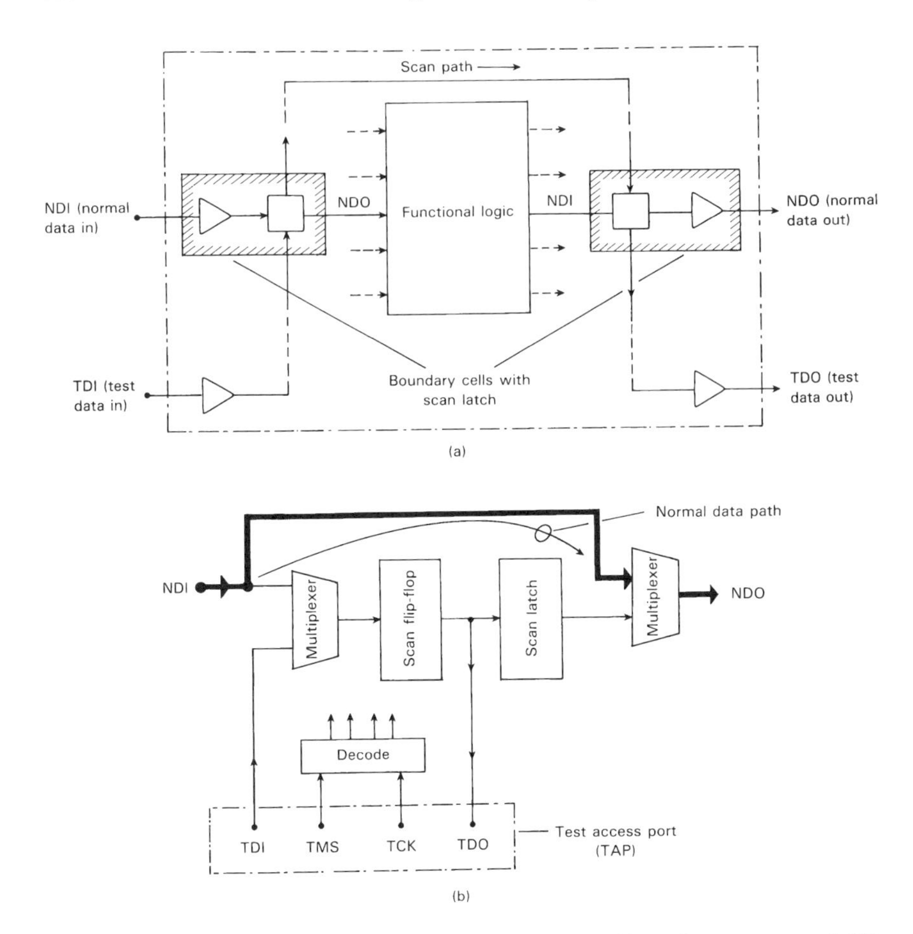

Fig. 6.41 The principles of boundary scan: (a) provision of boundary scan on each I/O so as to give a complete scan path within each IC; (b) simplified details of a boundary scan cell; exact details may be proprietary.

(f) test mode, use data (e) as a test vector to exercise the circuits or interconnect on the PCB surrounding the particular IC, and capture the resultant PCB response in input boundary cells
(g) test mode, scan out test results (f)

It will be seen that tests (b) to (d) are the usual load-and-test methods associated with scan-path IC testing, but the provision of a boundary scan cell on every I/O now enables the same load-and-test technique to be applied to the circuitry surrounding the IC, the internal logic signals within the IC effectively being isolated from the rest of the system. This is the power of boundary scan. Notice that modes (b), (d), (e) and (f) are identical scan modes with the boundary cells in cascade, and hence only two signals are necessary for boundary scan control purposes, these two control signals being decoded on-chip to give four test conditions (see Figure 6.41(b)).

Multiple ICs on an assembled PCB are all under the control of the JTAG four-wire test bus with its four lines TDI, TDO, TMS, and TCK. Figure 6.42(a) indicates how each IC has its own boundary scan which enables it to be tested independently, or to receive or transmit test data to other circuits or components via the PCB interconnect. Notice one further feature: although it is possible to test each individual IC by scanning in test vectors and scanning out the results, as the IC is effectively isolated from its surrounds it is also possible to have BILBO self-test on the chip as previously described, with some additional BILBO control signals, the I/O boundary scan scanning out a MISR test signature at the end of this self-test. A considerable flexibility of test is therefore possible.

Still further facilities are now being offered, with the availability of the off-the-shelf chip sets from IC vendors providing such facilities as:

- test bus selection, which can select certain scan paths on a complex PCB assembly and short-circuit the remainder
- test bus controller circuits for overall management of the scan path data
- scan test circuits for capturing data on busses, or loading test data on to busses in place to normal data (see Figure 6.42(b))

Considerable housekeeping may accrue in developing and keeping track of all the test data in a complex assembly, and therefore a further commercial development has been CAD resources designed specifically for the boundary scan protocols and system test data generation. This is illustrated in Figure 6.42(c).

Further details of boundary scan and IEEE Standard 1149.1 may be found in the published literature, including details of the sixteen-state algorithmic state machine (ASM) chart or state diagram, which defines the internal operation of the JTAG controller [7,75,95–102]. It is evident that the increasing complexity of both the individual ICs and the complete systems into which they are assembled have fostered these DFT dvelopments, with scan path being the key element in providing the internal controllability and observability without requiring an unacceptable number of additional test pins.

6.9 FURTHER TESTING CONCEPTS

Research into very many aspects of test for digital networks has been and continues to be undertaken, but may not be reflected in current practice. We will, however, review and reference some of this work herewith in order to give a more

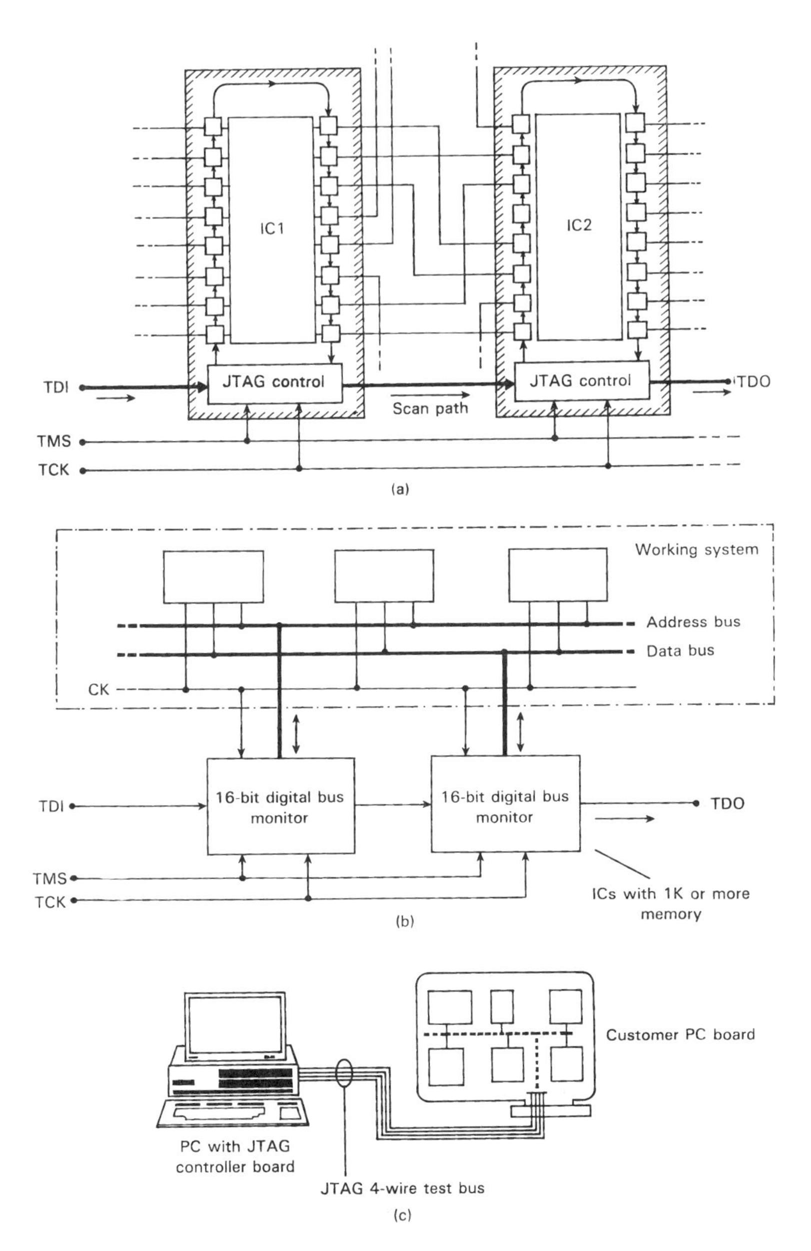

Fig. 6.42 Boundary scan applications: (a) the scan path through several on-board ICs; (b) scan-path principles applied to bus monitoring, with the monitoring ICs being able to capture system data or load in test data; (c) CAD resources to assist in the generation of test data and general housekeeping. (Source: based upon Texas Instruments technical data.)

complete overview of testing concepts. Notice that all of the following deals with digital circuits; the testing of analog circuits has not received the same amount of research in the past, and the procedures covered in Sections 6.6.2 and 6.6.3 remain the norm. The recent and continuing increase of mixed analog/digital custom circuits is having the effect of emphasizing the analog side of testing, and therefore we may see forthcoming developments such as the use of digital signal processing (DSP) for analog test [7]. However, we will not cover any of these ideas here, but instead consider the digital developments only. Remember also that with IC testing we are not concerned with fault location but only with fault detection.

As was seen in Section 6.1, one of the main problems with digital network testing is the volume of test data which has to be entered and its response checked (see Figure 6.5). One ideal test senario would be as shown in Figure 6.43(a), where one additional input pin is added to the network under test so as to switch it from normal mode to test mode, and one additional output pin is added to show the final test result, the latter being a simple pass/fail (fault-free/faulty) indication; all this to be done with the minimum of additional circuitry. An alternative ideal senario would be continuous on-line checking as shown in Figure 6.43(b), again with only one pass/fail output pin and the minimum of additional test circuitry. Unfortunately, these ideals cannot be fully realized.

Most of the research in digital logic test will be seen to be concerned almost exclusively with combinational logic only. This is because no realistic means

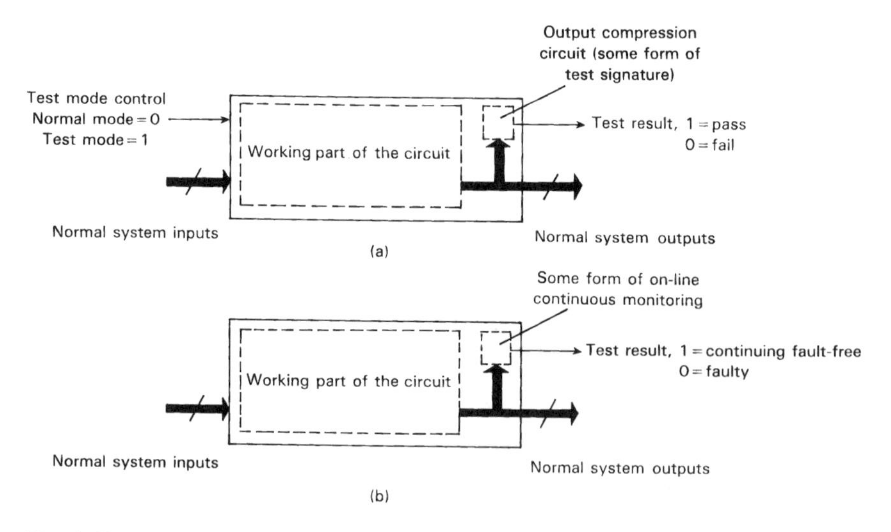

Fig. 6.43 Ideal test objectives: (a) one test mode input pin and one test result output pin; (b) continuous (on-line) testing, with one test monitoring output pin.

apart from some form of functional test can be formulated to test sequential networks with their possibly very long counting sequences and their inherent feedback connections, leaving us with the testing rules which we have already encountered, namely:

- partitioning long counters into manageable sections, as in Figure 6.22
- separating the sequential elements from the combinatinal logic, as in Figure 6.24
- reconfiguring the sequential elements into shift registers, which constitute the most readily testable form of sequential circuit

Beyond these simple ad hoc rules no further useful ideas can be envisaged apart from some form of self-checking (see Section 6.9.4). Test pattern generation for sequential networks has very little meaning.

6.9.1 Input and Output Compaction

A number of authors have observed that although the total number of primary inputs to a circuit may be large, typical combinational outputs often depend upon a restricted subset of these inputs. Thus, although there may be n inputs, where n is large, requiring 2^n input vectors for exhaustive global testing, individual outputs may be dependent upon $i, j, \ldots$ inputs, $i, j, \ldots < n$, giving $2^i, 2^j, \ldots$ input vectors for exhaustive test purposes.

It may also be observed that in many cases individual or local outputs depend upon a disjoint set of primary inputs. Partitioning of the overall system at the design stage to acheive this may also be relevant. Hence, although one might be limited by time considerations to an input test set of, e.g., 2^j vectors, by judicious sharing and combining of input signals when in test mode exhaustive testing of several outputs can simultaneously be undertaken.

A significant approach to the generation of reduced input test sets for exhaustive local testing is that of Akers [103], which continues on from earlier developments [104,105]. The system total of n primary inputs is generated from j independant binary test signals, $j < n$, by linear (Exclusive-OR) relationships, as shown in Figure 6.44. The value of j is the maximum number of the primary inputs involved in any one of the m primary output functions and the independency between these m output functions. In the case of a simple 3-bit test vector generator with three test outputs w_1, w_2, and w_3, a total of seven bit streams can be generated (see Table 6.5) from which 28 exhaustive 3-bit test vector sets can be selected. The total number of bit streams possible from a j-bit test vector generator is $2j - 1$.

Published details cover the derivation of appropriate exhaustive test sets for given multiple-output networks, with the emphasis being upon minimizing the width of the test vector generator w_i, and hence the test time.

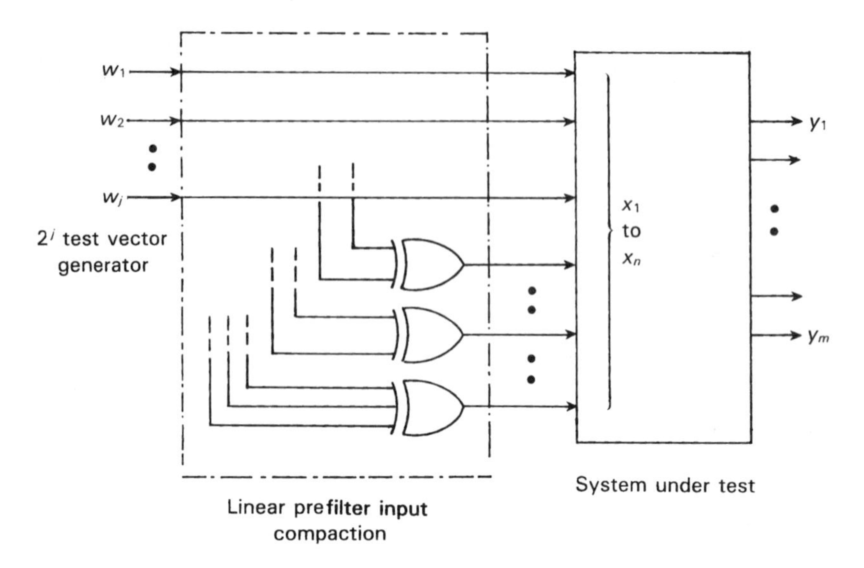

Fig. 6.44 The use of a linear prefilter to generate n binary inputs from the j-bit test vector generator; the circuit or system under test is combinational logic only (see text).

The several discrete primary outputs y_i, $i = 1, \ldots, m$, may also be compacted into fewer test output lines z_i, $i = 1, \ldots, l$, $l < m$, by the use of Exclusive-OR relationships (see Figure 6.45). (This compaction has been termed 'space compression,' since it does not modify the time for the exhaustive test of any individual output [106,107], in contrast to the previous input compaction which

Table 6.5 The Seven Possible Input Vectors Available from Three Primary Inputs by Linear Summation[a]

Primary inputs						
w_1	w_2	w_3	$w_1 \oplus w_2$	$w_1 \oplus w_3$	$w_2 \oplus w_3$	$w_1 \oplus w_2 \oplus w_3$
0	0	0	0	0	0	0
1	0	0	1	1	0	1
0	1	0	1	0	1	1
1	1	0	0	1	1	0
0	0	1	0	1	1	1
1	0	1	1	0	1	0
0	1	1	1	1	0	0
1	1	1	0	0	0	1

[a] Note that the choice of any three of the seven vectors will give an exhaustive test set provided that all the subscripts 1, 2 and 3 appear an odd number of times in the three chosen vectors.

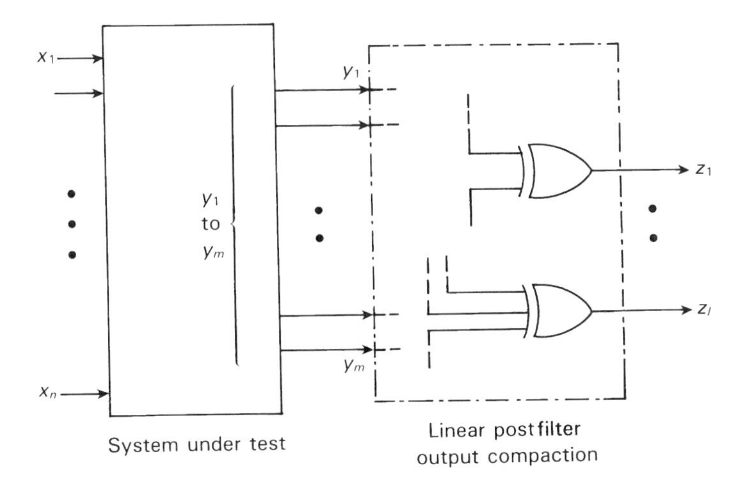

Fig. 6.45 The use of a linear postfilter to generate l binary outputs from the m primary circuit or system outputs.

has been termed 'time compression.') However, while exhaustive energization of the circuit under test is still a part of this approach, the principal thrust of output compaction has been to eliminate the need for individual evaluation of the several primary outputs when under test. Instead, the aim has largely been to consider some optimal combinations of the y_i outputs such that maximum fault cover will still be given by the reduced number of z_i test outputs (see Figure 6.45).

If the final fault detection mechanism is, for example, a simple counting procedure which simply counts the number of logic 1s (or 0s) appearing on an output line when under test (this is otherwise known as the syndrome count, and will be specifically considered in Section 6.9.2), then maximum fault coverage (minimum fault masking) will be obtained when this count tends to the maximum value of 2_j, where j is the number of binary inputs in the exhaustive test input vector set, or alternatively to zero count [106]. The usual objective of output compaction and other forms of output modification is therefore to maximize (or minimize) this value. A consideration of the choice of functions to combine to acheive this in the compaction method of Figure 6.45 is a nontrivial problem, since there are in total $2^m - 1$ possible combinations of the $y_1, \ldots, y_m$ outputs [106–108], but a further output compaction proposal is to combine *all* the y_i outputs under test with a further function y_{m+1}, y_{m+1} being chosen such that the final output count is some specific numerical signature. This is illustrated in Figure 6.46.

As in most forms of compaction, a correct output signature does not absolutely guarantee the correct functionality of the circuit under test. For example,

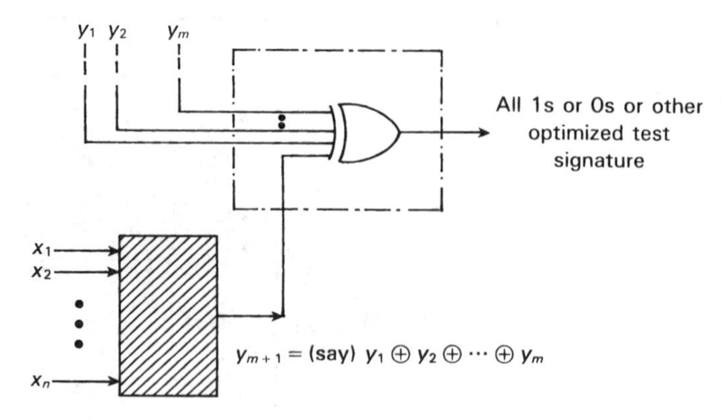

Fig. 6.46 Linear output compression of all y primary outputs, $i = 1, \ldots m$, with a further test function y_{m+1} generated from the primary inputs so as to give, e.g. all 1s in the compressed test output.

depending upon exactly how the circuit was designed and the signature chosen, the test configuration of Figure 6.46 may not be able to detect a stuck-at fault on a primary input line which causes outputs $y_1, \ldots, y_m$ to lose this input variable, or an internal fault which affects two (or any multiple of two) primary outputs identically. The more compaction involved, the more care needs to be taken in considering whether such compression is acceptable—a fault-free output signature count of a single 0 or a single 1 may be more secure against stuck-at primary inputs than a maximum count of 0s or 1s; the following is an example, if the test configuration of Figure 6.46 is used:

Fault-free output $f_1(X)$:	1 0 1 1 1 1 0 1
Fault-free output $f_2(X)$:	0 1 1 0 0 0 0 1
Fault-free output $f_3(X)$:	1 0 1 0 0 1 1 0
Test function output $f_{m+1}(X)$:	0 1 1 1 1 0 1 1
Final XOR test count signature:	0 0 0 0 0 0 0 1

Notice that if an even number of outputs lose the same one-valued minterms in their outputs, then this test signature will not detect such missing outputs, and hence this test and similar tests are not 100% secure.

Further details of input and output compaction techniques may be found in the references listed above, but such techniques remain largely ad hoc techniques rather than formal means to ease the testing problem.

6.9.2 Syndrome, Transition and Other Counting Concepts

Syndrome Counting

Mention of syndrome count has been made in the previous section. The syndrome of a combinational function or network is formally defined as

$$S = k/2^n$$

where k is the number of 1-valued minterms in the function output, and n is the number of independent primary input variables. With this definition, S ranges between 0 (no 1-valued minterms) and 1 (all minterms 1-valued). However, integer values are preferred in practice, and hence the normalizing factor of $1/2^n$ is usually omitted, leaving the syndrome S numerically equal to the number of ones in the output of a given function.

Syndrome testing was first proposed by Savir, and much work was undertaken to define 'syndrome-testable' networks [7,14,15,33,109–113]. The basic objective was to establish an unique syndrome signature from a network under test, usually with an exhaustive input test set, for all possible circuit failings, thus compressing the test output response into a simple ones-count as introduced above and as shown in Figure 6.47.

However, as indicated by the previous simple example, this simple ones-count is not foolproof. Certainly, if the syndrome count differs from the healthy count, then the circuit must be faulty, but it is always possible to postulate some circuit faults which will result in the same output. A trivial example is an multiple-input Exclusive-OR function which has an equal number of 0s and 1s in its output: if one of the inputs is stuck at 0 or at 1 the output still has an equal number of 0-valued and 1-valued minterms in its truthtable.

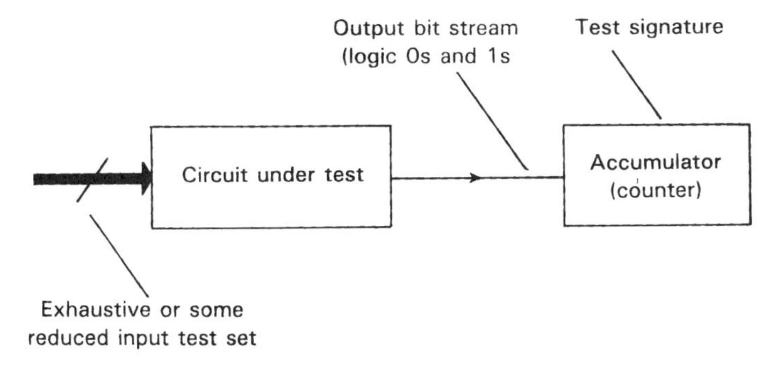

Fig. 6.47 Syndrome counting, where the output accumulator is a simple counter counting the number of 1s on the output bit stream (cf. the more complex and secure output count used in the single-input LFSR signature analysis of Figure 6.34(a).)

Hence, a simple 1s count is not generally satisfactory. Further work to improve this means ot test has included the following:

(a) checking to see whether a simple syndrome count for a given network will detect all given circuit faults such as all stuck-at faults
(b) modification of the network under test by the addition of test inputs and/or internal logic which modifies the syndrome count so that it is secure against all given circuit faults
(c) constrained syndrome testing, which involves more than one syndrome test, certain primary inputs being held constant during each such test
(d) weighted syndrome testing for multiple output networks where the individual outputs are given weighting factors, the final count being

$$\sum_{i=1}^{m} w_i S_i$$

where w_i is the weight given to the syndrome count S_i from output y_i

Further details of all these proposals may be found in the published literature [111–113]; we will refer to (d) again in Section 6.9.3, but in practice syndrome testing has not been actively adopted.

Transition Counting

An alternative to the syndrome count is a count of the number of transitions from 0 to 1 (or vice versa) in the output bit stream when under test. Like the syndrome count, this transition count usually considers a fully exhaustive input test set. However, unlike the syndrome test, where the count is independent of the order of application of the test vectors, a transition count is dependent upon the order of application. Thus, the exhaustive input test set may be ordered such that the number of transitions in the healthy output response is mimimized, which in the extreme is a single 0-to-1 or 1-to-0 transition [7,14,114–117].

An ordering of the exhaustive input test vectors such that there is only one transition in the output response (i.e., initially all 0s followed by all 1s, or vice versa) is not in itself 100% secure. The danger here is if a block of logic 0s at the end of the healthy logic 0 bit stream becomes all 1s, or if a block of 1s at the beginning of the healthy logic 1 bit stream becomes all 0s, then this mode of failure will not be detected by the single transition count. However, this shortcoming may be very elegantly overcome by repeating the initial test vector of the block of 0s and the initial test vector of the block of 1s at the end of their respective blocks, giving a total input test sequence of $2^n + 2$ vectors with a foolproof healthy count of one [117].

Accumulator Syndrome

Yet another output count procedure has been proposed by Saxena and Robinson based upon the syndrome count; this is the 'accumulator syndrome,' and is effectively the area under the syndrome count over the exhaustive test sequence of 2^n input vectors [118]. For example, if the output bit stream begins

0 0 1 1 0 0 1 0 . . .

then the accumulator syndrome count would increase as follows:

0 0 1 3 5 7 10 13 . . .

Note that this count, like the transition count, is dependent upon the order of application of the test vectors.

However, as shown by Table 6.6, the accumulator syndrome count is not inherently secure against all conceivable faults. Fault masking (aliasing) can occur in all these forms of output compression, except where the input test sequence can be specifically matched to the function under test so as to give some extreme count, for example, the single transistion count noted above.

Further details and analyses of the fault masking properties of these test compression methods may be found [14,118–120]. A combination of transition and syndrome counting has also been proposed [14], but the fault coverage is difficult to determine if random input test vector sequences are assumed. On the other hand, in all cases the use of a functionally dependent test vector sequence which may give a 100% secure test raises the practical problem of generating such a sequence, which cannot be done on-chip by any simple means other than possibly using a ROM or PROM for such a purpose.

Table 6.6 A Simple Example Showing the Failure of the Three Forms of Output Compression to Provide Either Individually or Collectively a Unique Fault-Free Signature[a]

	Output response on test	Syndrome count	Transition count	Accumulator syndrome count
(i) Healthy	0 1 1 0 0 1 0 0	3	2	16
(ii) Faulty	0 1 0 1 1 0 0 0	3	2	16
(iii) Faulty	0 1 1 0 0 0 1 0	3	2	15
(iv) Faulty	0 1 1 0 1 0 0 0	3	2	17
(v) Faulty	0 0 1 1 0 1 1 0	4	2	16
(vi) Faulty	1 0 0 1 0 1 0 0	3	3	16

[a] Specific re-ordering of the input test vector sequence can improve the coverage of the transistion and accumulator counts.

Arithmetic Coefficients

The canonic expansion for any combinational logic function is classically given by the Boolean minterm expansion, which for any three-variable function $f(X)$ is:

$$f(X) = a_0\bar{x}_1\bar{x}_2\bar{x}_3 + a_1x_1\bar{x}_2\bar{x}_3 + a_2\bar{x}_1x_2\bar{x}_3 + a_3x_1x_2\bar{x}_3 + a_4\bar{x}_1\bar{x}_2x_3 + a_5x_1\bar{x}_2x_3 + a_6\bar{x}_1x_2x_3 + a_7x_1x_2x_3$$

$$a_j,\ j = 0, \ldots, 2^n - 1,\ \varepsilon\ \{0,1\}$$

There is, however, an alternative canonic arithmetic expansion for any $f(X)$ [121,122], which for any three-variable function is:

$$f(X) = b_0 + b_1x_1 + b_2x_2 + b_3x_1x_2 + b_4x_3 + b_5x_1x_3 + b_6x_2x_3 + b_7x_1x_2x_3$$

$$b_j,\ j = 0, \ldots, 2^n - 1,\ \varepsilon \text{ positive and negative integer values}$$

For example, the parity function $\bar{x}_1\bar{x}_2x_3 + \bar{x}_1x_2\bar{x}_3 + x_1\bar{x}_2\bar{x}_3 + x_1x_2x_3$ may be written as:

$$f(X) = x_1 + x_2 - 2\,x_1x_2 + x_3 - 2\,x_1x_3 - 2\,x_2x_3 + 4\,x_1x_2x_3$$

which, if evaluated for each minterm, will be found to give the {0,1} truthtable for the function. The procedure for determining the arithmetic coefficients for any given function is a matrix multiplication of the normal {0,1} truthtable of the function, the transform being:

$$\mathbf{T}^n = \begin{bmatrix} \mathbf{T}^{n-1} & 0 \\ -\mathbf{T}^{n-1} & \mathbf{T}^{n-1} \end{bmatrix}$$

$$\mathbf{T}^0 = +1$$

Heidtmann has considered the possibility of using these arithmetic coefficients, which are unique for any given function, as test signatures for stuck-at faults [122], but no advantages have been found in their use in comparison with more conventional Boolean techniques.

Exclusive-OR Coefficients

There is also the positive Reed-Muller Exclusive-OR canonic expansion for any Boolean function, which for a three-variable function is:

$$f(X) = c_0 \oplus c_1x_1 \oplus c_2x_2 \oplus c_3x_1x_2 \oplus c_4x_3 \oplus c_5x_1x_3 \oplus c_6x_2x_3 \oplus c_7x_1x_2x_3$$

$$c_j,\ j = 0, \ldots, 2^n - 1,\ \varepsilon\ \{0,1\}$$

where the addition is modulo-two. The transform for determining the positive canonic Reed-Muller coefficients for any given function is also given by a matrix multiplication of the normal Boolean truthtable of the function, the matrix being:

$$\mathbf{T}^n = \begin{bmatrix} \mathbf{T}^{n-1} & 0 \\ \mathbf{T}^{n-1} & \mathbf{T}^{n-1} \end{bmatrix}_{\text{mod.2}}$$

$$\mathbf{T}^0 = +1$$

(Note that there are a total of 2^n possible Reed-Muller canonic expansions for any Boolean function of n input variables to take into account the possible combinations of complemented and uncomplemented literals in the expansion; the positive canonic expansion considered above has by definition all its x_i literals uncomplimented [121,123,124]. We will not concern ourselves with the alternatives here.) Again, certain research has been done in considering the possible use of R-M coefficients for testing purposes [125–127], but again, no lasting applications or attractions have been found to justify their use in test methodologies.

Spectral Coefficients

Although the binary values of 0 and 1 are the normally accepted way of defining the logic state in 2-valued digital logic, there are other mathematical relationships which can be employed. These alternatives, the spectral coefficients, have been the subject of considerable research, and offer certain academic advantages for both logic design and test in comparison with our discrete {0,1} Boolean representations and Boolean algebra.

It is possible to consider the transformation of any binary-valued output vector of a function $f(X)$ of n input variables into some other alternative domain by multiplying the binary-valued vector by a $2^n \times 2^n$ complete orthogonal transform matrix. The transformation produced by such a mathematical operation is such that the information content of the original vector is fully retained in the resulting 'spectral domain' vector, that is, the spectral data are unique for the given function, and that the original binary data can be recreated by the application of an inverse transform.

The theory of complete orthogonal transforms may be found in many mathematical textbooks [128–132]. However, the simplest to use for the transformation of discrete binary data, and the most well structured, is the Hadamard transform, defined by the following recursive structure:

$$\mathbf{Hd}^n = \begin{bmatrix} \mathbf{Hd}^{n-1} & \mathbf{Hd}^{n-1} \\ \mathbf{Hd}^{n-1} & -\mathbf{Hd}^{n-1} \end{bmatrix}$$

$$\mathbf{Hd}^0 = +1$$

For $n = 3$ this gives the following, shown here operating on the example function $f(X_1) = \bar{x}_1\bar{x}_2x_3 + \bar{x}_1x_2\bar{x}_3 + x_1\bar{x}_2\bar{x}_3 + x_1x_2x_3$:

$$
\underbrace{\begin{bmatrix}
1 & 1 & 1 & 1 & 1 & 1 & 1 & 1 \\
1 & -1 & 1 & -1 & 1 & -1 & 1 & -1 \\
1 & 1 & -1 & -1 & 1 & 1 & -1 & -1 \\
1 & -1 & -1 & 1 & 1 & -1 & -1 & 1 \\
1 & 1 & 1 & 1 & -1 & -1 & -1 & -1 \\
1 & -1 & 1 & -1 & -1 & 1 & -1 & 1 \\
1 & 1 & -1 & -1 & -1 & -1 & 1 & 1 \\
1 & -1 & -1 & 1 & -1 & 1 & 1 & -1
\end{bmatrix}}_{[\mathbf{Hd}]}
\underbrace{\begin{bmatrix} 0 \\ 1 \\ 1 \\ 0 \\ 1 \\ 0 \\ 0 \\ 1 \end{bmatrix}}_{\mathbf{X}]}
=
\underbrace{\begin{bmatrix} 4 \\ 0 \\ 0 \\ 0 \\ 0 \\ 0 \\ 0 \\ -4 \end{bmatrix}}_{\mathbf{R}]}
$$

The spectrum $\mathbf{R}]$ of $f(X)$ is thus given by the eight numbers

r_0	r_1	r_2	r_{12}	r_1	r_{13}	r_{23}	r_{123}
4	0	0	0	0	0	0	−4.

However, it is often preferable to convert the binary {0,1} truthtable values into {+1, −1} by the simple substitution of +1 for logic 0 and −1 for logic 1 before doing the above transform. It is left as an exercise for the reader to confirm that the spectrum for the above function is now given by 0 0 0 0 0 0 0 +8, which conventionally would be identified as

$$
\mathbf{S}] = \begin{matrix} s_0 & s_1 & s_2 & s_{12} & s_3 & s_{13} & s_{23} & s_{123} \\ 0 & 0 & 0 & 0 & 0 & 0 & 0 & +8 \end{matrix}
$$

A full consideration of the mathematics and meaning of the numbers which result in the spectral domain may be found in the literature and will not be covered here [113,125,131]. Each of the spectral coefficient values represents some global information about the function, and not discrete information as is the case with the normal {0,1} binary data—knowing that a function output is logic 1 on a particular input test vector tells us absolutely nothing about the rest of the function, but knowing the value of any single spectral coefficient tells us something about the whole function. It is straightforward to show that each spectral coefficient is a correlation coefficient (agreements—disagreements) which gives a measure of 'how like' the output function is to (a) logic 1, (b) each of the n primary x_i inputs, and (c) every possible Exclusive-OR combination of the inputs; the value of $s_0 = 0$ in the above example indicates that there are the same number of agreements and disagreements between the output and logic 1; the value of

$s_1 = 0$ indicates that there are the same number of agreements and disagreements between the output and the input variable $x1$, and so on up to $s_{123} = +8 = 2^n$, indicating that the function is in full agreement with the Exclusive-OR of all three input variables—the function is indeed $x_1 \oplus x_2 \oplus x_3$ if it is examined.

Any fault in a logic circuit will change its output spectum from the fault-free values. Hence, if we know the fault-free spectrum and determine all the spectral coefficients of the circuit under test, then if they differ in any way the circuit is faulty. However, rather than check all coefficients, it would be advantageous to be able to check only one in order to confirm the fault-free or faulty status of the circuit.

If we look at the first spectral coefficient r_0, it will be appreciated that it is the syndrome count previously considered. It is the weakest of the possible counting signatures unless it is maximally valued, but then the output is only logic 0 or 1. A great deal of research has been done in order to try to define the optimum coefficient(s) to act as a secure test signature [7,14,113,125,133–135], but the problem always remains that it is necessary to apply a lengthy test sequence (ideally exhaustive) in order to determine any test count. Figure 6.48 illustrates the commonality of these various output compression possibilities, with only the duplication in Figure 6.48(a) providing a 100% secure test for all possible fault conditions.

Undoubtedly, the spectral domain with its global information provides a much deeper insight into the functional structure of combinational circuits than is available from the discrete Boolean domain [113]. However, to date it has not proved successful in providing an alternative general-purpose test methodology [7], although work with alternative orthogonal transforms and new methods of calculating the coefficients is still an on-going academic activity. For further fundamental developments, see particularly Miller [14] and the many references contained therein.

6.9.3 Further PLA and BIST Testing Concepts

PLAs

PLA structures continue to be a prominent field of research and development, one of the objectives being to find alternative means of test which involve the minimum addition to the basic PLA architecture.

Syndrome testing, and in particular weighted syndrome testing, has been considered as a possible means of PLA test. The most definitive work in weighted syndrome testing is that of Serra and Muzio [136]. Following an earlier work by Yamada [137] which considered the individual outputs of PLAs under fault modeling, and that of Barzila et al. who first introduced the concept of a weighted syndrome summation for multiple-output networks [112], Serra and Muzio have

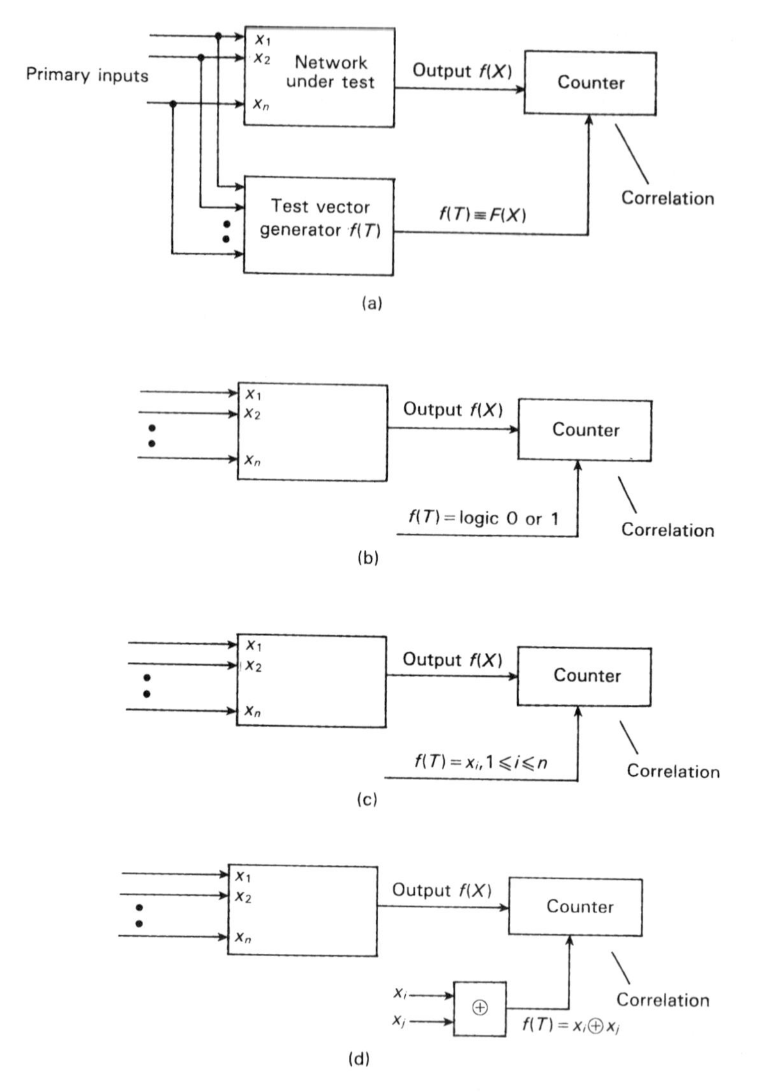

Fig. 6.48 Test signatures based upon the correlation between a primary output $f(X)$ and a test vector generator $f(T)$ over the exhaustive test vector sequence of 2^n vectors: (a) the ultimate situation of $f(X)$ compared against a copy of itself; (b) $f(X)$ compared against logic 0 or 1, equivalent to syndrome testing; (c) $f(X)$ compared against a single primary input variable x, = spectral coefficient s; (d) $f(X)$ compared against the Exclusive OR of two primary inputs, = spectral coefficient s. Note that in all cases a maximum count of, 2^n or a minimum count of zero is the extreme case.

shown that the weighted syndrome sum of all outputs is a sufficient test for all types of single faults within PLA architectures. The fault models considered are:

- stuck-at faults—one row or one column stuck at either logic 0 or logic 1
- bridging faults—two adjacent rows or columns shorted together taking the logical AND or OR of the individual signals
- crosspoint faults—the loss or spurious presence of a contact in the internal matrix of connections

These conditions are considered for all input and internal lines in any PLA dedication.

The analyses show that any single PLA fault cannot cause 1-valued minterms to be both removed and added in the sum-of-products outputs, due to the structure of the PLA itself. Thus, this fault effect is always unidirectional in a syndrome count, either increasing or decreasing its value. The only exception to this is where two internal lines are stuck, one at logic 0 and the other at logic 1; here it is now conceivable for 1-valued minterms to be both added and removed from two (or more) outputs, hence resulting in an unchanged total syndrome count across all outputs.

A simple analysis of the latter possibility can be made at the PLA dedication stage. If present, then the final weighting given to the individual syndrome counts to provide the final weighted-syndrome-summation (WSS) test signature can be chosen such that masking of this failing does not occur. Alternatively, it may be appropriate in some cases to use an unused output line connected to an already existing product to achieve a similar total test result. This is a once-and-for-all design activity for a given dedication.

The weighting suggested by Serra and Muzio is to select powers of two for the weights, so that the overall summation can be simply achieved by the bit position to which each output is connected in a single WSS accumulator. This is indicated in Figure 6.49, each m-bit binary word from the m individual outputs of the PLA adding to the number stored in the WSS accumulator, the final summation being interrogated by the output AND detector. The probability of multiple faults not being detected by this binary weighting has been considered [136], and it has been concluded that this possibility is small, the probability decreasing as the number of paths through the PLA increases.

Other PLA test developments include concurrent testing using checker codes (see Section 6.9.4), pseudo-exhaustive testing wherein the primary inputs and the checker lines are partitioned into subsets with exhaustive testing of each subset [138], and built-in self-test proposals which exploit the feature that simple nonlinear feedback shift registers can generate marching 0s and 1s appropriate for testing NOR array structures [139]. Other PLA built-in test proposals include those of Hassan and McCluskey [140] and Treur, Fujiwara and Agarwal [141]. Further details and additional references may be found in Refs. 14–16.

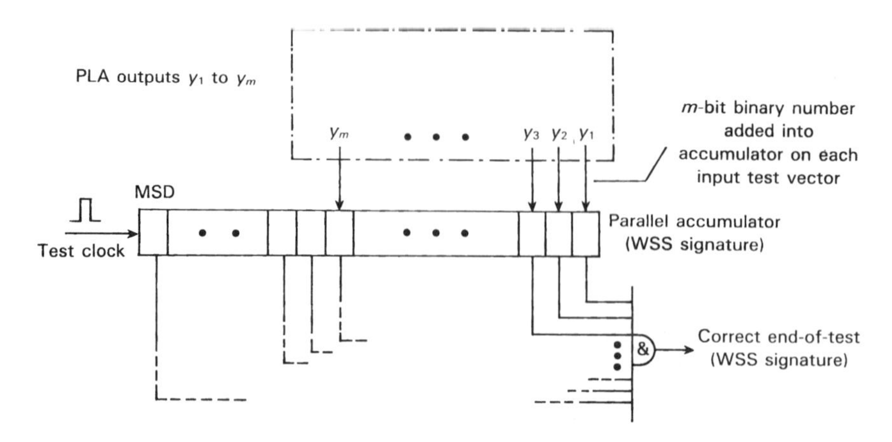

Fig. 6.49 The generation of a binary-weighted syndrome summation WSS from the y_i outputs of a PLA under test. Note that $y_1, \ldots y_n$ need not be routed in the strict power-of-two order shown here.

BIST

Finally, there is research interest in the possibilities of merging internal BIST-type structures with boundary scan for ordinary (non-PLA) digital circuits. A topology such as shown in Figure 6.50 has been proposed by Robinson [142], which claimes to minimize the sequence length necessary for test and to give a reduced silicon area overhead in comparison with previous proposals.

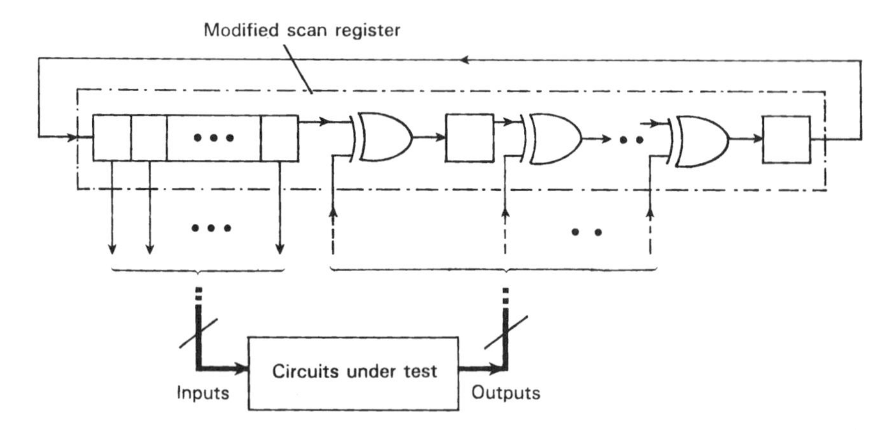

Fig. 6.50 The circular BIST methodology proposed by Robinson, wherein the initial stages of the scan register provide input test vectors and the later stages act as a multiple input signature register (MISR) in the overall test mode.

These and other BIST developments [63,99,143] will continue in order to refine this increasingly necessary form of design-for-test as chip complexity grows and device geometries continue to shrink to low submicron values.

6.9.4 On-Line Checking

As has previously been noted, some form of continuous on-line checking of a digital network avoids all the difficulties of having to generate test patterns, or reconfiguring the network into an alternative test mode. Also, the on-line tests are automatically carried out at the normal working speed of the system. However, the penalty to be paid is a noticeable increase in silicon area to accomodate the additional circuitry.

Duplication of networks has already been mentioned (see Figure 6.10, for example). Clearly, this involves more than 100% additional silicon area overhead, but it may be appropriate under critical circumstances. (Usually networks will be at least triplicated—*triple-modular-redundancy*, or TMR—so that majority voting to determine the correct output state is possible [15,19].) However, here we will briefly reference some on-line checker proposals which have a minimum silicon area overhead and which merely provide a flag output indicating faulty/fault-free conditions, that is, they are fault detecting but not fault correcting.

On-line (otherwise known as 'concurrent' or 'implicit') testing involves some form of information redundancy wherein more output lines are present than are required for normal system operation. The digital signals on the additional output lines are related to the expected healthy output signals by some chosen relationship, and any discrepancy in this relationship will be detected as a circuit fault. This is illustrated in Figure 6.51(a).

One of the commonly proposed coding schemes for the check bits is the Berger code [7,15,144,145]. The standard Berger code involves the use of $\lceil \log 2\ (m + 1) \rceil$ check bits for m data output bits, and where the value of the check bits is the binary count of the number of zeros in the data output, for example, as follows:

Healthy data output	Check bit value	Berger check code
0 0 0 0 0 0 0 0	8	1 0 0 0
0 1 0 1 0 1 0 1	4	0 1 0 0
1 1 1 1 1 1 1 0	1	0 0 0 1

The Berger code generator is designed to generate the appropriate binary numbers from the primary inputs, thus making the output check bits continuously available alongside the normal primary system outputs. The output checker circuit also

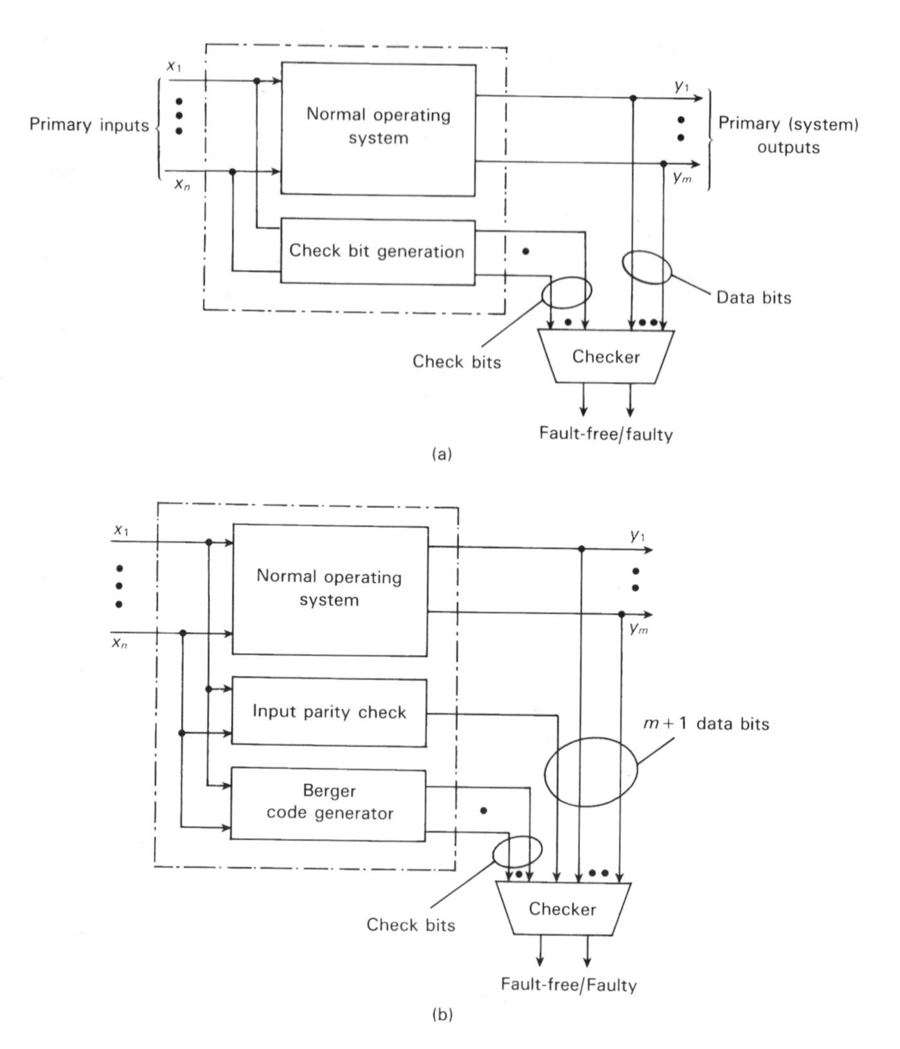

Fig. 6.51 The organization required for concurrent checking, with the check bits being related to the fault-free primary system outputs: (a) the basic concept; note that the check bit generation is ideally a very small percentage of the total circuit complexity; (b) a particular on-line check scheme with primary input parity checking and Berger code check bits related to the fault-free $m + 1$ output bits.

determines this number by simply counting the number of zeros in the data output, which it then compares with the check bits for agreement. Any disagreement indicates some internal circuit fault [7,15,19,144–146].

However, additional security has been proposed by including a parity check on the primary inputs (see Figure 6.51(b)). The Berger code generator now generates an output code which represents the number of zeros in the healthy data output plus this parity bit, which the output checker circuit continuously compares against the $m + 1$ data bits [147].

The particular strength of the Berger code as an on-line checker is for what are termed unidirectional faults in the circuit under test, that is, additional 1-valued minterms are either added to or subtracted from blocks of ones in the y_i outputs under fault conditions. This will be recognized as the classic failure characteristic of PLAs (growth faults, shrinkage faults, etc.), and hence Berger code checking has been particularly advocated for on-line PLA testing [146–148]. The Berger code generator is then made by using additional product and sum lines in the PLA—the dotted boundary in Figure 6.51 now represents a PLA—with the checker circuit as a separate small circuit at the PLA output. The checker circuit itself must be fail-safe [7,15].

There are, however, many other checker codes that can be proposed, particularly for arithmetic or signal processing circuits. Anong these codes are the following:

- residue codes, which involve a mathematical division or other operation considering the fault-free primary outputs as a m-bit number, using the remainder or overflow or other result of the mathematical calculation as the check data
- Reed-Muller codes, which involve Exclusive-OR (linear) relationships
- M-out-of-N codes, which contain N information bits of which M must be logic 0

and several others. This is a very sophisticated area in its own right, and extends from simple on-line go/no-go checking procedures, as in Figure 6.51, right across to error detection and correction procedures for data busses, data transmission in telecommunications, and fail-safe systems. The reader is therefore referred to further publications to amplify the introductory information which we can afford to give here [64,144,149]. For further discussions on the specific area of on-line digital circuit test, good coverage may be found in Russell [15] and also in Refs. 7 and 19.

On-line checking is of importance in an increasing number of applications, and is likely to become more prominent in custom ICs than it has been in the past. The silicon overheads involved in on-line checkers are usually less than may be involved with BIST and other off-line test methodologies, the latter being considered later in Section 6.10. However, as far as is known, no USIC vendor actively encourages the adoption of on-line testing in their available macros, and it is left to the OEM designer to consider how appropriate or otherwise such methodologies are to incorporate into his or her particular custom design.

6.9.5 Cellular Arrays and Special Macros

Cellular Arrays

Cellular arrays are two-dimensional assemblies of (ideally) identical cells which provide specific signal processing capability, or arithmetic or sorting or similar duties. Figure 6.52 illustrates the general architecture, with data flowing in parallel through the array to produce the required output information. A particular attraction of such architectures, which are also known as systolic arrays, is that they can usually be arranged in a very compact silicon layout, with the minimum

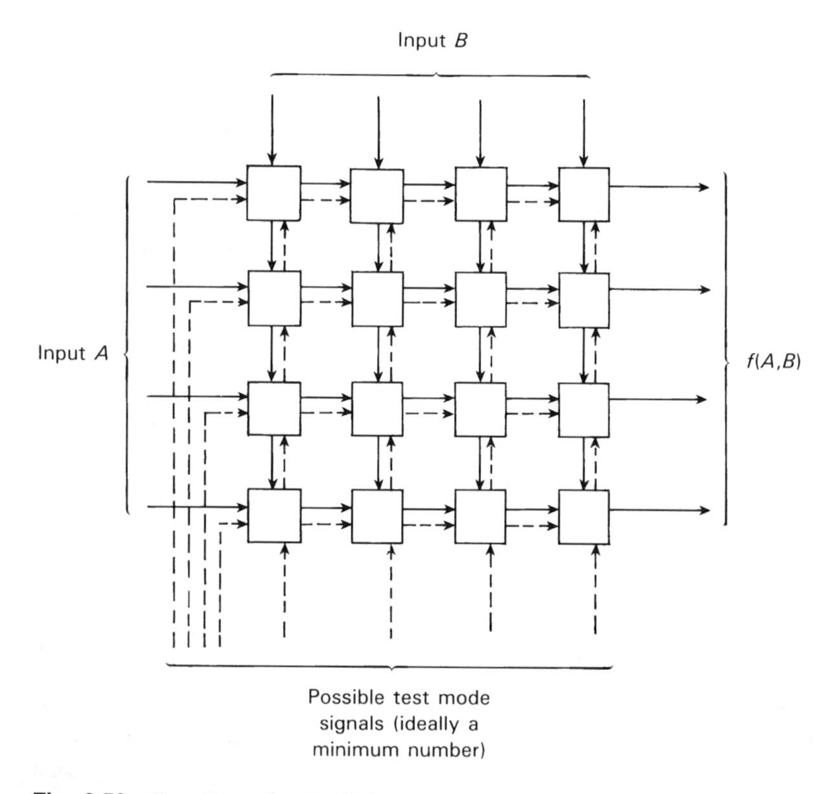

Fig. 6.52 Two-dimensional cellular arrays with parallel data flow through the array structure, and the possible test mode for testing purposes.

of area being necessary for interconnections, hence acheiving a high maximum operating speed.

To avoid exhaustive testing, there needs to be some means of providing controllability and observability of all the cells in the array in order to confirm their functionality. This requirement may be approached by being able to switch each cell from its normal mode to a test mode. In the test mode, the cell may, for example, be made 'transparent' so that test signals flow through unchanged to and from the cell, as shown in the dotted lines in Figure 6.52; other methods may use column and row bypassing logic so that test signals may be routed around cells to and from the appropriate test points.

This area of test remains very largely an ad hoc specialized area, greatly dependent upon the purpose of the array and the cell details. Every design requires to be considered on its own merits and logical structure. However, the regularity of the architecture and the relative simplicity of each cell often leads to the ablity to formulate relatively simple tests based upon the flow of signals through the array [7], which give exceptionally high fault coverage. No automatic test pattern generation (ATPG) is usually considered, and as far as is known, no built-in self-test has yet been used for the test of these strongly structured arrays.

Further details of this area may be found in the published literature [150–152], and in many regular international conferences on circuit design and signal processing which embrace this subject.

Special Macros

Recently, research has been given to combinational functions which have certain functional properties enabling them to be easily tested. Among such circuits have been *self-dual logic primitives* [113], a self-dual function being defined as

$$f(x_1, x_2, \ldots, x_n) \equiv \overline{f(\bar{x}_1, \bar{x}_2, \ldots, \bar{x}_n)}$$

where every x_i input variable is complemented in the right-hand side of the above expression together with overall complementation. This holds true for all 2^n input combinations. A simple example of a self-dual function is $x_1x_2 + x_1x_3 + x_2x_3$, as may readily be shown by determining that the minterms of the function and its self-dual are identical.

The particular attraction of the self-dual function is that if any input vector is applied to the function, such as $x_1\bar{x}_2x_3$ in the above case, followed by the vector with all literals complemented, for example, $\bar{x}_1x_2\bar{x}_3$ in this case, then by definition the function output will switch from logic 0 to logic 1 or vice versa. Further, if a network of self-dual functions is made, it is an easy exercise to show that if any complete input vector $\dot{x}_1\dot{x}_2 \ldots \dot{x}_n$ is applied to the primary inputs where $\dot{x}_i$ is x_i or $\bar{x}_i$, followed by the complemented vector $\bar{\dot{x}}_1\bar{\dot{x}}_2 \ldots \bar{\dot{x}}_n$, then under fault-

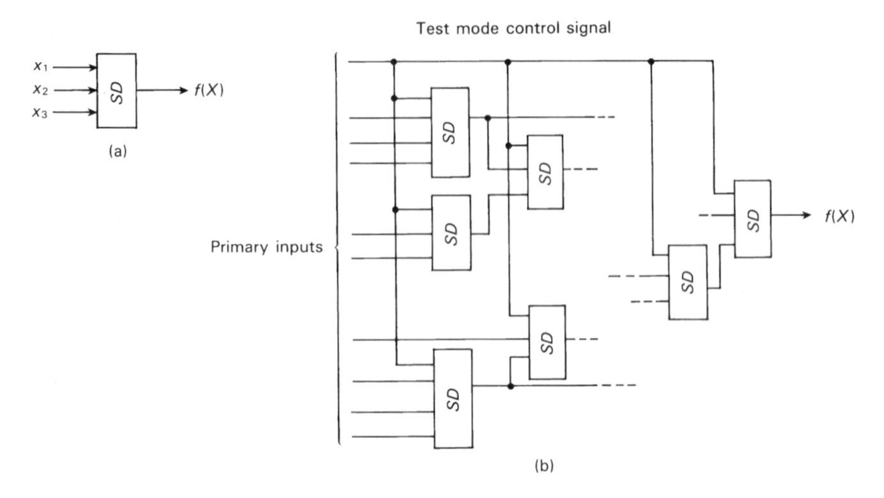

Fig. 6.53 The concept of using self-dual functions for test purposes: (a) example self-dual function e.g., $\overline{x_1x_2 + x_2x_3 + x_1x_3}$; (b) network of self-dual functions, where the logic value on all internal nodes and the output node changes when any existing input vector is replaced by its dual.

free conditions all the self-dual circuit outputs throughout the network will switch state, including the final output(s). Hence, if a combinational logic network is composed of self-dual circuits as shown in Figure 6.53, controllability and observability of the nodes in the network will be greatly eased.

A self-dual function such as $f(X) = \overline{x_1x_2 + x_1x_3 + x_2x_3}$ may therefore be suggested as an universal logic building-block, since 2-input NOR and NAND relationships can be obtained by setting one of the inputs to 0 or 1. The normal circuit requirements would be designed with all cells in this logic state. For test purposes, any one input test vector comprising all the n primary inputs and this common control mode signal could be applied, followed by a second test vector consisting of the complements of these $n + 1$ inputs. The stuck-at fault coverage with just two such vectors is reported to be high, but it may require further pairs of vectors to ensure 100% fault coverage. Details of this work may be found in Ref. 153, and is indicative of research work in logic primitives which are more complex than our more familiar Boolean logic gates. (A difficulty with more complex primitives is, of course, that the usual methods of logic design based upon Boolean algebra and relationships are generally inadequate to handle more complex primitives. This difficulty is partially resolved by the use of spectral design concepts which can manipulate complex logical relationships [113], but it currently remains outside the mainstream of digital logic design. However, the adoption recently of more complex cells in CPLDs by Zilinx/Algotronix and Motorola/Pilkington and others may add some impetus to these considerations.)

Additional work on totally self-checking (TSC) modules and their use in totally self-checking combinational circuits may also be found (for example, see Pagey and Al-Khalili and the further references contained therein [154]).

6.10 THE SILICON AREA OVERHEADS OF DFT

Because of the commercial difficulties in quoting actual financial costs, the 'cost' of DFT is usually measured in silicon area overheads for the given feature or methodology. Only the IC vendor can translate this into true money terms, and then it may vary widely depending upon a range of commercial factors.

Most of the published information on silicon overheads deals with scan-path and BIST techniques; ad hoc techniques are not amenable to formal analysis since by definition they may vary with each design. Therefore, overheads in formal DFT methodologies on the basis of additional devices (transistors) and/or logic gates form the basis of most of the calculated overhead estimates.

One of the first papers in this area was that of Ohletz, Williams and Mucha [155]. Like the majority of subsequent papers, it considers that the combinational logic in a network does not materially increase with the introduction of DFT, but the storage (sequential) part becomes more complex. The usual calculation for scan and self-test silicon overheads therefore introduces the factor K, where K is defined as:

$$K = \frac{\text{No. of combinational logic gates in the circuit}}{\text{No. of latches (flip-flops) in the circuit}}$$

The value of K generally varies between 5 for latch-intensive designs and 12–20 for less latch-intensive circuits. Alternative calculations for the silicon area overheads in PLA architectures may also be found, but we will specifically comment upon these later.

The % silicon overhead for scan-path and BIST may be calculated as follows. Assuming CMOS technology, let:

P = number of transistor-pairs per D-type circuit
L = number of additional transistor-pairs per D-type circuit to form a scan-path or a BILBO register stage
F = average number of transistor-pairs per combinational logic gate, where a gate is conventionally taken as a 2-input NAND or NOR circuit

The total number of transistor-pairs per D-type circuit without DFT is therefore $\{P + KF\}$, and with the introduction of DFT is increased by L to $\{P + L +$

$KF\}$. Therefore, the percentage transistor-pair overhead, which may be taken as roughly equal to the silicon area overhead, is given by:

$$\frac{L}{P + KF} \times 100\%$$

Some publications count the number of transistors rather than transistor-pairs, which does not affect the calculation, but may refer to a master-slave D-type register as two latches when considering K. Here we are taking a master-slave circuit as one circuit. The precise values taken for P and L strongly affect the final percentage values.

The count of the number of transistors or transistor-pairs will vary with the technology and perhaps with the detailed circuit and macro design. However, as an example, consider the details given in Figure 6.54, which shows a CMOS realization for a D-type master-slave circuit and the additional circuitry for the BILBO stage of Figure 6.32; from this circuit detail we may determine the following numbers:

Circuit	No. of transistor-pairs
D-type circuit (see Figure 6.54(a))	10
Additional BILBO complexity per stage (see Figure 6.54(b))	8
Additional one-off first-stage BILBO multiplexer (see Figure 6.32)	5
Additional m-input LFSR XOR feedback, = (m − 1) 2-input XORs in cascade (see Figure 6.32)	5 (m − 1)

Thus the total transistor-pair count for n D-type circuits is $10n$, and for n LFSR stages is $\{10n + 8n + 5 + 5\,(m - 1)\}$, $\simeq 18n$ as m is small in comparison with the remaining terms. A similar calculation for n CALBO stages (see Figure 6.36) gives a count of 26 n assuming an equal count of type 90 and type 150 circuits, but more recent work on CALBO has indicated that not more than two type 150 circuits are required for any length register, the remainder being type 90 circuits [156], and hence this result of 26 n is somewhat high. These results are shown plotted in Figure 6.55.

These results should not be taken as exact, because as detailed circuit design and layout will influence the precise results. However, they do indicate the form of analysis which may be carried out, and show that there is usually a considerable silicon area penalty involved in DFT techniques. Further figures may be found in Ref. 155 for scan-path circuits, and a frequent general quote is that design-for-test may involve somewhere between a 10% and a 30% silicon area penalty.

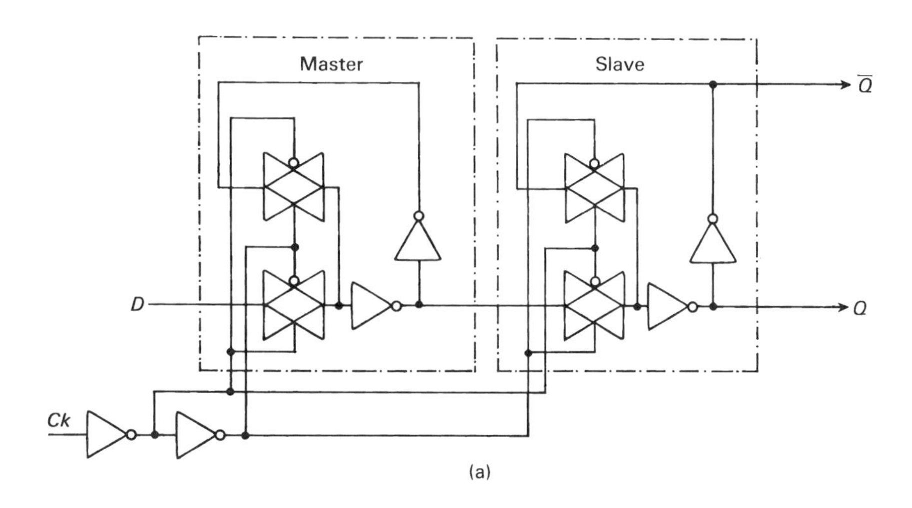

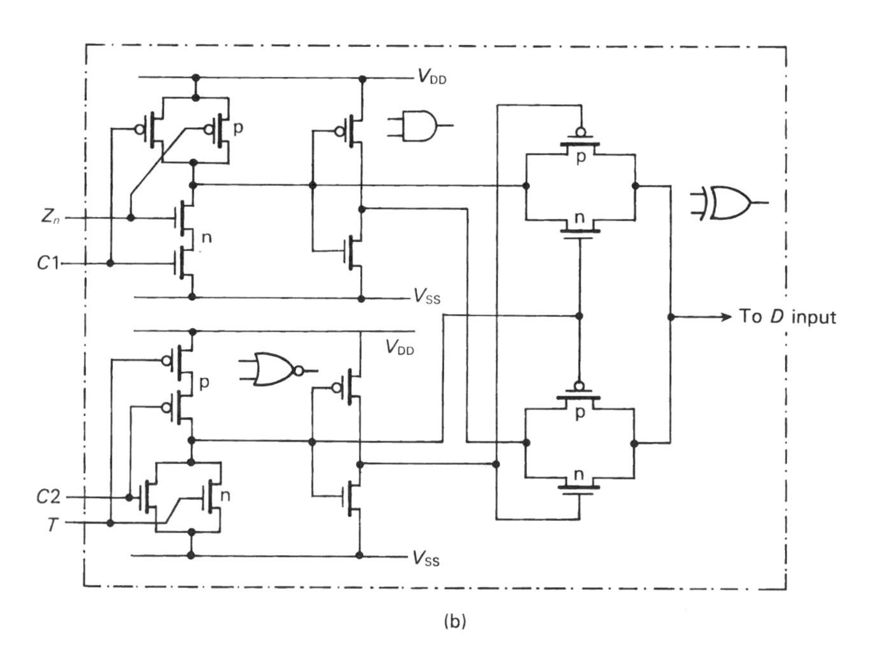

Fig. 6.54 CMOS circuit realizations: (a) master-slave flip-flop; (b) the addition to (a) necessary to form a BILBO LFSR stage.

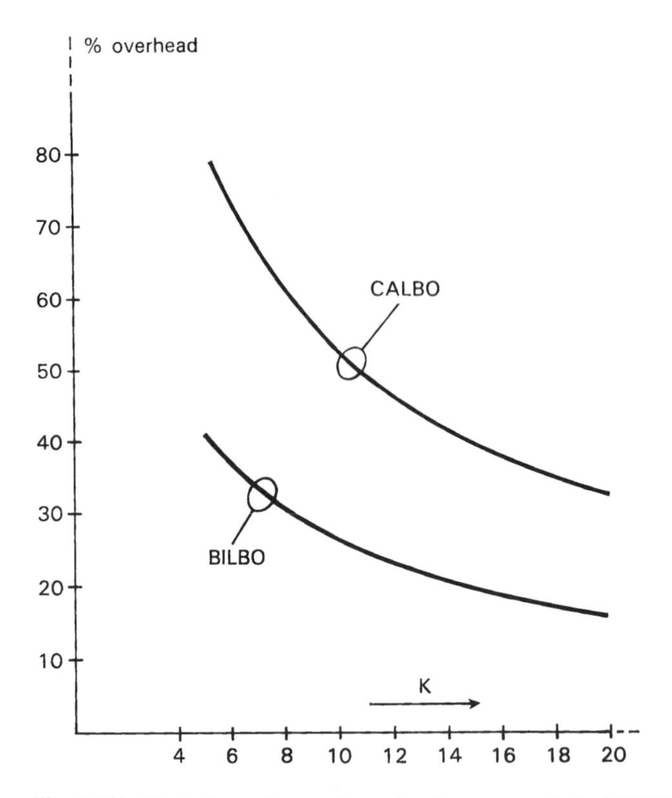

Fig. 6.55 Typical transistor-pair overheads necessary in both BILBO and CALBO compared with the use of D-type storage circuits in non-DFT architectures. Scan-path DFT may give somewhat lower overhead values than BILBO.

The overheads involved in incorporating some form of DFT in programmable logic array structures has also been considered. These usually involve a consideration of how many additional columns are necessary in the AND array and how many additional rows are necessary in the OR array, plus any extra check circuitry. Results such as those shown in Figure 6.56 may be found in the published literature [48,157–159].

Finally, there has also been research work on the development of knowledge-based CAD tools which can examine all the financial aspects of design and test at the early stages of a new design, in order to forecast the likely costs of incorporating different forms of built-in test, or even not to include any. This is the subject area termed *economics modelling*, and as well as considering the actual silicon costs it can, in the more comprehensive models, take into consideration the *lifetime costs* of the equipment, that is, to include the likely costs of

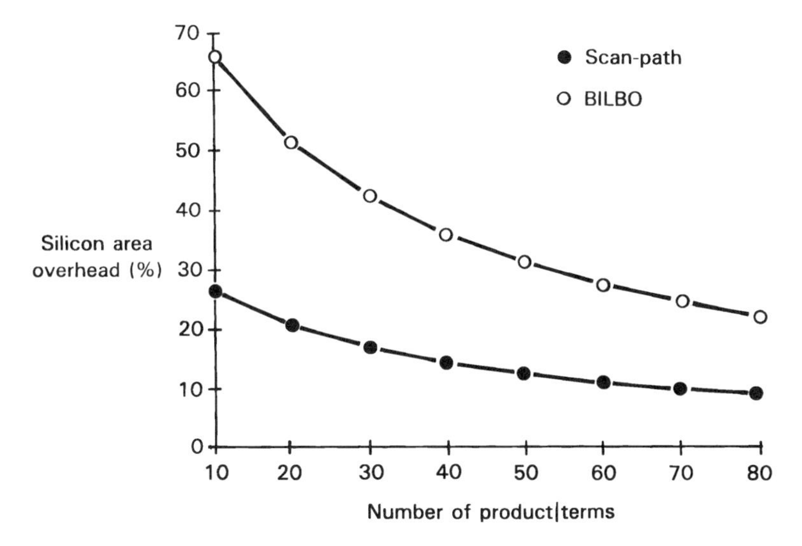

Fig. 6.56 Area overhead predictions for scan-path and BILBO methodologies applied to PLA architectures.

maintenance and repair over the useful lifetime of the circuit or system. An objective of this comprehensive work is to attempt to provide a test strategy planning tool which may be used at the initial design stage, to point to a financially optimum test strategy for the particular circuit or system under consideration. The work of Zhu and Breuer and of Ambler et al. may be noted in this context [160–163], but a problem with all such economic modelling remains the difficulty of finding accurate numbers to put into the (often complex) financial equations [7]. Nevertheless, such CAD tools should be useful to many companies when they have been honed to provide good financial forecasts.

6.11 SUMMARY

It will be apparent from preceding sections of this chapter that testing is a critical consideration for VLSI integrated circuits, due primarily to the limited accessibility provided by the available I/Os. The higher the value of internal gate count per I/O pin, the more difficult testing procedure will be unless some form of DFT is built in to the design.

For combinational circuits, the basic choice is between some fully exhaustive test sequence or a nonexhaustive test sequence. The former is a functional test which avoids having to consider the exact internal design of the circuit or the generation of some test pattern generation, but in general cannot be used unless the circuit is very small or is partitioned into accessible smaller parts for test purposes. The latter is a nonfunctional test where the test vectors are determined by some consideration of the likely faults within the circuit, automatic test pattern generation (ATPG) programs being used to determine some minimum set of test vectors which will give acceptable fault coverage. The stuck-at fault model has been the classic and most successful fault model in this work.

However, as has been seen, ATPG programs themselves begin to run into difficulties as circuit size escalates to the tens of thousands of logic gates, and also cannot easily accomodate sequential circuits. As a result, it has become necessary to consider problems of test at the very start of the circuit design process, and to introduce, where necessary, design-for-test circuit enhancements to ease the final test problem. Controllability and observability are the key requirements, and this is acheived by partitioning and/or by the use of some form of shift register which enables test data to be shifted into and out from the circuit when in test mode.

Since the majority of real-life circuits are not purely combinational circuits but contain sequential logic, all DFT techniques are designed to cater to this situation. The normal and possibly the only way to do so is to segregate the sequential (storage) elements from the combinational logic gates when in test mode, and to test each separately. We therefore have scan-test and built-in self-test as the now widely accepted means of test for complex digital circuits. Reconfiguration of the sequential elements into shift register configurations when under test is central to these methodologies.

Pseudo-random test sequences feature strongly in built-in self tests rather than binary or other similar sequences. As we have seen, autonomous maximum-length shift registers (LFSRs) are used in the built-in logic block observation (BILBO) test methodology to generate pseudo-random test sequences, the principal advantage being the ease of reconfiguration of the sequential elements into a LFSR configuration. An alternative to BILBO is CALBO, which uses an autonomous maximum-length cellular automaton reconfiguration instead of a LFSR to generate pseudo-random test vectors—this has not, to date, been used by custom IC vendors as far as is known, since it requires a somewhat greater silicon area to introduce compared with BILBO without a proven increase in testing efficiency. All pseudo-random test sequences tend to be almost if not completely exhaustive sequences for each circuit partition, and no test vector generation problems are therefore involved. Several variations of LFSR test vector generation have been introduced, including multiple-LFSR configurations.

Strongly structured architectures such as memory circuits, PLDs and other array architectures have been seen to have their own particular test methodologies based upon their structure, usually involving some consideration of their most likely internal faults rather than the stuck-at fault model. When built into custom ICs as large macros, these circuits must normally be provided with appropriate accessibility to enable the recognized form of test to be done.

Purely analog ICs do not generally present extreme testing difficulties, being testable with conventional analog test instrumentation, nowadays usually under some computer control. However, the problem of test of mixed analog/digital ICs is far more difficult, and remains the biggest problem area for circuit and system testing. Partitioning of the analog and the digital parts is normally the key requirement, with testing of the two parts being undertaken separately. However, ongoing research into the possibilities of some form of self-test using the two parts to form part of the test structure for the other may lead to some more formal means of mixed-signal testing; digital signal processing (DSP) forms a particular part of this research.

Continuous on-line testing is divided into two categories, namely, for fault detection and correction and for fault detection only. The former normally involves complete circuit or module redundancy, with triple-modular-redundancy (TMR) being a familiar case. The latter involves some additional outputs to the normal primary outputs, the signals on which have a fixed relationship to the fault-free signals on the primary outputs, any discrepancy in this relationship being indicative of a circuit fault or faults. This whole detailed subject area is the province of coding theory, and forms an essential part of critical and safety-critical systems.

Finally, we have seen that boundary scan has been introduced to solve the problems of accessibility of the I/O pins of ICs after assembly on the final printed-circuit board. This is now very specifically defined by the IEEE Standard 1149/1. Boundary scan provides not only accessibility to test the individual assembled ICs, but also and equally important the interconnect between all ICs and the final PCB I/Os. Again, it is the shift register configuration that provides the internal accessibility without the need for any appreciable number of additional test pins, but again, one has to accept an increased time to undertake all the required load-and-test operations as the penalty to be paid for any form of serial test procedure in comparison with the situation where all required test points (nodes) are freely available.

The OEM system designer involved in custom IC design activities will encounter many of these factors, and should therefore be knowledgeable on the subject. It is clearly a very broad area: here we have only been able to survey the subject and have not considered in any detail the underlying mathematics of many of the areas. For more detailed information and discussion several texts specifically dealing with test should be consulted [7,15,16,19,33,144,163,164]. Finally, continuing attention should be paid to journals such as *Computer Design, IEEE Design and Test of Computers, IBM Journal of Research, Integration—*

the VLSI Journal, and (particularly) the *Journal of Electronic Testing: Theory and Applications (JETTA)* for on-going developments in this area. Some final comments will be made later in Chapter 8.

REFERENCES

1. Parker, K.P., *Integrating Design and Test: Using CAE Tools for ATE Programming*, IEEE Computer Society Press, CA, 1987
2. Bateson, J., *In-Circuit Testing*, Van Nostrand Reinhold, NY, 1985
3. Bennetts, R.G., *Introduction to Digital Board Testing*, Crane-Russak, NY, 1985
4. Naish, P. and Bishop, P., *Designing ASICs*, Wiley, NY, 1988
5. Trontelj, J., Trontelj, L. and Shenton, G., *Analog Digital ASIC Design*, McGraw Hill, UK, 1989
6. The Open University, *Design for Testability*, Microelectronics for Industry Publication PT505 DFT, The Open University, UK, 1988
7. Hurst, S.L.,*VLSI Testing: Digital and Mixed Analogue/Digital Techniques*, Institution of Electrical Engineers (IEE) Publications, UK, 1997
8. Wilkins, B.R., *Testing Digital Circuits*, Van Nostrand Reinholt, UK, 1986
9. Bennetts, R.G., *Design of Testable Logic Circuits*, Addison-Wesley, UK, 1984
10. Kularatna, N., *Modern Electronic Test and Measuring Instruments*, Institution of Electrical Engineers (IEE) Publications, UK, 1996
11. Higgins, R.J., *Digital Signal Processing in VLSI*, Prentice Hall, NJ, 1990
12. Richardson, N., E-beam probing for VLSI circuit debug, *VLSI System Design*, Vol. 8, No. 9, 1987, pp. 24–29 and 58
13. Bate, J.A. and Miller, D.M., Exhaustive testing of stuck-open faults in CMOS combinational circuits, *Proc. IEE*, Vol. 135, 1988, pp. 10–16
14. Miller, D.M. (Ed.), *Developments in Integrated Circuit Testing*, Academic Press, UK, 1987
15. Russell, G. and Sayers, I.L., *Advanced Simulation and Test Methodologies for VLSI Design*, Van Nostrand Reinhold, UK, 1989
16. Williams, T.G. (Ed.), *VLSI Testing*, North-Holland, Amsterdam, 1986
17. Roth, J.P., Diagnosis of automata failure: a calculus and a method, *IBM J. Research and Development*, Vol. 10, 1976, pp. 278–291
18. Miczo, A., *Digital Logic Testing and Simulation*, Harper and Row, NY, 1986
19. Lala, P.K., *Fault Tolerant and Fault Testable Hardware Design*, Prentice Hall, UK, 1986
20. Kirkland, T. and Mercer, R., Algorithms for automatic test pattern generation, *IEEE Design and Test of Computers*, Vol. 5, No. 3, 1988, pp. 43–45
21. Goel, P., RAPS test pattern generation, *IBM Technical Disclosure Bulletin*, Vol. 21, 1978, pp. 2787–2791
22. Goel, P. and Rosales, B.C., PODEM-X: an automatic test pattern generation system for VLSI logic structures, *Proc. IEEE 18th. Design Automation Conf.*, 1981, pp. 260–268

23. Fujiwara, H. and Shimona, T., On the acceleration of test generation algorithms, *IEEE Trans. Computers*, Vol. C.32, 1983, pp. 1137–1144
24. Funatsu, S. and Kawasi, M., An automatic test generation system for large digital circuits, *IEEE Design and Test of Computers*, Vol. 2, No. 5, 1985, pp. 54–60
25. Ghosh, S., Behavioural-level fault simulation, *IEEE Design and Test of Computers*, Vol. 5, No. 3, 1988, pp. 31–42
26. McCluskey, E.J., Comparing causes of IC failures, *Center for Reliable Computing Report No. 86-2*, Stanford University, CA, 1986
27. Breuer, M.A. and Friedman, A.D., Functional level primitives in test generation, *IEEE Trans. Computers*, Vol. C.29, 1983, pp. 223–235
28. Kawai, M., Shibano, M., Funatso, S., Kato, S., Kurobe, T., Oakawa, K. and Sasaki, T., A high level test pattern generation algorithm, *Proc. IEEE Int. Test Conf.*, 1983, pp. 346–352
29. Lin, T. and Su, S.Y.H., The S-algorithm: a promising approach for systematical functional test generation, *IEEE Trans. Computer Aided Design*, Vol. CAD.4, 1985, pp. 250–263
30. Levendal, Y.H. and Menon, P.R., Test generation algorithms for non-procedural computer hardware description languages, *IEEE Digest Int. Symp. on Fault Tolerant Computers*, 1981, pp. 200–205
31. Sasaki, T., Yamada, A., Kato, S., Nakazawa, T., Tomita, K. and Namizu, N., MIXS: a mixed level simulator for large digital system verification, *Proc. IEEE 17th Design Automation Conf.*, 1980, pp. 626–633
32. Miczo, A., A sequential ATPG: a theoretical limit, *Proc. IEEE Int. Test Conf.*, 1983, pp. 143–147
33. Bardell, P.W., McAnney, W.H. and Savir, J., *Built-in Test for VLSI: Pseudo-random Techniques*, Wiley, NY, 1987
34. Jain, S.K. and Stroud, C.E., Built-in self-testing of embedded *memories, IEEE Design and Test of Computers*, Vol 3, No. 5, 1986, pp. 27–37
35. Sun, Z. and Wang, L-T., Self-testing embedded RAMs, *Proc. IEEE Int. Test Conf.*, 1984, pp. 184–186
36. Nair, R., Thatte, S.M. and Abraham, J.A., Efficient algorithms for testing semiconductor random access memories, *IEEE Trans. Computers*, Vol. C.27, 1978, pp. 572–576
37. Suk, D.S. and Reddy, S.M., A march test for functional faults in semiconductor random access memories, *IEEE Trans. Computers*, Vol. C.30, 1981, pp. 982–985
38. Marinescu, M., Simple and efficient algorithms for functional RAM testing, *Proc. IEEE Int. Test Conf.*, 1982, pp. 236–239
39. Prince, B., *Semiconductor Memories*, Wiley, UK, 1995
40. Bardell, P.H. and McAnney, W.H., Built-in self-test for RAMs, *IEEE Design and Test of Computers*, Vol. 5, No. 5, 1988, pp. 29–36
41. Grnodis, A.J. and Hoffman, D.E., 250 MHz advanced test system, *IEEE Design and Test of Computers*, Vol. 5, No. 2, 1988, pp. 24–35
42. Venkateswaran, R. and Mazumder, P., A survey of DA techniques for PLD and FPGA based systems, *Integration, the VLSI Journal*, Vol. 17, 1994, pp. 191–240
43. Smith, J.E., Detection of faults in programmable logic arrays, *IEEE Trans. Computers*, Vol. C.28, 1979, pp. 845–853

44. Ramanatha, K.S. and Biswas, N.N., A design for testability of undetectable cross-point faults in programmable logic arrays, *IEEE Trans. Computers*, Vol. C.32, 1983, pp. 551–557
45. Ostapka, D.L. and Hong, S.J., Fault analysis and test generation for programmable logic arrays (PLAs), *Proc. IEEE Fault Tolerant Computing Conf.*, 1978, pp. 83–89
46. Eichelberger, E.B. and Lindbloom, E., Heuristic test-pattern generator for programmable logic arrays, *IBM Journal of Research and Development*, Vol. 24, No. 1, 1980, pp. 15–22
47. Zhu, X.-A. and Breuer, M.A., Analysis of testable PLA designs, *IEEE Design and Test of Computers*, Vol. 5, No. 4, 1988, pp. 14–18
48. Agrawal, V.K., Multiple fault detection in programmable logic arrays, *IEEE Trans. Computers*, Vol. C.29, 1980, pp. 518–522
49. Jacob, J. and Biswas, N.N., Further comments on 'Detection of faults in programmable logic arrays,' *IEEE Trans. Computers*, Vol. C.39, 1990, pp. 155–156
50. Thatte, S.M. and Abraham, J.A., Test generation for microprocessors, *IEEE Trans. Computers*, Vol. C.29, 1980, pp. 429–441
51. Abraham, J.A. and Parker, K.P., Practical microprocessor testing: open and closed loop approaches, *Proc IEEE COMPCON*, 1981, pp. 308–311
52. Bellon, C., Automatic generation of microprocessor test patterns, *IEEE Design and Test of Computers*, Vol. 1, No. 1, 1984, pp. 83–93
53. Daniels, R.G. and Bruce, W.C., Built-in self-test trends in Motorola microprocessors, *IEEE Design and Test of Computers*, Vol. 2, No. 2, 1985, pp. 64–71
54. Sridhar, T. and Hayes, J.P., A functional approach to testing bit-sliced microprocessors, *IEEE Trans. Computers*, Vol. C.30, 1981, pp. 563–571
55. Perry, T.S., Intel's secret is out, *IEEE Spectrum*, Vol. 26, No. 4, 1989, pp. 21–27
56. Moore, E.F. (Ed.), *Sequential Machines: Selected Papers*, Addison-Wesley, UK, 1964
57. Eichelberger, E.B. and Williams, T.W., A logic design structure for LSI testability, *IEEE Proc. Design Automation Conf.*, 1977, pp. 462–468
58. Williams, T.W. and Parker, K.P., Design for testability—a survey, *Proc. IEEE*, Vol. 71, 1983, pp. 98–112
59. Fumatsu, S., Wakatsuki, N. and Yamada, A., Easily-testable design of larger digital networks, *NEC Journal of Research and Development*, No. 54, 1979, pp. 49–55
60. Stewart, J.H., Future testing of large LSI cards, *IEEE Digest Semiconductor Test Symp.*, 1977, pp. 6–15
61. Ando, H., Testing VLSI with random access scan, *IEEE Digest COMPCON*, Spring 1980, pp. 50–52
62. Agrawal, V.D., Cheng, K.T., Johnson, D.D. and Lin, T., Designing circuits with partial scan, *IEEE Design and Test of Computers*, Vol. 5, No. 2, 1988, pp. 8–15
63. Martlett, R.A. and Deer, J., RISP (Reduced Intrusion Scan Path) methodology, *J. Semiconductor ICs*, Vol. 6, No. 2, 1988, pp. 15–18
64. Anderson, A. and Lee, P., *Fault Tolerance: Principles and Practice*, Prentice Hall, NJ, 1980
65. von Neumann, J., Probabilistic logic and synthesis of reliable organisms from unreliable components, *Annals of Mathematical Studies*, No. 34, 1, pp. 43–98

66. Breuer, M.A. and Friedman, A.D., *Diagnosis and Reliable Design of Digital Systems*, Pitman, NY, 1977
67. Jensen, F., *Electronic Component Reliability*, Wiley, NY, 1995
68. Hewlett-Packard, *A Designer's Guide to Signature Analysis*, Application Note 222, Hewlett-Packard Co., CA, 1977
69. Williams, T.W., Daehn, W., Gruetzner, M. and Stark, C.W., Aliasing errors in signature analysis registers, *IEEE Design and Test of Computers*, Vol. 4, No. 2, 1987, pp. 39–45
70. Beucler, F. and Manner, M.J., 'HILDO,' a highly integrated logic device observer, *VLSI Design*, Vol. 5, No. 6, 1984, pp. 88–96
71. Cosgrave, B., The UK 5000 gate array, *Proc. IEE*, Vol. 132.G, 1985, pp. 90–92
72. Gelsinger, P.P., Built-in self-test of the 80386, *Proc. IEEE Conf. on Computer Design*, 1986, pp. 169–173
73. Paraskeva, M., Knight, W.G. and Burrows, D.F., A new test structure for VLSI self-test: the structured test register (STR), *IEE Electronic Letters*, Vol. 21, 1985, pp. 856–857
74. Williams, T.W., Walther, R.G., Bottorff, P.S. and DasGupta, S., Experiment to investigate self-testing techniques in VLSI, *Proc. IEE*, Vol. 132.G, 1985, pp. 105–107
75. Maunder, C.M., The status of IC design-for-tesability, *J. Semiconductor ICs*, Vol. 6, 1989, pp. 25–29
76. Stroud, C.E., An automated BIST approach for general sequential logic synthesis, *Proc. IEEE Design Automation Conf.*, 1988, pp. 3–8
77. Illman, R. and Clarke, S., Built-in self-test of the Macrolan chip, *IEEE Design and Test of Computers*, Vol. 7, 1990, pp. 29–40
78. Bardell, P.H., Design considerations for parallel pseudo random pattern generators, *J. Electronic Testing: Theory and Application*, Vol. 1, No. 1, 1990, pp. 73–87
79. Wolfran, S., Random sequence generation by cellular automata., *Advances in Applied Mathematics*, Vol. 7, 1986, pp. 127–169
80. Thanailakis, O.O. and Card, H.C., Group properties of cellular automata and VLSI applications, *IEEE Trans. Computers*, Vol. C.35, 1986, pp. 1013–1024
81. McCleod, R.D., Hortensius, P.D., Pries, W. and Card, H.C., Design for testability using logic block observers based upon cellular automata, *Proc. IEEE Trans. Computers*, Vol. C.38, 1989, pp. 1466–1473
82. Serra, M., Slater, T., Muzio, J.C. and Miller, D.M., The analysis of linear cellular automata and their aliasing properties, *IEEE Trans. Computer Aided Design*, Vol. CAD.9, 1990, pp. 769–778
83. Serra, M., Miller, D.M. and Muzio, J.C., Linear cellular automata and LFSRs are isomorphic, *Record 3rd. Technical Workshop: New Directions for IC Testing*, Canada, 1988, pp. 213–223
84. Serra, M., Slater, T., Muzio, J.C. and Miller, D.M., The analysis of one-dimensional linear cellular automata and their aliasing properties, *IEEE Trans. Computer Aided Design*, Vol. CAD.9, 1990, pp. 767–778
85. Slater, T. and Serra, M., *Tables of Linear Hybrid 90/150 Cellular Automata*, University of Victoria, Dept. of Computer Science Report No. DCS-105-IR, January 1989

86. Daehn, W. and Mucha, J., A hardware approach to self-testing of large programmable logic arrays, *IEEE Trans. Computers*, Vol. C.30, 1981, pp. 829–933
87. Tamir, Y. and Sequin, C.H., Design and application of self-testing comparitors implemented in MOS technology, *IEEE Trans. Computers*, Vol. C.33, 1984, pp. 493–506
88. Khakbaz, J. and McCluskey, E.J., Concurrent error detction and testing for large PLAs, *IEEE Trans. Electronic Devices*, Vol. ED.29, 1982, pp. 756–764
89. Chen, C.Y., Fuchs, W.K. and Abraham, J.A., Efficient concurrent error detection in PLAs and ROMs, *Proc. IEEE ICCD*, 1985, pp. 525–529
90. Fujiwara, H. and Kinoshita, K., A design of programmable logic arrays with universal tests, *IEEE Trans. Computers*, Vol. C.30, 1981, pp. 823–828
91. Saluja, K.K., Kinoshita, K. and Fujiwara, H., An easily testable design of programmable logic arrays for multiple faults, *IEEE Trans. Computers*, Vol. C.32, 1983, pp. 1036–1046
92. Hong, S.J. and Ostpako, D.L., FITPLA: a programmable logic array for function-dependent testing, *Proc. IEEE 10th. Fault Tolerant Computing Symp.*, 1980, pp. 131–136
93. Khabaz, J., A testable PLA design with low overhead and high fault coverage, *IEEE Trans. Computers*, Vol. C.33, 1984, pp. 743–745
94. LSI Logic, *Databook and Design Manual*, LSI Logic Corporation, CA, 1986
95. IEEE, *Standard 1149.1–1990, Standard Test Access Port and Boundary Scan Architecture*, Institute of Electrical and Electronics Engineers, NY, 1990, and *Supplement P.1149–1a*, 1993
96. Maunder C., Boundary scan: a framework for structured design-for-test, *Proc. IEEE Int. Test Conf.*, 1987, pp. 714–723
97. Bleeker, H., van den Eijnden, P. and de Jong, F., *Boundary-Scan Test: A Practical Approach*, Kluwer Academic Publishers, MA, 1993
98. Wagner, P.T., Interconnect testing with boundary-scan, *Proc. IEEE Int. Conf.*, 1987, pp. 52–57
99. Scholz, H.N., Tulloss, R.E., Yau, C.W. and Wach, W. ASIC implementations of boundary-scan and built-in self-test (BIST), *J. Semicustom ICs*, Vol. 6, No. 4, 1989, pp. 30–38
100. Plessey Semiconductors, *JTAG/BIST in ASICs*, Application Note AN.87, GEC Plessey Semiconductors, UK, October 1987
101. Dettmer, R., JTAG: setting the standard for boundary-scan testing, *IEE Review*, Vol. 32, 1989, pp. 49–52
102. Andrews, W., JTAG works to standardize chip, board and system self-test, *Computer Design*, Vol. 28, No. 13, 1989, pp. 28–31
103. Akers, S.B., In the use of linear sums in exhaustive testing, *Proc. IEEE Design Automation Conf.* 1895, pp. 148–153
104. Tang, T. and Cheng, C.L., Logic test patterns using linear codes, *IEEE Trans. Computers*, Vol. C.33, 1984, pp. 845–850
105. McCluskey, E.J., Verification testing: a pseudo exhaustive test technique, *IEEE Trans. Computers*, Vol. C.33, 1984, pp. 541–546
106. Agrawal, V.K., Increasing effectiveness of built-in testing by output data modification, *Proc. IEEE 13th. Int. Symp. on Fault Tolerant Computing*, 1983, pp. 227–233

107. Saluja, K.K. and Karpovsky, M., Testing computer hardware through data compression in space and time, *Proc. IEEE Int. Test Conf.*, 1983, pp. 83–88
108. Hurst, S.L., Use of linearisation and spectral techniques in input and output compaction testing of digital networks, *Proc. IEE*, Vol. 136-E, 1989, pp. 48–56
109. Savir, J., Syndrome testable design of combinational circuits, *IEEE Trans. Computers*, Vol. C.29, 1980, pp. 442–551 and correction pp. 1012–1013
110. Savir, J., Syndrome testing of 'syndrome untestable' combinational circuits, *IEEE Trans. Computers*, Vol. C.30, 1981, pp. 606–608
111. Markowsky, G., Syndrome-testability can be achieved by circuit modification, *IEEE Trans. Computers*, Vol. C.30, 1891, pp. 604–606
112. Barzilia, Z., Savir, J., Markoswsky, G. and Smith, M.G., The weighted syndrome sums approach to VLSI testing, *IEEE Trans. Computers*, Vol. C.30, 1981, pp. 996–1000
113. Hurst, S.L., Miller, D.M. and Muzio, J.C., *Spectral Techniques in Digital Logic*, Academic Press, NY, 1985
114. Hayes, J.P., Transition count testing of combinational circuits, *Trans. IEEE Computers*, Vol. C.25, 1976, pp. 613–620
115. Hayes, J.P., Generation of optimal transition count tests, *IEEE Trans. Computers*, Vol. C.27, 1978, pp. 36–41
116. Reddy, S.M., A note on logic circuit testing by transition counting, *IEEE Trans. Computers*, Vol. C.26, 1977, pp. 313–314
117. Fujiwara, H. and Kimoshita, K., Testing logic circuits with compressed data, *Proc. IEEE 8th. Int. Symp. on Fault Tolerant Computing*, 1978, pp. 108–113
118. Saxena, N.R. and Robinson, J.P., Accumulator compression testing, *IEEE Trans. Computers*, Vol. C.35, 1986, pp. 317–321
119. Robinson, J.P, and Saxena, N.R., A unified view of test compression methods, *IEEE Trans. Computers*, Vol. C.36, 1987, pp. 94–99
120. Savir, J. and McAnney, W.H., On the masking probabilty with one's count and transition count, *Proc. IEEE Int. Conf. on Computer Aided Design*, 1985, pp. 111–113
121. Davio, M., Deschamps, J.P. and Thayse, A., *Discrete and Switching Functions*, McGraw-Hill, NY, 1978
122. Heidtmann, K.D., Arithmetic spectrum applied to stuck-at fault detection for combinational networks, *IEEE Trans. Computers*, Vol. C.40, 1991, pp. 320–324
123. Wu, X., Chen, X. and Hurst, S.L., Mapping of Reed-Muller coefficients and the minimization of Exclusive-OR switching functions, *Proc. IEE*, Vol. 129-E, 1982, pp. 15–20
124. Besslich, P.W., Efficient computer method for ExOR logic design, *Proc. IEE*, Vol. 130-E, 1983, pp. 203–206
125. Karpovsky, M.G. (Ed.), *Spectral Techniques and Fault Detection*, Academic Press, NY, 1985
126. Damarla, T.R. and Karpovsky, M.G., Fault detection in combinational networks by Reed-Muller transformations, *IEEE Trans. Computers*, Vol. C.38, 1989, pp. 788–797
127. Muzio, J.C., Stuck fault sensitivity of Reed-Muller and arithmetic coefficients, *3rd.*

International Workshop on Spectral Techniques, Dortmund, Germany, 1988, pp. 36–45

128. Birhoff, G. and MacLane, S., *A Survey of Modern Algebra*, Macmillan, NY, 1977
129. Ayres, F., *Theory and Problems of Matrices*, Schaum's Outline Series, McGraw-Hill, NY, 1962
130. Hohn, F.E., *Elementary Matrix Algebra*, Macmillan, NY, 1964
131. Beauchamp, K.G., *Applications of Walsh and Related Functions*, Academic Press, UK, 1984
132. Beauchamp, K.G., *Transforms for Engineers*, Oxford University Press, UK and NY, 1987
133. Miller, D.M. and Muzio, J.C., Spectral fault signatures for single stuck-at faults in combinational networks, *IEEE Trans. Computers*, Vol. C.33, 1984, pp. 765–769
134. Lui, P.K. and Muzio, J.C., Spectral signature testing of multiple stuck-at faults in irredundant combinational networks, *IEEE Trans. Computers*, Vol. C.35, 1986, pp. 1088–1092
135. Muzio, J.C. and Serra, M., Spectral criteria for the detection of bridging faults, *Fault Detection Research Group Report No. 12–1986*, University of Victoria Dept. of Computer Science, Canada, 1986
136. Serra, M. and Muzio, J.C., Testing programmable logic arrays by sum of syndromes, *IEEE Trans. on Computers*, Vol. C.36, 1987, pp. 1097–1101
137. Yamada, T., Syndrome-testable design of programmable logic arrays, *Proc. IEEE Int. Test Conf.*, 1983, pp. 453–458
138. Ha, D.S. and Reddy, S.M., On the design of pseudo exhaustive testable PLAs, *IEEE Trans. Computers*, Vol. C.37, 1988, pp. 468–472
139. Daehn W. and Mucha, J., A hardware approach to self-testing of large PLAs, *IEEE Trans. Computers*, Vol. C.30, 1981, pp. 929–933
140. Hassan S.Z. and McCluskey, E.J., Testing PLAs using multiple parallel signature analysers, *Digest IEEE Int. Symp. on Fault Tolerant Computing*, 1983, pp. 422–425
141. Treur, R., Fujiwara, H. and Agarwal, V.K., Implementng a built-in self-test PLA design, *IEEE Design and Test of Computers*, Vol. 2, 1985, pp. 37–48
142. Robinson, J.P., Circular BIST test generation, *J. Semicustom ICs*, Vol. 7, 1990, pp. 12–16
143. Zorian, Y. and Agarwal, V.K., Optimizing error-masking in BIST by output data modification, *J. Electronic Testing: Theory and Applications*, Vol. 1, 1990, pp. 59–71
144. Johnson, B.W., *Design and Analysis of Fault Tolerant Systems*, Addison-Wesley, MA, 1989
145. Berger, J.M., A note on error detection codes for asymmetric channels, *Information and Control*, Vol. 4, 1961, pp. 68–73
146. Serra, M., Some experiments on the overhead for concurrent checking using Berger codes, *Record of 3rd. Technical Workshop: New Directions for IC Testing*, Canada, 1988, pp. 207–212
147. Wessels, D. and Muzio, J.C., Adding primary input error coverage to concurrent checking for PLAs, *Record of 4th. Technical Workshop: New Directions for IC Testing*, Canada, 1989, pp. 135–153

148. Abrahams, J.A, and Davidson, E.S., The design of PLAs with concurrent error detection, *Proc. IEEE Fault Tolerant Computing Symp.*, 1982, pp. 303–310
149. Pradham, D.K. (Ed.), *Fault Tolerant Computing: Theory and Techniques*, Prentice Hall, NJ, 1986
150. Massara, R.E. (Ed.), *Design and Test Techniques for VLSI and WLSI Circuits*, Institution of Electrical Engineers (IEE) Publications, UK, 1989
151. Moore, W.R., Maly, W. and Strojaw, A., *Yield Modelling and Defect Tolerance in VLSI*, Adam Hilger, UK, 1988
152. Mamane, W.P. and Moore, W.R., Testing of VLSI regular arrays, *IEE Int. Conf. on Computer Design*, 1988, pp. 145–148
153. Taylor, D., *Design of Certain Semicustomised Structures Incorporating Self-Test*, PhD Thesis, Huddersfield University, UK, 1990
154. Pagey, S. and A.J. Al-Khalili, Universal TSC module and its application to the design of TSC circuits, *Microelectronics Journal*, Vol. 28, 1977, pp. 29–39
155. Ohletz, M.J., Williams, T.W. and Mucha, J.P., Overheads in scan and self-test designs, *Proc. IEEE Int. Test Conf.*, 1987, pp. 460–470
156. Zhang, S., Miller, D.M. and Muzio, J.C., Determination of minimal cost one-dimensional linear hybrid cellular automata, *IEE Electronic Letters*, Vol. 27, 1991, pp. 1625–1627
157. Goel, P., Test generation cost analysis and projections, *Proc. IEEE 17th. Design Automation Conf.*, 1980, pp. 77–84
158. Miles, J.R., Ambler, A.P. and Totton, K.A.E., Estimation of area and performance overheads for testable VLSI circuits, *Proc. IEEE Int. Conf. on Computer Design*, 1988, pp. 213–223
159. IBM, Embedded array test with ECIPT, *IBM Technical Disclosure Bulletin*, Vol. 28, 1984, pp. 2376–2378
160. Zhu, X. and Breuer, M.A., A knowledge-based system for selecting design methodologies, *IEEE Design and Test of Computers*, Vol. 5, 1988, pp. 41–58
161. Varma, P., Ambler, A.P. and Baker, K., An analysis of the economics of self-test, *Proc. IEEE Int. Test Conf.*, 1984, pp. 20–30
162. Dear, I.D., Dislis, C., Ambler, A.P. and Dick, J.H., Economic effects in design and test, *IEEE Design and Test of Computers*, Vol. 8, 1991, pp. 64–77
163. Dislis, C., Dick, J.H., Dear, I.D. and Ambler, A.P., *Test Economics and Design for Testability*, Ellis Horwood, UK, 1995
164. Special Issue, *Electrical Testing*, IBM Journal of Research, Vol. 34, No. 2/3, 1990, pp. 137–448

7

The Choice of Design Style: Technical and Managerial Considerations

Every original-equipment manufacturing company has to make a number of initial decisions when considering the introduction of a new or enhanced product relating to the technology in which the product is to be designed and manufactured. Such decisions involve managerial, technical and economic aspects, and the final choice may not be a clear-cut decision.

The main diversifications of design and manufacture are shown in Figure 7.1. Obviously, there are many manufactured products which cannot incorporate any microelectronics, for example, fabrics, mechanical fixtures, adhesives and so on, and equally there are products which by definition must be built around microelectronics, for example, lap-top computers and radar, but between the two extremes is a vast range of products which already do or (more importantly) could incorporate electronics in order to improve their performance or sales appeal.

The breadth of present-day electronics is indicated in Figure 7.2, from which it is clear that microelectronics generally covers the low-voltage, low-power areas, with high voltage, high power or special applications being covered by discrete devices. However, the microelectronics itself is invariably only part of a complete product: there must always be other devices or fixtures associated with the microelectronics to make the product relevant for its intended use. Very frequently, weaknesses in these peripheral items make the end product unreliable or unsuitable for the end-user, and therefore the nonmicroelectronic design and manufacturing activities are equally if not more important than the microelectronics part. It is no use having a superb IC if the input and output peripherals are cheap and unsatisfactory.

The remaining sections of this chapter will deal largely with the items con-

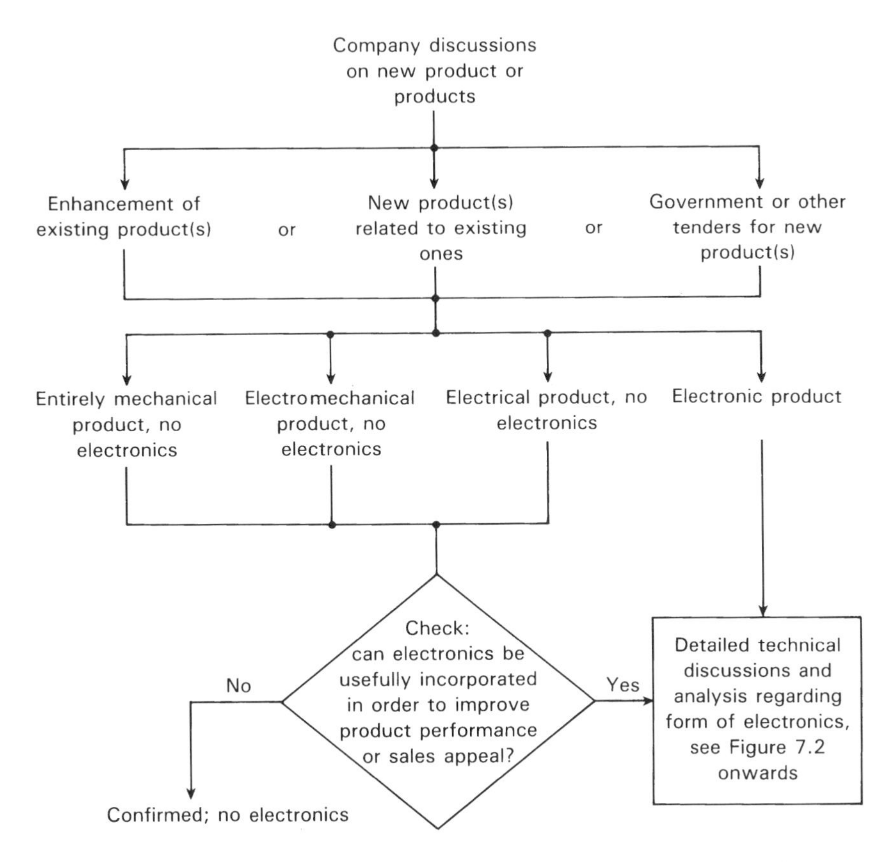

Fig. 7.1 Initial OEM company considerations relating to the introduction of any new engineering product.

tained within the dotted rectangle of Figure 7.2, and will consider the factors and costs involved in choosing a particular microelectronic design style. However, there is one parameter which the majority of digital ICs have in common, and that is the d.c. supply voltage. The most common voltage still remains 5V, which was initially established in the early days of SSI and MSI TTL circuits and has remained so ever since. There is no particular merit in 5V, having originally been established by the then maximum reliable reverse-voltage rating without breakdown of pn junctions, but there is now a growing use of lower voltages for VLSI circuits in order to reduce chip dissipation; some recently announced off-the-shelf standard products have quoted voltage supplies of 2.5V and 3.3V, PLD devices in particular being for 3.3V working. A problem with such lower voltages

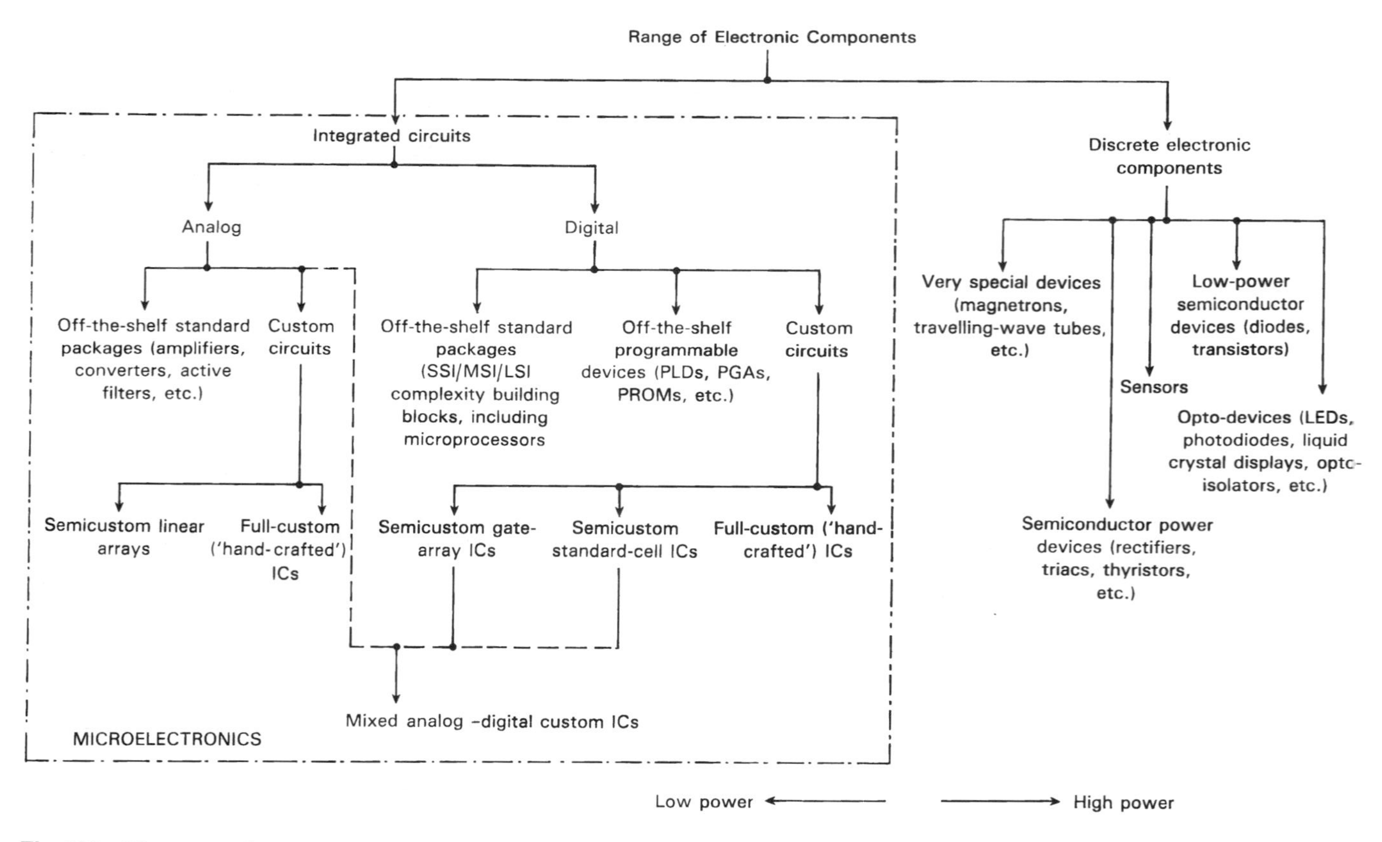

Fig. 7.2 The range of available electronic components ranging from high-power discrete devices to low-power VSLI microelectronics. The areas of specific concern in this text are those marked with an asterisk.

is that there are still a majority of other circuits requiring a 5V supply, and therefore a 5V interface may have to be provided. However, the majority of custom ICs still continue to use 5V, although there is freedom for the OEM to specify alternatives if the specific product so requires. One advantage of CMOS technology is that voltages can be varied fairly easily, allowing an OEM to specify, e.g., 1.5V or 3V working for small battery-operated products if necessary.

On the other hand, analog circuits have not standardized on 5V. Instead, a number of voltage ratings will be found in off-the-shelf devices, for example, ±9V and ±15V, although as in the digital area there are lower supply voltages being pursued for some products [1]. This is yet another problem confronting the manufacture of a mixed analog/digital product, and may require the provision of two or more power supplies. The adoption of custom microelectronics may simplify this mixed supply requirement, but will not eliminate it entirely since it is not practical to run sensitive analog circuits from the same supply rails as digital circuits due to the interference which can be caused by fast-changing logic signals.

Unless otherwise specified, the following sections will assume that no unusual power supply requirements are required. Power dissipation, however, may be a crucial factor, and will be referred to where relevant.

7.1 THE MICROELECTRONIC CHOICES

The first choice which automatically arises from the definition of a new microelectronics-based product is whether the circuits are:

- entirely digital, with binary input and output signals
- entirely analog, with no digital signals present but with a need for accurate voltage amplification or other signal processing
- mixed analog/digital

The last category almost invariably involves some form of analog-to-digital (A-to-D) and/or digital-to-analog (D-to-A) conversion, for example, as is shown in Figure 7.3, with the digital part performing some logic processing on the digital data.

In the following sections we will consider the decisions which face the OEM designer in each of the above categories. As will be seen, the last category often has to be regarded as a mixture of separate analog and digital parts, with the best design style having to be chosen for each part independently.

7.1.1 Digital-Only Products

The categories of digital circuits available to the OEM for a new product range from full custom, designed in every detail to match the required specification,

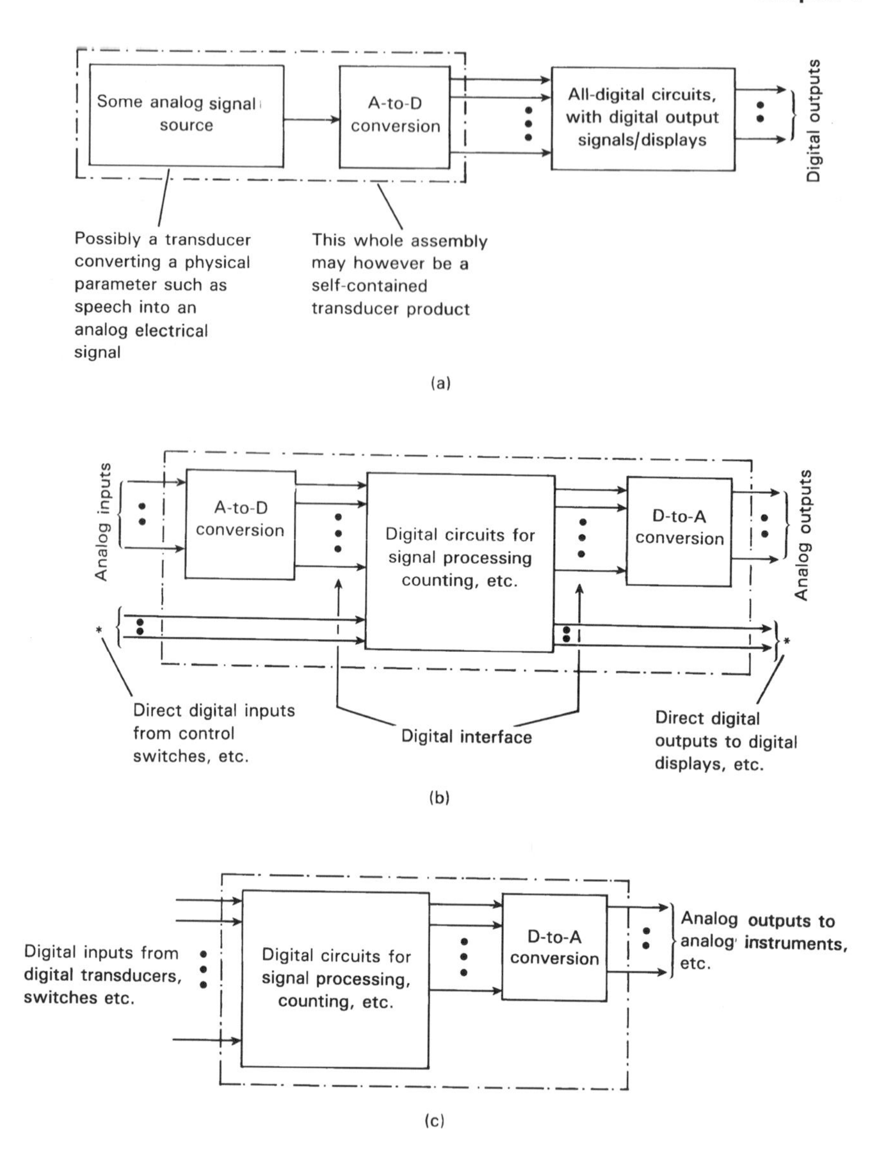

Fig. 7.3 The general structure of a mixed analog/digital product: (a) analog source signal(s) with some A-to-D conversion; (b) analog inputs and analog outputs, with some internal signal processing; (c) digital inputs with final analog outputs.

through the various standard-cell and gate-array choices, to programmable logic devices and finally standard off-the-shelf nonprogrammable parts. The principal characteristics of these various groups have been considered in preceding chapters, but in order to summarize them, Table 7.1 gives the main design attributes as far as the OEM is concerned.

Which type of circuit to use for a new product depends upon the anticipated production quantitiy, since it is clearly uneconomical to incur the expense of designing a custom IC if only very small quantities are required [2,3]. There will be exceptions to this generalization, for example, where product weight or size or absolute performance takes precedence over all economic factors, as may arise in avionics, military or other specialist fields. But for the majority of products involving microelectronics, the initial choice of design style is likely to be influenced by the broad guidelines shown in Figure 7.4. Further considerations of the economic factors, including nonrecurring engineering (NRE) costs, will be covered in Section 7.4.

Microprocessor Choice

The new product's technical requirements must, however, play the dominant role in the choice of design style. The off-the-shelf microprocessor or microcontroller IC, with its selling price of a few dollars, is a strong contender in the choice of production method for a new product, and must enter into this initial decision-making. Simple microprocessors and microcontrollers for OEM use range from 4-bit processors such as the Texas Instrument's 1000 series, 8-bit such as the Zilog Z8* and Motorola 68** series, to 16-bit such as the Intel 80** and the Motorola 68*** series. The more powerful 32-bit and 64-bit μPs are, of course, also available, but the smaller processors are entirely appropriate for a vast range of product designs which do not require massive computational power [4].

The general decision tree of when or when not to use a microprocessor is given in Figure 7.5. In comparison with later decision routes, factors such as space and power constraints, and to some extent cost, are not deciding factors in the decision to adopt a microprocessor-based design. Rather, the dominant factors are as follows:

- the architecture and the computational power of the processor
- the flexibility whereby several versions of the same product can be made by changing the software only, leaving the hardware standardized for all the variants

The last item is the supreme advantage of software-based equipments.

Outside product areas such as word processors, etc., it is unlikely that production quantities of software-based products will be high, possibly thousands rather than hundreds of thousands. However, if very large production require-

Table 7.1 The Control Which the OEM Designer Has Over the Detailed IC Design and Manufacture

Category of circuit	Control over the chip design and fabrication	CAD availability
1 Hand-crafted Full custom All mask levels unique, no fixed chip size	Full control; anything can be done within the limits of the technology and budget	Yes
2 Standard cells, parameterized (or 'soft') cells/macros All mask levels unique, no fixed chip size	Can only use the functional cells in the CAD library, but can (i) optimize cell performance for the particular application and (ii) place and route in any manner	Yes
3 Standard cells, fixed (or 'hard') cells/macros All mask levels unique, no fixed chip size	Can only use the pre-designed cells in the CAD library but can place and route in any manner	Yes
4 Uncommitted arrays, component-level cells Final interconnection mask(s) only unique, fixed chip sizes	Fixed chip layout of components. Full flexibility to interconnect components in each cell and route cells in any manner, subject to constraint of available wiring space	Yes
5 Uncommitted arrays, functional cells Final interconnection mask(s) only unique, fixed chip size	Fixed chip layout of fixed functional cells (gates or other primitives). Full flexibility to route cells in any manner, subject to constraint of wiring space	Yes
6 Programmable devices, field-programmable Completely prefabricated, no unique mask, fixed chip sizes	Fixed chip layout and circuit topology. Can only destroy existing internal connections or active wanted connections to dedicate the chip	Yes
7 Programmable devices, mask-programmable Final interconnection mask only unique, fixed chip sizes	A vendor's version of (6) above should volume production of a field-programmable device be required; still the same fixed chip layout and circuit topology as (6)	Same as for (6) above
8 Fixed standard functions Completely prefabricated standard off-the-shelf products, e.g. microprocessor, memory. TTL 7000 series, CMOS 4000 series, etc.	Absolutely standard products, with optimized chip layouts. Customer/user has no control or knowledge of the detailed chip design or fabrication	Not specifically available

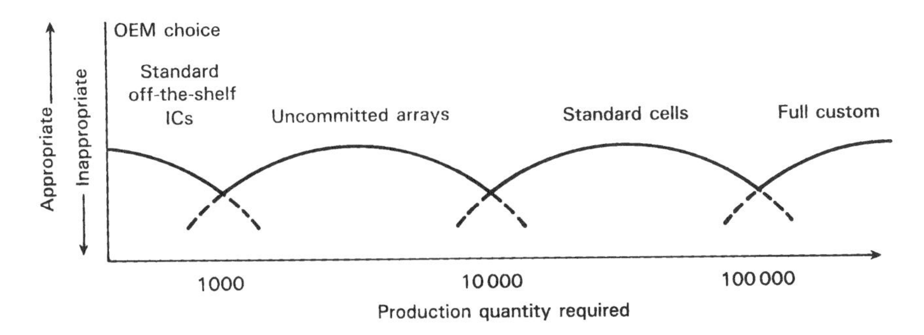

Fig. 7.4 A repeat of Figure 4.33, giving a general guide to the most likely economic design style for varying production quantities; however, the increasing availability of complex programmable logic devices (CPLDs) may compete in the middle range of this generalization.

ments are anticipated, then the possibility of using a standard-cell design route with the microprocessor/microcontroller as an on-chip macro may be justified. This will be seen later in connection with Figure 7.8. It is unlikely that any OEM would reach this level of sophistication without appreciable prior in-house experience in the design and manufacture of microelectronic-based products.

Other Standard Parts

For other digital-only applications where relatively small production quantities are anticipated, other widely available off-the-shelf standard parts, including programmable devices, may be relevant. The general decison tree leading to the choice of standard parts is given in Figure 7.6.

However, the further choices between standard digital logic ICs or PLDs and between bipolar technology or CMOS technology are not clear-cut decisons: standard ICs allow ready second-sourcing, but vendor-specific PLDs and particularly the newer CPLDs can reduce the number of IC packages and hence the size of the printed-circuit board(s) in the final equipment. Also, PLDs can provide some degree of design security that is not available with standard parts, since the exact programming is hidden from the end-user. On the technicalities of bipolar vs. CMOS, the latest advanced low-power Schottky TTL series ICs will provide a higher system speed than CMOS, but the lower power 74HC or 74HCT CMOS series are good all-around choices, also providing a wider operating temperature range than bipolar. For wide operating voltage requirements, the older 4000 series CMOS ICs provide a working voltage range of +3 to +15V, but price now generally favors the 74 series CMOS circuits. Programmable logic

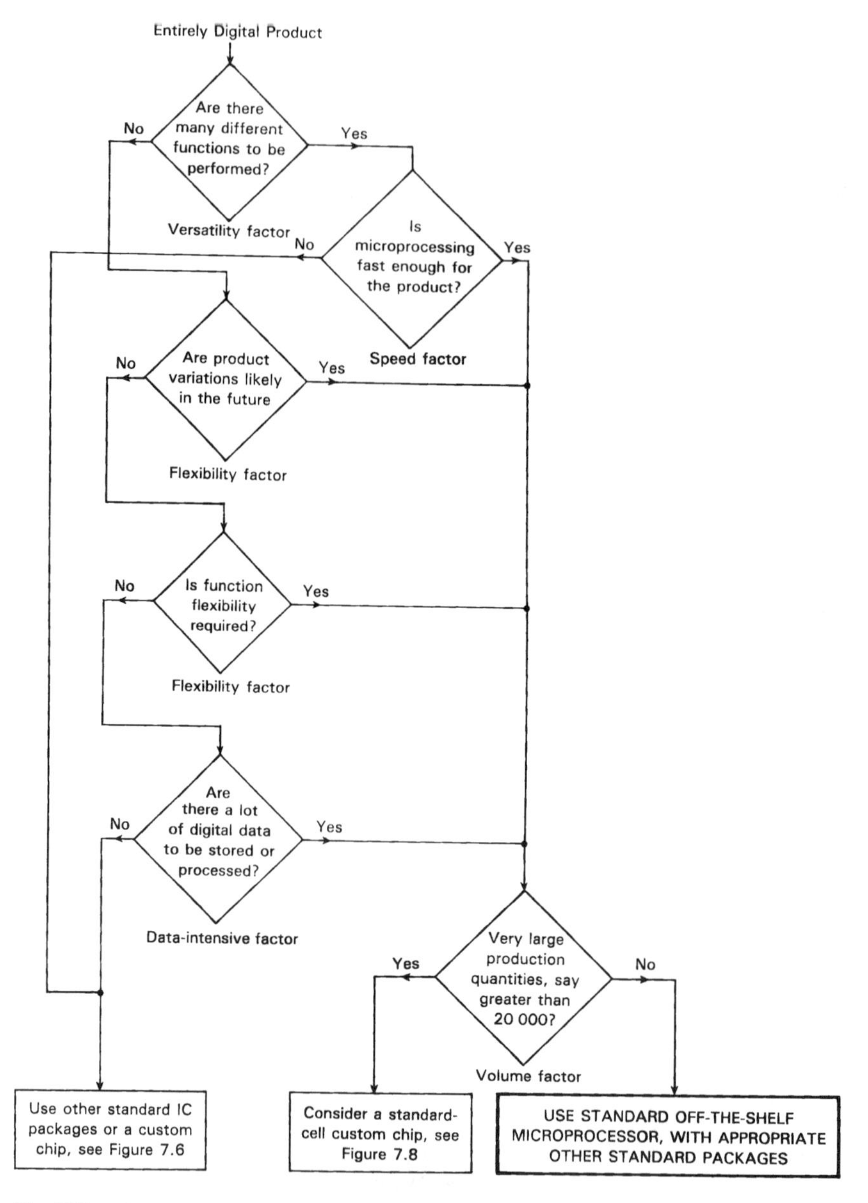

Fig. 7.5 The decision tree of when to consider the use of an off-the-shelf microprocessor or microcontroller. (Source: based upon Ref. 3)

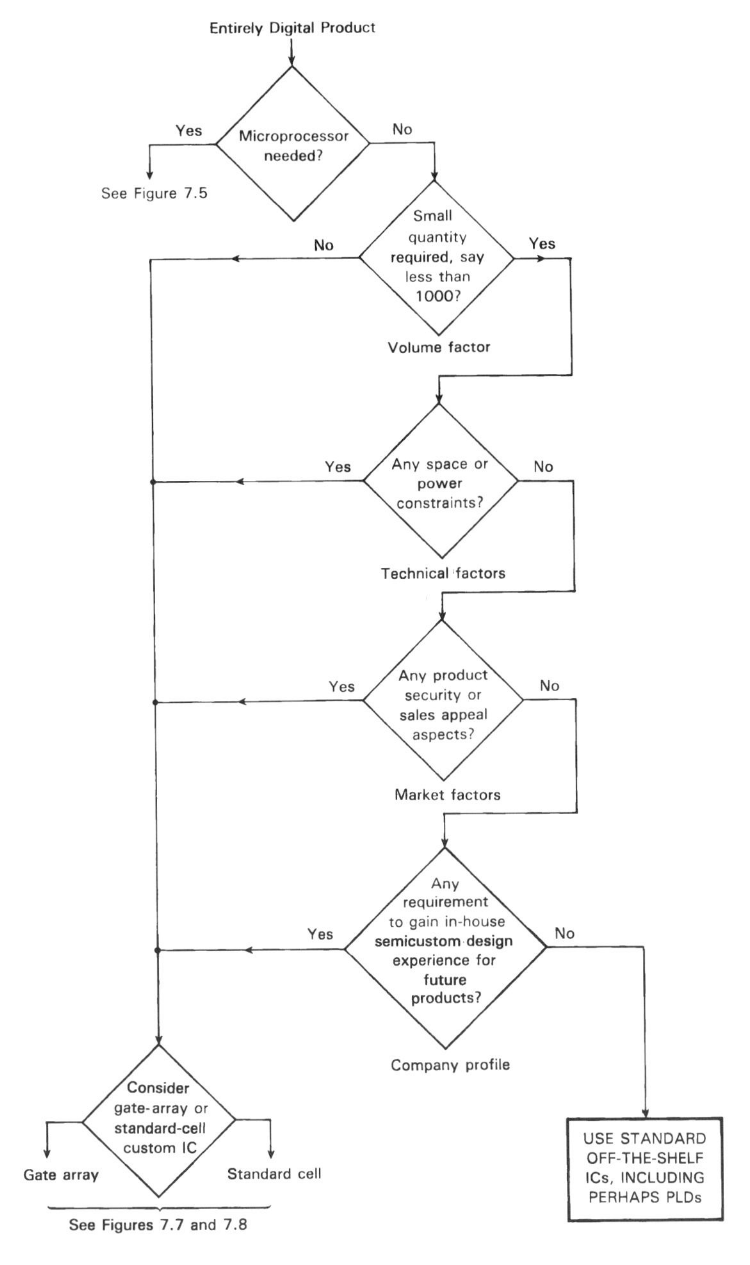

Fig. 7.6 The decision tree of when to consider the use of off-the-shelf standard digital ICs, including programmable logic, devices if they can provide the required circuit complexity. (Source: based upon Ref. 3.)

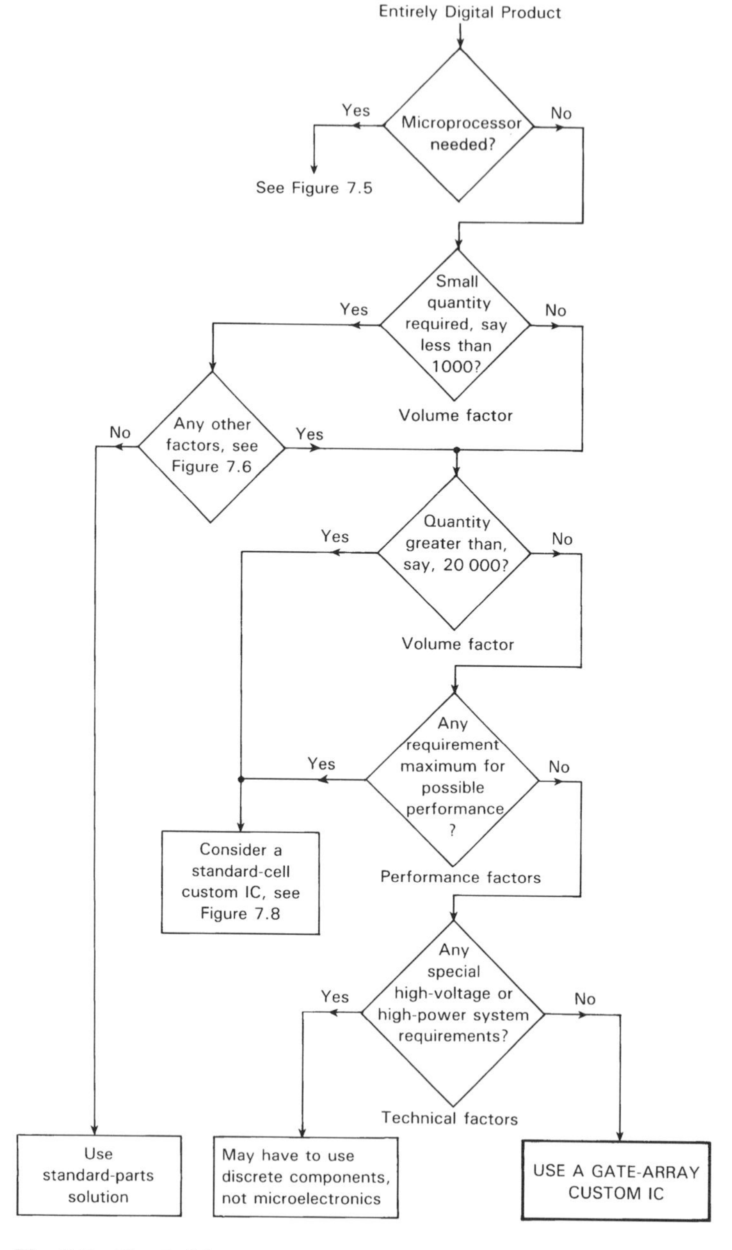

Fig. 7.7 The decision tree of when to consider the use of a channelled-architecture gate-array USIC for digital applications. The place of the channel-less gate-array IC is less sure, but is currently confined to sophisticated custom circuits with on-chip memory requirements. (Source: based upon Ref. 3.)

device costs are generally decreasing; the cost of CAD resources for PLD design work will be referred to in Section 7.4.2.

Gate-Array Choice

For somewhat greater production quantities, the next choice of possible design style is the gate array. For very high-speed applications, this may require ECL or GaAs products, but for the majority of applications it will currently be CMOS, usually a channelled architecture for general-purpose applications but perhaps channel-less for more sophisticated needs. However, we will here concentrate upon the decisions leading to the adoption of a relatively simple CMOS custom circuit.

Figure 7.7 illustrates the general decision tree leading to the choice of a gate array. This tree should be read in conjunction with both the preceding and the subsequent diagrams, since there is no absolute division between them. However, a number of additional practical considerations are involved when considering the gate-array route which are not shown in Figure 7.7, among them being the following:

- the number of available gates and I/Os on the vendor's products compared with the OEM's needs
- I/O interface requirements, particularly if nonstandard
- the technology—bipolar or MOS
- power dissipation if critical
- the number of mask levels necessary to customize the array
- the comprehensiveness of the vendor's standard library of cell interconnections and supporting CAD tools
- how much of the custom design is to be carried out in-house by the OEM and how much by the vendor or independent design house

These are all factors requiring close consideration when it seems that a gate-array choice may be appropriate for the new product. On the choice of technology, CMOS will be the usual decision, with bipolar if very high-speed operation is required; BiCMOS technology does not seem to have become widely accepted, and has been dropped by certain IC manufacturers.

IC design and manufacturing costs, which are not present when using standard parts, will inevitably be involved in the adoption of a gate-array solution. These cost relationships (see Section 7.4) must therefore be considered, unless the particular attributes of a custom design such as design security or minimum size or power dissipation are overriding considerations.

Standard-Cell Choice

If large production needs without any product design changes are envisaged, then a standard-cell choice may be preferable to the gate array; if complex macros

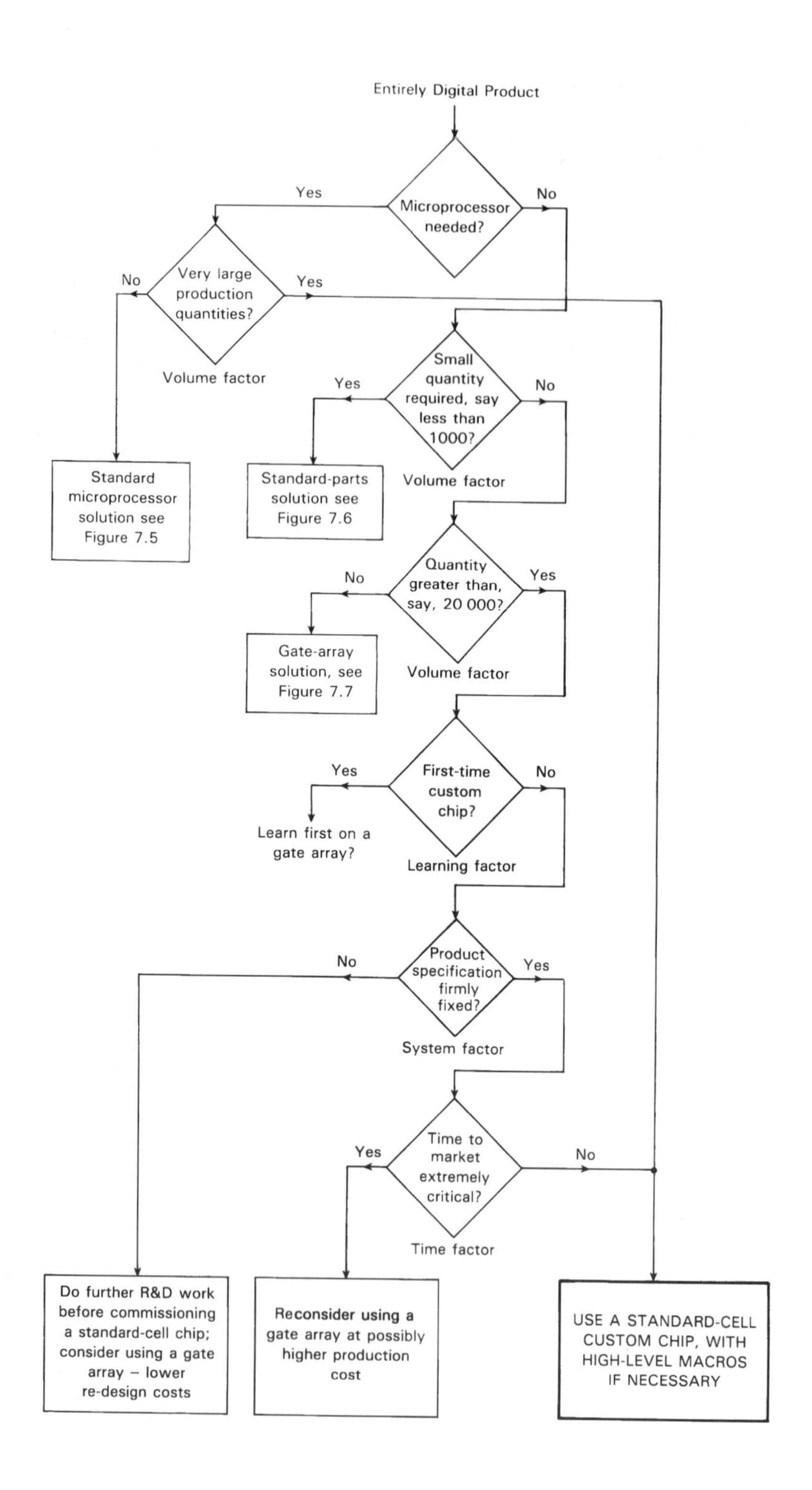
Entirely Digital Product
Microprocessor needed?
Yes
No
Very large production quantities?
No
Yes
Volume factor
Standard microprocessor solution see Figure 7.5
Small quantity required, say less than 1000?
Yes
No
Volume factor
Standard-parts solution see Figure 7.6
Quantity greater than, say, 20 000?
No
Yes
Volume factor
Gate-array solution, see Figure 7.7
First-time custom chip?
Yes
No
Learn first on a gate array?
Learning factor
Product specification firmly fixed?
No
Yes
System factor
Time to market extremely critical?
Yes
No
Time factor
Do further R&D work before commissioning a standard-cell chip; consider using a gate array – lower re-design costs
Reconsider using a gate array at possibly higher production cost
USE A STANDARD-CELL CUSTOM CHIP, WITH HIGH-LEVEL MACROS IF NECESSARY

and memory requirements are present, then this choice may provide these means much more readily.

Figure 7.8 illustrates the decision routes to a standard-cell design. However, design costs, time to market, and the possibility of product design revisions involving iteration of the custom design and the full mask set, makes this a choice which perhaps needs more careful attention than a gate-array decision.

The availability of microprocessors and RAM and ROM macros in most vendors' standard-cell libraries may be a deciding factor in choosing the standard-cell design solution, since, as we have seen in previous chapters, it is not practical to compile these on a channelled-architecture gate-array IC. However, this decision is not without alternatives, in particular the following:

- the use of cheap off-the-shelf microprocessors and memory ICs, leaving the custom IC to accommodate the remaining less structured digital logic
- the use of a channel-less gate array, which may be able to provide complex macro requirements if the vendor has included them in his interconnect library

The latter has not yet established a well-defined place in the range of design styles adopted by the majority of OEMs, but may increasingly do so. They are, however, most appropriate for the more sophisticated high-speed product requirements rather than the average OEM needs.

Full-Custom Choice

A full-custom digital design solution has not been listed in the preceding decision diagrams. The increasing availability of efficient high-level macros in standard-cell form and the forecast increasing availability of more digital signal processing (DSP) macros diminishes the need for full-custom designs. Hence, it is not until extremely large production quantities of a new digital product are envisaged—for example, video recorders or other high-volume home entertainment products with global sales—that full-custom becomes appropriate, or unless complex DSP requirements are needed which cannot readily be acheived with existing standard parts or standard-cell libraries.

Full-custom designs may, however, be more frequently required for analog circuits than for digital circuits. The following section will therefore refer more significantly to the full-custom design style.

Fig. 7.8 The decision tree of when to consider the use of a standard-cell USIC for digital applications. (Source: based upon Ref. 3.)

7.1.2 Analog-Only Products

In contrast to VLSI digital circuits where hundreds of thousands of transistors may be involved, particularly if memory macros are present, analog-only systems tend to remain relatively small, with the whole system design capable of being undertaken by one competent system designer. The individual building blocks—the operational amplifiers, comparitors, active filters, etc.—may be much more complex than individual logic gates, so that the majority of OEM system designers would not have the necessary skills or experience to undertake their detailed design, and therefore analog-only systems are almost invariably built using predesigned circuits in one form or another. The many excellent textbooks on analog circuit design, such as that by Gayakwad [5], cover the circuit theory and analysis, but cannot cover the specialist translation of this design data into a vendor's silicon layout.

Figure 7.9 shows the decision tree for purely analog systems. The majority are still produced using standard off-the-shelf parts due to their wide availability and low cost, so that it is extremely difficult for any form of custom IC to compete on purely financial terms. Both bipolar and MOS analog products are available, although bipolar generally gives better performance due to the higher value of g_m for bipolar in comparison with MOS.

However, if an appreciable quantity of a product is planned, or if space, design security or even sales appeal are significant factors, then, as indicated in Figure 7.9, a custom IC may be a suitable design choice. A further advantage, not specifically listed in Figure 7.9, is that a wider range of packaging options (see Section 7.2) may be available in the custom route compared with standard parts, which may have advantages.

The most widely available and used custom product for purely analog purposes is the uncommitted component array, with an increasing range of tile-based architectures available from vendors. Standard-cell products for analog purposes tend to be combined in a library of predominantly digital parts, and are therefore more relevant for mixed analog/digital purposes (see Section 7.1.3). Perhaps the greatest risk in adopting the custom analog IC path is that it is clearly more expensive to correct or modify a custom circuit compared with a standard parts design, and since analog system design is prone to require 'fine-tuning' or other last-minute circuit adjustments, it is much more important to have a right-first-time design with the custom route than perhaps it is when using standard parts. The saving factor is that SPICE simulation is usually capable of performing a complete custom circuit simulation before any manufacturing costs are incurred, and should therefore highlight errors or omissions in the design activity.

Ignoring these design considerations, on purely financial grounds and omitting any NRE costs (see Section 7.4.1), the choice between standard parts and custom is generally summarized in Figure 7.10. The dotted line indicating hybrid

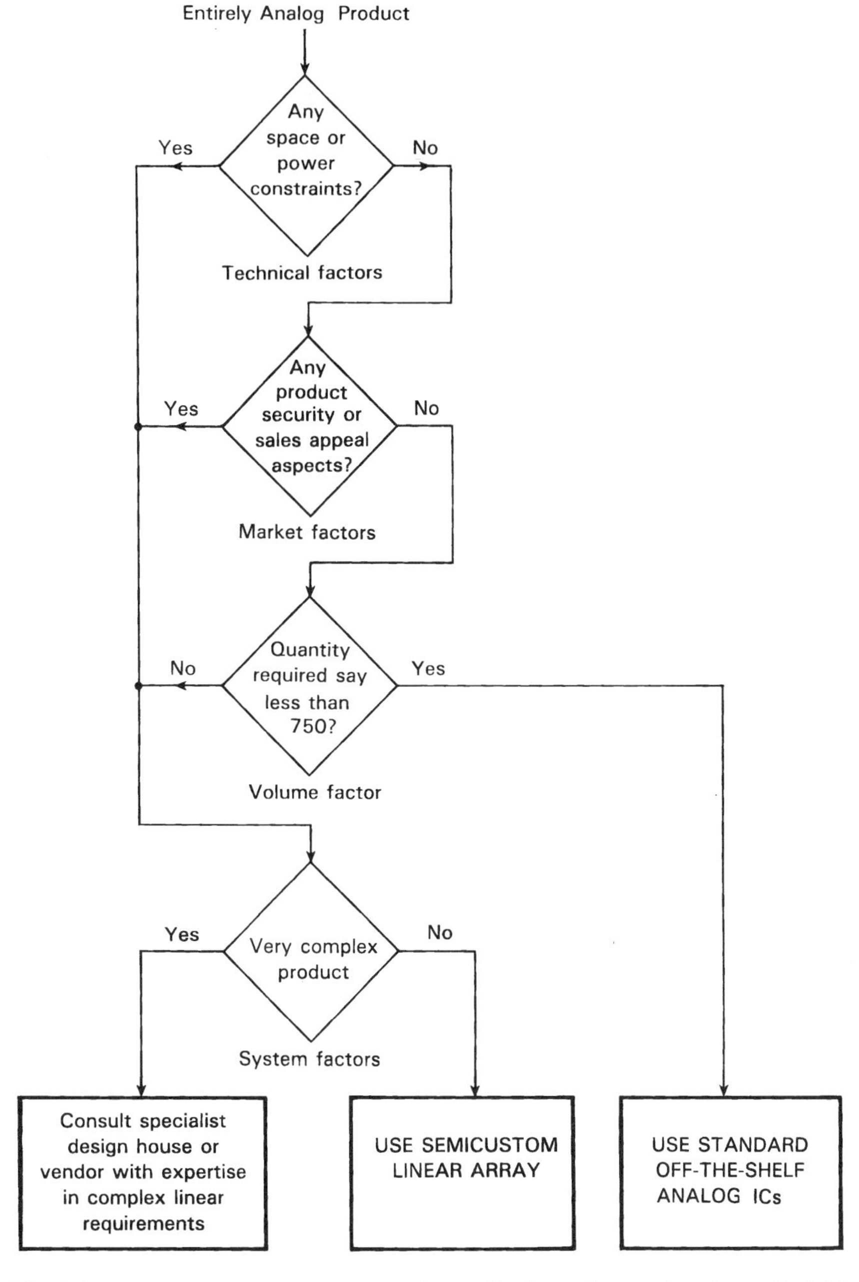

Fig. 7.9 The decision tree for purely analog applications. (Source: based upon Ref. 3.)

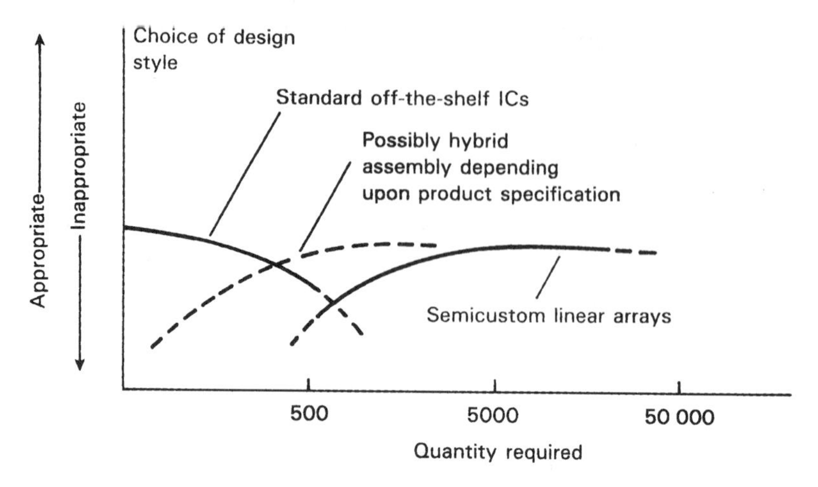

Fig. 7.10 A general guide to the most economic design style for analog-only microelectronic circuits.

assemblies (see Section 7.2) is not an area which we are able to detail, since it varies greatly from one application to another, but it is nevertheless one which may offer both financial and technical attractions, particularly where unusual voltages or other parameters are involved.

Possibly the greatest problem that faces the OEM designer with a custom analog requirement is his or her level of experience. For simple design requirements, there should be little problem provided SPICE simulation is available, but for more complex requirements involving, e.g., precise filtering requirements, the services of a specialist design house may be necessary. Such independent specialists may adopt a tile-based design strategy, or for particularly complex requirements may adopt a full-custom approach to meet the OEM's specification. This with the following mixed analog/digital area remains a major field where full-custom designs may still be appropriate.

7.1.3 Mixed Analog/Digital Products

The design style for mixed analog/digital products (see Figure 7.3) is perhaps the most difficult to summarize, since it depends heavily upon the size of the analog part(s) and the complexity of the digital part(s). If the main design component is the digital part with the analog merely, e.g., A-to-D input and/or D-to-A output, then it is most easy to purchase the A-to-D and D-to-A parts as multi-sourced off-the-shelf items, leaving the digital design to be decided on its own merits.

However, if the digital part is small and the analog part dominates, then an uncommited analog array which can provide some minor digital capability may be appropriate; alternatively, the digital requirement may possibly be readily realized with off-the-shelf ICs, leaving the analog part to be the main consideration. For complex combined analog and digital requirements, then a mixed standard-cell design strategy may be considered.

These possibities are summarized in Figure 7.11. Yet another extreme situation is when the product requirements are beyond the capabilities of the OEM to design, e.g., some complex telecommunications or digital signal processing requirement where the OEM can accurately specify what is required but not how to realize it; in this case, we may find that the services of a specialist design house are once again called in to produce the microelectronic design, and if the complexity and sales value of the product warrants, then a full-custom USIC may be justified.

7.2 PACKAGING

The assembly of semiconductor dice into appropriate packages is normally necessary so that the IC may be handled by the OEM and assembled into the final product. The only exception to this is in forms of 'direct-die-mounting,' where the individual chips—often termed 'naked dice'—scribed from a tested production wafer are used in thick-film or thin-film hybrid assemblies [6–8] or a chip-on-board (COB) assembly, each assembly containing one or more IC plus further discrete components as appropriate. Each individual die is fixed to the substrate with a gold/eutectic solder or similar means, the die I/Os then being wire-bonded to the substrate to complete the interconnections. Other forms of die-to-substrate interconnect, such as 'bump-on-die' soldering, may be used. The whole assembly is then usually given some hermetic protective covering [8–13].

Such assembly methods are only undertaken by companies who specialize in hybrid assembly techniques. For the OEM it can offer a very compact assembly for new products, but usually the work has to be subcontracted to a specialist company who is not the IC manufacturer. Hence, an additional outside agency becomes involved in the OEM's project management, with a corresponding loss in direct control of all the manufacturing steps.

The most usual situation is for the OEM to use off-the-shelf and custom ICs already assembled on receipt into appropriate packaging, with one IC per package. The OEM has no choice in the packaging when using standard ICs, the majority still being packaged in the familiar plasic-encapsulated dual-in-line ('DIP' or 'DIL') packages. Analog or other special off-the-shelf ICs with relatively few leads may have alternative forms of standard packaging, but the DIL is the most common. However, for custom ICs the situation is completely differ-

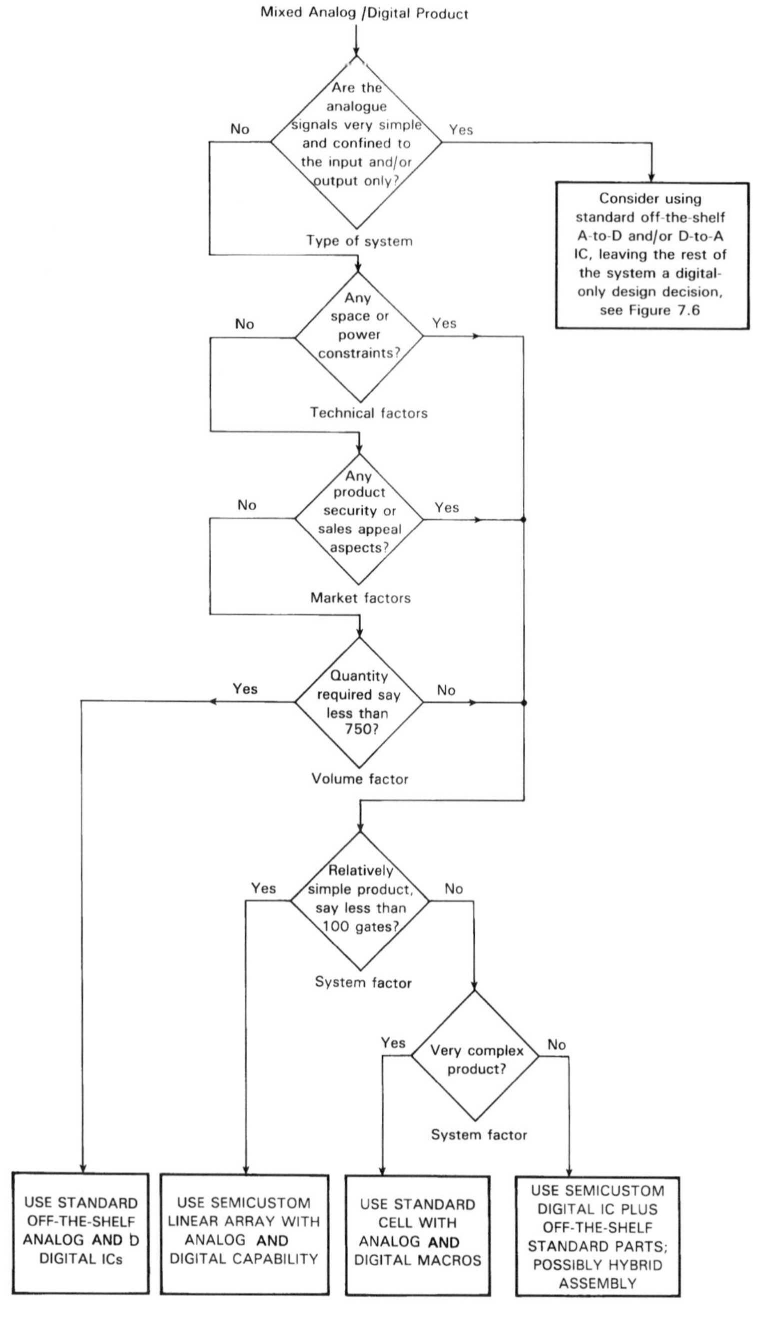

Fig. 7.11 The decision tree for mixed analog/digital circuit applications. (Source: based upon Ref. 3.)

ent; the OEM now comes into the decision making on the form of package, and in consultation with the custom IC vendor or outside design house has to agree upon the type of packaging to be used for his or her circuit. This is, therefore, a further technical and financial aspect to be actively considered by the OEM during the custom circuit adoption procedure.

The technology of packaging is complex. The package into which an IC is mounted has to provide many features, including the following:

- mechanical robustness, so that it may be handled in automatic testers and final OEM product assembly without damage
- thermal properties, so that the heat dissipated on the dice can be removed without an unacceptable temperature rise
- electrical properties, so that on-chip signals can be connected to the package I/O pins without unacceptably degrading the available performance

The thermal design [6,14] is complicated by the need to ensure that expansion stresses are minimized as package temperature varies, which usually means some small degree of flexibility between package and printed-circuit board or other mounting. The electrical design is equally if not more difficult [10], since die-to-package lead inductances and capacitances mean that the fastest on-chip performance now available cannot be brought out without some small loss of performance; also, the electrical characteristics of all die-to-package interconnections should be identical, which implies that all such connections must be of identical length within the package.

Package design is thus a complicated specialist field [9–16]. However, the OEM can only accept what is commercially available, and make an informed choice between the alternatives that are offered.

Looking at the categories of IC packing that are available, we have two main types, namely:

(i) through-hole mounting packages, in which the package I/O pins pass through holes in the PCB or other mounting, being soldered on the underside
(ii) surface-mount packages, which do not use any through-holes but are mounted directly on the face of the PCB, etc.

The former is the most common, but the latter has many-electrical and other advantages at the cost of the OEM of mastering more difficult attachment/connection/inspection techniques, with possibly higher tooling costs for the product assembly line. Removal of faulty or incorrectly assembled surface-mount devices may be particularly awkward, and due to the difficulty of easy inspection of connections may need boundary scan (see Section 6.8.2) for product testing.

Table 7.2 The Principal Range of Commercial IC Packages[a]

Type of package P = plastic encapsulation C = ceramic	Designation	Through-hole mounting (T) or surface mounting (S)	Connector pitch	No. of I/Os[b]
Standard dual-in-line P	DIL or DIP	T	0.1″	16–64
Shrink-size dual-in-line P	SDIP	T	0.07″	40–64
Small-outline dual-in-line P	SOP	S	0.5″	8–32
Quad-in-line P	QUIP	T	0.1″ (two rows each side)	16–40
Flat plastic package P	FPP	S	1 mm or less	54–120
Plastic leaded chip carrier P	PLCC	S	0.05″	24–156
Standard dual-in-line C	DIL	T	0.1″	16–64
Ceramic leadless chip carrier C	LCC	S	0.5″ OR 0.4″	16–156
Pin-grid arrays C (also available in plastic encapsulation = PPGA)	PGA	T	0.1″ (nested squares of pins)	64–324

[a] Mounting sockets are available for many of these types, but this introduces additional interconnection risks in the final product. More sophisticated packages have to more pins per package and/or smaller 'footprint' size.
[b] Minimum and maximum number of pins may vary from different vendors.

The range of widely available packages is given in Table 7.2, with Figure 7.12 illustrating some of these possibilities. The overall dimensions and pin spacings are defined by internationally accepted JEDEC standards, with 0.100 ± 0.010 inch pin spacings being common in DIL packages, although other types of packages employ spacings of 0.050 inch, 0.040 inch, 1.0 mm, 0.8 mm, 0.65 mm, etc. (JEDEC is the Joint Electronic Devices Engineering Council, now part of the Electronic Industry Association [EIA]. However, JEDEC is still the term most commonly used in connection with packaging standards.)

As far as custom ICs are involved, there are several factors which strongly influence the most appropriate type of package [17–19]. These include the following considerations:

- The volume of custom circuits required may not be large, and therefore the set-up cost and possible delay in plastic encapsulation may not be justified; ceramic packages which can be held in stock and 'hand-assembled' may be more economic and give a much faster turn-around time for small quantities.

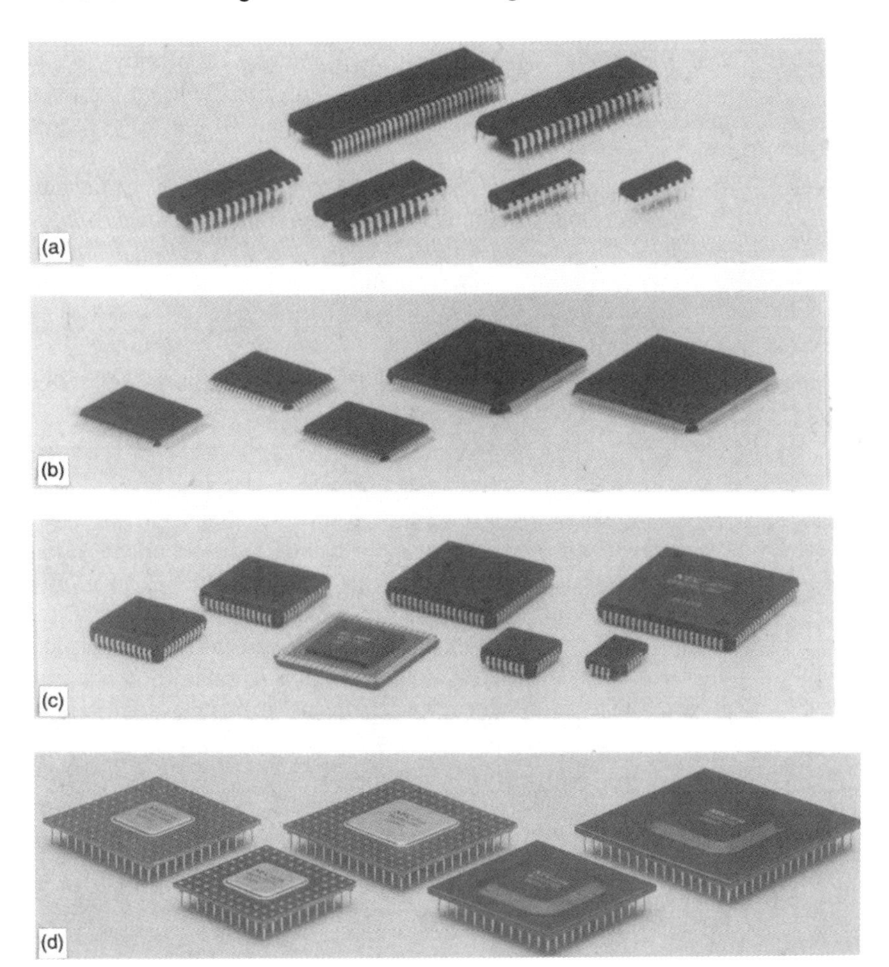

Fig. 7.12 Some of the common range of IC packaging methods: (a) dual-in-line (DIL) plastic packaging; (b) flat plastic packaging (FPP); (c) plastic and ceramic leadless chip carriers (PLCC and LCC); (d) pin-grid array and plastic encapsulated pin-grid array (PGA and FPGA) packaging. Newer packages with smaller 'footprint' area and closer pin spacings are also available. (Courtesy NEC Electronics, UK.)

- Custom ICs often require a larger number of I/Os in comparison with standard off-the-shelf ICs, and hence the relatively small number of pins on dual-in-line and other standard packages is inadequate (a typical provision is 25 pins per 1000 logic gates in a VLSI circuit, but many custom designs require substantially more than this).

- One of the reasons for adopting custom microelectronics may be the space factor, and thus a package with the smallest 'footprint' area may be particularly significant—in this respect, the dual-in-line package is inefficient.
- There may be circumstances where the circuit dissipation is particularly high, and thus the thermal properties of certain packages may become important.
- The size of the actual custom chip may require a large die size cavity in the package, and hence a larger package is required in which to assemble it than is necessary from the actual number of I/Os required—thus, spare package I/Os may be available which perhaps could profitably be used for additional test purposes if appreciated at an early design stage.

Details of many packaging parameters may be found in vendors' literature [17]. The OEM should be aware of these technicalities and implications as part of the decision mechanism in choosing or not choosing a custom circuit design style. If a custom design is adopted, then it is most likely that the final packaging choice will be between ceramic dual-in-line packaging, as shown in Figure 7.13, for relatively simple USICs, and leadless chip carriers or pin grid arrays for more complex requirements. Plastic encapsulation (see Figure 7.12) may become justified for very large volume requirements, but ceramic packaging may remain manditory for very high reliability and/or military and avionics applications.

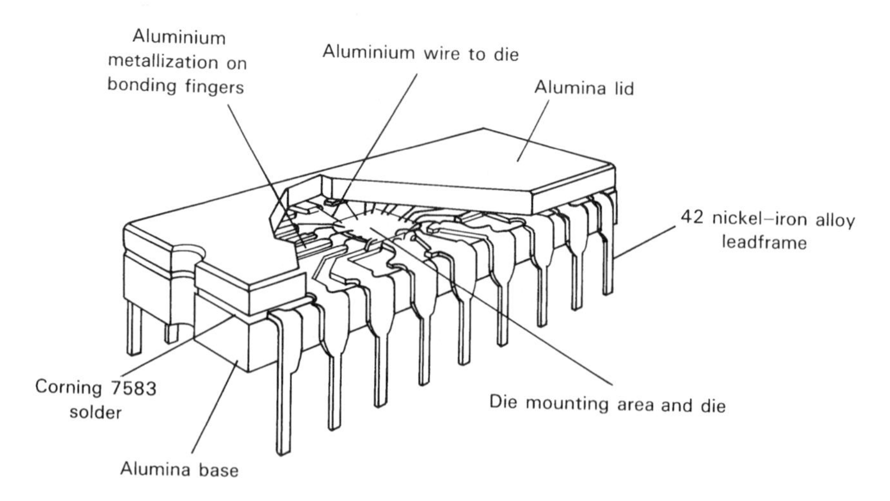

Fig. 7.13 The CERDIP ceramic dual-in-line package which is appropriate for many OEM applications. A higher reliability ceramic DIL package, the ceramic sidebrazed package, is also available, which is an industry standard for very high reliability applications.

7.3 TIME TO MARKET

The time to market, that is, the design, manufacturing and testing time for a new product, has been mentioned in previous chapters as an important parameter, particularly so in microelectronics-based products where there is a continuous evolution and hence a rapid obsolescence of products. This is particularly important in designs involving custom ICs, since any error or omission in design may require a rerun of the fabrication and hence an appreciable delay to market of the final product.

The typical market life cycle for a microelectronic product is shown in Figure 7.14. The time scale on the x-axis varies with product, but may typically be as shown. However, market research has indicated that if a microelectronic-based product suffers an unforseen delay of six months in its time to market, then there will be a 30% loss of sales over the lifetime of the product [20].

To attempt to put some illustrative figures on this loss, consider a product with a selling price of $1000 giving a 20% profit margin, with a target production of 10,000 units over a three year period. With correct time to market, this gives a company profit of $\$1000 \times 0.2 \times 10{,}000 = \$2M$. However, in the event of an unforseen six months' delay to market, there will be a loss of sales of 3000 units, which represents a loss of company income of $600,000 over the product lifetime. This feature is discussed in several publications [3,21–23], and necessitates a consideration of the design styles which provide:

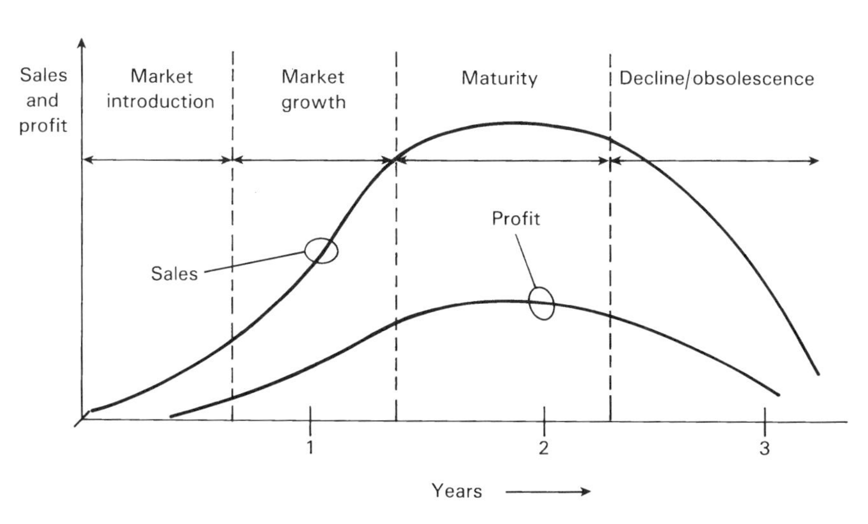

Fig. 7.14 The typical product life cycle of a microelectronic-based product, with possibly a two-year useful market life before obsolescence. Any delay in market introduction will reduce the areas under these curves.

- the shortest design-to-prototype time
- the shortest time to correct initial design errors or omissions
- the lowest risk factor of encountering any delays

To quantify the above factors would require a full analysis of each particular product, company expertise and outside suppliers' stability and reliability. However, the general trends are as shown in Figure 7.15, taking the adoption of multisource off-the-shelf standard parts as 1.0 in each case. These generalizations also assume the following:

- The OEM has experience in the use of standard parts before adopting a PLD or custom design style, and would not attempt complex circuits without considerable in-house experience—if this is not the case the risks may escalate considerably.
- Full in-house supporting CAD is available for the PLD design choice, which can provide comprehensive design and simulation resources and automatic down-loading into the programming unit, thus making the OEM self-sufficient in all the design activities.

The greatest risk with the gate-array, standard-cell and full-custom approach remains that of some error in the specification or design stage unnoticed by the OEM at the simulation approval stage, which then necessitates a rerun of the mask and wafer fabrication, with possible queueing for these outside rerun activities. There is, therefore, considerable attraction in being able to implement any design corrections in-house, as is possible with standard parts or PLD/CPLD design methodologies, but this may not always be able to meet the complexity, performance, space or cost requirements.

7.4 FINANCIAL CONSIDERATIONS

Financial considerations are possibly the most difficult parameters to survey, since they are clearly dependent upon moving technical and market factors. All we can attempt here is to illustrate certain underlying concepts, and provide broad information. Any financial figures quoted here are for illustrative purposes only, and must not be read as detailed information which remains continuously valid. In real terms, the cost of silicon and CAD/CAE tools is likely to continue to fall, but manpower costs may increase. Regrettably, there is no regular publication which concentrates upon these financial aspects, and what figures are randomly published are likely to date very rapidly.

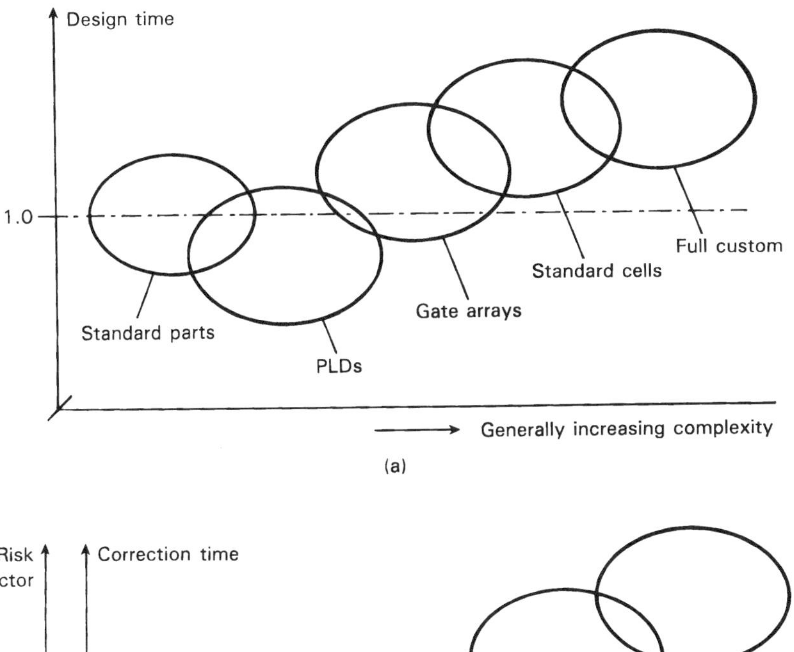

(a)

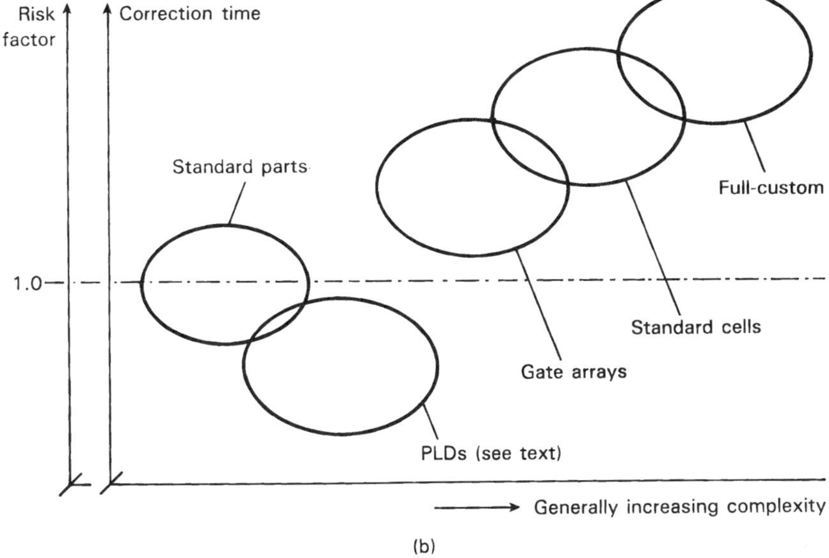

(b)

Fig. 7.15 A relative comparison of design times and correction times and the risk factor with different design styles; exact quantification is product, company, vendor and cost dependent: (a) broad comparison of design-to-prototype times; (b) broad comparison of risk and correction times, which are roughly similar.

7.4.1 NRE Costs

The design of all products involves nonrecurring engineering (NRE) costs, which include designers' time and overheads and other factors which are ideally completely paid for by the time the product is on the production line.

In the case of PLDs and custom microelectronics, the OEM may have to consider costs arising from the factors listed in Table 7.3. Some of these may not arise, and others, such as training, may not be considered by some companies to be NRE expenditure directly costed to a specific new product. Nevertheless, they are costs which must be covered somewhere within the company finances. Note that external PLD costs should not arise, since the whole objective of PLDs is to undertake all design activities in-house so as to give maximum control over the design and prototype manufacture.

We will examine certain of these NRE costs more closely in the following sections, but let us first illustrate how such costs ought to be considered at the early product planning stage. Let:

$E1$ = the total microelectronic NRE cost involved with a particular design style
$E2$ = the remaining OEM NRE costs for the rest of the product, e.g., mechanical design, PCB design, tooling, etc.
A = the per-unit cost to the OEM of the PLD or custom IC
B = the per-unit cost of all the remaining product components
C = the per-unit assembly and test costs

Then, for a production run of N units, the total OEM costs are

$$E1 + E2 + N(A + B + C)$$

Table 7.3 The Microelectronic NRE Costs Which May Be Incurred by the OEM When Using PLD or Custom IC Design Styles[a]

NRE item	PLD	Custom IC
In-house NRE costs		
Purchase or hire of CAD tools	✓	✓
Designer training	✓	✓
Design time and overheads	✓	✓
Special assembly and test resources	✓	✓
Purchase of PLD programmer	✓	–
Outside NRE costs		
Subcontract part or all of the IC design work	–	✓ [b]
Cost of mask-making	–	✓ [b]
Cost of custom test jigs if necessary to interface with standard tester	–	✓ [b]
Iteration costs	–	✓

[a] The individual item costs are not necessarily the same between the two columns.
[b] Possibly a single lump sum NRE cost to OEM.

giving a per-unit cost of

$$\frac{E1 + E2}{N} + (A + B + C)$$

Clearly, if $E1$ is large and/or N is small, then the microelectronic NRE costs will have a profound effect on per-unit costs, and the OEM should consider a design style which reduces this factor as much as possible. On the other hand, NRE costs become much less significant for large quantity production, as the $E1 + E2$ costs can now be amortized over a much larger production number N, with $A + B$ now becoming the most significant parameters.

These are simple equations to give the most economic form of design style; the often insuperable problem, however, is to obtain accurate numbers to put into the equations for different design styles with different suppliers and cost structures.

7.4.2 CAD Costs

The CAD costs possibly represent the largest single NRE expenditure when considering a design style other than one using standard off-the-shelf parts, particularly for first-time adoption by the OEM. In Chapter 5, Section 5.6 indicated some typical hardware costs which may be incurred for both IC design and PCB design activities. Software costs are more elusive, since vendors may often supply software relating to their particular products at preferential rates, and also the OEM may elect to purchase a minimum or a more comprehensive software suite.

A broad generalization of CAD costs which the OEM may consider is shown in Figure 7.16, to which software costs and hardware and software maintenance costs should be added. Hence, a total of at least twice the initial hardware costs may be involved when an OEM builds up in-house resources for design purposes. (This is broadly true for all computer-based activities within a company.)

Thus, the major CAD decisions facing the OEM when a design style other than one using off-the-shelf components is under consideration are as follows:

(i) Adopt a PLD/CPLD or a custom IC design?
(ii) Decide not to purchase any CAD tools, but instead subcontract all microelectronic design activity to an appropriate outside contractor (not really sensible for the PLD/CPLD route as this destroys the PLD advantage of being able to do everything in-house)?
(iii) Purchase some minimum CAD resource just sufficient for the new project, and write off the cost against the first project?
(iv) Alternatively purchase a more comprehensive CAD resource appropriate for future possibly more complex projects, but risk obsolescence before fully utilized?

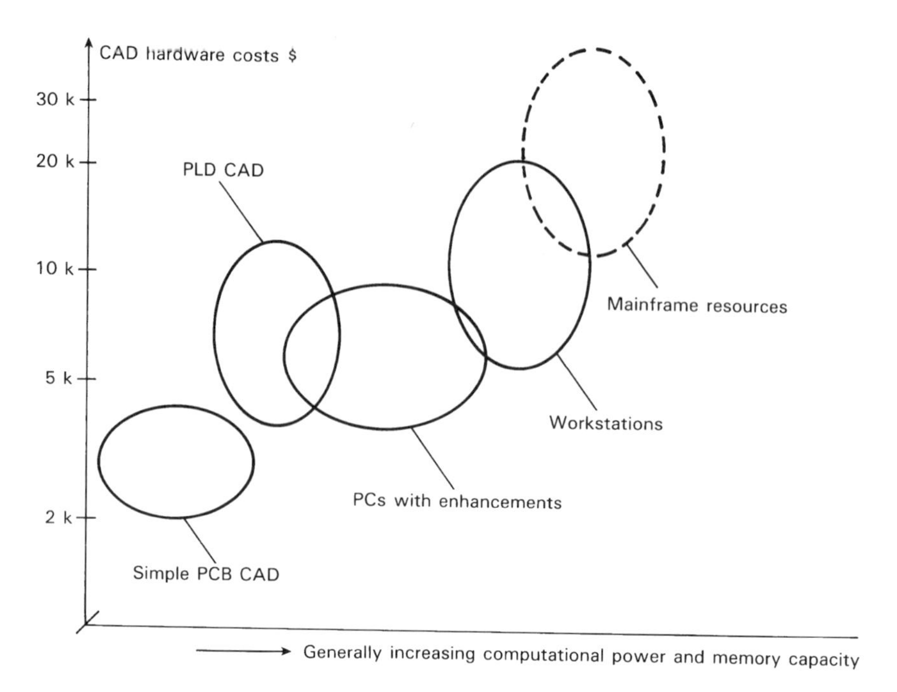

Fig. 7.16 The broad picture of CAD hardware costs; software costs are additional, except possibly with PLD CAD, which may be marketed as a complete vendor's package. (See also Table 5.6 for additional information.)

For first-time company adoption of custom ICs, it is perhaps (iii) which is most appropriate, with, e.g., a PC-based CAD resource on which to gain company experience (see also Section 7.4.3). For the PLD/CPLD route, then the CAD tools recommended or possibly supplied by the PLD vendor may be the best choice, but again, with the risk of perhaps not being fully appropriate for subsequent designs or other vendors' devices. These are therefore decisions to be made by informed managerial judgements, and cannot here be realistically quantified.

7.4.3 Learning Costs

The learning time required by OEM designers to become conversant with new CAD tools is not inconsiderable. Indeed, for a manufacturing company embarking on microelectronic design activities for the first time, it may be a traumatic experience, with considerable company expenditure in salaries during this period over and above that incurred in profitable design work.

Because of this, there is a school of thought which says that the OEM designer can never be as efficient as a designer in a specialist design office who spends 100% of his or her time on custom design [24]. In terms of the productivity parameter 'number of transistors or gates committed per hour,' this is probably true, but if the OEM subcontracts all this design activity, then the savings in in-house learning costs have to be set against the commercial rates charged by the outside agency. In purely financial terms, the latter may be cheaper, but this does not give the OEM any increased in-house experience for future design activities, and there are always problems of potentially incomplete or misunderstood interface communications and staff mobility [24–28].

In broad terms, it takes perhaps two years for an OEM designer to become fully conversant and 'up-to-speed' with CAD tools for IC design activities, of which time perhaps one quarter is spent in nonprofitable activities. With a salary plus company overhead figure of around $100,000 per annum, this implies a company learning cost per designer from scratch of $50,000. For new CAD tools the subsequent learning/retraining times will be considerably less, perhaps $5,000–$10,000 per new system in nonprofitable activity.

There are also the even more intangible learning costs for the company as a whole if no previous microelectronic-based products have been produced and marketed; it may well take five years for a company to build up expertise in source of components, reliable and unreliable suppliers and subcontractors, new shop floor assembly and testing skills, and new sales and marketing requirements.

The general trend of design costs is thus as shown in Figure 7.17. As before, the difficulty is to put real figures into this picture, although some figures may be extracted from existing publications, noticeably by Fey and Paraskevopoulos [29–31].

7.4.4 Testing Costs

The total testing costs involved in a microelectronic-based product includes the following:

- OEM cost of incoming IC testing (if done)
- OEM cost of final product testing
- the vendor's cost of wafer probing and packaged IC testing if custom circuits are adopted

The first two costs occur whatever the chosen microelectronic design style, but the last is particularly significant when custom circuits are involved.

The significance of test has been considered in many sections of the preceding chapters. It is one of the most critical technical factors to discuss with the vendor once the custom IC design style has been chosen. However, it is likely

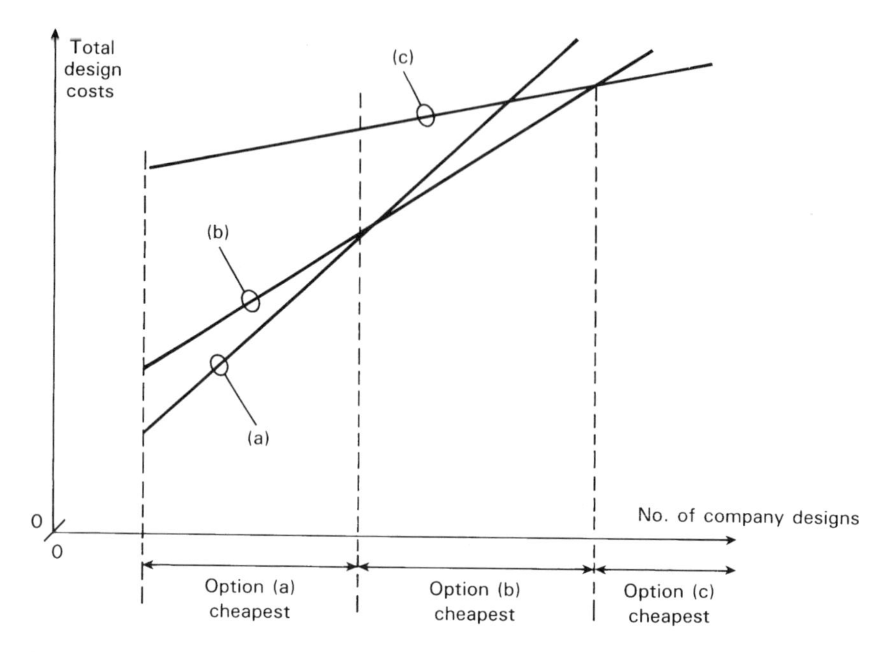

Fig. 7.17 A broad picture of OEM custom microelectronic design costs, which includes the capital cost of CAD tools and learning costs, where: (a) all design work is subcontracted to an outside agency; (b) minimum in-house CAD resources are purchased, e.g., up to schematic capture and netlist preparation; (c) full in-house CAD capability is available. The exact crossover points of these three categories cannot be quantified without detailed product analysis.

that the vendor will not separately charge the OEM for testing unless very long or special test sequences are required, but instead will include the necessary test jigs, etc., in an overall NRE charge (see Table 7.3), and unit test costs in the per-unit price of the packaged circuit. Some details of test cost modeling may be found in Ref. 32, but these relate more to quantity production tests and yeild than to small-quantity situations.

There is, however, one practical difference between the test of a gate-array USIC and a standard-cell or full-custom USIC. In the former, the die size and the number and positions of all the I/Os are fixed, and d.c. supply inputs always use the same I/Os regardless of custom dedication. In the latter case, both the die size and the allocation of the I/Os are fully flexible, and may (probably will) vary between each custom design.

For the vendor, this means that standard wafer-probe cards and test-set interface boards can be used on the computer-controlled IC tester for each size of gate array, but a new probe card and a new interface board may have to be specially manufactured for each individual standard-cell and full-custom circuit. (A probe card is a small card containing a number of spring-loaded probes which bear down on the I/O pads of a die during the wafer test, the wafer being stepped and each die being probed in turn; the interface board provides the necessary signal conditioning between the circuit under test, whether at wafer-probe or subsequent packaged stage, and the [usually large mainframe] tester which provides the test vectors and monitors the test response.) Clearly, if the probe card and the interface board are unique for a particular OEM's circuit, then this cost will be passed on to the OEM in some guise or another.

Hence, the OEM may not know the precise breakdown of costs involved in the vendor's production and testing activities, but the cost of the required testing and any special jigs, etc., will be reflected somewhere in the vendor's pricing structure.

7.4.5 Overall Product Costs

Let us finally attempt to bring these factors together to illustrate the costs facing an OEM considering different design styles. The following figures should be taken as relative or illustrative only, exact figures being dependent upon (a) circuit complexity, (b) existing OEM expertise and CAD/CAM resources, (c) the vendor's need to secure new orders, and (d) the continuing evolution of the market. More detailed sources of costing may be found in Refs. 2, 3, 21, and 22, and in the comprehensive analyses of Refs. 29–31. Reference 31 is reprinted and commented upon in Ref. 3

PLD Vs. Gate Arrays

One of the most widely publicized costings relate to the PLD vs. gate-array choice, since PLDs have strongly attacked the market previously held by gate arrays. Taking 2000 and 5000 gate circuit requirements, cost such as below have been published:

	2000-gate gate array	5000-gate gate array	2000-gate PLD	5000-gate PLD
Total NRE (including design) cost, $\$ \times 1000$	30	50	10	15
Manufacturing cost per gate, cents	0.15	0.2	0.4	0.5

Using these illustrative values, the total costs for a production run of N units are as follows:

(i) 2000-gate array, \$30k + $(2000 \times 0.0015 \times N)$
(ii) 5000-gate array, \$50k + $(5000 \times 0.002 \times N)$
(iii) 2000-gate PLD, \$10k + $(2000 \times 0.004 \times N)$
(iv) 5000-gate PLD, \$15k + $(5000 \times 0.005 \times N)$

Assuming a linear relationship with N (which is not necessarily valid), then the economics of the gate-array vs. the PLD choice is therefore as shown in Figure 7.18.

Care should always be taken to ensure that like-for-like is being compared in such comparisons. For example, in some figures it may be assumed that for

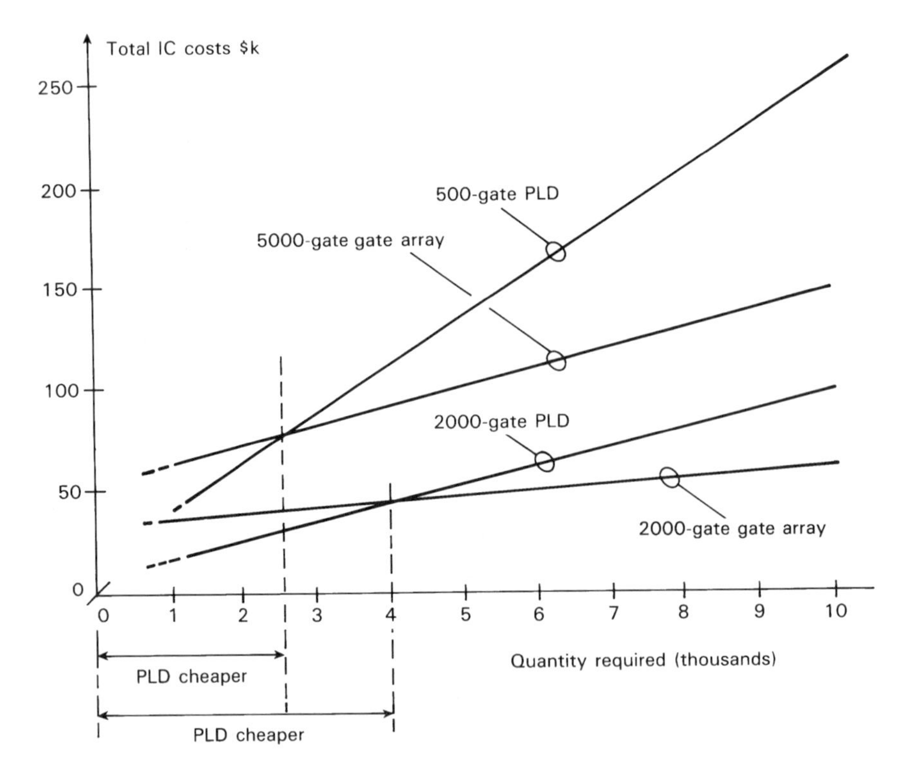

Fig. 7.18 Total PLD and gate-array costs for 2000-gate and 5000-gate applications. Note that falling PLD prices will move the crossover points to higher production values. The increasing availability of CPLDs is unlikely to materially change this relatively simple circuit situation.

the gate-array choice all the circuit design costs are lumped into a vendor's global NRE cost, but for the PLD choice, as there is no subcontracted vendor's design work, the NRE design costs are zero. Similarly, the IC packaging cost may be significant, at least as much as an untested die, which will tend to smooth out differences in the cost per (packaged) gate in different design styles. And, finally, there may be a propensity to adjust values in order to produce neat graphical representations. Nevertheless, there is good evidence that a PLD may be more cost effective than a gate array over an appreciable band of production quantity requirements.

Gate Arrays Vs. Standard Parts

Cost comparisons between gate arrays and off-the-shelf standard parts usually assume that some in-house CAD must be available for the former, but not for the latter. Hence, higher NRE charges are involved in the first-time adoption of the gate-array design style than when using standard parts, but it is then often assumed that this capital expenditure on CAD is freely available for subsequent new product designs without further cost.

An example breakdown of NRE and other costs is given in Table 7.4. The significant feature here is that due to higher initial NRE costs, the first custom IC product is not cheaper than when using standard parts. Subsequent designs, however, when the CAD tools are available and the OEM has gained experience in custom design activities, now begin to show cost advantages.

Table 7.4 Design and Manufacturing Costs for a Typical Digital Product Using Off-the-Shelf Parts and Custom ICs, Showing No Saving to the OEM in First-Time Adoption of the Latter

Activity or expenditure	Off-the-shelf standard parts ($)	First custom circuit ($)	Second custom circuit ($)
Global NRE costs up to the approval of prototype stage			
Total CAD costs	–	**40 000**	–
OEM design time costs	**40 000**	**50 000**	**30 000**
Simulation costs	–	5000	4000
Prototype costs	5000	10 000	10 000
Test costs	**7500**	**7500**	**7500**
Design iteration costs	**5000**	**10 000**	**3000**
Non-IC design and documentation costs	10 000	7500	7500
Company overheads	**15 000**	**15 000**	**15 000**
Sub-totals	82 500	145 000	77 000
Manufacturing costs of 2000 units	140 000	106 000	106 000
Total costs	**222 500**	**251 000**	**183 000**
Ratio	1.0	1.12 (12% increase)	0.82 (18% saving)

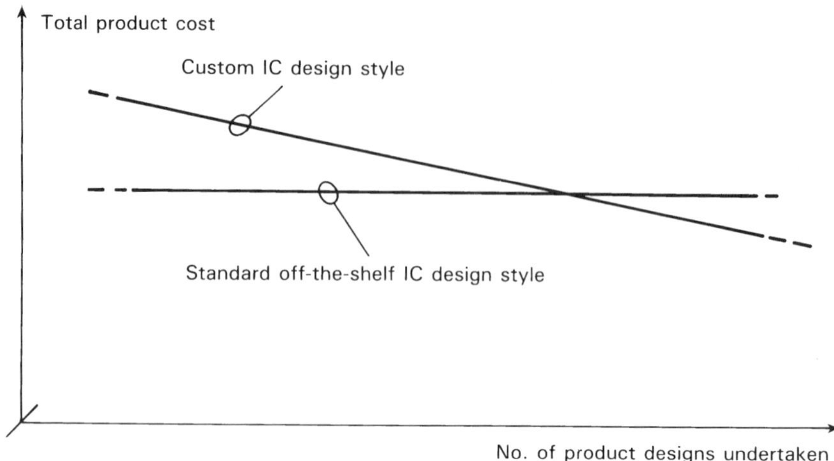

Fig. 7.19 The usual cost characteristic between using standard parts and a custom IC when either is technically feasible for the product. The crossover point where OEM expertise and resources makes the custom choice less expensive is difficult to quantify since it depends upon many company and product factors, but may be the third company design provided that no new CAD tools have to be purchased and learned.

Irrespective of how realistic are the cost figures put into these calculations, it is almost always true that the first-time adoption of custom microelectronics does not save money where an alternative off-the-shelf design style is technically possible. This law is illustrated in Figure 7.19, and is a natural result of new CAD costs and OEM learning time.

Standard Cells Vs. Gate Arrays

Let us now consider standard-cell vs. gate-array design styles. While it is generally true that more powerful CAD resources may be required for standard-cell design activities compared with corresponding gate-array design, the actual design time may be shorter due to less constraints and difficulties in placement and routing. Thus, if we assume that overall NRE cost roughly balance, the major cost difference between the two design styles is in the mask-making and fabrication. In particular, the full mask set of the standard-cell USIC compared with the reduced mask set of the gate array will be the major cost distinction.

Since mask cost up to $3000–$10,000 per mask depending upon area and resolution, a complex standard-cell USIC requiring fifteen or more mask levels involves a considerable initial expenditure, being a major component of the vendor's NRE charges. If a design alteration should, unfortunately, be necessary, then this cost will be repeated.

Considering a relatively simple 1.5 μm circuit, mask costs may be:

- gate-array choice, two commitment masks @ $5000 per mask = $10,000
- standard-cell choice, ten masks @ $5000 per mask = $50,000

Hence, a saving of $40,000 has to be made on the production costs of the latter compared with the former before the standard-cell approach becomes financially advantageous. The position of the break-even point is very vendor-dependent, but could be in the region of 20,000 or more production circuits. Notice also that if, as some vendors suggest, an original gate-array design is moved over to a standard-cell design because of increasing product sales, the full mask set costs have to be recovered before the change becomes financially worthwhile; in practice, due to the limited life cycle (see Figure 7.14) and the possibility of some unforeseen hold-up in the standard-cell replacement, it is extremely unlikely that an OEM would finance and risk such a change unless exceptionally good market prospects prevail.

A final rule-of-thumb of all these deliberations is that on financial grounds, the optimum choice of design style for digital circuits is generally as shown in Figure 7.20. However, note that (a) full-custom has not been included in this survey, since this is usually confined to very specialized and/or extremely large volume requirements, and (b) the growing capabilities of complex programmable logic devices may take over an appreciable part of the gate-array and standard-cell market if vendors' forecasts are correct, although the precise finances in the adoption of CPLDs is still somewhat unclear.

Analog and Mixed Analog/Digital Circuits

Analog and mixed analog/digital custom IC finances cannot be analyzed in the same detail as the above digital-only USICs, since vendors do not generally publish the costs of such custom products. This is largely because of the broad breadth of OEM requirements, making it difficult to give general design, manufacturing and testing costs, allied with the fact that it is very flexible how much the vendor or the OEM does in the actual circuit design activity.

Little additional information can thus be given on these costs. The financial equations for analog USICs may be found discussed in more detail in the report by Trontelj et al. [33] but few hard figures can be quoted.

Lifetime Costs

Finally, as was noted at the end of Chapter 6 in Section 6.10, when dealing with the silicon overheads involved in different test strategies, research work is being

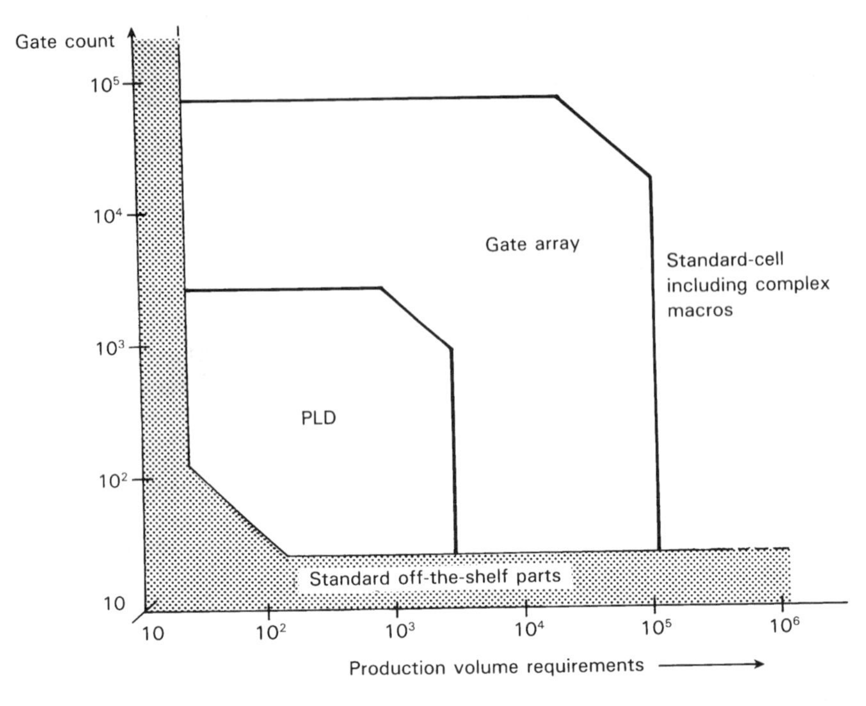

Fig. 7.20 The overall picture for digital-only products, where each design style has its band of gate count/number required which is financially advantageous. Full custom may be considered as a higher continuation of the standard-cell design choice.

done on economics modeling, covering the whole lifetime of a product from design, through manufacture and test, to in-service maintenance and repair. This is still very much an ongoing research area involving knowledge-based computer support and other means, the final objective being a provide a tool which at the initial design stage can point to an optimum design and test strategy for a new microelectronics product. As was noted in Section 6.10, a practical difficulty in the application of such models is to find the figures to put in the equations, without which no firm results and guidance can be generated. Readers are therefore recommended to study the development of this subject to date [34–41], and to be aware of continuing active work in this area.

7.5 SUMMARY

From all the preceding discussions it will be apparent that it is difficult to give firm guidelines on the optimum choice of microelectronic design style for a new company product. The technical decisions on what is *not* appropriate for a given product are much easier than deciding on the 'best' choice; financial considerations are even more diffuse and often indeterminate when all costs, both OEM's and vendor's, are taken into account.

Analyses and cost models are still being developed to assist in design style choice. Mention has previously been given to the definitive work of Fey and Paraskevopoulos [29–31]; a further disclosure, specifically for gate arrays, is that of Edward [42], but to obtain accurate costs representing the continuously changing state of the industry remains the difficulty. The cost of design-for-test covered in the previous chapter is also of supreme importance as custom circuits reach the complexity of VLSI.

Management and design engineers in original-equipment manufactuting companies should be aware of all these aspects in order to be able to take informed decisions on design styles. Information such as covered in the literature [2,3,29,32,33,35–40,43,44], some of which include case studies, should be appreciated. However, in the end it is most probable that if a USIC design style is chosen, it will be for technical and other reasons rather than cost (see the list of potential advantages given at the beginning of Chapter 4); cost is unlikely to be the dominant factor unless very high production volumes are anticipated.

Time to market perhaps remains the most crucial of all the parameters for the success of a new microelectronic product—witness the rapid obsolescence of PCs which do not have this year's latest microprocessor—and must be very high on the list of factors requiring detailed managerial attention.

REFERENCES

1. Vargha, D., Design trade-offs for low voltage single-supply op-amps, *Electronic Product Design*, Vol. 18, 1997, pp. 47–54
2. ICE, *ASIC Outlook 1992*, Integrated Circuit Engineering, Scottsdale, AZ, 1991
3. The Open University, *Microelectronic Decisions*, Microelectronics for Industry Publication PT505MED, The Open University, UK, 1988
4. Wilson, R., 8-bit MCUs wage a counterattack, *Computer Design*, Vol. 29, No. 11, 1990, pp. 73–82
5. Gayakwad, R.A., *Op-Amps and Linear Integrated Circuits*, Prentice Hall, NJ, 1988
6. DiGiacoma, J. (Ed.), *VLSI Handbook*, McGraw-Hill, NY, 1989
7. Geiger, R.L., Allen, P.E. and Strader, N.R., *VLSI Design Techniques for Analog and Digital Circuits*, McGraw-Hill, NY, 1990
8. Haskard, M.R., *Thick Film Hybrids: Manufacturing and Design*, Prentice Hall, NJ, 1989
9. Griffiths, G.W., A review of semiconductor packaging and its role in electronic manufacture, *J. Semicustom ICs*, Vol. 7, No. 1, 1989, pp. 31–39

10. Bakaglu, H.G., *Circuits, Interconnections and Packaging for VLSI*, Addison-Wesley, MA, 1990
11. Sinnadurai, F.N. (Ed.), *Handbook of Microelectronic Packaging and Interconnect Technologies*, Electrochemical Publications, Scotland, 1985
12. Christou, A., *Integrating Reliability into Microelectronics Manufacturing*, Wiley, NY, 1994
13. Editorial, Increasing efficiency with flip-chip on board, *Embedded Systems Engineering*, Vol. 5, No. 1, 1997, p. 8
14. O'Flaherty, M., Cahill, C., Rogers, K. and Slattery, O., Validation of numerical models of ceramic pin grid array packages, *Microelectronics Journal*, Vol. 28, 1997, pp. 229–238
15. VLSI Support Technologies, *Computer-Aided Design, Testing and Packaging*, IEEE Computer Society Press, CA, 1982
16. Booth, W., *VLSI Era Packaging*, Electronic Trend Publications, CA, 1987
17. National Semiconductor, *ASIC Packaging*, Booklet No. 101185, National Semiconductor Corporation, Germany, 1987
18. Freeman, E., Cost, device speed, size and reliability determine the best package for an ASIC, *EDN*, Vol. 32, No. 9, 1987, pp. 77–82
19. Fahr, G.K., Logic array packaging, *Application Report Note No. A23*, LSI Logic Corporation, CA, 1982
20. Reinertsen, D.G., Whodunit? The search for the new product killers, *Electronic Business*, Vol. 9, 1983, pp. 62–66
21. Xilinx, *The Programmable Gate Array Data Book*, Xilinx Corporation, CA, 1989
22. Altera, *PLDs vs. ASIC Economics*, Altera Corporation Product Information Bullitin No. 6, CA, 1988
23. Hurst, S.L., Custom microelectronics: the risk factor, *Microprocessors and Microsystems*, Vol. 14, No. 4, 1990, pp. 197–203
24. Kutzin, M., Advantages of ASIC design centres, *J. Semicustom ICs*, Vol. 5, No. 1, 1987, pp. 12–15
25. Heberling, C., The modern silicon foundary and CAD software, *J. Semicustom ICs*, Vol. 6, No. 1, 1988, pp. 19–24
26. Horne, N.W., The case for strategic alliances, *J. Semicustom ICs*, pp. 25–28
27. Shenton, G., ASIC procurement: is getting married a good idea? *J. Semicustom ICs* pp. 29–32
28. White, A., The painless path to ASIC design, *J. Semicustom ICs*, pp. 33–36
29. Fey, C.F. and Paraskevopoulos, D.E., Studies in LSI technology: a comparison of product cost using MSI gate arrays, standard cells and full custom, *IEEE J. Solid State Circuits*, Vol. SC.21, 1986, pp. 297–303
30. Fey, C.F. and Paraskevopoulos, D.E., A model of design schedules for application specific ICs, *J. Semiconductor ICs*, Vol. 4, No. 1, 1986, pp. 5–12
31. Fey, C.F. and Paraskevopoulos, D.E., A techno-economic assessment of application-specific circuits: current status and future trends, *Proc. IEEE*, Vol. 75, 1987, pp. 829–841 (also reprinted in [3] and [6])
32. Teradyne, *The Economics of VLSI Testing*, Teradyne, Inc., CA, 1984
33. Trontelj, J., Trontelj, L. and Shenton, G., *Analog Digital ASIC Design*, McGraw-Hill, UK, 1989

34. Goel, P., Test generation costs, analysis and projections, *Proc. IEEE Int. Test Conf.*, 1980, pp. 77–84
35. Turino, J., *Design to Test: A Definitive Guide for Electronic Design, Manufacture and Service*, Van Nostrand Reinholt, NY, 1990
36. Ambler, A.P., Abidir, M. and Sastry, S. (Eds.), *Economics of Design and Test for Electronic Circuits and Systems*, Ellis Horwood, UK, 1992 (extended version of papers presented at the First International Workshop on the Economics of Design and Test, Austin, TX, 1992)
37. Davis, B., Economic modeling of board test strategies, *Proc. IEEE 2nd. Int. Workshop on the Economics of Design and Test*, Austin, TX, May 1993
38. Dislis, C., Dear, I.D. and Ambler, A.P., The economics of chip level testing and DFT, *IEEE Int. Test Conf., Digest Test Synthesis Seminar*, 1994, pp. 2.1.1–2.1.8
39. Special Issue, Economics of Electronic Design, Manufacture and Test, *J. of Electronic Testing: Theory and Applications (JETTA)*, Vol. 5, Nos. 2/3, 1994, pp. 127–312
40. Dislis, C., Dick, J.H., Dear, I.D. and Ambler, A.P., *Test Economics and Design for Testability*, Ellis Horwood, UK, 1994
41. Hurst, S.L., *VLSI Testing: Digital and Mixed Analogue/Digital Techniques*, IEE Publications, UK, 1997
42. Edward, L.M.N., USIC cost simulation: a new solution to an old problem, *J. Semicustom ICs*, Vol. 8, No. 2, 1990, pp. 3–12
43. The Open University, *Microelectronic Matters*, Microelectronics for Industry Publication No. PT504MM, The Open University, UK, 1988
44. Altera, *A Manager's Perspective*, Altera Corporation, CA, 1990

8

Conclusions

8.1 THE PRESENT STATUS

From the information which we have attempted to survey in the preceding chapters, it is apparent that the present status of custom microelectronics is one of considerable expertise and achievement, particularly in the following respects:

- Silicon fabrication technology has reached a very high level of maturity, with submicron geometry now being in full production for state-of-the-art requirements.
- Extensive gate-array and standard-cell developments have provided the OEM with a wide range of products and choice of vendors for digital applications.
- CAD resources have also developed to the stage where most digital circuits can be readily designed and simulated for a target realization in gate-array or standard-cell form.
- Costs of both design and fabrication have fallen in real terms over the past two decades to the point where—assuming no start-up/learning time expenditure—a custom realization may become financially viable for production quantities as low as a few thousand units.
- The problems of testing large networks are now widely appreciated and freely discussed between OEM and vendor, with the adoption of DFT strategies where appropriate.
- Recent growth in the capabilities of complex programmable logic devices has introduced a new player in the field of VLSI digital design,

which allows the OEM to maintain in-house all the design and development work for an increasing size of digital network.

However, as we have seen, there are still shortcomings and risks which the OEM should appreciate when deciding to adopt a custom design style, among these factors being the following:

- OEM analog and mixed analog/digital custom requirements are not as comprehensively served by vendors' products and the supporting CAD design tools as the digital-only area—this is partly a reflection of the much greater range of needs found in the analog side of OEM activities.
- CAD for the very highest levels of design, namely, the architectural and behavior levels involving hardware description languages (HDLs and VHDL, etc.), have yet to reach the level of maturity and widespread acceptance of the lower-level CAD tools.
- First-time adoption of custom-design techniques by an OEM inevitably involves a learning period and possible risk of errors and omissions in first designs.
- Obsolescence of products and CAD tools, difficulties of second-sourcing without incurring further NRE costs, and take-overs and mergers between vendors are serious factors which an OEM has to consider if long-term product availability is required.

Silicon will remain the dominant semiconductor technology in unipolar and bipolar form for both custom and off-the-shelf microelectronic circuits. The challenge of BiCMOS and GaAs does not seem to have been successful; BiCMOS has recently been dropped by some vendors, and GaAs has been discontinued in favor of fifth-generation bipolar transistors in certain mobile telephone products [1]. CMOS with bipolar for very high frequencies thus remain the present market leaders.

Optical lithography and the use of mask sets remains and is likely to remain the principal method of patterning silicon during the wafer fabrication stages (see Section 8.2). E-beam direct-write-on-wafer has not proved to be a viable alternative for the custom field, largely due to the equipment capital costs and the time to process, but may retain an important niche where very fast turn-around (a day or two rather than weeks) and only one wafer needs to be processed. Other fast turn-around techniques also do not seem to have been overtly successful.

8.2 FUTURE DEVELOPMENTS

The past and present developments relating to semiconductor technology and custom microelectronics are likely to remain the underlying principles and practices for the future. Future developments are unlikely to be anything radically new over and above that which is already known or under active consideration, but continuing developments are to be expected in semiconductor fabrication, design and test strategies, intelligent CAD support and other areas such as neural networks and optoelectronics. Looking at these in turn, some further comments can be made.

Semiconductor Technologies

The emphasis upon smaller geometries will be a continuing activity, with deep submicron fabrication becoming a standard production capability. Standard-cell CMOS products with a 0.35 μm geometry and a usable gate density of 14000 logic gates per mm^2 are already available on the commercial market, with a quoted maximum complexity of over 3,000,000 equivalent gates. The library functions include microcontroller and digital signal processing macros. However, it is probable that the majority of OEM custom requirements will continue not to require the maximum possible capabilities, and perhaps around a 1 or 1.5μm geometry will suffice for most OEM circuits.

However, the limits of optical lithography and the escalating capital cost of production lines to fabricate very small geometry microelectronic circuits is putting a constraint upon further reduction in device size. Figure 8.1 indicates an optical limit of around 0.2 μm, below which electron beam lithography is indicated. Previous forecasts of the limit of optical lithography have, however, been overcome with continuing development, and therefore the possible final limit of optical means will probably prove to be around 0.5 to 1.0 μm. The economic problem of increasing production line costs is such that multinational and/or multicompany funding is being pursued, such that the 10^8 or more dollar costs are shared over several financial sources. The present interest in 'cluster tools' in which the wafer fabrication steps are divided over a number of smaller specific processing tools may help to control these capital costs [2].

Silicon-on-insulator (SOI), gallium-arsenide and other III/V semiconductor compounds [3–5] will continue to receive research attention, but it remains most probable that these technologies will remain for specialist applications, e.g., radar, rather than for the wider custom IC market. The increase in performance of CMOS and bipolar silicon technologies should mean that these other semiconductor technologies will not prove to be worthwhile for mainstream VLSI circuit adoption. Absolute maximum fabrication capabilities will continue to be driven by the memory field, with MOS remaining the dominating technology.

The semiconductor technologies for analog circuits should not see any radical changes in the future compared with the present, but will benefit from the general research and improvements in both bipolar and unipolar device develop-

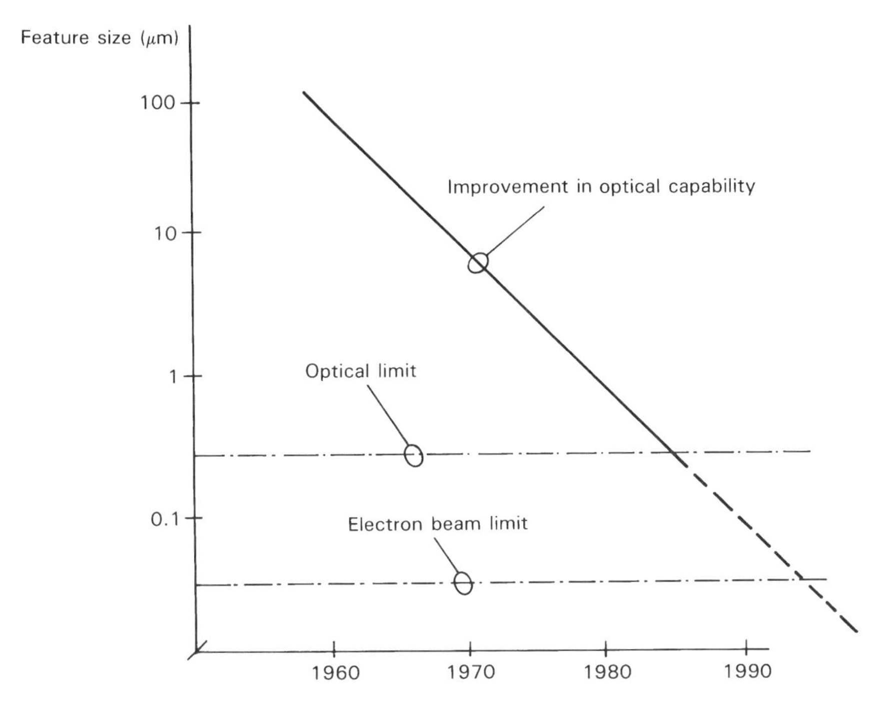

Fig. 8.1 The limits of optical lithography (see text), with electron-beam lithography providing about a further order of magnitude reduction in pattern resolution. X-ray patterning may also be considered. (Note, mask alignment and etching accuracy are additional practical limitations.)

ment. The greatest evolution may be in the CAD tools used for analog circuit design (see below) rather than the semiconductor technologies.

Design, Test and CAD

The adopted principals of digital design have not materially changed for several decades, being very largely based upon Boolean algebra. Research activity in the 1980s on spectral design techniques [6,7] has not proved advantageous to date, and is considered to be unlikely to be so in the future. However, there has recently been an upsurge in research work on fundamental logic design, including work on binary decision diagrams (BDDs), state machine optimization, structural optimization, simulated annealing, high-level synthesis using genetic algorithms, stochastic techniques and other concepts [8–16]. Previous constraints due to computer-intensive computational requirements have, to some extent, been reduced due to the increasing desk-top computing power now becoming available, including what has been termed 'high-performance computing' (HPC), which is defined as computing resources which provide more than an order of magnitude more computing power than is normally available on a desktop computer [17]. Parallel computation may also be a future means to handle computationally intensive simulation and other design requirements.

Similarly, on the digital signal processing side and on the analog side, there are ongoing design research and development activities [12,18–20]. The digital signal processing field, in particular, is showing many design developments in 'one-time-programmable' (OTM) microcontrollers, integrated analog input/analog output DSP-based control processors, in-system reconfigurable FPGA digital signal processors and other off-the-shelf products which may filter down to the custom circuits area as future library macros. Another recent development which may foreshadow a new approach to embedded system design is the mixed-mode USIC illustrated in Figure 8.2, which combines a processor solely designed for use within a custom IC; this processor has a powerful range of fast arithmetic instructions but omits the complex architecture for rapid interrupts which are a feature of most general-purpose off-the-shelf μPs [21]. Such developments may allow efficient microprocessor macros to be more readily incorporated in custom ICs than overpowerful general-purpose processors.

On the CAD side, present research and development on high-level design, hardware description languages and knowledge-based systems will continue, and will cover analog as well as digital circuit synthesis [22–30]. Structured approaches to the design and automated test of analog and mixed-signal systems are of particular potential significance; innovative new companies such as Opmaxx in Beaverton, OR, with their DesignMax, FaultMax and TestMax software developments are pointing the way to a possible new generation of design and test tools [31]. CAD vendors will increasingly package the best software currently available so as to provide a 'seamless' set of tools for OEM use from behavioral level downwards. However, there is no present indication that any radically new concepts for design-for-testability will emerge in the foreseeable future, but boundary scan and self-test will increasingly be part of the CAD capability.

In connection with the latter, the existence of IEEE Standard 1149.1–1990 was covered in Chapter 6, Section 6.8.2. This will continue to be the recognized standard for digital boundary scan. However, the further IEEE specification for analog scan tests, namely, P.1149.4, which is currently in provisional form awaiting finalization, will cover the boundary scan requirements for analog ICs, and should be part of the OEM's design responsibilities in the future. Other standard specifications, such as the IEEE Standard 1394–1995, which covers high-speed interconnect between peripherals and PCs, may also have to become part of the knowledge required by the OEM designer in custom design activities [32–35].

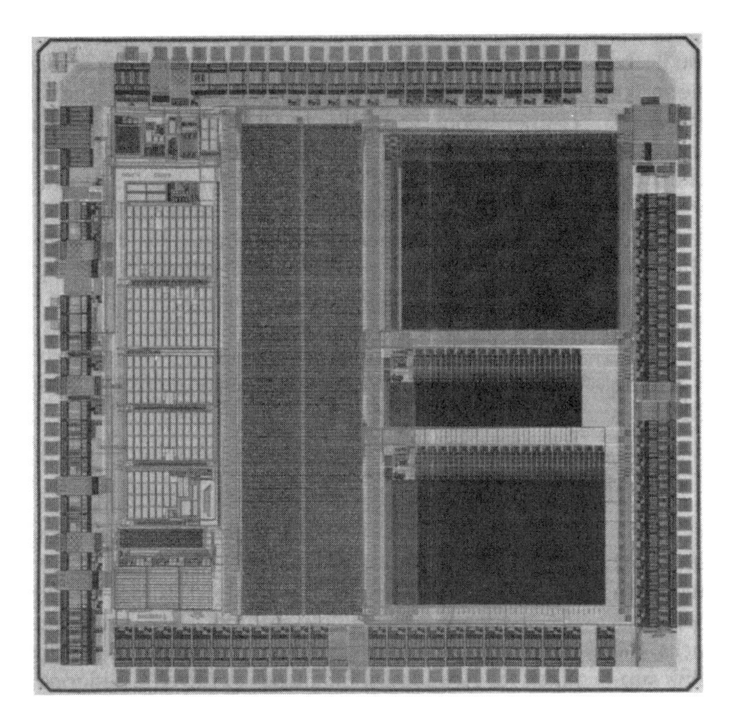

Fig. 8.2 A mixed-mode custom design for a mobile telecomms product incorporating a XAP processor, implemented on a 0.5 μm CMOS process. The analog cells are on the left, the central column includes the 3000-gate XAP processor, and the three right-hand blocks are RAM and ROM memory. The analog cells are full-custom with the digital cells being taken from a standard-cell library. (Courtesy of Cambridge Consultants Ltd., Cambridge, UK.)

The future may also see the commercial availability of economics modeling software as introduced in the preceding chapter, thus to allow the OEM to take informed decisions on the optimum choice of design style at an early stage in the design process [36]. However, the viability of this tool will ultimately depend upon the cooperation and ability of vendors in providing good and up-to-date cost data to include in the OEM calculations, and this may be the difficulty in its global usefulness.

Neural Networks

All the presently available commercial digital computers are binary with von Neuman architectures. There has, however, been a considerable amount of research work over the past decades on multiple-valued logic, where the radix is not confined to two. This work on ternary and higher valued logic has never reached commercial viability, except in one or two special cases such as certain memory arrays, the major difficulty being the absence of any microelectronic device which can readily and stably take more than two stable states [37–40]. Unless there is some breakthrough in physical devices, there is little future prospect of a higher radix than 2 becoming available. (The only foreseeable possibility may be in the optoelectronics area [see below], but the global dominance of binary will be difficult to replace.)

However, the work on neural networks, sometimes and more correctly termed *artificial neural networks* (ANNs), has reached a stage where neural network circuits in microelectronic form have been fabricated, and the performance of learning and self-adaptive networks has been demonstrated [41–45]. There is, therefore, a strong possibility that neural networks may enter the OEM custom microelectronics area for control engineering, pattern recognition and similar applications, although full-custom is likely to be the first in this field in advance of IC vendors providing library data.

Optoelectronics

The final area of future possibilities in the custom microelectronics field which we will briefly reference is the optoelectronics field, where the system information is carried by photons rather than by electron flow. Again, a very great deal of research has been done in this field, the one area where it has become firmly established being the telecommunications area with optical fiber transmission and optical amplifiers being widely used on metropolitan area networks upwards.

The present commercial status of optical telecoms is largely that of discrete optical devices, with the digital logic processing being done with very high speed conventional semiconductor devices. The system circuits may therefore be regarded as full-custom hybrid assembly products, although the conventional electronics may be considered as very special standard macros (or even off-the-shelf products) in many cases. However, considerable academic research has also been pursued on optical logic, with optical signals being used to realize Boolean logic relationships, and on the use of optical signals for multiple-valued logic, threshold logic and neural network realizations. The theoretical advantages of optical signals are considerable, the greatest one being the absence of cross-talk between optical paths and freedom from the effects of any outside electromagnetic or electrostatic inteference.

Details of the research to date may be found in the published literature [46–56], but although some small monolithic circuits with optical logic have been produced, it remains very much in the future for any possible adoption of this technology by OEMs to meet their product requirements. Silicon-based monolithic circuits as at present are therefore likely to continue as the main technology for OEM custom applications.

8.3 FINAL SUMMARY

To return from the possible distant future to the immediate future, for the majority of OEMs the capabilities of silicon fabrication will therefore continue to provide the required performance requirements. The most need is for the continuing development in user-friendly supporting CAD tools, particularly in (a) increased

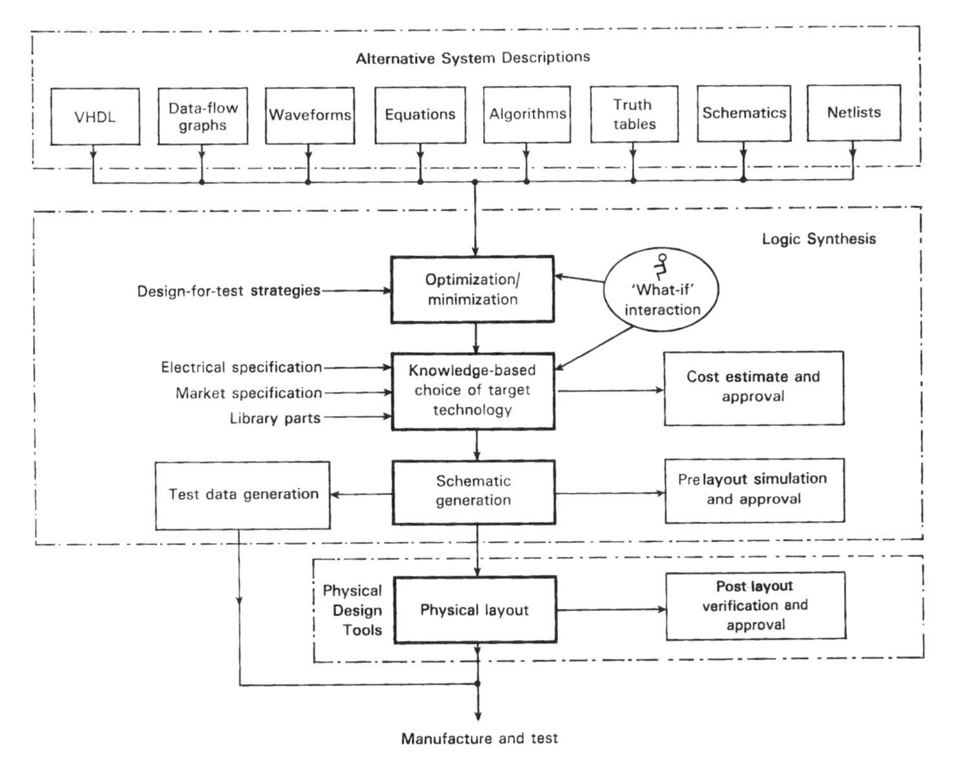

Fig. 8.3 A possible future integrated CAD resource for OEM design activities, with the vendor taking over after successful completion of the hierachical design activities for other than PLD designs. Some additional feedback loops (not shown) must be present to accomodate failure to meet target requirements at lower levels of design. Several commercial HDLs may be involved or chosen in the top hierarchical level, and PLDs/CPLDs should be present in the choice of design style.

standardization of data exchange formats between design styles and vendor's products, (b) seamless transfer of design data up and down the hierarchy of levels of design, (c) increased ease to consider and incorporate appropriate design-for-test strategies in a new product design, and (d) increased ease to handle analog and mixed analog/digital custom designs. Figure 8.3 shows an ideal CAD structure, allowing easy 'what-if' interaction by the OEM designer at the early design stages. However, whatever the precise style of custom design adopted, there will always remain the learning curve for new circuit designers, and therefore it is a wise policy to undertake reasonably straightforward designs in the first place in order to gain confidence and expertise.

Future developments will continue to be presented, particularly at the IEEE Design Automation Conference (DAC), the Custom Integrated Circuits Conference (CICC) and the International Conference on Computer Design (ICCD), and published in their resulting Proceedings, with testing aspects covered specifically in the specialist *Journal of Electronic Testing: Theory and Application (JETTA)*.

REFERENCES

1. Design news, Silicon transistors replace GaAs in mobile phones, *Electronic Product Design*, Vol. 18, No. 3, 1997, p. 14
2. Atherton, L.F. and Atherton, R.W., *Wafer Fabrication: Factory Performance and Analysis*, Kluwer Academic Publishers, MA, 1995
3. Liou, J.J., *Principles and Analysis of AlGaAs/GaAs Hetrojunction Bipolar Transistors*, Artech House, MA, 1996
4. Ploog, K.H. (Ed.), *III/V Quantum System Research*, Institution of Electrical Engineers Publications, UK, 1995
5. Misugi, T. and Shibatomi, A. (Eds.), *Compound and Josephson High-Speed Devices*, Plenum Press, NY, 1993
6. Beauchamp, K.G., *Applications of Walsh and Related Functions*, Academic Press, UK, 1984
7. Hurst, S.L., Miller, D.M. and Muzio, J.C., *Spectral Techniques in Digital Logic*, Academic Press, UK, 1985
8. Murgai, R., Brayton, R.K. and Vincentelli, A.S., *Logic Synthesis for Field Programmable Gate Arrays*, Kluwer Academic Publishers, MA, 1996
9. Sait, S.M., Ali, S. and Benton, M.S.T., Scheduling and allocation in high-level synthesis using stochastic techniques, *Microelectronics Journal*, Vol. 27, 1996, pp. 693–712
10. Lanchares, J., Hidalgo, J.I. and Sanchez, J.M., A method for multiple-level logic synthesis based on the simulated annealing algorithm, *Microelectronics Journal*, Vol. 28, 1997, pp. 143–150
11. Lee, M.T. -C., *High-Level Synthesis of Digital VLSI Circuits*, Artech House, MA, 1997

12. Litovski, V. and Zwolinski, M., *VLSI Circuit Simulation and Optimization*, Chapman and Hall, UK, 1997
13. Kam, T., Villa, T., Brayton, R. and Vincentelli, A.S., *Synthesis of Finite State Machines: Functional Optimization*, Kluwer Academic Publishers, MA, 1997
14. Monteiro, J. and Devadas, S., *Computer-Aided Design Techniques for Low Power Sequential Logic Circuits*, Kluwer Academic Publishers, MA, 1997
15. Saucier, G. and Mignotte, A. (Eds.), *Logic and Architecture Synthesis: State-of-the-Art and Novel Approaches*, Chapman and Hall, UK, 1995
16. Sasao, T. and Fujita, M. (Eds.), *Representations of Discrete Functions*, Kluwer Academic Publishers, MA, 1996
17. Hey, A.J.G., High-performance computing—past, present and future, *IEE Computing and Control Journal*, Vol. 8, No. 1, 1997, pp. 33–42
18. Richards, M.A., Gadient, A.J. and Frank, G.A. (Eds.), *Rapid Prototyping of Application Specific Processors*, Kluwer Academic Publishers, MA, 1997
19. Meng, T.H. and Malik, S. (Eds.), *Asynchronous Circuit Design for VLSI Signal Processing*, Kluwer Academic Publishers, MA, 1994
20. Chang, H., Charbon, E., Choudhury, U., Demir, A., Felt, E., Liu, E., Malavasi, E., Vincentelli, A.S. and Vassiliou, I., *A Top-Down Constraint-Driven Methodology for Analog Integrated Circuits*, Kluwer Academic Publishers, MA, 1997
21. Dettmer R., Getting XAPped: the XAP processor, *IEE Review*, Vol. 43, No. 3, 1997, pp. 113–115
22. Mantooth, A. and Fiegenbaum, M., *Modeling with Analog Hardware Description Language*, Kluwer Academic Publishers, MA, 1994
23. Analogy, *AHDL Tutorial*, Analogy, Inc., Beaverton, OR, 1993
24. Saleh, R.A., Rhodes, D.L., Christen, E. and Antao, B.A.A., Analog hardware description languages, *Proc. IEEE Custom Integrated Circuits Conf.*, 1994, pp. 349–356
25. Civardi, L., Gatti, U. and Torelli, G., An AHDL-based methodology for computer simulation of switched-capacitor filters, *Microelectronics Journal*, Vol. 27, 1996, pp. 485–497
26. Harjani, R.A., Rutenbar, R.A. and Carley, L.R., A prototype framework for knowledge-based analog circuit synthesis, *Proc. IEEE Design Automation Conf.*, 1987, pp. 42–49
27. David, G., Boute, R.T. and Shriver, B. (Eds.), *Declarative Systems*, North Holland, Amsterdam, 1990
28. Fatchey, E.-T. and Perry, E.E., BLADES: an artificial intelligence approach to analog circuit design, *IEEE Trans. Computer Aided Design*, Vol. CAD.8, 1989, pp. 680–692
29. Berckan, E., From analog design description to layout: a new approach to analog silicon compilation, *Proc. IEEE Custom Integrated Circuits Conf.*, 1989, pp. 4.4.1–4.4.4
30. Bourai, Y., Izeboudjen, Y., Bouhabel, Y. and Tafit, A., A new approach for an AHDL based on system semantics, *Proc. IEEE Design Automation Conf.*, 1996, pp. 201–206
31. Opmaxx, *Commercial literature, DesignMax, FaultMax and TestMax*, Opmaxx, Beaverton, OR, 1997

32. IEEE, Standard 1149.1-1990, *Standard Test Access Port and Boundary Scan Architecture*, Institute of Electrical and Electronics Engineers, NY, 1990
33. IEEE, *Supplement 1149.1b to IEEE Std. 1149.1*, Institute of Electrical and Electronics Engineers, NY, 1994
34. IEEE, *Standard P.1149.4, Mixed Signal Test Bus Framework Proposals*, Institute of Electrical and Electronics Engineers, NY, 1992
35. IEEE, *Standard P.1149.5, Standard Module Test and Maintenance Bus*, Institute of Electrical and Electronics Engineers, NY, 1994
36. Dislis, C., Dick, J.H., Dear, I.D. and Ambler, A.P., *Test Economics and Design for Testability*, Ellis Horwod, UK, 1995
37. Hurst, S.L., Multiple-valued logic: its status and its future, *IEEE Trans. Computers*, Vol. C.33, 1984, pp. 1160–1179
38. Rine, D.C. (Ed.), *Computer Science and Multiple Valued Logic*, North Holland, NY, 1984
39. Muzio, J.C. and Wesselkamper, T.C., *Multiple-Valued Switching Theory*, Adam Hilger, UK, 1986
40. Butler, J.T., *Multiple-Valued Logic in VLSI*, IEEE Computer Science Press, CA, 1991
41. Picton, P.D., *Introduction to Neural Networks*, Macmillan, UK, 1994
42. Fakhraie, S.M. and Smith, K.C., *VLSI-Compatable Implementations for Artificial Neural Networks*, Kluwer Academic Publishers, MA, 1996
43. Jabri, M.A., Coggins, R.J. and Flower, B.G., *Adaptive Analog VLSI Neural Systems*, Chapman and Hall, UK, 1996
44. Mead, C., *Analog VLSI and Neural Systems*, Addison-Wesley, MA, 1989
45. Grossburg, S., *Neural Networks and Natural Intelligence*, MIT Press, MA, 1988
46. Journal, *International Journal of Optical Computing*, Wiley, UK, published quarterly
47. Special Issue on Optoelectronics, *The GEC Journal of Research*, Vol. 10, No. 2, 1993, pp. 66–128
48. Cvijetic, M., *Coherent and Nonlinear Lightwave Communications*, Artech House, MA, 1996
49. Chuang, S.L., *Physics of Optoelectronic Devices*, Wiley, NY, 1995
50. Riaziat, M.L., *Introduction to High-Speed Electronics and Optoelectronics*, Wiley, NY, 1996
51. Caufield, H.J., Kinser, J. and Rogers, S.K., Optical neural networks, *Proc. IEEE*, Vol. 77, 1989, pp. 1573–1583
52. Feitelson, D.G., *Optical Computing*, MIT Press, MA, 1988
53. Zappe, H.P., *Introduction to Semiconductor Integrated Optics*, Artech House, MA, 1995
54. Psaltis, D., Optical realization of neural network models, *Proc. SPIE Int. Optical Computing Conf.*, 1986, pp. 278–282
55. Hsu, K., Brady, D. and Psaltis, D., Experimental demonstration of optical neural computers, *Neural Information Processing Systems*, 1988, pp. 377–386
56. Kupiec, S.A. and Caulfiefd, H.J., Massively parallel optical PLA, *Int. J. Optical Computing*, Vol. 2, 1991, pp. 49–62

Appendix A
The Elements and Their Properties

To supplement the specific information covered in the main chapters of this text, this and the following appendices give some further information which may provide some additional background material. This first appendix gives the familiar periodic table of the elements, followed by some factors of significance for microelectronic fabrication or circuit performance.

The periodic table is shown in Table A.1, with the atomic number of each element given in parentheses. The elements between Group II and III are the transition elements, while the elements from lanthanum (La), atomic number 57, to lutetium (Lu), atomic number 71, are the lathanons or rare earth elements. The higher end of the table constitutes the naturally radioactive and increasingly unstable elements.

The arrangement of electrons and their energies determines virtually all the mechanical, chemical and electrical properties of the elements. Table A.2 lists some of the elements with the number of atoms in the various electron shells surrounding the nucleus in their lowest energy state, that is, at 0°K. Note that each shell has a principal quantum number 1, 2, 3, ..., and may have several orbital quantum numbers s, p, d,....For example, within the N-shell there may be four quantum numbers: 4s, 4p, 4d, and 4f.

Good electrical conductors such as copper (Cu, 29), silver (Ag, 47), platinum (Pt, 78) and gold (Au, 79) are characterized by a single loosely bound electron in their outer (valency) orbit, with inner orbits full or almost full. Poor conductors (insulators), on the other hand, such as the inert gases neon (Ne, 10), argon (Ar, 18), etc., are characterized by an atomic structure having no easily available electrons for conduction purposes. Insulating compounds such as silicon

Table A.1 The Periodic Table of the Elements with the Atomic Number of Each Element Given in Parentheses

Group I	Group II											Group III	Group IV	Group V	Group VI	Group VII	Group VIII
H (1)																	He (2)
Li (3)	Be (4)											B (5)	C (6)	N (7)	O (8)	F (9)	Ne (10)
Na (11)	Mg (12)	TRANSITION ELEMENTS										Al (13)	Si (14)	P (15)	S (16)	Cl (17)	Ar (18)
K (19)	Ca (20)	Sc (21)	Ti (22)	V (23)	Cr (24)	Mn (25)	Fe (26)	Co (27)	Ni (28)	Cu (29)	Zn (30)	Ga (31)	Ge (32)	As (33)	Se (34)	Br (35)	Kr (36)
Rb (37)	Sr (38)	Y (39)	Zr (40)	Nb (41)	Mo (42)	Tc (43)	Ru (44)	Rh (45)	Pd (46)	Ag (47)	Cd (48)	In (49)	Sn (50)	Sb (51)	Te (52)	I (53)	Xe (54)
Cs (55)	Ba (56)	La (57*)	Hf (72)	Ta (73)	W (74)	Re (75)	Os (76)	Ir (77)	Pt (78)	Au (79)	Hg (80)	Tl (81)	Pb (82)	Bi (83)	Po (84)	At (85)	Rn (86)
Fr (87)	Ra (88)	Ac (89†)															

Lanthanons *(Rare earths)	Ce (58)	Pr (59)	Nd (60)	Pm (61)	Sm (62)	Eu (63)	Gd (64)	Tb (65)	Dy (66)	Ho (67)	Er (68)	Tm (69)	Yb (70)	Lu (71)
Actinons †(Radioactive elements)	Th (90)	Pa (91)	U (92)	Np (93)	Pu (94)	Am (95)	Cm (96)	Bk (97)	Cf (98)	Es (99)	Fm (100)	Md (101)	No (102)	Lr (103)

dioxide (SiO_2) likewise have no electrons readily available for conduction purposes in their molecular structure.

The semiconductor elements and semiconductor compounds fall between these two extremes. Table A.3 gives a section of the periodic table which is of interest for microelectronic purposes, specifically the elements which have four electrons in their valency band and which can be made in a crystal lattice structure, and the elements in the immediately adjoining bands.

The two naturally occuring elements for practical semiconductor purposes are silicon and germanium, both of which can be made in a crystal lattice form and which have an intrinsic conduction midway between conductors and insulators. Carbon in its crystal form (diamond) has too high an electron energy gap in its valency band to be useful—it is effectively an insulator—while tin and lead have too low an energy gap and are effectively conductors at normal ambient temperatures. This leaves silicon and germanium as the useful elements, with silicon having the the preferable fabrication and performance properties under normal operating conditions. The possible n-type dopants are given by the group V elements, contributing additional electrons for conduction purposes when introduced into a germanium or silicon crystal structure, while the possible p-type dopants are from the group III elements. Semiconductor compounds such as gal-

Table A.2 An Extract from the Periodic Table Expanded To Give the Electron Shell Arrangements.

Atomic number	Element	Electron arrangement																				
		K	L		M			N				O					P				Q	
		1s	2s	2p	3s	3p	3d	4s	4p	4d	4f	5s	5p	5d	5f	5g	6s	6p	6d	6f	7s	7p
1	H	1																				
2	He	2																				
3	Li	2	1																			
4	Be	2	2																			
5	B	2	2	1																		
6	C	2	2	2																		
7	N	2	2	3																		
8	O	2	2	4																		
9	F	2	2	5																		
10	Ne	2	2	6																		
11	Na	2	2	6	1																	
12	Mg	2	2	6	2																	
13	Al	2	2	6	2	1																
14	Si	2	2	6	2	2																
15	P	2	2	6	2	3																
16	S	2	2	6	2	4																
17	Cl	2	2	6	2	5																
18	Ar	2	2	6	2	6																
19	K	2	2	6	2	6		1														
20	Ca	2	2	6	2	6		2														
21	Sc	2	2	6	2	6	1	2														
22	Ti	2	2	6	2	6	2	2														
23	V	2	2	6	2	6	3	2														
24	Cr	2	2	6	2	6	5	1														
25	Mn	2	2	6	2	6	5	2														
26	Fe	2	2	6	2	6	6	2														
27	Co	2	2	6	2	6	7	2														
28	Ni	2	2	6	2	6	8	2														
29	Cu	2	2	6	2	6	10	1														
30	Zn	2	2	6	2	6	10	2														
31	Ga	2	2	6	2	6	10	2	1													
32	Ge	2	2	6	2	6	10	2	2													
33	As	2	2	6	2	6	10	2	3													
34	Se	2	2	6	2	6	10	2	4													
35	Br	2	2	6	2	6	10	2	5													
36	Kr	2	2	6	2	6	10	2	6													
37	Rb	2	2	6	2	6	10	2	6			1										
38	Sr	2	2	6	2	6	10	2	6			2										
39	Y	2	2	6	2	6	10	2	6	1		2										
40	Zr	2	2	6	2	6	10	2	6	2		2										
41	Nb	2	2	6	2	6	10	2	6	4		1										
42	Mo	2	2	6	2	6	10	2	6	5		1										
43	Tc	2	2	6	2	6	10	2	6	6		1										
44	Ru	2	2	6	2	6	10	2	6	7		1										
45	Rh	2	2	6	2	6	10	2	6	8		1										
46	Pd	2	2	6	2	6	10	2	6	10												
47	Ag	2	2	6	2	6	10	2	6	10		1										
48	Cd	2	2	6	2	6	10	2	6	10		2										

Table A.2 Continued

Atomic number	Element	Electron arrangement																				
		K	L		M			N				O					P				Q	
		1s	2s	2p	3s	3p	3d	4s	4p	4d	4f	5s	5p	5d	5f	5g	6s	6p	6d	6f	7s	7p
49	In	2	2	6	2	6	10	2	6	10		2	1									
50	Sn	2	2	6	2	6	10	2	6	10		2	2									
51	Sb	2	2	6	2	6	10	2	6	10		2	3									
52	Te	2	2	6	2	6	10	2	6	10		2	4									
78	Pt	2	2	6	2	6	10	2	6	10	14	2	6	9			1					
79	Au	2	2	6	2	6	10	2	6	10	14	2	6	10			1					
80	Hg	2	2	6	2	6	10	2	6	10	14	2	6	10			2					
81	Tl	2	2	6	2	6	10	2	6	10	14	2	6	10			2	1				
82	Pb	2	2	6	2	6	10	2	6	10	14	2	6	10			2	2				
83	Bi	2	2	6	2	6	10	2	6	10	14	2	6	10			2	3				
84	Po	2	2	6	2	6	10	2	6	10	14	2	6	10			2	4				
85	At	2	2	6	2	6	10	2	6	10	14	2	6	10			2	5				
86	Rn	2	2	6	2	6	10	2	6	10	14	2	6	10			2	6				
87	Fr	2	2	6	2	6	10	2	6	10	14	2	6	10			2	6			1	
88	Ra	2	2	6	2	6	10	2	6	10	14	2	6	10			2	6			2	
89	Ac	2	2	6	2	6	10	2	6	10	14	2	6	10			2	6	1		2	
90	Th	2	2	6	2	6	10	2	6	10	14	2	6	10			2	6	2		2	
91	Pa	2	2	6	2	6	10	2	6	10	14	2	6	10	2		2	6	1		2	
92	U	2	2	6	2	6	10	2	6	10	14	2	6	10	3		2	6	1		2	

lium-arsenide (GaAs) will be seen to be formed from group III/V elements. The values of significant parameters for silicon, germanium and gallium-arsenide are given in Table A.4.

The effect on the intrinsic resistivity of silicon with n-type and p-type doping is shown in Figure A.1. It will be seen that an impurity concentration of, e.g., one atom of impurity per 10^6 atoms of silicon will give an electrical resistivity of about 10^2 ohm-meters compared with the intrinsic resistivity of pure silicon of about 2.3×10^7 ohm-meters, a reduction of over 10^5. However, note how these values compare with conductors and insulators, which at 300°K have values as shown on page 445.

Table A.3 An Extract from the Periodic Table of the Elements Closely Associated with Semiconductors.

Group II	III	IV	V	VI
	Boron (5)	Carbon (6)	Nitrogen (7)	Oxygen (8)
Magnesium (12)	Aluminium (13)	Silicon (14)	Phosphorus (15)	Sulphur (16)
Zinc (30)	Gallium (31)	Germanium (32)	Arsenic (33)	Selenium (34)
Cadmium (48)	Indium (49)	Tin (50)	Antimony (51)	Tellurium (52)
Mercury (80)	Thallium (81)	Lead (82)	Bismuth (83)	

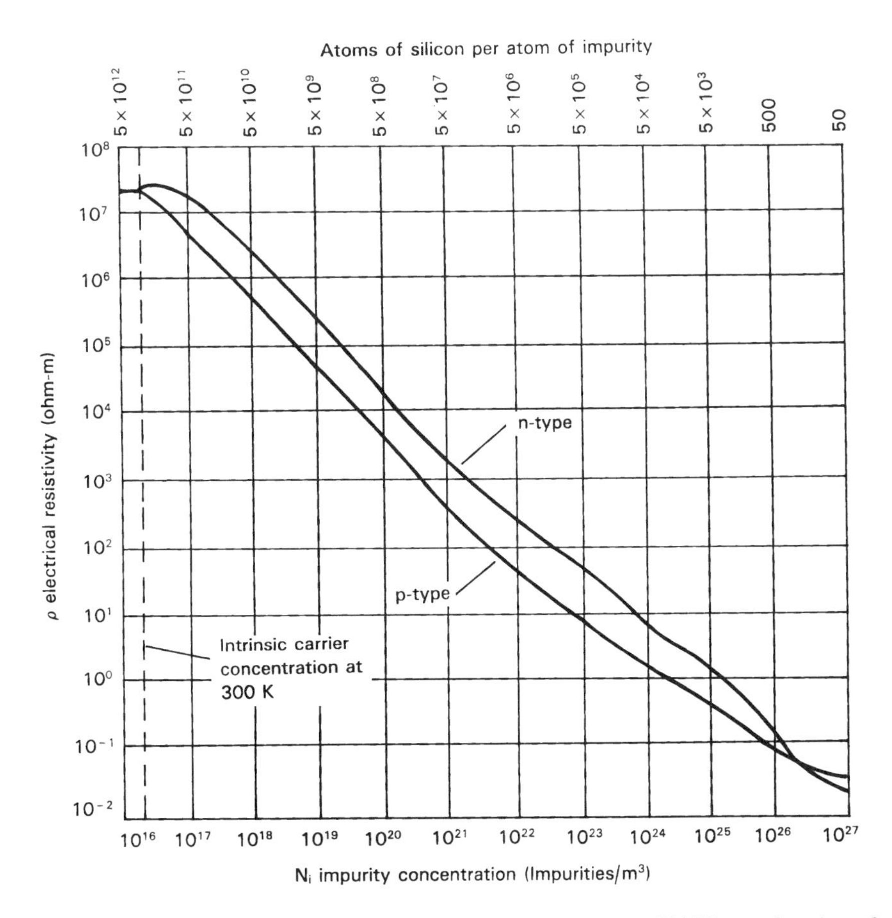

Fig. A.1 The variation of electrical resistivity ρ for silicon at 300°K as a function of the impurity doping concentration. Crystalline silicon has approximately 5×10^{22} atoms per cubic meter.

Material	Resistivity ρ (Ωm)
Copper	1.7×10^{-8}
Aluminium	2.7×10^{-8}
Gold	11×10^{-8}
Intrinsic silicon	2.3×10^{7}
Glass, mica, etc.	10^{16} to 10^{22}

Further physical data may be found in many appropriate texts, for example, Refs. 1 to 8 given in Chapter 2, and elsewhere.

Table A.4 Significant Physical Features for Silicon, Germanium and Gallium-Arsenide at 300°K

	Si	Ge	GaAs
Energy gap Eg (eV)	1.1	0.7	1.4
Intrinsic carrier density n_i (carriers m^{-3})	1.45×10^{16}	2.5×10^{19}	1×10^{13}
Relative permittivity e_r	12	16	12
Hole mobility μ_P ($m^2 V^{-1} s^{-1}$)	0.05	0.19	0.5
Electron mobility μ_N ($m^2 V^{-1} s^{-1}$)	0.14	0.39	0.85
Hole diffusion constant D_P ($m^2 s^{-1}$)	1.3×10^{-3}	5×10^{-3}	1.3×10^{-3}
Electron diffusion constant D_n ($m^2 s^{-1}$)	3.6×10^{-3}	10×10^{-3}	22×10^{-3}

Appendix B
Fabrication and Yield

Lithography is at the heart of almost all microelectronic fabrication processes. Figure B.1 summarizes the steps which are conventially involved in the production of masks for photolithography or for direct-write-on-wafer (see also Chapter 4, Section 4.4).

The procedures which conventionally take place at each mask level during wafer fabrication are summarized in Figure B.2. From this it will be appreciated that at each mask stage, a number of separate fabrication and inspection processes must be done, all of which are critical to the production of defect-free final circuits. With E-beam direct-write-on-wafer, the steps are similar except that the wafers have to be loaded into and aligned in the E-beam machine for each patterning of the resist material. Further details of these various processes may be found in Ref. 3 of Chapter 2.

The number of whole dice (chips) which can be obtained from wafers of increasing diameter is given in Table B.1. The extreme edge of a wafer is invariably kept free of circuits, since there is likely to be an increasing number of defects near its circumference due to minor mechanical damage as well as processing variations at the edges of the wafers as they go through the many stages of fabrication. Table B.1 gives the die size in microns (10^{-6} m) and the wafer diameter in centimeters (10^{-2} m); imperial measure with die size in mils (10^{-3} inches) and wafer diameter in inches may be encountered in some texts, the relationships between these measures being:

- 1 μm = 0.04 mils
- 1 mm = 40 mils
- 1 mil = 25.4 μm (2.54×10^{-5} m)

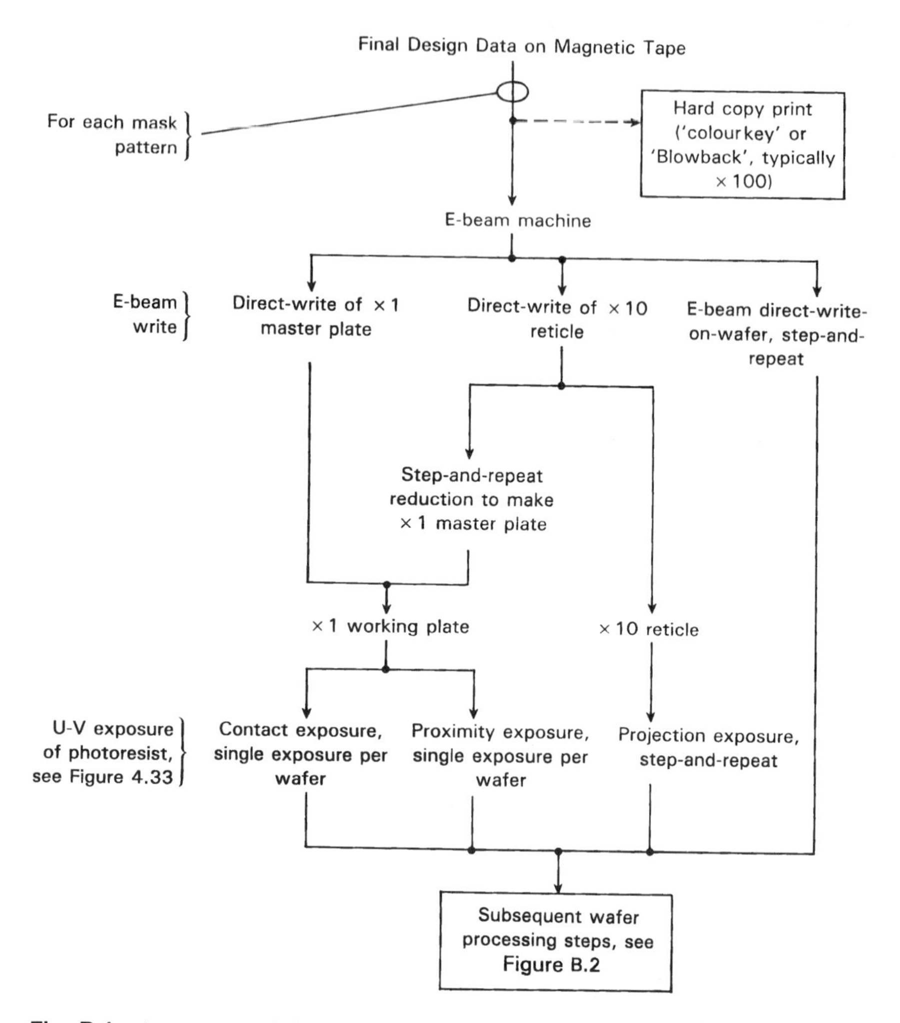

Fig. B.1 A summary of the principal alternative ways of patterning resist on silicon wafers.

- 100 mils = 2.54 mm
- 1 inch = 2.54 cm (2.54×10^{-2} m)
- 1 mil^2 = 645 μm^2 (6.45×10^{-10} m^2)

Modeling of the die yield of a wafer, which is principally dependent upon die size and the processing 'goodness' and not upon the circuit complexity on the die, has been extensively studied, since with accurate modeling, the yield and

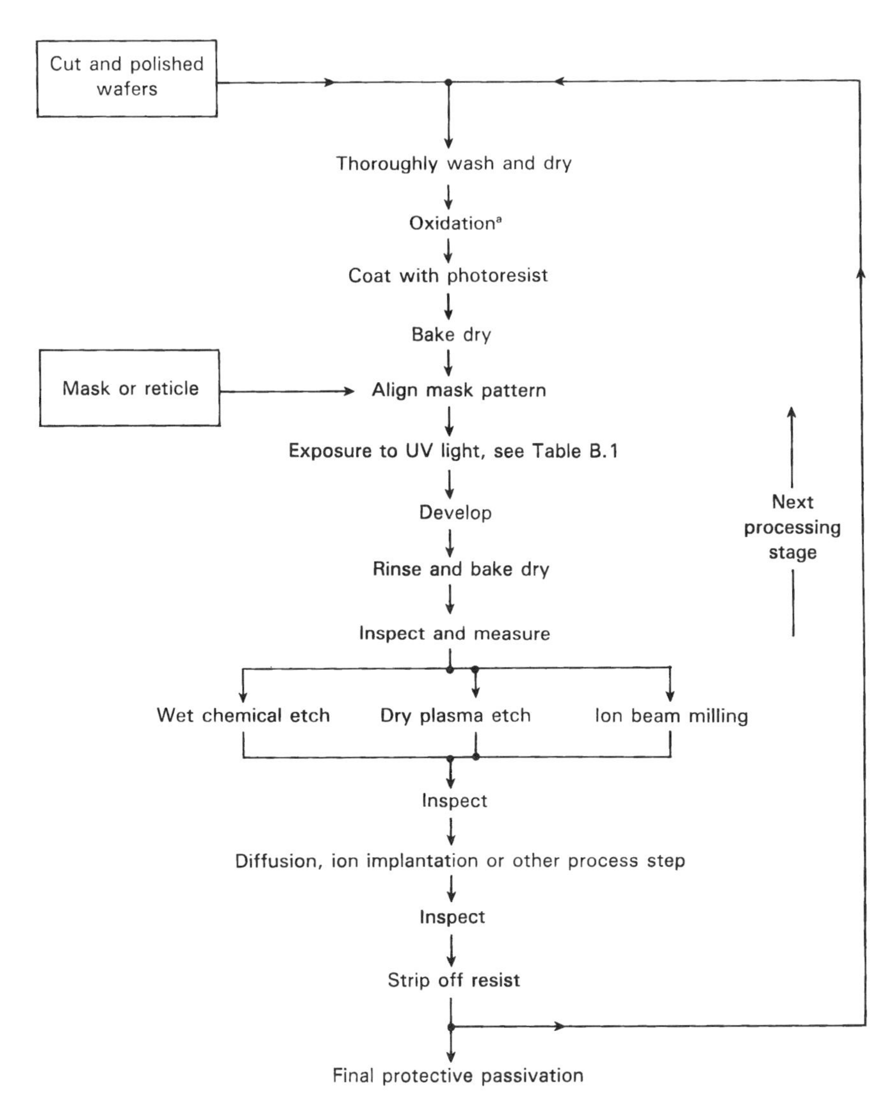

[a] Not at every stage.

Fig. B.2 A summary of the steps at each mask level of a wafer fabrication process.

Table B.1 The Approximate Number of Whole Dice Available from Wafers of Increasing Diameter, Neglecting the Presence of Any Process Monitoring Circuits (see Figure 6.2)

Square die size (μm × μm)	Die area (μm^2)	Approximate no. of dice per wafer			
		Wafer diameter			
		7.5 cm	10 cm	12.5 cm	15 cm
2000	4×10^6	1035	1840	2920	4800
2250	5.06×10^6	835	1480	2290	3300
2500	6.25×10^6	655	1180	1800	2600
2750	7.56×10^6	545	980	1490	2200
3000	9×10^6	450	820	1250	1900
3250	10.56×10^6	372	703	1090	1650
3500	12.25×10^6	324	601	935	1400
3750	14.06×10^6	284	510	810	1200
4000	16×10^6	256	448	710	1000
4250	18.06×10^6	214	400	625	880
4500	20.25×10^6	190	350	550	800
4750	22.56×10^6	168	313	490	730
5000	25×10^6	148	282	445	660
5250	27.56×10^6	135	264	395	580
5500	30.25×10^6	123	230	360	500
5750	33.06×10^6	110	212	316	450
6000	36×10^6	98	190	300	410
6250	39.06×10^6	90	172	273	370
6500	42.25×10^6	82	158	250	340
6750	45.56×10^6	75	146	232	310
7000	49×10^6	68	134	217	290
7250	52.56×10^6	67	124	197	270
7500	56.25×10^6	60	115	184	250
7750	60.63×10^6	52	105	173	235
8000	64×10^6	52	100	162	220
8250	68.06×10^6	51	90	152	200
8500	72.25×10^6	43	86	142	185
8750	76.56×10^6	42	81	132	175
9000	81×10^6	40	80	125	160

therefore the cost of new circuits may be forecast by the IC manufacturer. Also, once the detailed modeling parameters for a production process have been established, the quality of the production line may be maintained by appropriate parameter measurements. However, published theory and published parameter values usually lag considerably behind vendors' current achievements, such data being very company sensitive, and as a result such information as is available has historically been out of date as far as the current state-of-the-art is concerned.

In theory, wafer yield Y, $0 \leq Y \leq 1$, is normally considered to be a function of the density of defects per unit area of the wafer D_0, the area of the die A, and parameters α which are specific to the particular model and process. Many detailed equations have been proposed for Y, all of which have a statistical basis. The many variants try to take into account factors such as the defect density D_0

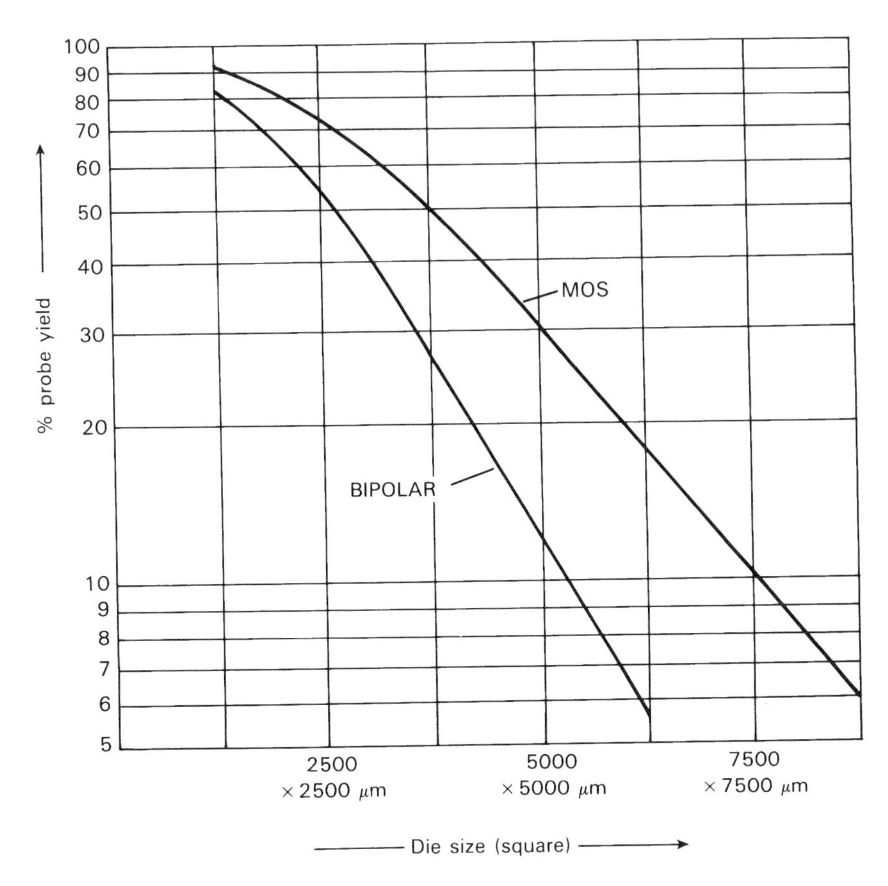

Fig. B.3 The general characteristic of wafer yield from a 20 cm wafer as a function of die size.

not being uniform across a wafer, but all predict that the yield of fault-free circuits decreases as the die size increases. Further details may be found in Ref. 3 of Chapter 2.

One of the simple modeling equations that has been used is:

$$Y = \left[\frac{1 - e^{-D_0 A}}{D_0\, A}\right]^2 \times 100\%$$

The difficulty here is to determine a realistic value for D_0, which has to vary with technology, wafer diameter and production expertise. Some 1980s values for the above equation are:

- $D_0 = 10^5$ defects per m^2 for bipolar technology
- $D_0 = 5 \times 10^4$ defects per m^2 for MOS technology

These values are shown plotted in Figure B.3. Corresponding current values are not available, but are likely to be at least an order of magnitude better than these published figures.

Hence, while the absolute values indicated in Figure B.3 must not be taken as representative of late 1990s practice, the general characteristics of wafer yield with die size shown by this model is likely to remain valid.

Appendix C
The Principal Equations Relating to Bipolar Transistor Performance

The basic equation which underpins all bipolar transistor performance is the theoretical pn junction diode equation. Considering a simple pn junction such as shown in Figure C.1(a), the forward direction of current is when the p-region is made positive with respect to the n-region, and the reverse direction of current is when the p-region is negative with respect to the n-region. The expression relating the current I_D through the diode to the voltage V_D across it is given by the nonlinear equation

$$I_D = I_S\ (e^{qV_D/kT} - 1) \tag{C.1}$$

where

I_S = the reverse saturation current
q = electron charge, $= 1.6 \times 10^{-9}$ C
k = Boltzmann's constant, $= 1.38 \times 10^{-23}$ JK^{-1}
T = temperature (°K)

The term q/kT has the dimensions of V^{-1}, and has a numerical value of between 20 and 40 at 300°K depending upon the detailed properties of the pn junction. However, for the majority of bipolar transistors the value of this term is taken to be 40 V^{-1} for emitter-base junctions. Hence, the diode equation (C.1) may be rewritten as

$$I_D = I_S(e^{40V} - 1) \tag{C.2}$$

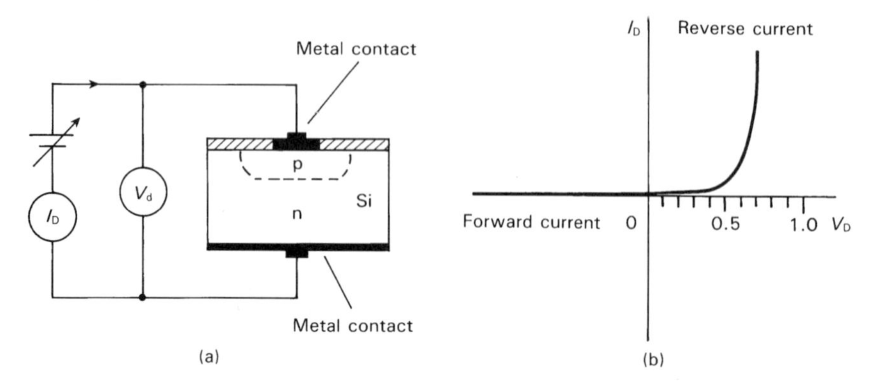

Fig. C.1 The silicon pn junction diode: (a) the measuring circuit: (b) the forward and reverse current vs. voltage characteristics.

where V is in volts.

The current I_S for a typical silicon pn junction at room temperature may be less than a picoamp (10^{-12} A). Thus when the reverse voltage V_D across the junction is, e.g., -1 V, the exponential term in Equation (C.2) is e^{-40} which is negligible, giving $I_D = -I_S$. However, when V_D is made positive, the forward current I_D increases rapidly, being about 1 mA when V_D is $+0.63$ V and continuing to increase by a factor of about 50 (e^4) for every 0.1 V increase in V_D. This is indicated in Figure C.1(b). The -1 term in Equation (C.2) may now be neglected.

In the normal bipolar junction transistor (BJT) it is the emitter-base pn junction which is forward biased by the transistor input conditions (see Figure C.2(a)). Hence, Equation (C.2) can be rewritten in terms of the transistor parameters, giving:

$$I_E = I_S(e^{40V_{EB}} - 1) \tag{C.3}$$

for pnp transistors, where V_{EB} is positive for forward biasing, and

$$I_E = I_S(e^{40V_{BE}} - 1) \tag{C.4}$$

for npn transistors where V_{BE} is positive for forward biasing.

The remaining d.c. current relationships for a transistor operating in common-base configuration are:

$$I_C = h_{FB}\, I_E$$

and

$$\begin{aligned} I_B &= I_E - I_C \\ &= (1 - h_{FB})\, I_E \end{aligned} \tag{C.5}$$

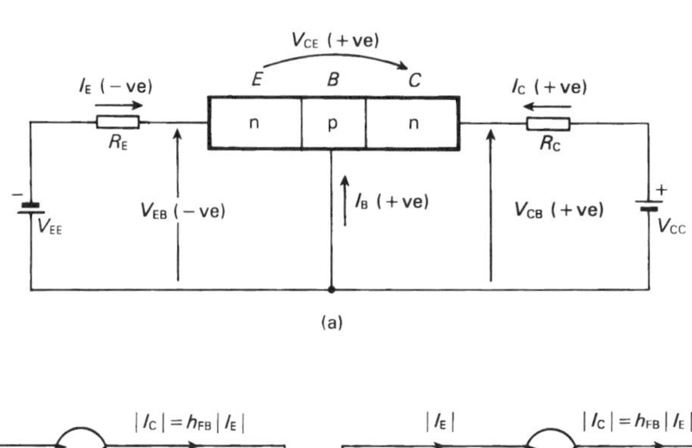

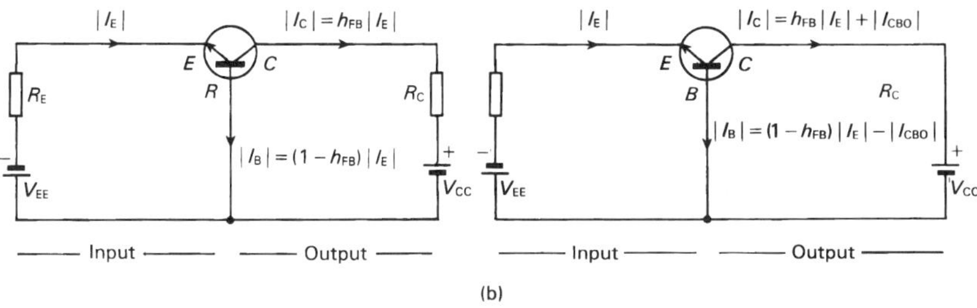

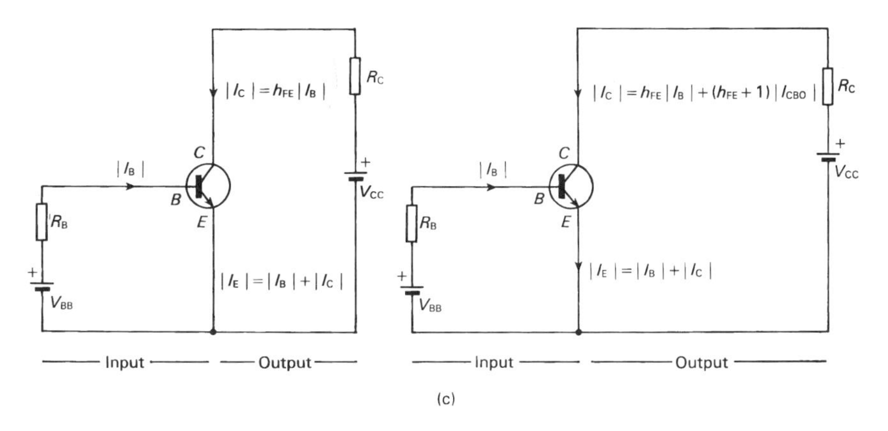

Fig. C.2 The silicon npn bipolar junction transistor (BJT): (a) the normal operating polarities; (b) current relationships in common-base working with and without the collector leakage current I_{CBO}; (c) current relationships in common emitter working also with and without I_{CBO}.

(see Figure C.2(b)), where h_{FB}, sometimes termed α, is the large-signal (or d.c.) common-base current gain. Note that as h_{FB} has a value very close to, but not exceeding unity, we may assume that $I_C \simeq I_E$, giving us:

$$\begin{aligned} I_C &= I_S(e^{40V_{BE}} - 1) \\ &\simeq I_S\,(e^{40V_{BE}}) \end{aligned} \tag{C.6}$$

when the transistor is conducting.

For the BJT operating in common-emitter configuration where the base current is now regarded as the input current, the above d.c. relationships are rearranged to become:

$$\begin{aligned} I_C &= \frac{h_{FB}}{1 - h_{FB}} I_B \\ &= h_{FE}\, I_B \end{aligned} \tag{C.7}$$

where h_{FE}, sometimes termed β, is the large-signal common-emitter current gain.

These relationships do not take into account the small but temperature-sensitive leakage current I_{CBO} flowing through the reverse-biased collector-to-base junction (see Figure C.2(b)). Inclusion of this current modifies the current distribution to:

$$I_C = h_{FB}\, I_E + I_{CBO}$$

and

$$\begin{aligned} I_B &= I_E - (h_{FB}\, I_E + I_{CBO}) \\ &= (1 - h_{FB})I_E - I_{CBO} \end{aligned}$$

whence

$$I_E = \frac{1}{1 - h_{FB}} I_B + \frac{1}{1 - h_{FB}} I_{CBO} \tag{C.8}$$

Substituting for I_E in the equation for I_C now gives:

$$\begin{aligned} I_C &= \frac{h_{FB}}{1 - h_{FB}} I_B + \frac{h_{FB}}{1 - h_{FB}} I_{CBO} + I_{CBO} \\ &= h_{FE}\, I_B + (h_{FE} + 1)\, I_{CBO} \end{aligned} \tag{C.9}$$

where h_{FE} is as previously. However, for silicon transistors operating at normal temperatures the leakage term in this equation may usually be ignored.

The small-signal (or a.c.) parameters of the junction transistor, which represent the transistor performance about some fixed (quiescent) d.c. operating point, are determined by evaluating the slope of the various current/voltage relationships. Consider the transconductance g_m of the BJT, where g_m is the rate of change of output current with input voltage. In common-emitter mode this is:

$$g_m = \mathrm{d}\, I_C / \mathrm{d}\, V_{BE}$$

which by differentiation of Equation (C.6) gives

$$g_m = 40\, I_S\, e^{40 V_{BE}}$$

which using Equation (C.6) again gives

$$g_m = 40\, I_C \qquad \text{(C.10)}$$

Thus the g_m of a bipolar transistor is *proportional to the d.c. current flowing in the collector*, and is often quoted as 40 mA V^{-1} per milliamp of collector current. Thus if I_C is 100 μA, g_m = 4 mA per volt; if I_C is 10 mA, then g_m is 400 mA, per volt, and so on.

Of the many other small-signal parameters which may be found, the hybrid or *h* parameters are often used for low/medium frequency modeling. The four *h* parameters for common-emitter working are as follows:

$$\text{a.c. input impedance } h_{ie} = \left.\frac{\Delta V_{BE}}{\Delta I_B}\right|_{V_{CE}\text{constant}}$$

$$\text{a.c. output conductance } h_{oe} = \left.\frac{\Delta I_C}{\Delta V_{CE}}\right|_{I_B\text{constant}}$$

$$\text{a.c. forward current gain } h_{fe} = \left.\frac{\Delta I_C}{\Delta I_B}\right|_{V_{CE}\text{constant}}$$

$$\text{a.c. reverse voltage feedback } h_{re} = \left.\frac{\Delta V_{BE}}{\Delta V_{CE}}\right|_{I_B\text{constant}}$$

The a.c. equivalent circuit using *h* parameters is shown in Figure C.3 in comparison with the low-frequency hybrid π model based on the transconductance parameter *gm*.

The significance of the four *h* parameters is that they can be readily measured individually without requiring inconvenient operating conditions on the transistor. The two parameters h_{oe} and h_{re} are often neglected, leaving h_{ie} and h_{fe} as the parameters of major significance. Comparing the two simplified equivalent circuits shown in Figure C.3, the two current generators $g_m\, v_{be}$ and $h_{fe} i_b$ should give the same output results, and thus we have the relationships

$$g_m\, v_{be} = h_{fe} i_b$$

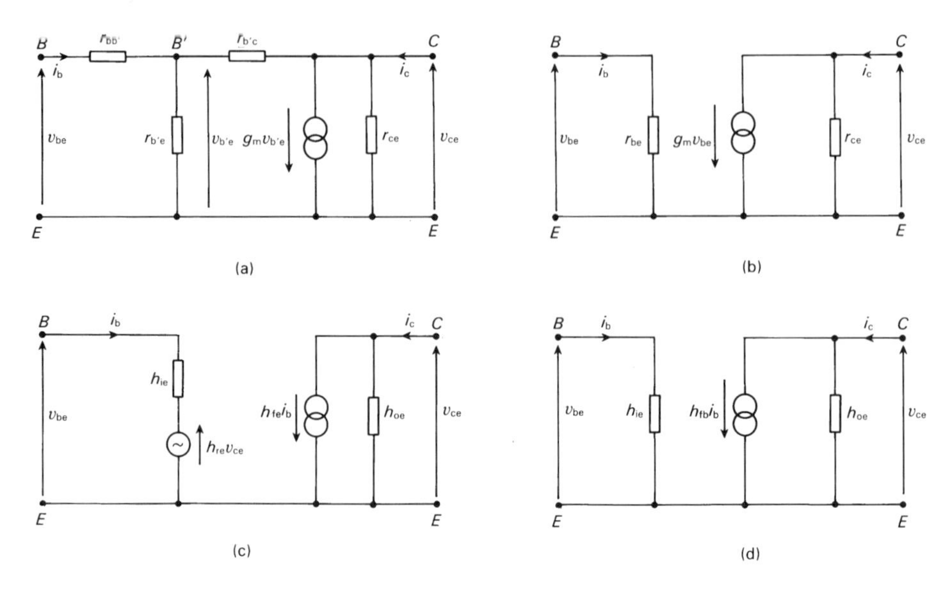

Fig. C.3 BJT small-signal equivalent circuits for common-emitter working: (a) the low-frequency hybrid π model; (b) the simplified hybrid π model with only the dominant parameters; (c) the low-frequency h-parameter model: (d) the simplified h-parameter model.

whence

$$h_{fe} = g_m \, h_{ie}$$

giving

$$h_{ie} = \frac{h_{fe}}{g_m} = \frac{h_{fe}}{40 \, I_C} = \frac{25 \, h_{fe}}{I_C} \text{ where } I_C \text{ is in mA}$$

This final equation is particularly significant since it states that the common-emitter a.c. input impedance is approximately given by *the current gain times* 25 Ω *per mA of the collector current* I_C. These and many other BJT relationships may be more rigorously and accurately determined, for example, see Ref. 1 below and elsewhere.

However, consider a little further the a.c. performance of the BJT: from

Figure C.3(b) it will be noted that the a.c. gain performance of the BJT is very largely governed by the value of g_m, which as we have seen has a predetermined value proportional to the value of the collector current I_C. But from Figure C.3(d), it would appear that the gain performance is largely governed by the value of the forward current gain term h_{fe}, which is a parameter that is dependent upon the expertise of the manufacturer in making high-gain devices. To reconcile these two aspects, it must be appreciated that h_{ie} in the h parameter model is also dependent upon h_{fe} and the transistor g_m value as shown above, and therefore the effect of high current gain is to increase the input impedance term h_{ie}, thus reducing the input current and hence maintain $h_{fe}\, i_b$ sensibly constant for any given value of I_C. The transconductance term thus remains the dominant parameter which largely determines the a.c. gain performance of a BJT, except at high frequencies where further parameters come into prominance, being independent of device dimensions. High gain, however, is very desirable in order to increase the input impedance in common-emitter working.

Further details of transistor action and small-signal equivalent circuits may be found in the texts cited below and in many other sources. It will be found that the SPICE simulation program uses a very comprehensive model to represent the transistor, although in use it is common to use default values (often zero) for a number of the parameters—the significance of SPICE for device modeling was introduced in Section 5.2.3 of Chapter 5 (see also the Chapter 5 references 112–115).

REFERENCES

1. Goodge, M., *Semiconductor Device Technology*, Macmillan, UK, 1983
2. Sedra, A.S. and Smith, K.C., *Microelectronic Circuits*, Holt, Reinhart and Wilson, NY, 1987
3. Glazer, A.B. and Subak-Sharpe, G.E., *Integrated Circuit Engineering*, Addison-Wesley, MA, 1979
4. Sze, S.M. (Ed.), *VLSI Technology*, McGraw-Hill, NY, 1983
5. Neudeck, G.W., et al. (Eds.), *Modular Series on Semiconductor Fundamentals*, Volumes 1 to 5, Addison-Wesley, MA, 1987–1989

Appendix D
The Principal Equations Relating to Unipolar (MOS) Transistor Performance

Unlike the bipolar junction transistor, most of the MOS device parameters are dependent upon device geometry, particularly the length L and the width W of the conducting channel between source and drain. It is this characteristic of MOS devices that allows scaling and the freedom for the system designer to become involved at the silicon design level in certain circumstances.

Here we will consider only enhancement-mode, insulated-gate field-effect transistors (IGFETs), sometimes still ambiguously referred to as MOSFETs (metal-oxide-silicon field-effect transistors) even though the gate is now invariably polysilicon rather than metal. Enhancement-mode means that the FET is nonconducting when V_{GS} is zero, a forward gate-to-source voltage greater than a threshold value being necessary to allow source-to-drain conduction; the expression $(V_{GS} - V_{TH})$ will be found to be prominent in almost all the following equations. Junction FETs (JFETs) and depletion-mode types have generally similar equations for their performance; full details may be found in very many texts such as Refs. 1 to 9 of Chapter 2, Ref. 8 of Chapter 4 and elsewhere.

The initial development of FET characteristics classically begins with a consideration of the gate input capacitance, with its dielectric of silicon dioxide between gate and substrate. This capacitance is given by:

$$C_{OX} = \frac{\epsilon_0 \, \epsilon_r \, A}{x} \tag{D.1}$$

where

C_{OX} = the capacitance of the physical gate-substrate capacitance (F)
ϵ_0 = absolute permittivity ($F \, m^{-1}$)

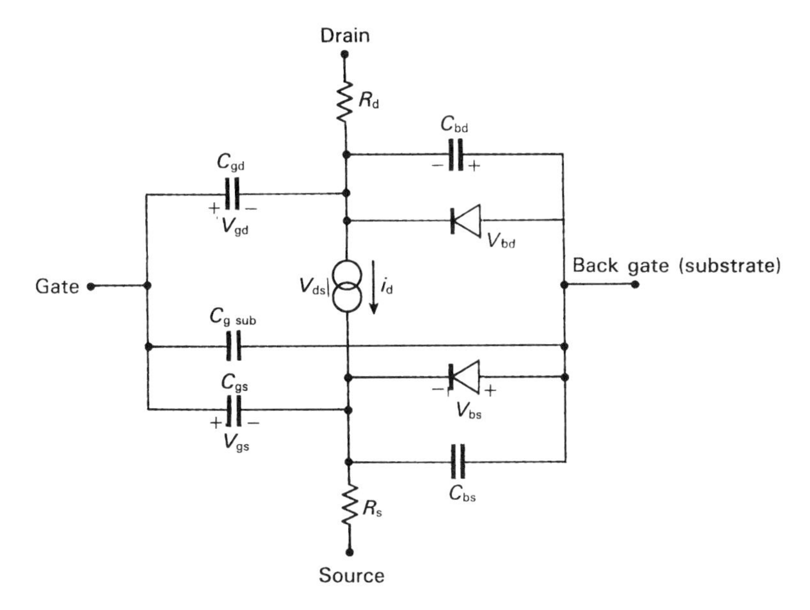

Fig. D.1 The SPICE model of an insulated-gate field-effect transistor.

ϵ_r = relative permittivity of the gate oxide
x = gate oxide thickness (m)

This may be given in per-unit area, namely

$$C_{OX} = \frac{\epsilon_0\, \epsilon_r}{x}\ \mathrm{F\ m^{-2}} \tag{D.2}$$

The importance of capacitance in understanding the action of FETs is well illustrated by consideration of the model used in SPICE simulation, which is shown in Figure D.1. As will be seen, capacitive terms between gate, source, drain and substrate are very conspicuous; they are, however, together with the drain and source diodes largely nonlinear terms, so that a considerable complexity accrues in undertaking SPICE simulations over a range of operating conditions.

From a consideration of the charge induced by the gate in the channel between source and drain, the expression for the drain current may be derived, namely:

$$I_D = \frac{W}{L}\, \mu_N\, C_0 \left[(V_{GS} - V_{TH})\, V_{DS} - \frac{V_{DS}^{\,2}}{2} \right] \tag{D.3}$$

where

I_D = drain current (A)
W = width of gate (m)
L = length of gate (m)
μ_N = drift mobility of channel electrons ($m^2\ V^{-1}\ s^{-1}$)
C_0 = gate to channel capacitance per unit area ($F\ m^{-2}$)
V_{GS}, V_{TH}, V_{DS} = gate-to-source, threshold, and drain-to-source voltages as previously (V)

The above equation is used for n-channel devices; for p-channel devices, μ_N is replaced by μ_P, the drift mobility of channel holes. Further, $\mu_N\ C_0$ (or $\mu_P\ C_0$) may be referred to as the process gain factor K', ($A\ V^{-2}$), and in turn W/L. K' may be considered as a single parameter β, termed the transistor gain factor. Hence:

$$I_D = \beta \left[(V_{GS} - V_{TH})\ V_{DS} - \frac{V_{DS}^2}{2} \right] \tag{D.4}$$

It will be appreciated that because the mobility μ_N for electrons is higher than the hole mobility μ_P (in silicon 0.15 $m^2\ V^{-1}\ s^{-1}$ compared with 0.05 $m^2\ V^{-1}\ s^{-1}$), β for n-channel devices will be higher than for p-channel devices. However, note that the geometric aspect ratio L/W of the transistor is also present in the transistor gain factor and therefore in the overall expression for the resulting drain current I_D.

The above expression for I_D contains certain simplifications, particularly bulk substrate effects which tend to act as a second 'back' gate to modify the drain current, but for most practical purposes equations (D.3) and (D.4) are appropriate. However, if the term $[(V_{GS} - V_{TH})\ V_{DS} - \frac{1}{2}\ (V_{DS}^2)]$ is examined more closely, it will be seen that for $V_{GS} - V_{TH}$ constant the resultant current I_D will increase with increasing V_{DS} up to some maximum ('saturated') value. The value for V_{DS} at this point is given by differentiating I_D with respect to V_{DS} and equating to zero, that is:

$$\frac{d\ I_D}{d\ V_{DS}} = (V_{GS} - V_{TH}) - \frac{1}{2}\ (2\ V_{DS}) = 0$$

whence

$$V_{DS(sat)} = (V_{GS} - V_{TH}) \tag{D.5}$$

The point at which this occurs is known as the saturation (or 'pinch-off') point. Below this point the FET is operating in its unsaturated mode, sometimes termed its 'triode' or 'ohmic' mode, where I_D increases rapidly with increasing $V_{DS,}$ V_{GS} constant. The value of I_D at $V_{DS(sat)}$ is given by:

$$I_{D(sat)} = \beta \left[\frac{(V_{GS} - V_{TH})^2}{2} \right] = \beta \left[\frac{(V_{DS(sat)})^2}{2} \right] \tag{D.6}$$

When V_{DS} is small compared with $(V_{GS} - V_{TH})$, the $\frac{1}{2}\, V_{DS}{}^2$ term in the expression for I_D may be neglected, which gives:

$$I_D = \frac{W}{L} \mu_N\, C_0\, \{(V_{GS} - V_{TH})V_{DS}\} \tag{D.7}$$

whence

$$\frac{V_{DS}}{I_D} = \frac{L}{W} \frac{1}{\mu_N\, C_0\, (V_{GS} - V_{TH})} = \frac{L}{W} \frac{1}{K'\, (V_{GS} - V_{TH})} \tag{D.8}$$

The FET is now behaving as a linear voltage-controlled resistor, the resistance value being determined by the aspect ratio L/W of the channel and the gate voltage $V_{GS.}$

At values of voltage V_{DS} greater than $V_{DS(sat)}$, increasing V_{DS} increases the length of the conducting channel which is in the pinch-off condition (known as the channel length modulation effect), a factor which can be accounted for by incorporating a new factor λ in the expression for I_D, giving:

$$I_D = I_{D(sat)}\, (1 + \lambda V_{DS}) = \beta \left[\frac{(V_{GS} - V_{TH})^2}{2} \right] (1 + \lambda V_{DS}) \tag{D.9}$$

for the current in the saturated region, where λ has a small value typically between 0.01 and 0.05. The classic I_D/V_{DS} characteristics for increasing V_{GS} are therefore as shown in Figure D.2.

The expression for the threshold voltage V_{TH} depends heavily upon the fabrication process and to a negligible extent upon the device geometry, as would be expected. The usual equation is:

$$V_{TH} = V_{FB} + 2\Psi_B + \gamma\, \{2\Psi_B\}^{1/2} \tag{D.10}$$

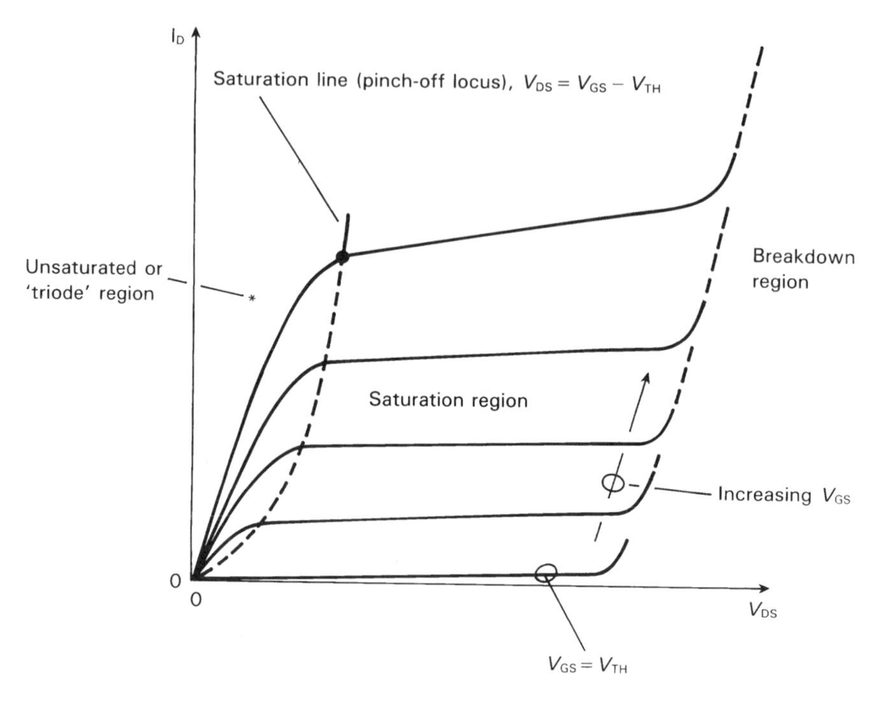

Fig. D.2 The I_D vs. V_{DS} output characteristics of an enhancement-type FET for increasing values of gate voltage V_{GS}.

where

$$V_{FB} = \text{the transistor flat-band voltage}$$

$$\Psi_B = \frac{k\,T}{q} \log n \left[\frac{N_A}{n_I}\right]$$

that is, a parameter dependent upon the acceptor level concentration N_A in the substrate and the free electron concentration n_I of intrinsic silicon (assuming an n-channel device),

and

$$\gamma = \frac{\{2\epsilon_0\epsilon_r q\, N_A\}^{1/2}}{C_0}$$

i.e., a further parameter dependent upon the acceptor level concentration together with the gate capacitance parameters.

V_{TH} is not, therefore, a parameter which can be freely varied with device

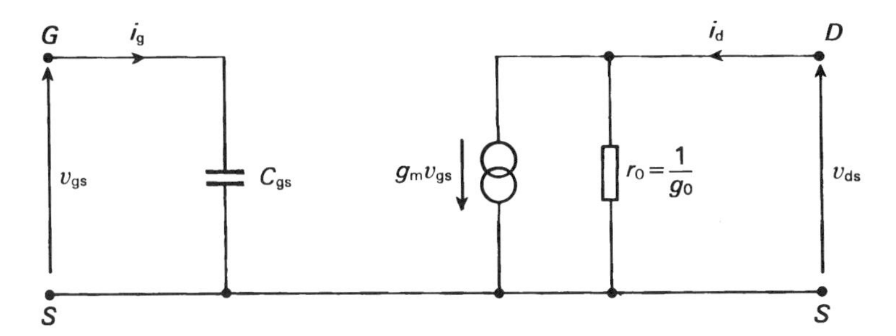

Fig. D.3 The low-frequency small-signal a.c. equivalent circuit for the insulated-gate FET. At higher frequencies additional capacitive terms need to be introduced.

geometry as can the drain current I_D, and hence is not a flexible parameter for the circuit designer's use at the silicon level. Its value for enhancement-mode FETs may be taken very roughly as 0.2 V_{DD}, so for +5 V operation V_{TH} is about 1 V. For depletion-mode devices, V_{TH} may roughly be taken as −0.7 V, giving V_{TH} about −3.5 V for +5 V operation. Note that V_{TH} roughly tracks with V_{DD}, thus giving good circuit tolerance to variations in V_{DD}.

Turning to the small-signal a.c. performance of the IGFET, because of its very high input impedance, with leakage currents perhaps as low as 10^{-5} A, the usual low-frequency a.c. equivalent circuit is as shown in Figure D.3. For higher frequency operation additional capacitive terms have to be added to this simple a.c. equivalent circuit.

The small-signal gate input capacitance C_{gs} depends upon the gate area $W \times L$ and the gate oxide capacitance C_0 per unit area (see equation (D.2)). Typically, C_{gs} may be taken as 0.7 $C_0 WL$.

The value of the transconductance g_m may be found by differentiating the preceding equations for I_D with respect to V_{GS}, the remaining paramters being held constant. This therefore gives:

$$g_m = \frac{W}{L} \mu_N C_0 V_{DS}$$

$$= \beta V_{DS} \tag{D.12}$$

from equations (D.3) and (D.4) for the nonsaturated region of operation, that is, when $V_{DS} \leq (V_{GS} - V_{TH})$, and:

$$g_m = \beta (V_{GS} - V_{TH})(1 + \lambda V_{DS})$$

$$\simeq \beta (V_{GS} - V_{TH})$$

$$= \beta V_{DS(sat)} \tag{D.13}$$

from Equation (D.9) for the saturated region of operation. Hence, g_m increases to a maximum value when in saturation.

The small-signal a.c. output conductance, g_0 is given by differentiating I_D with respect to $V_{GS,}$ the remaining parameters being held constant. This therefore gives:

$$g_0 = \beta \{(V_{GS} - V_{TH}) - V_{DS}\} \tag{D.14}$$

for the nonsaturated region of operation, and:

$$g_0 = \beta \left[\frac{(V_{GS} - V_{TH})^2}{2} \right] \lambda \tag{D.15}$$

from Equation (D.9) for the saturated region of operation, which is a small value conductance equal to a high value impedance.

An alternative expression for g_m in the saturated region can be obtained by substituting for $(V_{GS} - V_{TH})$ in terms of $I_{D,}$ which yields:

$$g_m = [\sqrt{(2\ \mu_N\ C_0)}] \cdot [\sqrt{(W/L)}] \cdot [\sqrt{I_D}] \tag{D.16}$$

which emphasizes the following:

- g_m is proportional to the square root of the d.c. bias current I_D in an amplifier circuit
- is proportional to the square root of the channel dimensions W/L

Similarly, an alternative expression for the output conductance in terms of I_D can be obtained, giving:

$$g_0 = \frac{\lambda\ I_D}{1 + \lambda V_{DS}} \tag{D.17}$$

whence the output impedance r_0 is

$$r_0 = \frac{1 + \lambda\ V_{DS}}{\lambda\ I_D} \tag{D.18}$$

The above Equation (D.16) for g_m should be contrasted with that for the bipolar junction transistor, where g_m is independent of the geometrical size of the transistor but is proportional to the collector current (see Equation (C.10)). From Equation (D.16), an IGFET having $\mu_N\ C_0 = 20\ \mu A\ V^{-2}$ and equal width and length dimensions, $g_m = 0.2$ mA V^{-1} at $I_D = 1$ mA. With $W/L = 100$, which is an extremely wide transistor for a given length, g_m is still only 2 mA V^{-1}. This should be contrasted with the value of 40 mA V^{-1} for a BJT at an operating current of 1 mA. Hence, IGFETs are generally characterized by a low value of g_m in comparison with other active devices.

Finally, from Figure D.3 it will be seen that the maximum possible voltage gain is g_m/g_0, $= g_m\, r_0$, which from Equations (D.16) and (D.18) gives:

$$A_{max} = \left[\frac{2\ \mu_N\ C_0\ W\ I_D}{L}\right]^{1/2} \left[\frac{1 + \lambda\ V_{DS}}{\lambda\ I_D}\right]$$

$$= \frac{k}{\sqrt{I_D}} \tag{D.19}$$

Hence, A_{max} is inversely proportional to the square of the drain current, decreasing with increasing values of I_D. Therefore, for maximum gain performance, FET currents should be kept low, which implies a difficulty in using FETs for high current, high amplification duties.

Further details may be found in many standard textbooks, for example, see the references suggested at the end of Appendix C.

Symbols and Abbreviations

ACT	application configurable technology
ALU	arithmetic logic unit
ANN	artificial neural network
ASIC	application specific integrated circuit (see also CSIC and USIC)
ASM	algorithmic state machine
ASSP	application specific standard part
ASTTL	advanced Schottky transistor-transistor logic
A-to-D	analog-to-digital
ATE	automatic test equipment
ATPG	automatic test pattern generation
BDD	binary decision diagram
BiCMOS	bipolar-CMOS fabrication
BFL	buffered FET logic
BILBO	built-in logic block observation
BJT	bipolar junction transistor
C	capacitor
CAD	computer-aided design (see also CAE and CAM)
CAE	computer-aided engineering
CALBO	cellular automaton logic block observation
CAM	computer-aided manufacture
CCD	charge-coupled device
CDI	collector-diffusion isolation
CIF	Caltech intermediate form
CML	current-mode logic

CMOS	complementary MOS (see also nMOS and pMOS)
CPLD	complex programmable logic device
CPU	central processing unit
CSIC	custom- (or customer-) specific integrated circuit
DA	design automation
DFT	design-for-testability
DIL or DIP	dual-in-line package or dual-in-line plastic package
D-MESFET	depletion-mode metal-semiconductor field-effect transistor (GaAs device, see also E-MESFET)
DMOS	dynamic MOS
DRAM	dynamic read-only memory (see also SRAM)
DRC	design rules checking
D-to-A	digital-to-analog
E^2PROM	see EEPROM
EAROM	electrically alterable ROM
ECAD	electronic computer-aided design
ECL	emitter-coupled logic
EDA	electronic design automation
EE PROM	electrically erasable ROM
E_g	energy band gap of an electron
E-MESFET	enhancement-mode field-effect transistor (see also D-MESFET)
EPROM	erasable-programmable ROM
ERC	electrical rules checking
eV	electron-volt
FAST™	advanced Schottky TTL technology, originated by Fairchild
FET	field-effect transistor
FPGA	field programmable gate array
FPLA	field programmable logic array
FPLS	field programmable logic sequencer
FTD	full technical data (see also LTD)
GaAs	gallium arsenide
GDS_{II}	graphics design system, version II (a mask pattern standard)
Ge	germanium
GHz	gigahertz (10^9 Hz)
GPIB	general purpose instrumentation bus, electrically identical to Hewlett Packard instrumentation bus HPIB
HCMOS	high-speed CMOS
HDL	hardware description language
HILO	a commercial simulator for digital circuits

HPC	high performance computing
IC	integrated circuit
IGFET	insulated-gate FET
I^2L	integrated injection logic
I^3L	oxide isolated integrated injection logic
I/O	input/output
ISL	integrated Schottky logic
JFET	junction FET
K	1024 (as used in defining memory size, etc.)
kHz	kilohertz (10^3 Hz)
L	inductor, or length in MOS technology geometry
LAN	local area network
λ	lambda, a unit of length in MOS technology geometry
L/W	length/width ratio, used in MOS design layouts
LCA	logic cell array
LOCMOS	local oxidation CMOS
LSI	large scale integration (see also SSI, MSI, VLSI, ULSI and WSI)
LSTTL	low power Schottky TTL
LTD	limited technical data (see also FTD)
MESFET	metal-semiconductor FET
MgCMOS	metal-gate CMOS (see also SgCMOS)
MHz	megahertz (10^6 Hz)
mil	one thousandth of an inch
MIPS	million instructions per second
MOS	metal-oxide-silicon
MOST	metal-oxide-silicon transistor
ms	millisecond (10^{-3} s)
MSI	medium scale integration (see also SSI, LSI, VLSI, ULSI and WSI)
nMOS	n-channel metal-oxide-silicon
NRE	nonrecurring expenditure or engineering
ns	nanosecond (10^{-9} s)
OEM	original equipment manufacturer
op. amp.	operational amplifier
PAL	programmable array logic
PCB	printed circuit board
PDM	process device monitor (see also PED and PMC)
PED	process evaluation device (see also PDM and PMC)
PG	pattern generator
PGA	pin grid array (packaging), or programmable gate array

PLA	programmable logic array
PLD	programmable logic device
PLS	programmable logic sequencer
PMC	process monitoring circuit (see also PDM and PED)
pMOS	p-channel metal-oxide-silicon
PMUX	programmable multiplexer
ps	picosecond (10^{-12} s)
PROM	programmable ROM
QIL	quad-in-line (package)
R	resistor
RAM	random-access memory (see also DRAM and SRAM)
RAS	random-access scan
ROM	read-only memory
RTL	resistor-transistor logic, or register transfer language
SBC	standard buried collector (fabrication)
SDFL	Schottky-diode FET logic
SgCMOS	silicon-gate CMOS (see also MgCMOS)
Si	silicon
S-a-0	stuck at 0
S-a-1	stuck at 1
SiO_2	silicon dioxide
Si_3N_4	silicon nitride
SMD	surface mount device
SOI	silicon-on-isolator
SOS	silicon-on-sapphire
SOSMOS	silicon-on-sapphire CMOS
SPICE	simulation program, integrated circuit emphasis (public domain software)
SRAM	static read-only memory (see also DRAM)
SSI	small scale integration (see also MSI, LSI, VLSI, ULSI and WSI)
STL	Schottky transistor logic
STTL	Schottky transistor-transistor logic
TPG	test pattern generation (see also ATPG)
TTL	transistor-transistor logic
UCA	uncommitted component array
ULA™	uncommitted logic array (original name used by Ferranti, UK)
ULSI	ultra large scale integration (see also SSI, MSI, LSI, VLSI and WSI)
μP	microprocessor

μs	microsecond (10^{-6} s)
USIC	user specific integrated circuit, used in this text in preference to ASIC or CSIC
V_{BB}	d.c. base supply voltage
V_{CC}	d.c. collector supply voltage
V_{DD}	d.c. drain supply voltage
V_{EE}	d.c. emitter supply voltage
V_{GG}	d.c. gate supply voltage
V_{SS}	d.c. source supply voltage
VDU	visual display unit
VHDL	a particular HDL
VHSIC	very high speed integrated circuit
VLSI	very large scale integration (see also SSI, MSI, LSI, ULSI and WSI)
VMOS	a form of MOS fabrication
W	width (in MOS technology fabrication)
WSI	wafer scale integration (see also SSI, MSI, LSI, VLSI and ULSI)

Index